D1447286

To Steven Krantz

With compliments of

the author.

Garret Sobczyk

Sept 30, 2013

New Foundations in Mathematics

Garret Sobczyk

# New Foundations in Mathematics

## The Geometric Concept of Number

Garret Sobczyk
Departamento de Física y Matemáticas
Universidad de Las Américas
Puebla, Mexico

ISBN 978-0-8176-8384-9 ISBN 978-0-8176-8385-6 (eBook)
DOI 10.1007/978-0-8176-8385-6
Springer New York Heidelberg Dordrecht London

Library of Congress Control Number: 2012948769

Mathematics Subject Classification (2010): 11A05, 15A16, 15A18, 15A21, 15A63, 15A63, 15A66, 15A69, 15A75, 17B45, 22E60, 51N10, 51N15, 53A04, 53A05, 53A30, 58A05, 65D05, 83A05, 20B30



Printed on acid-free paper

Springer is part of Springer Science+Business Media (www.birkhauser-science.com)

*I dedicate this book to those who search for truth and beauty no matter how treacherous and narrow that path may be.*

# Preface

This book provides an introduction to geometric algebra and its application to diverse areas of mathematics. It maintains the spirit of its predecessor, *Clifford Algebra to Geometric Calculus: A Unified Language for Mathematics and Physics*, and as such it has many unique features not seen in any other undergraduate textbook. It provides many innovative ways of looking at geometrical ideas and topics for student research and thesis projects.

The material has been developed over the many years that the author has taught undergraduate courses at the Universidad de Las Américas-Puebla, Mexico, in linear algebra, vector calculus, differential geometry, numerical analysis, modern algebra, and number theory. Whereas this book cannot be considered a textbook for all of these different subjects, there is a common theme they all share: they can all be efficiently formulated using the unified geometric number system advocated here. Geometric algebra, which has undergone extensive development in the second half of the twentieth Century, has its origins in the seminal works of Grassmann, Hamilton, and Clifford in the nineteenth century.

The book begins with the introduction of the spectral basis in modular number systems and in modular polynomials. This often overlooked concept provides insight into and greatly simplifies the proofs of many basic theorems in number theory and the corresponding closely related structure theorems of a linear operator. Since geometric numbers obey exactly the same algebraic rules as square matrices of real numbers, the languages are completely compatible and structure theorems that are valid for one are equally valid for the other.

The concept of a matrix as an array of numbers with an unintuitive multiplication rule hardly provides a geometric way of looking at things. Nevertheless, matrices have proven to be an extremely effective computational tool and have played a major role in the development of diverse areas of mathematics. Geometric algebra rectifies this defect by providing a geometric perspective, and many new algebraic tools. Combining both of these powerful systems by simply considering matrices whose elements are geometric numbers adds much needed geometric content and flexibility to both languages. The author hopes that this book captures both the idea and the spirit of the powerful geometric number system that has kept him going since he learned about the subject as a graduate student at Arizona State University many years ago.

We assume readers to have had undergraduate differential and integral calculus, a first course in modern algebra, and the mathematical maturity that an upper-level mathematics or physics undergraduate student might be expected to have. The many topics covered in the book should also appeal to first-year graduate students in mathematics, physics, engineering and computer science. Any unfamiliarity that a reader might have regarding mathematical terminology can be quickly overcome by a quick reference to the unlimited resources on the internet. We also recommend that the reader has knowledge of and access to symbolic mathematical software such as Mathematica or Maple. Such software considerably lightens the computational work required and makes for easy verification of results. A simple Mathematica package is provided for calculating the spectral basis for a modular polynomial.

There are three main groupings of interrelated core chapters:

- Chapters 1–5 introduce the fundamental concepts of a spectral basis of modular numbers and modular polynomials with applications in number theory, numerical analysis, and linear algebra. The hyperbolic numbers, introduced alongside the well-known complex numbers, are used to solve the cubic equation and provide a mathematical foundation for the theory of special relativity. The geometric extension of the real numbers is achieved by introducing new *anticommuting* square roots of plus or minus one which represent orthogonal directions in successively higher dimensions.
- Chapters 7–10 lay down the ideas of linear and multilinear algebra. Matrices of geometric numbers are considered throughout. New proofs of the Cayley–Hamilton Theorem, Gram–Schmidt orthogonalization, and the spectral decomposition of a linear operator are given in geometric algebra, as well as a comprehensive geometric interpretation of complex eigenvalues and eigenvectors in an Hermitian (definite or indefinite) inner product space.
- Chapters 13–16 develop the basic ideas of vector calculus and differential geometry in the context of geometric algebra. The classical integration theorems are derived from a single fundamental theorem of calculus. Manifolds are embedded in Euclidean or pseudo-Euclidean spaces and consequently have both intrinsic and extrinsic curvature, characterized by the projection and shape operators. Highlighted is a special treatment of conformal mappings and the conformal Weyl tensor, which have applications in physics and engineering.

Chapter 6 covers some of the more traditional topics in linear algebra which are not otherwise used in this book. Chapters 11, 12, 17, and 18 provide additional breadth and scope by treating the symmetric group, by giving a novel look at the concept of space-time in special relativity, by laying down the basic ideas of projective geometry, and by giving an introduction to Lie algebras and Lie groups, topics which are not usually covered in an undergraduate course. In the Table of Contents, a "*" is used to indicate those sections which are considerably more technical and may be omitted on first reading.

The author is indebted to many students and collegues from around the world who have contributed much during the many stages of the development of these ideas. Foremost he is indebted to David Hestenes, who first introduced him to geometric algebra many years ago as a graduate student at Arizona State University during the years 1965–1971. The author is indebted to Roman Duda of the Polish Academy of Sciences, Bernard Jancewicz, Jan Łopuszański, and Zbigniew Oziewicz of the Institute of Theoretical Physics in Wrocław, Poland, and Stony Brook University, for encouragement and support during these difficult early years. He wants to thank Rafal Ablamowicz (USA), Timothy Havel (USA), William Baylis (Canada), and Pertti Lounesto (Finland), who contributed in different ways to the writing of this book. In addition, he wants to thank Jaime Keller for inviting him to Mexico, Luis Verde-Star (Mexico), Waldyr Rodrigues (Brazil), Josep Parra (Spain), and Eduardo Bayro-Corrochano (Mexico). Among former students and now sometimes collaborators, he wants to thank José María Pozo (Spain), Marco Antonio Rodríguez (Mexico), Omar Leon Sanchez (Mexico), and Alejandra C. Vicente (Mexico). Graphics design artist Ana Sánchez Stone was a great help with all of the figures in Chap. 11 and in particular Fig. 11.6. The author is greatly indebted to Universidad de Las Américas-Puebla and Sistemas Nacionales de Investigadores de México, for many years of support. This book could not have been written without the constant support and encouragement over the years by my wife, Wanda, my mother and father and is a present for my mother's 100th birthday.

Cholula, México Garret Sobczyk

# Contents

# Chapter 1
# Modular Number Systems

*For out of olde feldes, as men seith,*
*Cometh al this newe corne fro yeere to yere;*
*And out of olde bokes, in good feith,*
*Cometh al this new science that men lere.*

–Chaucer

We begin by exploring the algebraic properties of the modular numbers, sometimes known as *clock arithmetic*, and the modular polynomials. The modular numbers and modular polynomials are based upon the *Euclidean algorithm*, which is simply the idea of dividing one integer into another or one polynomial into another polynomial, which we first learned in secondary school. Studying the modular number system leads us to introduce the concept of a *spectral basis*. This fundamental concept, which is largely neglected in elementary mathematics, will serve us well in our study of linear algebra and other topics in later chapters.[1]

## 1.1 Beginnings

In Euclid's *Elements*, Book VII, we find

> **Proposition 2:** *Given two numbers not prime to one another, to find their greatest common measure.*

Then follows what mathematicians refer to as the Euclidean algorithm [32]. We shall need the following consequence of this venerable algorithm. Given $r$ positive

[1]This chapter is based upon an article by the author that appeared in the American Mathematical Monthly [80].

G. Sobczyk, *New Foundations in Mathematics: The Geometric Concept of Number*, DOI 10.1007/978-0-8176-8385-6_1,

integers $h_1, h_2, \ldots, h_r \in \mathbb{N}$ whose *greatest common divisor* ($gcd$) is $1 \in \mathbb{N}$, then there exist integers $b_1, b_2, \ldots, b_r \in \mathbb{Z}$ with the property that

$$b_1 h_1 + b_2 h_2 + \cdots + b_r h_r = 1. \tag{1.1}$$

The justified fame of the Euclidean algorithm arrives from the fact that it has a much larger realm of applicability than just the integers. In particular, Let $\mathbb{K}$ be any *field* and let $\mathbb{K}[x]$ be the corresponding *integral domain* of polynomials over $\mathbb{K}$ [28, p.248, 250]. Given $r$ polynomials $h_1(x), h_2(x), \ldots, h_r(x) \in \mathbb{K}[x]$ whose greatest common divisor ($gcd$) is $1 \in \mathbb{K}$ (no common zeros), then there exist polynomials $b_1(x), b_2(x), \ldots, b_r(x) \in \mathbb{K}[x]$ with the property that

$$b_1(x)h_1(x) + b_2(x)h_2(x) + \cdots + b_r(x)h_r(x) = 1 \tag{1.2}$$

The identities (1.1) and (1.2), and the striking analogy between them, provide the basis for what follows.

## Examples

1. $gcd(4,15) = 1 \quad \Longrightarrow \quad 4 \cdot 4 + (-1) \cdot 15 = 1,$
2. $gcd(4,15,7) = 1 \quad \Longrightarrow \quad (-24) \cdot 4 + 6 \cdot 15 + (+1) \cdot 7 = 1,$
3. $gcd(x+1, x^2+1) = 1 \quad \Longrightarrow \quad (-1/2)(x-1)(x+1) + (1/2)(x^2+1) = 1,$
4. $gcd(x+1, x^2+1, x+2) = 1 \quad \Longrightarrow \quad x^2(x+1) + (x^2+1) - x^2(x+2) = 1.$

## 1.2 Modular Numbers

Given any integer $n \in \mathbb{Z}$ and any positive integer $h \in \mathbb{N}$, the Euclidean algorithm tells us that there is a unique integer $q \in \mathbb{Z}$ and a nonnegative integer $r$, $0 \le r < h$, such that

$$n = qh + r.$$

The set $\mathbb{Z}_h = \{0, 1, 2, \ldots, h-1\}$ of all possible remainders $r$, after division by $h$, denotes the *modular number system* modulo($h$) where $h \in \mathbb{N}$. The numbers $b \in \mathbb{Z}_h$ represent equivalence classes, and addition, multiplication, and equality in $\mathbb{Z}_h$ are defined modulo($h$). We write $b + c \overset{h}{=} d$ and $bc \overset{h}{=} d$ to mean that $b + c \equiv d \bmod(h)$ and $bc \equiv d \bmod(h)$. The modular number system $\mathbb{Z}_h$ is *isomorphic* to the *factor ring* $\mathbb{Z}/<h>$ for the *ideal*

$$<h> = \{0, \pm h, \pm 2h, \pm 3h, \ldots\} = \{nh\}_{n \in \mathbb{Z}}$$

over the integers $\mathbb{Z}$. In terms of the ideal $< h >$, the equivalence classes of $\mathbb{Z}_h$ are explicitly expressed by

$$\mathbb{Z}_h \stackrel{\mathrm{h}}{=} \{0+ < h >, 1+ < h >, \dots, h-1+ < h >\}. \tag{1.3}$$

The technical details, in the framework of modern algebra, can be found in [28, p.261, 262].

For any positive integer $h \in \mathbb{N}$, by *unique prime factorization*, we can write $h = p_1^{m_1} p_2^{m_2} \dots p_r^{m_r}$, where each $p_i$ is a distinct prime factor of $h$. We can also order the factors $p_i^{m_i}$ so that their multiplicities satisfy $1 \le m_1 \le m_2 \le \dots \le m_r$. Now define $h_i = h/p_i^{m_i}$ for $i = 1, \dots, r$. Since the $h_i$ have no common factor other than 1, (1.1) holds, and we have

$$b_1 h_1 + b_2 h_2 + \dots + b_r h_r = 1,$$

for an appropriate choice of the integers $b_i \in \mathbb{Z}$. Whereas this equation holds in $\mathbb{Z}$, it is just as valid when interpreted as an identity in $\mathbb{Z}_h$. Defining the numbers $s_i \stackrel{\mathrm{h}}{=} b_i h_i \in \mathbb{Z}_h$, we can rewrite the above identity as

$$s_1 + s_2 + \dots + s_r \stackrel{\mathrm{h}}{=} 1. \tag{1.4}$$

When interpreted as an identity among the numbers $s_i \in \mathbb{Z}_h$, the following additional important properties are easily verified by multiplying (1.4) on both sides by $s_i$ and simplifying modulo $h$. We find that

$$s_i^2 \stackrel{\mathrm{h}}{=} s_i \quad \text{and} \quad s_i s_j \stackrel{\mathrm{h}}{=} 0 \tag{1.5}$$

for $i, j = 1, 2, \dots, r$, and $i \neq j$. We say that the $s_i \in \mathbb{Z}_h$ are *mutually annihiliating idempotents* that partition unity. The set of numbers $\{s_1, s_2, \dots, s_r\}$ make up what we call the *spectral basis* of $\mathbb{Z}_h$.

Now suppose that $c \in \mathbb{Z}_h$. Multiplying both sides of the identity (1.4) by $c$ gives

$$cs_1 + cs_2 + \dots + cs_n \stackrel{\mathrm{h}}{=} c.$$

Since the $s_i$ are idempotents, they act as *projections* onto the modular numbers $\mathbb{Z}_{p_i^{m_i}}$; this is clear because from the definition of $s_i = b_i h_i$, it follows that $p_i^{m_i} s_i \stackrel{\mathrm{h}}{=} 0$. Thus, any number $c \in \mathbb{Z}_h$ can be written in the spectral basis as the unique *linear combination*

$$c \stackrel{\mathrm{h}}{=} \sum_{i=1}^{r} (c \bmod p_i^{m_i}) s_i \quad \text{in} \quad \mathbb{Z}_h, \tag{1.6}$$

of the basis elements $s_1, s_2, \dots, s_r$. This last identity is also known as the famous *Chinese Remainder Theorem*, dating back to the fourth century A.D. The interested reader may check out the web site

http://en.wikipedia.org/wiki/Chinese_remainder_theorem

The modular number systems $\mathbb{Z}_{p^m}$, modulo a power of a prime, play a particularly important role in Number Theory in that most modular problems reduce to problems involving a power of a prime. In dealing with such problems, it is best to represent numbers $a \in \mathbb{Z}_{p^m}$ in terms of the *p-adic number basis*

$$a = (a_{m-1}\, a_{m-2}\, \dots a_1\, a_0)_p = \sum_{i=0}^{m-1} a_i p^i \tag{1.7}$$

where each digit $a_i \in \mathbb{Z}_p$. Using (1.7) in (1.6), we find that

$$c \overset{h}{=} \sum_{i=1}^{r} (c \bmod p_i^{m_i}) s_i \overset{h}{=} \sum_{i=1}^{r} \sum_{j=0}^{m_i-1} c_{i,j}\, q_i^j, \tag{1.8}$$

where $q_i^0 = s_i$ and $q_i^j \overset{h}{=} p_i^j s_i$ for $i = 1,\dots,r$ and $j = 1,\dots,m_i - 1$. The set

$$\cup_{i=1}^{r}\{s_i, q_i, \dots, q_i^{m_i-1}\}$$

is called the *complete spectral basis* of $\mathbb{Z}_h$.

We are now in a position to directly solve for the idempotents $s_i$. Multiplying each side of the identity (1.4) by $h_i$ gives

$$h_i s_i \overset{h}{=} h_i,$$

which can be easily solved in $\mathbb{Z}_h$, getting $s_i = (h_i^{-1} \bmod p_i^{m_i})h_i$ for each $i = 1,2,\dots,r$. The $q_i \overset{h}{=} p_i s_i$ are *nilpotent* in $\mathbb{Z}_h$, for $i = 1,2,\dots,r$. The nilpotents $q_i$ have the *index of nilpotency* $m_i$, since $q_i^{m_i-1} \neq 0$ but $q_i^{m_i} \overset{h}{=} 0$ in $\mathbb{Z}_h$.

Let us calculate the complete spectral basis $\{s_1, s_2, q_2\}$ for $\mathbb{Z}_{12}$ where $h = 12 = 3 \cdot 2^2$, so that $p_1 = 3$ and $p_2 = 2$. By (1.4), we must have $s_1 + s_2 \overset{h}{=} 1$. By multiplying this equation by

$$h_1 = \frac{h}{p_1} = 2^2 \quad \text{and} \quad h_2 = \frac{h}{p_2^2} = 3,$$

we get with the help of (1.6)

$$4s_1 + 4s_2 \overset{h}{=} 4, \quad \text{or} \quad s_1 = (4 \bmod 3)s_1 \overset{h}{=} 4,$$

and

$$3s_1 + 3s_2 \overset{h}{=} 3, \quad \text{or} \quad 3s_2 = (3 \bmod 4) \overset{h}{=} 3 \implies s_2 \overset{h}{=} 9,$$

respectively. From $s_2 \overset{h}{=} 9$, we easily calculate $q_2 = 2s_2 = 18 \overset{h}{=} 6$, so the complete spectral basis for $z_{12}$ is

$$\{s_1 = 4, s_2 = 9, q_2 = 6\}. \tag{1.9}$$

Much of the power of the spectral basis is a consequence of the simple rules for multiplication of its idempotent and nilpotent elements. We give here the table of multiplication for the spectral basis (1.9) of $\mathbb{Z}_{12}$.

| $\cdot \bmod 12$ | $s_1$ | $s_2$ | $q_2$ |
|---|---|---|---|
| $s_1$ | $s_1$ | 0 | 0 |
| $s_2$ | 0 | $s_2$ | $q_2$ |
| $q_2$ | 0 | $q_2$ | 0 |

For a second example, consider $h = 360 = 5 \cdot 3^2 \cdot 2^3$ for which $h_1 = 3^2 \cdot 2^3$, $h_2 = 5 \cdot 2^3$ and $h_3 = 5 \cdot 3^2$. The spectral basis satisfying (1.4) and (1.5) is found to be

$$\{s_1 = 216, s_2 = 280, s_3 = 225\},$$

as we now show. To find $s_1$, multiply $s_1 + s_2 + s_3 = 1$ by $h_1 = 3^2 \cdot 2^3 = 72$, and use (1.6) to get

$$72 s_1 = 2 s_1 = 72 \text{ in } \mathbb{Z}_{360}$$

or $16 s_1 = s_1 = 8 \cdot 72 = 216$. Similar calculations give $s_2$ and $s_3$. An arbitrary $c \in \mathbb{Z}_{360}$ can now be written

$$c \stackrel{h}{=} c s_1 + c s_2 + c s_3 \stackrel{h}{=} (c_1)_5 s_1 + (c_2 c_3)_3 s_2 + (c_4 c_5 c_6)_2 s_3,$$

where $c_1 \in \mathbb{Z}_5$, $c_2, c_3 \in \mathbb{Z}_3$, and $c_4, c_5, c_6 \in \mathbb{Z}_2$. The complete spectral basis of $\mathbb{Z}_{360}$ is

$$\{s_1 = 216, s_2 = 280, q_2 = 120, s_3 = 225, q_3 = 90, q_3^2 = 180\}. \tag{1.10}$$

The multiplication table for spectral basis of $\mathbb{Z}_{360}$ is given below:

| $\cdot \bmod 360$ | $s_1$ | $s_2$ | $q_2$ | $s_3$ | $q_3$ | $q_3^2$ |
|---|---|---|---|---|---|---|
| $s_1$ | $s_1$ | 0 | 0 | 0 | 0 | 0 |
| $s_2$ | 0 | $s_2$ | $q_2$ | 0 | 0 | 0 |
| $q_2$ | 0 | $q_2$ | 0 | 0 | 0 | 0 |
| $s_3$ | 0 | 0 | 0 | $s_3$ | $q_3$ | $q_3^2$ |
| $q_3$ | 0 | 0 | 0 | $q_3$ | $q_3^2$ | 0 |
| $q_3^2$ | 0 | 0 | 0 | $q_3^2$ | 0 | 0 |

Employing the spectral basis, we also have an easy formula for finding the inverse $b^{-1}$ of $b \in \mathbb{Z}_h$. We have

$$b^{-1} \stackrel{h}{=} \sum_{i=1}^{r} (b_{i(m_i-1)} \dots b_{i0})_{p_i}^{-1} s_i,$$

so the problem of finding $b^{-1} \in \mathbb{Z}_h$ is reduced to the problem of finding the inverse in a prime power modular number system $\mathbb{Z}_{p^m}$. For example, using the spectral basis (1.9) for $\mathbb{Z}_{12}$, we can easily calculate the inverse $7^{-1}$ of $7 \in \mathbb{Z}_{12}$. We first write

$$7s_1 + 7s_2 = s_1 + 3s_2 \stackrel{h}{=} 7 \iff \frac{1}{1}s_1 + \frac{1}{3}s_2 \stackrel{h}{=} \frac{1}{7}.$$

But $\frac{1}{3} \bmod 4 = -1 \bmod 4$, so $\frac{1}{7} \stackrel{h}{=} s_1 - s_2 = 4 - 9 \stackrel{h}{=} 7$. We can easily check that $7 \cdot 7 = 49 = 1 \bmod 12$ as required.

Using $p$-adic notation, it is easy to understand that a number $b \in \mathbb{Z}_{p^m}$ will be divisible by $p^k$ if and only if $b = (b_{m-1} \dots b_k 0 \dots 0)_p$. Also, $b$ will not be divisible by $p^{k+1}$ if $b_k \neq 0$. The *Euler phi function* or *totient function* http://en.wikipedia.org/wiki/Euler's_totient_function $\phi(c)$, for $c \in \mathbb{N}$, is defined to be the number of positive integers $b$, $1 \leq b < c$, *relatively prime* to $c$, i.e., such that $gcd(b,c) = 1$, [61, p.51]. It follows that if $p \in \mathbb{N}$ is a prime, then $\phi(p^m) = (p-1)p^{m-1}$.

Since a number

$$c \stackrel{h}{=} \sum_{i=1}^{r} (c_{i(m_i-1)} \dots c_{i0})_{p_i} s_i$$

will be *relatively prime* to $h$ if and only if $c_{i0} \neq 0 \bmod (p_i)$ for $i = 1, \dots, r$, it follows that

$$\phi(h) = \prod_{i=1}^{r} (p_i - 1) p_i^{m_i - 1}$$

for the composite number $h = \prod_{i=1}^{r} p_i^{m_i}$. Since the product of any two elements $x, y \in \mathbb{Z}_h$ which are relatively prime to $h$ is also relatively prime to $h$, it follows that all the elements in $\mathbb{Z}_h$ which are relatively prime to $h$ form a multiplicative group of order $\phi(h)$, called the *U-group* $U(h)$ [28]. Once again, appealing to the Theorem of Lagrange for groups, we have Euler's generalization of Fermat's theorem that $b^{\phi(h)} = 1 \bmod (h)$ for each $b \in \mathbb{Z}_h$ such that $gcd(b,h) = 1$.

A very basic result of number theory is *Fermat's theorem* that states $b^{\phi(p)} = 1 \bmod (p)$, where $p$ is any prime number and $gcd(p,b) = 1$. The Fermat theorem is an immediate consequence of the fact that the nonzero elements of the finite field $\mathbb{Z}_p$ under multiplication make up a *group* of order $\phi(p) = p - 1$ and the *Theorem of Lagrange* which tells us that the order of each element of a group must divide the order of the group.

## Exercises

1. Calculate

   (a) $3 + 8 \bmod 11$ and $3 \cdot 8 \bmod 11$.
   (b) $3 + 4 =$ in $\mathbb{Z}_{11}$, $3 \cdot 4 =$ in $\mathbb{Z}_{11}$.
   (c) Find $3^{-1} =$ and $5^{-1} =$ in $\mathbb{Z}_{11}$.
   (d) Find $3^{-1} =$ and $5^{-1} =$ in $\mathbb{Z}_7$.

2. (a) Write $87 = (a_1 a_0)_5$ as a 5-adic integer in $\mathbb{Z}_{25}$.
   (b) Write $53 = (b_1 b_0)_5$ as a 5-adic integer in $\mathbb{Z}_{25}$.

(c) Use (a) and (b) to find $87+53$ and $87\cdot 53$ in $\mathbb{Z}_{25}$.
(d) Find $87^{-1}$ and $53^{-1}$ in $\mathbb{Z}_{25}$.

3. (a) Write $35=(p_1p_0)_3$ as a 3-adic integer in $\mathbb{Z}_9$.
   (b) Write $53=(b_1b_0)_3$ as a 3-adic integer in $\mathbb{Z}_9$.
   (c) Use (a) and (b) to find $35+53$ and $35\cdot 53$ in $\mathbb{Z}_9$.
   (d) Find $35^{-1}$ and $53^{-1}$ in $\mathbb{Z}_9$.
4. (a) Write $87=(p_2p_1p_0)_2$ as a 2-adic integer in $\mathbb{Z}_8$.
   (b) Write $53=(b_2b_1b_0)_2$ as a 2-adic integer in $\mathbb{Z}_8$.
   (c) Use (a) and (b) to find $87+53$ and $87\cdot 53$ in $\mathbb{Z}_8$.
   (d) Find $87^{-1}$ and $53^{-1}$ in $\mathbb{Z}_8$.
5. (a) Write $35=(p_2p_1p_0)_3$ as a 3-adic integer in $\mathbb{Z}_{27}$.
   (b) Write $53=(b_2b_1b_0)_3$ as a 3-adic integer in $\mathbb{Z}_{27}$.
   (c) Use (a) and (b) to find $35+53$ and $35\cdot 53$ in $\mathbb{Z}_{27}$.
   (d) Use (a) and (b) to find $35^{-1}$ and $53^{-1}$ in $\mathbb{Z}_{27}$.
6. (a) Find the complete spectral basis $\{s_1,s_2,q_2\}$ of $h=2\cdot 3^2=18$ in $\mathbb{Z}_h$.
   (b) Express $a=35$ and $b=51$ in terms of the spectral basis in $\mathbb{Z}_{18}$.
   (c) Find $35+51$ in $\mathbb{Z}_{18}$ using the spectral basis.
   (d) Find $35\cdot 51$ in $\mathbb{Z}_{18}$ using the spectral basis.
   (e) Find $35^{-1}$ and $51^{-1}$ in $\mathbb{Z}_{18}$ if they exist using the spectral basis. If an inverse does not exist, justify your answer.
7. (a) Find the complete spectral basis $\{s_1,s_2,q_2,s_3,q_3\}$ of $h=3^2\cdot 2^2\cdot 5^2=900$ for $\mathbb{Z}_h$.
   (b) Express $a=351$ and $b=511$ in terms of the spectral basis in $\mathbb{Z}_{900}$.
   (c) Find $351+511$ in $\mathbb{Z}_{900}$ using the spectral basis.
   (d) Find $351\cdot 511$ in $\mathbb{Z}_{900}$ using the spectral basis.
   (e) Find $351^{-1}$ and $511^{-1}$ in $\mathbb{Z}_{900}$ if they exist. If an inverse does not exist, justify your answer.
8. Referring back to the previous problems, find the Euler function $\phi(h)$ for $h=$ 7, 11, 25, 35, 8, 27, 18, and $h=900$.
9. (a) Use problem 6 (e) to solve $35x=51 \bmod(18)$.
   (b) Use problem 7 (e) to solve $351x=511 \bmod(900)$.
10. (a) Use the spectral basis found in problem 6 to find all solutions to the equation $x^2=1 \bmod(18)$.
    (b) Use the spectral basis found in problem 6 to find all solutions to the equation $x^2=-1 \bmod(18)$.
11. Solve the system of congruences

$$x=3 \text{ modulo } 13$$

$$x=4 \text{ modulo } 16$$

$$x=5 \text{ modulo } 21$$

    for their common solution $x$.
12. Verify that the $s_i$ in (1.5) are mutually annihiliating idempotents as claimed.

## 1.3 Modular Polynomials

We now treat the *modular polynomial ring* $\mathbb{K}[x]_h \equiv \mathbb{K}[x]/\!<\!h(x)\!>$ of the polynomial $h \equiv h(x)$ over an arbitrary field $\mathbb{K}$ in an exactly analogous way to the modular numbers $\mathbb{Z}_h$. In this case, the ideal $< h(x) >$ over the polynomial ring $\mathbb{K}[x]$ is defined by

$$<\!h(x)\!>= \{f(x)h(x)| \quad f(x) \in \mathbb{K}[x]\}.$$

In terms of the ideal $< h(x) >$, the equivalence classes of $\mathbb{K}[x]_h$ are explicitly expressed by

$$\mathbb{K}[x]_h = \{0+<\!h(x)\!>, x+<\!h(x)\!>, \ldots, x^{m-1}+<\!h(x)\!>\}, \tag{1.11}$$

where $m = \deg(h(x))$ is the degree of the polynomial $h(x)$, [84].

Addition of polynomials in $\mathbb{K}[x]_h$ is just the ordinary addition of polynomials, and the multiplication of polynomials in $\mathbb{K}[x]_h$ is done $\mathrm{mod}(h(x))$ by using the Euclidean algorithm for polynomials. Thus, for $f(x), g(x) \in \mathbb{K}[x]_h$, $f(x)\cdot g(x) \equiv r(x) \bmod(h(x))$ if

$$f(x)\cdot g(x) = q(x)h(x) + r(x)$$

where $r(x) = 0$ or $\deg(r(x)) < \deg(h(x))$. More simply, we write

$$f(x)g(x) \stackrel{\mathrm{h}}{=} r(x).$$

Addition and multiplication in $\mathbb{K}[x]_h$ is always done modulo the polynomial $h(x)$. We are particularly interested in the number fields $\mathbb{K}$ defined by $\mathbb{R}$ the real numbers, $\mathbb{C}$ the complex numbers, and $\mathbb{Z}_p$ the *finite modular number fields* defined for prime numbers $p \in \mathbb{N}$.

Now let $h(x) = p_1^{m_1} p_2^{m_2}, \ldots, p_r^{m_r}$, where each $p_i^{m_i} \equiv (x - x_i)^{m_i}$ for distinct $x_i \in \mathbb{K}$. Also, we order the factors of the $m \equiv \sum_{i=1}^{r} m_i$ degree polynomial $h(x)$ so that the multiplicities of the roots $x_i$ satisfy the inequalities $1 \le m_1 \le m_2 \le \cdots \le m_r$. Defining the polynomials $h_i(x) = h(x)/p_i^{m_i}(x)$, we see that the greatest common divisor of the $h_i(x)$ is $1 \in \mathbb{K}[x]$, and therefore we can invoke (1.2) of the Euclidean algorithm to conclude that there exist polynomials $b_i(x) \in \mathbb{K}[x]$ which satisfy

$$b_1(x)h_1(x) + b_2(x)h_2(x) + \cdots + b_r(x)h_r(x) = 1.$$

Whereas the above equation holds in $\mathbb{K}[x]$, it remains equally valid when interpreted as an identity in the modular polynomial ring $\mathbb{K}[x]_{h(x)}$. Defining the polynomials $s_i(x) \stackrel{\mathrm{h}}{=} b_i(x)h_i(x) \in \mathbb{K}[x]_h$, we can rewrite (1.2) as

$$s_1(x) + s_2(x) + \cdots + s_r(x) \stackrel{\mathrm{h}}{=} 1 \tag{1.12}$$

The following additional important properties easily follow from (1.2):

$$s_i^2(x) \stackrel{h}{=} s_i(x) \ \text{ and } \ s_i(x)s_j(x) \stackrel{h}{=} 0, \tag{1.13}$$

for $i, j = 1, 2, \ldots, r$, and $i \neq j$. We say that the $s_i(x) \in \mathbb{K}[x]_h$ are mutually annihiliating idempotents that partition unity. The set of polynomials $\{s_1(x), s_2(x), \ldots, s_r(x)\}$ make up the *spectral basis* of the polynomial ring $\mathbb{K}[x]_h$. By the *complete spectral basis* of $\mathbb{K}[x]_h$, we mean the set of $m$ polynomials

$$\left\{s_1, q_1, \ldots, q_1^{m_i-1}, \ldots, s_r, q_r, \ldots, q_r^{m_r-1}\right\},$$

where $q_i^0 = s_i(x)$ for $i = 1, \ldots, r$, and $q_i^j \stackrel{h}{=} (x - x_i)^j s_i(x) \neq 0$ for $j = 1, \ldots, m_i - 1$. We shall discuss properties of the *nilpotents* $q_i(x)$ shortly.

Every element $b(x) \in \mathbb{K}[x]_h$ has the form

$$b(x) = b_1 + b_2 x + \cdots + b_{m-1} x^{m-1} \in b(x) + < h(x) >$$

for $b_i \in \mathbb{K}$. For this reason, we refer to the elements of $\mathbb{K}[x]_h$ as *polynomial numbers*. The modular polynomials $\mathbb{K}[x]_h$, under addition, have the structure of an $m$-dimensional *vector space* over $\mathbb{K}$ with the *standard basis* $\{1, x, x^2, \ldots, x^{m-1}\}$. We will formally give the definition of a vector space in Chap. 4, Definition (4.1.1). Note that the standard basis of $\mathbb{K}[x]_h$ has exactly the same number of elements as the complete spectral basis of $\mathbb{K}[x]_h$, namely,

$$m = \sum_{i=1}^{r} m_i = \deg(h(x)).$$

Every polynomial $b(x) \in K[x]_h$ can be expressed as a unique linear combination of the powers of $x$ in the standard basis. If we now multiply both sides of the identity (1.12) by $b(x)$, we get

$$b(x)s_1(x) + b(x)s_2(x) + \cdots + b(x)s_r(x) \stackrel{h}{=} b(x).$$

Since the $s_i(x)$ are idempotents in $\mathbb{K}[x]_h$, they act as *projections* onto the polynomial rings $\mathbb{K}[x]_{(x-x_i)^{m_i}}$. Thus, any polynomial $b(x) \in \mathbb{K}[x]_h$ can be written in the unique form

$$b(x) \stackrel{h}{=} \sum_{i=1}^{r} (b(x) \bmod (x - x_i)^{m_i}) s_i(x). \tag{1.14}$$

This is known as the *Chinese Remainder Theorem* for the modular polynomials in $\mathbb{K}[x]_h$.

The modular polynomial ring $\mathbb{K}[x]_{(x-x_0)^m}$ for $x_0 \in \mathbb{K}$ plays exactly the same role in the theory of modular polynomials that numbers modulo the power of a prime play in number theory; most modular polynomial problems reduce to problems involving a power of $x-x_0$. In dealing with such problems, it is best to express the modular polynomial $f(x) \in \mathbb{K}[x]_{(x-x_0)^m}$ in terms of the $(x-x_0)$-adic number basis

$$f(x) = \sum_{i=0}^{m-1} a_i(x-x_0)^i \equiv (a_{m-1}a_{m-2}\dots a_1a_0)_{x-x_0} \tag{1.15}$$

where each coefficient $a_i \in \mathbb{K}$. Addition and multiplication is done in $\mathbb{K}[x]_{(x-x_0)^m}$ by taking advantage of our proficiency in working with decimal digits and truncating after $m$-digits, just as we worked in the $p$-adic number base (1.7). We will give an example of such a calculation shortly.

We are now in a position to directly solve for the idempotents $s_i(x)$ in (1.12). Multiplying each side of the identity (1.12) by $h_i(x)$ gives

$$h_i(x)s_i(x) = h_i(x) \quad \text{in} \quad \mathbb{K}[x]_h \tag{1.16}$$

which can be easily solved in $\mathbb{K}[x]_h$, getting

$$s_i(x) = [h_i(x)^{-1} \bmod (x-x_i)^{m_i}]h_i(x)$$

for each $i = 1,2,\dots,r$. The nilpotents

$$q_i(x) \stackrel{\mathrm{h}}{=} (x-x_i)s_i(x) \tag{1.17}$$

for $i = 1,2,\dots,r$, defined earlier, have the *indexes of nilpotency* $m_i$ in $\mathbb{K}[x]_h$ since $q_i^{m_i} = (x-x_i)^{m_i}s_i(x) \stackrel{\mathrm{h}}{=} 0$. For $x \in \mathbb{K}[x]_h$, using (1.14), we get the important *spectral decomposition formula*

$$x \stackrel{\mathrm{h}}{=} \sum_{i=1}^{r}(x-x_i+x_i)s_i \stackrel{\mathrm{h}}{=} \sum_{i=1}^{r} x_is_i + q_i \stackrel{\mathrm{h}}{=} \sum_{i=1}^{r}(x_i+q_i)s_i \text{ in } \mathbb{K}[x]_h. \tag{1.18}$$

As our first example, consider the modular polynomial ring $\mathbb{R}[x]_h$, where

$$h(x) = p_1(x)p_2(x) = (x+1)(x-1)^2 \in \mathbb{R}[x]$$

for $p_1(x) = (x+1)$ and $p_2(x) = (x-1)^2$. We calculate

$$h_1(x) = \frac{h(x)}{p_1(x)} = (x-1)^2, \quad \text{and} \quad h_2(x) = \frac{h(x)}{p_2(x)} = x+1.$$

To find $s_1(x)$, we multiply (1.12) with $r=2$ by $h_1(x)$ to get

$$(x-1)^2s_1(x) + (x-1)^2s_2(x) \stackrel{\mathrm{h}}{=} (x-1)^2 \implies ((x+1)-2)^2s_1(x) + 0s_2(x) \stackrel{\mathrm{h}}{=} (x-1)^2$$

or

$$((x+1)^2-4(x+1)+4)s_1(x)\stackrel{h}{=}(x-1)^2 \implies s_1(x)=\frac{1}{4}(x-1)^2$$

since $(x+1)^k s_1(x)\stackrel{h}{=}((x+1)^k \bmod (x+1))s_1(x)=0$ for all $k\geq 1$. Similarly, to find $s_2(x)$, we multiply $s_1(x)+s_2(x)\stackrel{h}{=}1$ by $h_2(x)$ to get

$$(x+1)s_1(x)+(x+1)s_2(x)\stackrel{h}{=}(x+1) \quad\implies\quad (x+1)s_2(x)\stackrel{h}{=}(x+1).$$

Noting that $x+1=(x-1)+2$ and multiplying $(x+1)s_2(x)\stackrel{h}{=}(x+1)$ by $(x-1)-2$ gives

$$((x-1)^2-4)s_2(x)\stackrel{h}{=}(x-3)(x+1) \implies s_2(x)\stackrel{h}{=}-\frac{1}{4}(x-3)(x+1).$$

Finally, we calculate

$$q_2(x)\stackrel{h}{=}(x-1)s_2(x)=-\frac{1}{4}(x-1)(x-1-2)(x+1)\stackrel{h}{=}\frac{1}{2}(x-1)(x+1),$$

so the complete spectral basis for $\mathbb{R}[x]_h$ is

$$\{s_1(x)=\frac{1}{4}(x-1)^2,\ s_2(x)\stackrel{h}{=}-\frac{1}{4}(x-3)(x+1),\ q_2(x)=\frac{1}{2}(x-1)(x+1)\}. \quad (1.19)$$

Since the multiplication table for the spectral basis (1.19) of $\mathbb{R}[x]_h$ is exactly the same as for the spectral basis (1.9) of $\mathbb{Z}_{12}$, we will not reproduce it here.

For our second example, consider the modular polynomial ring $\mathbb{R}[x]_h$ for the polynomial $h(x)=(x-1)(x+1)^2x^3$. We calculate

$$h_1(x)=(x+1)^2x^3,\ \ h_2(x)=(x-1)x^3,\ \text{ and }\ h_3=(x-1)(x+1)^2.$$

To find $s_1(x)$, multiply (1.12) by $h_1(x)$ to get

$$h_1(x)s_1(x)\stackrel{h}{=}(x-1+2)^2(x-1+1)^3s_1(x)\stackrel{h}{=}(x+1)^2x^3 \ \text{ in } \ \mathbb{R}[x]_h,$$

or $s_1(x)=\frac{(x+1)^2x^3}{4}$. Similarly, from

$$h_2(x)s_2(x)\stackrel{h}{=}(x-1)x^3s_2(x)\stackrel{h}{=}(x-1)x^3 \ \text{ in } \ \mathbb{R}[x]_h,$$

we get $(x+1-2)(x+1-1)^3s_2(x)=(x-1)x^3$ in $\mathbb{R}[x]_h$. This last relation can be expressed in the $(x+1)$-adic form $(1,-2)_{(x+1)}\cdot(1,-1)^3_{(x+1)}s_2(x)=(x-1)x^3$, from which we calculate successively

$$(1,-2)_{x+1}\cdot(3,-1)_{(x+1)}s_2(x)=(x-1)x^3$$

$$(-7,2)_{(x+1)}s_2(x)=(x-1)x^3,$$

and finally, by multiplying both sides of this last equation by $7(x+1)+2=(7,2)_{(x+1)}$ and dividing by 4, we get

$$s_2(x)\overset{h}{=}\frac{(7x+9)(x-1)x^3}{4}.$$

Since

$$s_3(x)\overset{h}{=}1-s_1(x)-s_2(x)\overset{h}{=}-(x-1)(x+1)^2(2x^2-x+1)\ \text{in}\ \mathbb{R}[x]_h$$

our calculations of the spectral basis of $\mathbb{R}[x]_h$ for $h(x)=(x-1)(x+1)^2x^3$ is complete.

However, for the complete spectral basis, we must also calculate the corresponding nilpotent elements and their powers. We find that

$$q_2(x)\overset{h}{=}(x+1)s_2(x)\overset{h}{=}\frac{(x-1)(x+1)x^3}{2},$$

$$q_3(x)\overset{h}{=}xs_3(x)\overset{h}{=}(x-1)^2x(x+1)^2x^2$$

and

$$q_3^2(x)\overset{h}{=}x^2s_3(x)\overset{h}{=}-(x-1)(x+1)^2x^2.$$

The complete spectral basis of $\mathbb{R}[a]_h$ for $h(x)=(x-1)(x+1)^2x^3$ is thus

$$\{s_1(x),s_2(x),q_2(x),s_3(x),q_3(x),q_3^2(x)\}. \tag{1.20}$$

Again, since the multiplication table for the spectral basis of $R[x]_h$ is the same as the multiplication table for the spectral basis (1.10) of $\mathbb{Z}_{360}$, we need not repeat the multiplication table here.

An arbitrary polynomial $f(x)\in\mathbb{R}[x]_h$ for $h=(x-1)(x+1)^2x^3$ can now be written in the form

$$\begin{aligned}f(x)&=f(x)s_1+f(x)s_2+f(x)s_3\\&=(a_0)_{x-1}s_1(x)+(b_1b_0)_{x+1}s_2(x)+(c_2c_1c_0)_{x^3}s_3(x),\end{aligned}$$

where $(a_0)_{x-1}=f(x)\text{mod}(x-1)$, $(b_1,b_0)_{x+1}=f(x)\text{mod}(x+1)^2$ and $(c_2,c_1,c_0)_x=f(x)\text{mod}(x^3)$. The inverse $[f(x)]^{-1}$ of $f(x)$ in $\mathbb{R}[x]_h$ will exist iff $a_0,b_0,c_0\in\mathbb{R}$ are all nonzero.

For example, for $f(x) = (x+3)^2$, we calculate

$$\begin{aligned}(x+3)^2 &= (x-1+4)^2 = (x-1)^2 + 8(x-1) + 16 \\ &= (x+1+2)^2 = (x+1)^2 + 4(x+1) + 4 = x^2 + 6x + 9,\end{aligned}$$

from which it follows that

$$(x+3)^2 \stackrel{h}{=} 16s_1 + (4+4q_2)s_2 + (9+6q_3+q_3^2)s_3.$$

The inverse of $(x+3)^2 \bmod ((x-1)(x+1)^2x^3)$ is given by

$$\frac{1}{(x+3)^2} \stackrel{h}{=} \frac{1}{16}s_1 + \frac{1}{4+q_2}s_2 + \frac{1}{9+6q_3+q_3^2}s_3$$

which simplifies to

$$\frac{1}{(x+3)^2} \stackrel{h}{=} \frac{1}{16}s_1 + \frac{4-q_2}{16}s_2 + \left(\frac{1}{9} - \frac{2}{27}q_3 + \frac{1}{27}q_3^2\right)s_3.$$

Other examples can be found in [79].

## Exercises

1. Let $h = x(x-1)(x+1)$.
   (a) Find the spectral basis $\{s_1, s_2, s_3\}$ for $\mathbb{R}[x]_h$.
   (b) Find the spectral decomposition formula (1.18) for $x \in \mathbb{R}[x]_h$.
   (c) Verify that $x+3 \stackrel{h}{=} 3s_1 + 4s_2 + 2s_3$ and $\frac{1}{x+3} \stackrel{h}{=} \frac{1}{3}s_1 + \frac{1}{4}s_2 + \frac{1}{2}s_3$.
   (d) Find all the square roots: $\sqrt{x+3}$ of $x+3 \in \mathbb{R}[x]_h$.
   (e) Fill in the following multiplication table for the spectral basis of $\mathbb{R}[x]_h$.

| $\cdot \bmod h(x)$ | $s_1$ | $s_2$ | $s_3$ |
|---|---|---|---|
| $s_1$ | | | |
| $s_2$ | | | |
| $s_3$ | | | |

2. Let $h = (x-1)(x+1)^2$.
   (a) Find the spectral basis $\{s_1, s_2, q_2\}$ for $\mathbb{R}[x]_h$.
   (b) Find the spectral decomposition formula (1.18) for $x \in \mathbb{R}[x]_h$.
   (c) Verify that $x+3 \stackrel{h}{=} 2s_1 + (4+q_2)s_2$ and $\frac{1}{x+3} \stackrel{h}{=} \frac{1}{2}s_1 + \frac{4-q_2}{16}s_2$.
   (d) Find all the square roots: $\sqrt{x+3}$ of $x+3 \in \mathbb{R}[x]_h$.
   (e) Fill in the following multiplication table for the spectral basis of $\mathbb{R}[x]_h$.

| $\cdot \bmod h(x)$ | $s_1$ | $s_2$ | $q_2$ |
|---|---|---|---|
| $s_1$ | | | |
| $s_2$ | | | |
| $q_2$ | | | |

## 1.4 Interpolation Polynomials

One important application of the spectral basis of the modulo polynomials $\mathbb{R}[x]_h$ is that it allows us to easily calculate the *Lagrange–Sylvester interpolation polynomial* of any function $f(x)$, which is *analytic* (smooth and continuous) at the spectral points $\{x_1, x_2, \ldots, x_r\} \in \mathbb{R}$, simply by replacing $x$ in $f(x)$ by (1.18) and reducing modulo $h(x)$. The Lagrange–Sylvester interpolation polynomial $g(x) \in \mathbb{R}[x]_h$ of $f(x)$ is defined by

$$\begin{aligned} g(x) &\overset{h}{=} f\Big(\sum_{j=1}^{r}(x_j+q_j)s_i\Big) \overset{h}{=} \sum_{j=1}^{r} f(x_j+q_j)s_j \\ &\overset{h}{=} \sum_{j=1}^{r}\Big[f(x_j)+\frac{f'(x_j)}{1!}q_j(x)+\cdots+\frac{f^{(m_j-1)}(x_j)}{(m_j-1)!}q_j^{m_j-1}(x)\Big]s_j(x), \\ &\overset{h}{=} \sum_{j=1}^{r}\Big[f(x_j)+\frac{f'(x_j)}{1!}(x-x_j)+\cdots+\frac{f^{(m_j-1)}(x_j)}{(m_j-1)!}(x-x_j)^{m_j-1}\Big]s_j(x) \\ &\overset{h}{=} \sum_{j=1}^{r}\left(\frac{f^{(m_j-1)}(x_j)}{(m_j-1)!},\ldots,\frac{f'(x_j)}{1!},f(x_j)\right)_{x-x_j} s_j(x), \end{aligned} \tag{1.21}$$

[29, p. 101]. Note that to find the interpolation polynomial $g(x)$, we are using the spectral decomposition of $x \in \mathbb{R}[x]_h$, (1.18), and evaluating $f(x)$ by expanding in Taylor series about the spectral points $x_j$.

For example, let $h(x) = (x-1)(x+1)^2x^3$ as above, and let $f(x)$ be any function for which the values

$$f(1),\ f'(-1),\ f(-1),\ f^{(2)}(0),\ f'(0),\ f(0)$$

are defined. In terms of the spectral basis $\{s_1(x), s_2(x), s_3(x)\}$ of $\mathbb{R}[x]_h$, the interpolation polynomial $g(x) \in \mathbb{R}[x]_h$ of $f(x)$ is given by

$$g(x) = \left(\frac{f(1)}{0!}\right)_{x-1} s_1(x) + \left(\frac{f^{(1)}(-1)}{1!}, f(1)\right)_{x+1} s_2(x) + \left(\frac{f^{(2)}(0)}{(2)!}, \frac{f'(0)}{1!}, f(0)\right)_{x} s_3(x).$$

Shown in the Figs. 1.1 and 1.2 are the graphs of $f(x)$ and its interpolation polynomial $g(x)$, for $f(x) = \frac{1}{x+1/2}$, and $f(x) = x\sin 5x$. Note that the interpolation polynomials agree with the given functions and their derivatives evaluated at the spectral points $x_i$ of the respective polynomials $h(x)$.

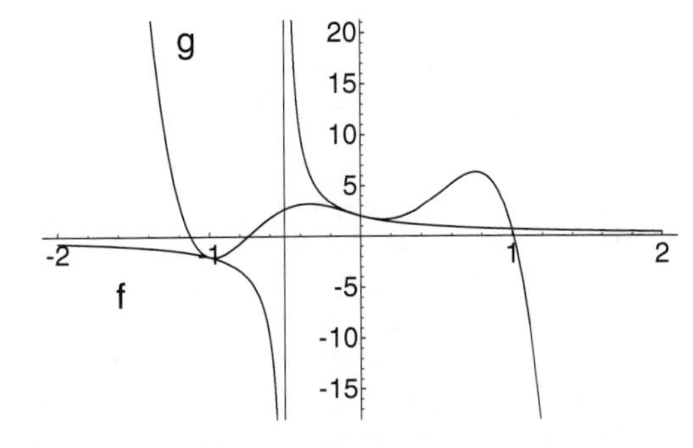


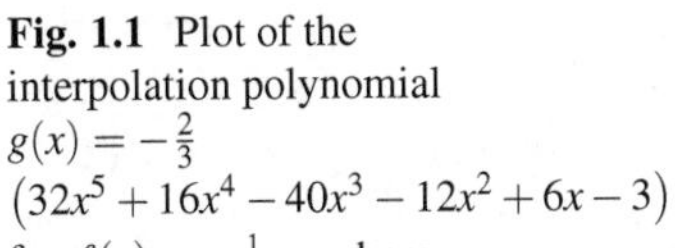
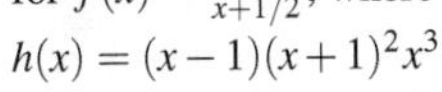

**Fig. 1.1** Plot of the interpolation polynomial $g(x) = -\frac{2}{3}(32x^5 + 16x^4 - 40x^3 - 12x^2 + 6x - 3)$ for $f(x) = \frac{1}{x+1/2}$, where $h(x) = (x-1)(x+1)^2x^3$

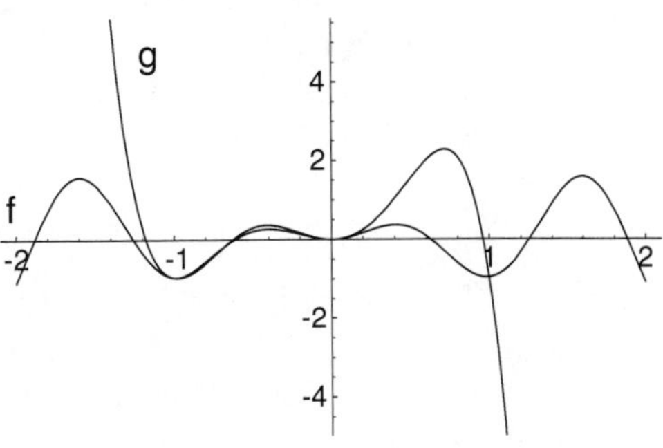


**Fig. 1.2** Plot of the interpolation polynomial $g(x) = 7.148x^5 - 5.959x^4 + 7.148x^3 + 5x^2$ for $f(x) = x\sin 5x$, where $h(x) = (x-1)(x+1)^2x^3$

The graphs of the figures were made using Mathematica. The following package in Mathematica can be used to calculate the spectral basis.

```
(* A Mathematica Package *)
(* POWER SERIES METHOD FOR EVALUATING SPECTRAL
    BASIS *)
(* OF POLYNOMIALS *)
(* Calculates p[i]=pqs[m,i,0] and q[i]^k=pqs[m,i,k]
 using power series for i=1,2, ... , Length[m],
 where m={m1,m2,...mr}*)
 psi[m_List]:=Product[(t - x[j])^(m[[j]]),
   {j,Length[m]}]
 psi[m_List,i_]:=psi[m]/((t-x[i])^(m[[i]]))
 pqs[m_,i_,k_]:=
 Together[Normal[Series[1/psi[m,i],{t,x[i],m[[i]]-
 (1+k)}]]]*(t-x[i])^k *psi[m,i]
(* Calculates p[i] and q[i] for i=1,2, ... ,
  Length[m] *)
 pqs[m_List]:=Do[p[i]=pqs[m,i,0];q[i]=pqs[m,i,1];
   Print[i],
 {i,Length[m]}]
(*end*)
```

There are interesting possibilities for the generalization of interpolation polynomials to apply to polynomials in several variables or to a function defined by a convergent power series such as the trigonometric or exponential functions, [83].

## Exercises

1. Let $h = (x+1)(x-1)^2$ in $\mathbb{Z}_3[x]_h$.
   (a) Make a multiplication table for the basis elements of the standard basis $\{1,x,x^2\}$. (Note that addition of the basis elements is just ordinary addition of polynomials.)
   (b) Find the spectral basis $\{s_1(x), s_2(x), q_2(x)\}$ of $\mathbb{Z}_3[x]_h$.
   (c) Express each element of the standard basis of $\mathbb{Z}_3[x]_h$ in terms of the spectral basis of $\mathbb{Z}_3[x]_h$.
   (d) Use part (c) to find the inverse $x^{-1}$ of $x$ and the inverse of $x^{-2}$ of $x^2$ in $\mathbb{Z}_3[x]_h$.
2. Let $h(x) = (x+1)(x-1)^2$ in $\mathbb{R}[x]_h$.
   (a) Make a multiplication table for the basis elements of the standard basis $\{1,x,x^2\}$. (Note that addition of the basis elements is just ordinary addition of polynomials.)
   (b) Find the spectral basis $\{s_1(x), s_2(x), q_2(x)\}$ of $\mathbb{R}[x]_h$.
   (c) Express each element of standard basis of $\mathbb{R}[x]_h$ in terms of the spectral basis of $\mathbb{R}[x]_h$.
   (d) Use part c) to find the inverse $x^{-1}$ of $x$ and the inverse of $x^{-2}$ of $x^2$ in $\mathbb{R}[x]_h$.
   (e) Find the Lagrange–Sylvester interpolation polynomial $g(x) \in \mathbb{R}[x]_h$ which best approximates the function $f(x) = \sqrt{x+2}$ in the interval $(-2,2)$. Plot the functions $f(x)$ and $g(x)$ in a graph.
   (f) Find the Lagrange–Sylvester interpolation polynomial $g(x) \in \mathbb{R}[x]_h$ which best approximates the function $f(x) = 1/x$ in the interval $(-2,2)$. Plot the functions $f(x)$ and $g(x)$ on a single graph. What can you say about $f(x)$ and $g(x)$ at the point $x = 0$?
3. Let $h(x) = (x+1)(x-1)^2$ in $\mathbb{Z}_5[x]_h$.
   (a) Make a multiplication table for the basis elements of the standard basis $\{1,x,x^2\}$.
   (b) Find the spectral basis $\{s_1(x), s_2(x), q_2(x)\}$ of $\mathbb{Z}_5[x]_h$.
   (c) Express each element of the standard basis of $\mathbb{Z}_5[x]_h$ in terms of the spectral basis of $\mathbb{Z}_5[x]_h$.
   (d) Use part (c) to find the inverse $x^{-1}$ of $x$ and the inverse of $x^{-2}$ of $x^2$ in $\mathbb{Z}_5[x]_h$.
4. Let $h(x) = (x-x_1)(x-x_2)^2$ in $\mathbb{R}[x]_h$.
   (a) Make a multiplication table for the basis elements of the standard basis $\{1,x,x^2\}$.
   (b) Find the spectral basis $\{s_1(x), s_2(x), q_2(x)\}$ of $\mathbb{R}[x]_h$.

(c) Express each element of standard basis of $\mathbb{R}[x]_h$ in terms of the spectral basis of $\mathbb{R}[x]_h$.
(d) Use part (c) to find the inverse $x^{-1}$ of $x$ and the inverse of $x^{-2}$ of $x^2$ in $\mathbb{R}[x]_h$.
(e) When will $\sqrt{x}$ exist in $\mathbb{R}[x]_h$?
(f) When will $x^{-1} = 1/x$ exist in $\mathbb{R}[x]_h$?

## *1.5 Generalized Taylor's Theorem

The interpolation polynomial of a function is a generalized Taylor's expansion about spectral points in the domain of the function. What is so far missing is an expression for the remainder, which is useful in numerical analysis. We begin our discussion with a *generalized Rolle's theorem.*

Let $h(x) = \prod_{i=1}^{r}(x-x_i)^{m_i}$ for distinct $x_i \in [a,b] \subset \mathbb{R}$ with multiplicity $m_i \geq 1$, and let $n = \deg(h(x))$. Given two functions $f(x)$ and $g(x)$, we say that $f(x) = g(x) \bmod(h(x))$ or $f(x) \overset{h}{=} g(x)$ if for each $1 \leq i \leq r$ and $0 \leq k < m_i$

$$f^{(k)}(x_i) = g^{(k)}(x_i). \tag{1.22}$$

If $f(x)$ and $g(x)$ are polynomials, then $f(x) \overset{h}{=} g(x)$ is equivalent to saying that if $f(x)$ and $g(x)$ are divided by the polynomial $h(x)$ (the Euclidean algorithm), they give the same remainder, as was discussed in Sect. 1.3. We denote the *factor ring of polynomials* modulo $h(x)$ over the real numbers $\mathbb{R}$ by

$$\mathbb{R}[x]_h := \mathbb{R}[x]/ < h(x) >,$$

see [28, p.266].

**Generalized Rolle's Theorem 1.5.1.** *Let $f(x) \in C[a,b]$ and $(n-1)$-times differentiable on $(a,b)$, an open interval that contains all the zeros of $h(x)$. If $f(x)$=0 $\bmod(h(x))$ , then there exists a $c \in (a,b)$ such that $f^{(n-1)}(c) = 0$.*

*Proof.* Following [53, p.38], define the function $\sigma(u,v) := \begin{pmatrix} 1, & u < v \\ 0, & u \geq v \end{pmatrix}$.

The function $\sigma$ is needed to count the *simple zeros* of the polynomial $h(x)$ and its derivatives.

Let $\#h^{(k)}$ denote the number of simple zeros that the polynomial equation $h^{(k)}(x) = 0$ has. Clearly, $\#h = r = \sum_{i=1}^{r} \sigma(0,m_i)$. By the classical Rolle's theorem,

$$\#h' = \sum_{i=1}^{r} \sigma(1,m_i) + (\#h) - 1 = \sum_{i=1}^{r} \sigma(1,m_i) + \sum_{i=1}^{r} \sigma(0,m_i) - 1.$$

Continuing this process, we find that

$$\#h'' = \sum_{i=1}^{r} \sigma(2,m_i) + (\#h') - 1 = \sum_{i=1}^{r} \sigma(2,m_i) + \sum_{i=1}^{r} \sigma(1,m_i) + \sum_{i=1}^{r} \sigma(0,m_i) - 2,$$

and more generally that

$$\#h^{(k)} = \sum_{i=1}^{r} \sigma(k, m_i) + (\#h^{(k-1)}) - 1 = \sum_{i=1}^{r} \sum_{j=0}^{k} \sigma(j, m_i) - k$$

for all integers $k \geq 0$.

For $k = n-1$, we have

$$\#h^{(n-1)} = \sum_{i=1}^{r} \sum_{j=0}^{n-1} \sigma(j, m_i) - (n-1) = \sum_{i=1}^{r} m_i - (n-1) = 1.$$

The proof is completed by noting that $\#f^{(k)} \geq \#h^{(k)}$ for each $k \geq 0$, and hence $\#f^{(n-1)} \geq 1$. □

### 1.5.1 Approximation Theorems

We can now prove

**Generalized Taylor's Theorem 1.5.2.** *Let $f(x) \in C[a,b]$ and $n$ times differentiable on the open interval $(a,b)$. Suppose that $f(x) = g(x) \bmod (h(x))$ for some polynomial $g(x)$ where $deg(g) < deg(h)$. Then for every $x \in [a,b]$, there exists a $c \in (a,b)$ such that*

$$f(x) = g(x) + \frac{f^{(n)}(c)}{n!} h(x).$$

*Proof.* For a given $x \in [a,b]$, define the function

$$p(t) = f(t) - g(t) - [f(x) - g(x)] \frac{h(t)}{h(x)}. \tag{1.23}$$

In the case that $x = x_i$ for some $1 \leq i \leq r$, it is shown below that $p(t)$ has a removable singularity and can be redefined accordingly. Noting that

$$p(t) = 0 \bmod (h(t)) \quad \text{and} \quad p(x) = 0$$

it follows that $p(t) = 0 \bmod (h(t)(t-x))$. Applying the generalized Rolle's theorem to $p(t)$, there exists a $c \in (a,b)$ such that $p^{(n)}(c) = 0$. Using (1.23), we calculate $p^{(n)}(t)$, getting

$$\begin{aligned} p^{(n)}(t) &= f^{(n)}(t) - g^{(n)}(t) - [f(x) - g(x)] \left(\frac{\mathrm{d}}{\mathrm{d}t}\right)^n \frac{h(t)}{h(x)} \\ &= f^{(n)}(t) - [f(x) - g(x)] \frac{n!}{h(x)}, \end{aligned}$$

so that

$$0 = p^{(n)}(c) = f^{(n)}(c) - [f(x) - g(x)]\frac{n!}{h(x)}$$

from which the result follows. □

Applying the theorem to the case when $x = x_i$, we find by repeated application of L'Hospital's rule that

$$\lim_{x \to x_i} \frac{f(x) - g(x)}{h(x)} = \frac{f^{(m_i)}(x_i) - g^{(m_i)}(x_i)}{h^{(m_i)}(x_i)} = \frac{f^{(n)}(c)}{n!}.$$

There remains the question of how do we calculate the polynomial $g(x)$ with the property that $f(x) \stackrel{h}{=} g(x)$ where $\deg(g(x)) < \deg(h(x))$? The *brute force* method is to impose the conditions (1.22) and solve the resulting system of linear equations for the unique solution known as the *osculating polynomial* approximation to $f(x)$, see [18], [93, p.52]. A far more powerful method is to make use of the special algebraic properties of the *spectral basis* of the factor ring $\mathbb{R}[x]_h$, as explained in Sect. 1.3. See also [81,85].

Much time is devoted to explaining the properties of Lagrange, Hermite, and other types of interpolating polynomials in numerical analysis. In teaching this subject, the author has discovered that many of the formulas and theorems follow directly from the above theorem. For rational approximation, we have the following refinement:

**Rational Approximation Theorem 1.5.3.** *Let $f(x) \in C[a,b]$ and $n$ times differentiable on $(a,b)$. Let $u(x)$ and $v(x)$ be polynomials such that $v(0) = 1$ and $deg(u(x)v(x)) < deg(h(x))$, and suppose that $f(x)v(x) - u(x) = 0 \bmod(h(x))$. Then*

$$f(x) = \frac{u(x)}{v(x)} + \frac{1}{n!v(x)}[f(t)v(t)]^{(n)}(c)h(x)$$

*for some $c \in (a,b)$.*

*Proof.* Define the function

$$p(t) = f(t)v(t) - u(t) - [f(x)v(x) - u(x)]\frac{h(t)}{h(x)}$$

where $x \in [a,b]$. Clearly, $p(t) = 0 \bmod(h(t)(t-x))$. Applying the generalized Rolle's theorem to $p(t)$, it follows that there exists a $c \in (a,b)$ such that

$$f(x)v(x) - u(x) = \frac{1}{n!}\left(\frac{\mathrm{d}}{\mathrm{d}t}\right)^n [f(t)v(t)]_{t \to c} h(x),$$

from which it follows that

$$f(x) = \frac{u(x)}{v(x)} + \frac{1}{n!v(x)}[f(t)v(t)]^{(n)}(c)h(x). \qquad \square$$

### 1.5.2 Hermite–Pade Approximation

Hermite–Pade Approximation is a rational approximation of a function of the kind

$$f(x) \widetilde{=} \frac{a(x)}{b(x)} = \frac{\sum_{i=0}^{n} a_i x^n}{\sum_{i=0}^{m} b_i x^m}. \tag{1.24}$$

In what follows, we only consider rational approximations where $b_0 = 1$. Let

$$h(x) = \prod_{i=1}^{r} (x - x_i)^{m_i}$$

where the zeros $x_i$ of $h(x)$ are all distinct and where $\deg(h(x)) = m+n+1$ is exactly the number of unknown coefficients of $a(x)/b(x)$ which determine the rational approximation to $f(x)$. Let $\{s_1, q_1, \ldots, q_1^{m_1-1}, \ldots, s_r, q_r, \ldots, q_r^{m_r-1}\}$ be the spectral basis of the spectral algebra of polynomials $\mathbb{R}[x]_{h(x)}$ of $h(x)$.

The approximation equation (1.24) above is clearly equivalent to

$$\frac{f(x)b(x) - a(x)}{b(x)} \widetilde{=} 0,$$

and if we assume that none of the zeros $x_i$ of $h(x)$ are zeros of $b(x)$, then the approximation equation becomes simply

$$f(x)b(x) - a(x) \widetilde{=} 0.$$

The *osculating-Pade approximation* $a(x)/b(x)$ to $f(x)$ is defined by choosing the $m+n+1$ unknowns $a_i, b_j$ so that they satisfy

$$f(x)b(x) - a(x) \equiv 0 \text{ modulo } (h(x)). \tag{1.25}$$

This will be the case if and only if the resulting system of $m+n+1$ linear equations in $m+n+1$ unknowns has a solution.

As an example, we will find the Hermite–Pade rational approximation for a given function $f(x)$ at the points $x_1 = 0, x_2 = 1$. Recall that the spectral basis for $R[x]_{h(x)}$ for $h(x) = t^2(t-1)^2$ is

$$\{s_1 = (2t+1)(t-1)^2, q_1 = t(t-1)^2, s_2 = (3-2t)t^2, q_2 = t^2(t-1)\}.$$

In terms of the spectral basis, the resulting linear equations from (1.25) are

$$[(f(0)+f'(0)q_1)(b(0)+b'(0)q_1)-(a(0)+a'(0)q_1)]s_1 \\ +[(f(1)+f'(1)q_2)(b(1)+b'(1)q_2)-(a(1)+a'(1)q_2]s_2,$$

or

$$f(0)-a(0)=0,\ f'(0)+f(0)b_1-a_1=0,\ (1+b_1)f(1)-(a_0+a_1+a_2)=0, \\ (1+b_1)f'(1)+f(1)b_1-(a_1+2a_2)=0.$$

The unique solution is

$$a_0=f(0),\ a_1=f'(0)-\frac{f(0)[2f(0)-2f(1)+f'(0)+f'(1)]}{f(0)-f(1)+f'(1)}$$

$$a_2=-\frac{f(0)^2-2f(0)f(1)+f(1)^2-f'(0)f'(1)}{f(0)-f(1)+f'(1)},$$

$$b_1=-\frac{2f(1)-2f(0)-f'(0)-f'(1)}{f(0)-f(1)+f'(1)},$$

provided that $f(0)-f(1)+f'(1)\neq 0$.

## Exercises

1. A rational approximation to a function $f(x)$ has the form

$$f(x)\cong\frac{a(x)}{b(x)}=\frac{\sum_{i=0}^{n}a_ix^n}{\sum_{i=0}^{m}b_ix^m}.$$

   where $b_0=1$. For $h(x)=x^2(x-1)^2$, determine the formulas to approximate $f(x)$ by the rational forms

   (a) $\frac{a_0+a_1x}{1+b_1x+b_2x^2}$.
   (b) $\frac{a_0}{1+b_1x+b_2x^2+b_3x^3}$.

2. Find all Hermite–Pade approximations of the form (a) and (b) to $f(x)=\ln(x+2)$ over the closed interval $[-1,1]$. Which one gives the best approximation over that interval?
3. Find all Hermite–Pade approximations to $f(x)=\exp(x)$ of the form (a) and (b) over the closed interval $[-1,1]$. Which one gives the best approximation over that interval?

# Chapter 2
# Complex and Hyperbolic Numbers

*The laws of nature are but the mathematical thoughts of God.*
–Euclid

The complex numbers were grudgingly accepted by Renaissance mathematicians because of their utility in solving the cubic equation.[1] Whereas the complex numbers were discovered primarily for algebraic reasons, they take on geometric significance when they are used to name points in the plane. The complex number system is at the heart of *complex analysis* and has enjoyed more than 150 years of intensive development, finding applications in diverse areas of science and engineering.

At the beginning of the twentieth century, Albert Einstein developed his theory of special relativity, built upon Lorentzian geometry, yet at the end of that century, almost all high school and undergraduate students are still taught only Euclidean geometry. At least part of the reason for this state of affairs has been the lack of a simple mathematical formalism in which the basic ideas can be expressed.

The hyperbolic numbers are blood relatives of the popular complex numbers and deserve to be taught alongside the latter. They not only serve to put Lorentzian geometry on an equal mathematical footing with Euclidean geometry but also help students develop algebraic skills and concepts necessary in higher mathematics. Whereas the complex numbers extend the real numbers to include a new number $\mathrm{i} = \sqrt{-1}$, the hyperbolic numbers extend the real numbers to include a *new* square root $\mathrm{u} = \sqrt{+1}$, where $\mathrm{u} \neq \pm 1$.[2] Whereas each nonzero complex number has a multiplicative inverse, this is no longer true for all nonzero hyperbolic numbers.

[1] A historical account of this fascinating story is told by Tobias Dantzig in his book *NUMBER: The Language of Science*, [17]. See also Struik's *A Concise History of Mathematics* [94], and [59].

[2] This chapter is based upon an article by the author that appeared in the College Mathematical Journal [77].

G. Sobczyk, *New Foundations in Mathematics: The Geometric Concept of Number*, DOI 10.1007/978-0-8176-8385-6_2,

## 2.1 The Hyperbolic Numbers

Whereas the algebraic equation $x^2 - 1 = 0$ has the real number solutions $x = \pm 1$, we assume the existence of a new number, the *unipotent* u, which has the algebraic property that $\mathrm{u} \neq \pm 1$ but $\mathrm{u}^2 = 1$. In terms of the *standard basis* $\{1, \mathrm{u}\}$, any hyperbolic number $w \in \mathbb{H}$ can be written in the form $w = x + \mathrm{u}y$ where $x, y$ are real numbers. Thus, the hyperbolic numbers $\mathbb{H} \equiv \mathbb{R}(\mathrm{u})$ are just the real numbers extended to include the unipotent u in the same way that the complex numbers $\mathbb{C} \equiv \mathbb{R}(\mathrm{i})$ are the real numbers extended to include the imaginary i where $\mathrm{i}^2 = -1$.

It follows that multiplication in $\mathbb{H}$ is defined by $(x + \mathrm{u}y)(r + \mathrm{u}s) = (xr + ys) + \mathrm{u}(xs + yr)$. Just as the complex numbers can be identified with skew-symmetric $2 \times 2$ matrices with equal diagonal entries, $a + \mathrm{i}b \leftrightarrow \begin{pmatrix} a & b \\ -b & a \end{pmatrix}$, the hyperbolic numbers correspond to the symmetric $2 \times 2$ matrices with equal diagonal entries: $x + \mathrm{u}y \leftrightarrow \begin{pmatrix} x & y \\ y & x \end{pmatrix}$. This correspondence is an *isomorphism* because the operations of addition and multiplication in $\mathbb{H}$ correspond to the usual matrix operations, as is explained later in Chap. 4.

The complex numbers $\mathbb{C}$ of the form $z = x + \mathrm{i}y$ where $\mathrm{i}^2 = -1$ and the hyperbolic numbers $\mathbb{H}$ are two-dimensional vector spaces over the real numbers, so each can identified with points in the plane $\mathbb{R}^2$. Using the standard basis $\{1, \mathrm{u}\}$, we identify $w = x + \mathrm{u}y$ with the point or vector $(x, y)$, see Fig. 2.1. The hyperbolic number $w = x + \mathrm{u}y$ names the corresponding point $(x, y)$ in the coordinate plane. Also pictured is the conjugate $w^- = x - \mathrm{u}y$.

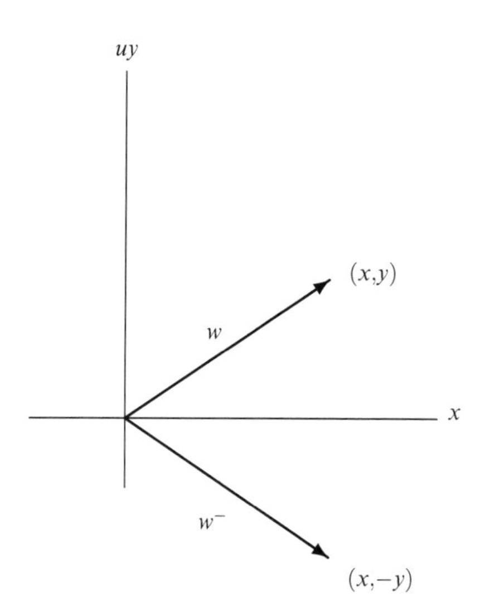


**Fig. 2.1** The hyperbolic number plane. The hyperbolic number $w$ and its conjugate $w^-$ are shown

The real numbers $x$ and $y$ are called the *real* and *unipotent* parts of the hyperbolic number $w$, respectively. The *hyperbolic conjugate* $w^-$ of $w$ is defined by

$$w^- = x - \mathrm{u}y.$$

The *hyperbolic modulus* or *length* of $w = x + \mathrm{u}y$ is defined by

$$|w|_h \equiv \sqrt{|ww^-|} = \sqrt{|x^2 - y^2|} \tag{2.1}$$

and is the *hyperbolic distance* of the point $w$ from the origin.

Note that the points $w \neq 0$ on the lines $y = \pm x$ are *isotropic*, meaning that they are nonzero vectors with $|w|_h = 0$. Thus, the hyperbolic distance yields a geometry, *Lorentzian geometry*, on $\mathbb{R}^2$ quite unlike the usual Euclidean geometry of the complex plane where $|z| = 0$ only if $z = 0$. It is easy to verify that

$$w^{-1} = \frac{w^-}{ww^-} = \frac{x - \mathrm{u}y}{x^2 - y^2}, \tag{2.2}$$

is the *hyperbolic inverse* of $w$, whenever $|w|_h \neq 0$.

## Exercises

1. Given the complex numbers $z_1 = 2 + 3\mathrm{i}$ and $z_2 = 3 - 5\mathrm{i}$, and the hyperbolic numbers $w_1 = 2 + 3\mathrm{u}$ and $w_2 = 3 - 5\mathrm{u}$,
   (a) Calculate $z_1 + z_2$ and $z_1 z_2$, and $w_1 + w_2$ and $w_1 w_2$.
   (b) Calculate the Euclidean norms $|z_1|, |z_2|$ of $z_1$ and $z_2$ and the hyperbolic norms $|w_1|, |w_2|$ of $w_1$ and $w_2$.
   (c) Calculate the inverses $z_1^{-1}$ and $z_2^{-1}$, and $w_1^{-1}$ and $w_2^{-1}$.
   (d) More generally, show that a complex number $z = x + \mathrm{i}y$ for $x, y \in \mathbb{R}$ will have an inverse iff $z \neq 0$. Show that a hyperbolic number $w = x + \mathrm{u}y$ for $x, y \in \mathbb{R}$ will have an inverse iff $x \neq \pm y$.
2. Find the matrices $[z_1]$ and $[z_2]$ representing the complex numbers $z_1$ and $z_2$ and the matrices $[w_1]$ and $[w_2]$ representing the hyperbolic numbers $w_1$ and $w_2$ given in Problem 1. Show that $[z_1][z_2] = [z_1 z_2]$ and $[w_1][w_2] = [w_1 w_2]$.
3. (a) Find all complex numbers $z = x + \mathrm{i}y$ with the property that $z^2 = \mathrm{i}$.
   (b) Show that there is no hyperbolic number $w = x + y\mathrm{u}$ with the property that $w^2 = \mathrm{u}$. In the next section, we learn which complex numbers and which hyperbolic numbers have square roots.

## 2.2 Hyperbolic Polar Form

Let us recall some basic properties of the exponential function. The exponential function is defined by

$$e^x = \sum_{j=0}^{\infty} \frac{x^j}{j!} = 1 + x + \frac{x^2}{2!} + \frac{x^3}{3!} + \frac{x^4}{4!} + \frac{x^5}{5!} + \cdots = \cosh x + \sinh x,$$

where

$$\cosh x = \sum_{j=0}^{\infty} \frac{x^{2j}}{(2j)!} = 1 + \frac{x^2}{2!} + \frac{x^4}{4!} + \frac{x^6}{6!} + \frac{x^8}{8!} + \cdots$$

are the even terms of the series and

$$\sinh x = \sum_{j=0}^{\infty} \frac{x^{2j+1}}{(2j+1)!} = x + \frac{x^3}{3!} + \frac{x^5}{5!} + \frac{x^7}{7!} + \cdots$$

are the odd terms of the series. The $\cosh x$ and $\sinh x$ are called the trigonometric *hyperbolic cosine* and *hyperbolic sine* functions, respectively.

Calculating $e^{ix}$, we find that

$$e^{ix} = \sum_{j=0}^{\infty} \frac{x^j}{j!} = 1 + ix - \frac{x^2}{2!} - i\frac{x^3}{3!} + \frac{x^4}{4!} + i\frac{x^5}{5!} + \cdots = \cos x + i \sin x, \tag{2.3}$$

where

$$\cos x = \sum_{j=0}^{\infty} (-1)^j \frac{x^{2j}}{(2j)!} = 1 - \frac{x^2}{2!} + \frac{x^4}{4!} - \frac{x^6}{6!} + \frac{x^8}{8!} + \cdots$$

are the even terms of the series and

$$\sin x = \sum_{j=0}^{\infty} (-1)^j \frac{x^{2j+1}}{(2j+1)!} = x - \frac{x^3}{3!} + \frac{x^5}{5!} - \frac{x^7}{7!} + \cdots$$

are the odd terms of the series. The $\cos x$ and $\sin x$ are called the trigonometric *cosine* and *sine* functions, respectively.

Calculating $e^{ux}$, we find that

$$e^{ux} = \sum_{j=0}^{\infty} \frac{x^j}{j!} = 1 + ux + \frac{x^2}{2!} + u\frac{x^3}{3!} + \frac{x^4}{4!} + u\frac{x^5}{5!} + \cdots = \cosh x + u \sinh x. \tag{2.4}$$

Every nonzero complex number $z \in \mathbb{C}$ can be written in the *polar form*

$$z = r(\cos\theta + i\sin\theta) \equiv r \exp i\theta \tag{2.5}$$

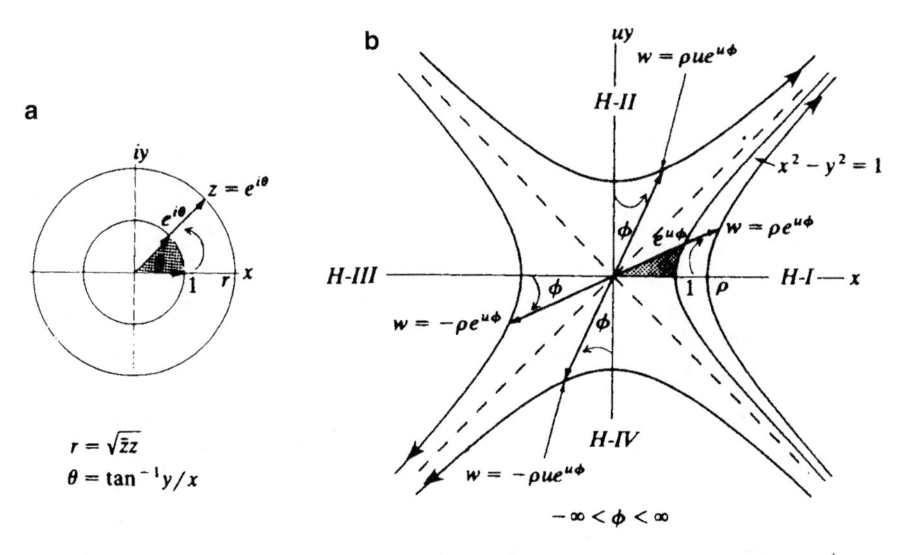


**Fig. 2.2** (**a**) In the $r$-circle the shaded area $= \frac{\theta}{2}$. (**b**) In the $\rho$-hyperbola the shaded area $= \frac{\phi}{2}$

for $0 \leq \theta < 2\pi$, where $\theta = \tan^{-1}(y/x)$ is the angle that the vector $z$ makes with the positive $x$-axis and $r = |z| \equiv \sqrt{\bar{z}z}$ is the Euclidean distance of the point $z$ to the origin. The set of all points in the complex number plane that satisfy the equation $|z| = r$ is a circle of radius $r \geq 0$ centered at the origin, see Fig. 2.2a.

Each time the parameter $\theta$ increases by $2\pi$, the point $z = r\exp i\theta$ makes a complete *counterclockwise* revolution around the $r$-circle. In the case of the $\rho$-hyperbola, the points on the branches, given by $w = \pm\rho\exp u\phi$ in the hyperbolic quadrants *H-I* and *H-III*, and $w = \pm\rho u\exp u\phi$ in the hyperbolic quadrants *H-II* and *H-IV*, respectively, are covered exactly once in the indicated directions as the parameter $\phi$ increases, $-\infty < \phi < \infty$. The hyperbolic quadrants are demarcated by the asymtotes $y = \pm x$ and are not the usual quadrants demarcated by the $x$- and $y$-axes.

Similarly, the set of all points in the hyperbolic plane that satisfy the equation $|w|_h = \rho > 0$ is a *four-branched* hyperbola of hyperbolic radius $\rho$. Such hyperbolic numbers $w = x + uy$ can be written in the *hyperbolic polar form*

$$w = \pm\rho(\cosh\phi + u\sinh\phi) \equiv \pm\rho\exp u\phi \tag{2.6}$$

when $w$ lies in the *hyperbolic quadrants H-I* or *H-III* or

$$w = \pm\rho(\sinh\phi + u\cosh\phi) \equiv \pm\rho u\exp u\phi \tag{2.7}$$

when $w$ lies in the hyperbolic quadrants *H-II* or *H-IV*, respectively. The *hyperbolic quadrants* of the hyperbolic plane are demarcated by the isotropic lines $|w|_h = 0$, which are the asymtotes of the $\rho$-hyperbolas $|w|_h = \rho > 0$. See Fig. 2.2b.

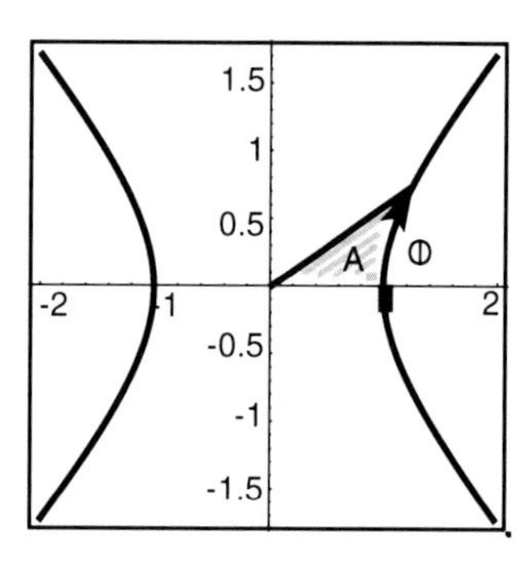


**Fig. 2.3** The hyperbolic angle $\phi = 2A$, where $A$ is the shaded area

The *hyperbolic angle* $\phi$ is defined by $\phi \equiv \tanh^{-1}(y/x)$ in the quadrants *H-I* and *H-III* or $\phi \equiv \tanh^{-1}(x/y)$ in *H-II* and *H-IV*, respectively. Just as the area of the sector of the unit circle with central angle $\theta$ is $\frac{1}{2}\theta$, the area of the unit hyperbolic sector determined by the ray from the origin to the point $\exp \mathrm{u}\phi = \cosh\phi + \mathrm{u}\sinh\phi$ (shaded in Fig. 2.2b) is $\frac{1}{2}\phi$.

Let us consider a simple example. Given the hyperbolic number $w = 5 + 3\mathrm{u}$, its hyperbolic magnitude is $|w|_h = \sqrt{|5^2 - 3^2|} = 4$, and its hyperbolic polar form is given by

$$w = 4\frac{5+3\mathrm{u}}{4} = 4\mathrm{e}^{\phi\mathrm{u}}$$

where $\phi = \tanh^{-1} 3/5 = \log 2 \doteq 0.693147$. Alternatively, we can calculate the hyperbolic angle $\phi$ by calculating the *hyperbolic arc length* along the hyperbola $x^2 - y^2 = 1$, getting

$$\int \mathrm{d}s_h = \int \sqrt{|\mathrm{d}x^2 - \mathrm{d}y^2|} = \int_1^{\frac{5}{4}} \sqrt{\left|1 - \left(\frac{\mathrm{d}y}{\mathrm{d}x}\right)^2\right|}\mathrm{d}x = \log 2$$

where $\frac{\mathrm{d}y}{\mathrm{d}x} = x/y = \frac{x}{\sqrt{x^2-1}}$. We can also calculate the hyperbolic angle $\phi$ by calculating $2A$ where $A$ is the area enclosed in Fig. 2.3. We find

$$2A = \frac{5}{4} \cdot \frac{3}{4} - 2\int_1^{5/4} \sqrt{x^2 - 1}\mathrm{d}x = \log 2.$$

The polar form of complex numbers provides the familiar geometric interpretation of complex number multiplication

$$r_1 \exp \mathrm{i}\theta_1 \cdot r_2 \exp \mathrm{i}\theta_2 = r_1 r_2 \exp\big(\mathrm{i}(\theta_1 + \theta_2)\big)$$

Similarly, the hyperbolic polar form gives a geometric interpretation of hyperbolic number multiplication, but because the hyperbolic plane is divided into four quadrants separated by the isotropic lines, we must keep track of the quadrants of the

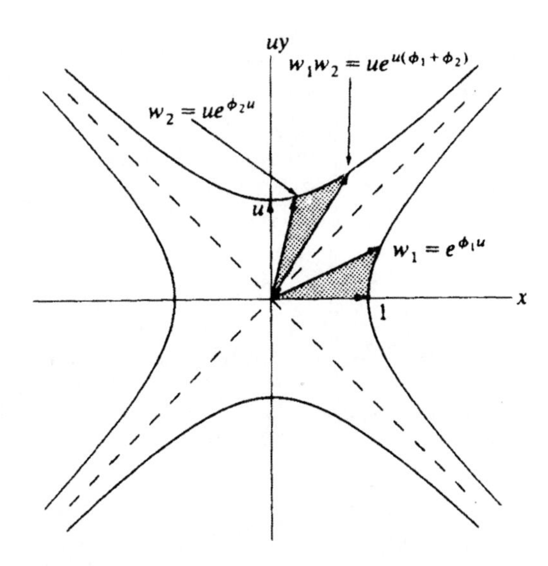


**Fig. 2.4** Shaded regions have equal area $\frac{1}{2}\phi_1$

factors. For example, if $w_1 = \rho_1 \exp \mathrm{u}\phi_1$ lies in quadrant *H-I* and $w_2 = \rho_2 \mathrm{u} \exp \mathrm{u}\phi_2$ lies in *H-II*, then

$$w_1 w_2 = \rho_1 \rho_2 \mathrm{u} \exp \mathrm{u}\phi_1 \exp \mathrm{u}\phi_2 = \rho_1 \rho_2 \mathrm{u} \exp\left(\mathrm{u}(\phi_1 + \phi_2)\right)$$

lies in quadrant *H-II* and is located as shown in Fig. 2.4.

## Exercises

1. Given the complex numbers $z_1 = 2 + 3\mathrm{i}$ and $z_2 = 3 - 5\mathrm{i}$, and the hyperbolic numbers $w_1 = 2 + 3\mathrm{u}$ and $w_2 = 3 - 5\mathrm{u}$,
   (a) Calculate the polar forms of $z_1$ and $z_2$. (b) Calculate the hyperbolic polar forms of $w_1$ and $w_2$.
   (c) Use the polar forms to calculate $\sqrt{z_1}$ and $\sqrt{z_2}$.
   (d) Use the hyperbolic polar forms to calculate $\sqrt{w_1}$ and $\sqrt{w_2}$.
2. (a) Given that $z = r\exp(\mathrm{i}\theta)$, show that $z^{-1} = \frac{1}{r}\exp(-\mathrm{i}\theta)$.
   (b) Given that $w = \rho\exp(\mathrm{u}\phi)$, show that $w^{-1} = \frac{1}{\rho}\exp(-\mathrm{u}\phi)$.
3. Find the matrix representation $[z]$ of $z = r\exp(\mathrm{i}\theta)$ and $[w]$ of $w = \rho\exp(\mathrm{u}\phi)$.
4. Plot the points $z = r\exp(\mathrm{i}\theta)$ and $\mathrm{i}z = r\mathrm{i}\exp(\mathrm{i}\theta)$ in the complex number plane, and show that $\mathrm{i}z = r\mathrm{i}\exp(\mathrm{i}\theta) = r\exp(\mathrm{i}(\theta + \pi/2))$.
5. Plot the points $w = \rho\exp(\mathrm{u}\phi)$ and $\mathrm{u}w = \rho\mathrm{u}\exp(\mathrm{u}\phi)$ in the hyperbolic number plane. Explain why $\mathrm{u}w$ does not have a square root in the real hyperbolic plane.

## 2.3 Inner and Outer Products

Let us now compare the multiplication of the hyperbolic numbers $w_1^- = x_1 - \mathrm{u}y_1$ and $w_2 = x_2 + \mathrm{u}y_2$ with the multiplication of the corresponding complex numbers $\overline{z_1} = x_1 - \mathrm{i}y_1$ and $z_2 = x_2 + \mathrm{i}y_2$. The *conjugate products* are

$$\overline{z_1}z_2 = (x_1 - \mathrm{i}y_1)(x_2 + \mathrm{i}y_2) = (x_1x_2 + y_1y_2) + \mathrm{i}(x_1y_2 - x_2y_1), \tag{2.8}$$

and

$$w_1^- w_2 = (x_1 - \mathrm{u}y_1)(x_2 + \mathrm{u}y_2) = (x_1x_2 - y_1y_2) + \mathrm{u}(x_1y_2 - x_2y_1). \tag{2.9}$$

The *real* and *imaginary* parts of the conjugate product $\overline{z_1}z_2$, denoted by $\langle\overline{z_1}z_2\rangle_0$ and $\langle\overline{z_1}z_2\rangle_\mathrm{i}$, respectively, are called the *inner* and *outer products* of the complex numbers $z_1$ and $z_2$. Likewise, the *real* and *unipotent* parts of the conjugate product $w_1^- w_2$, denoted by $\langle w_1^- w_2\rangle_0$ and $\langle w_1^- w_2\rangle_\mathrm{u}$, respectively, are called the *hyperbolic inner* and *outer products* of the hyperbolic numbers $w_1$ and $w_2$. The vectors $z_1$ and $z_2$ in the complex number plane and $w_1$ and $w_2$ in the hyperbolic number plane are said to be, respectively, *Euclidean orthogonal* or *hyperbolic orthogonal* if their respective inner products are zero.

From (2.8) and (2.9), it is seen that the components of the respective Euclidean and hyperbolic outer products are identical and give the *directed area* of the parallelogram with $w_1$ and $w_2$ as adjacent edges. Rather surprisingly, the concept of area is identical in Euclidean and Lorentzian geometry.

The conjugate products (2.8) and (2.9) are also nicely expressed in polar form. Letting $w_1 = \rho_1 \exp(\mathrm{u}\phi_1) \in H\text{-}I$ and $w_2 = \rho_2 \exp(\mathrm{u}\phi_2) \in H\text{-}I$, and $z_1 = r_1 \exp(\mathrm{i}\theta_1)$ and $z_2 = r_2 \exp(\mathrm{i}\theta_2)$, we find that

$$w_1^- w_2 = \rho_1\rho_2 \exp\big(\mathrm{u}(\phi_2 - \phi_1)\big) = \rho_1\rho_2\big(\cosh(\phi_2 - \phi_1) + \mathrm{u}\sinh(\phi_2 - \phi_1)\big)$$

where $\phi_2 - \phi_1$ is the hyperbolic angle between $w_1$ and $w_2$ and

$$\overline{z_1}z_2 = r_1r_2 \exp\big(\mathrm{i}(\theta_2 - \theta_1)\big) = r_1r_2\big(\cos(\theta_2 - \theta_1) + \mathrm{i}\sin(\theta_2 - \theta_1)\big)$$

where $\theta_2 - \theta_1$ is the Euclidean angle between $z_1$ and $z_2$. The special cases when $w_1 = s + \mathrm{u}s$ and/or $w_2 = t + \mathrm{u}t$ for $s, t \in \mathbb{R}$ must be considered separately since no hyperbolic polar forms of these numbers exists. This happens when either or both of $w_1$ and $w_2$ lie on the isotropic lines $y = \pm x$.

Multiplication by $\exp(\phi\mathrm{u})$ is a linear transformation that sends the standard basis $\{1, \mathrm{u}\}$ to the new basis $\{\exp(\phi\mathrm{u}),\ \mathrm{u}\exp(\phi\mathrm{u})\}$. The product

$$\exp(\phi\mathrm{u})^- \mathrm{u}\exp(\phi\mathrm{u}) = \mathrm{u},$$

shows that the new basis is hyperbolic orthogonal like the standard basis, since the hyperbolic inner product (real part) is zero. The mapping $w \to w\exp(\phi\mathrm{u})$ is naturally

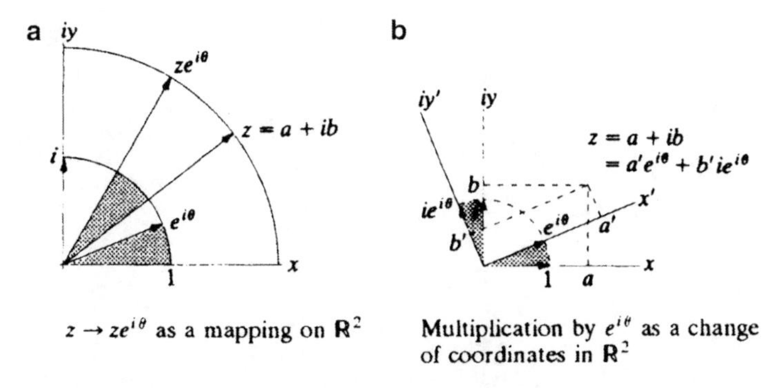


**Fig. 2.5** Euclidean rotation. The shaded areas each equal $\frac{1}{2}\theta$

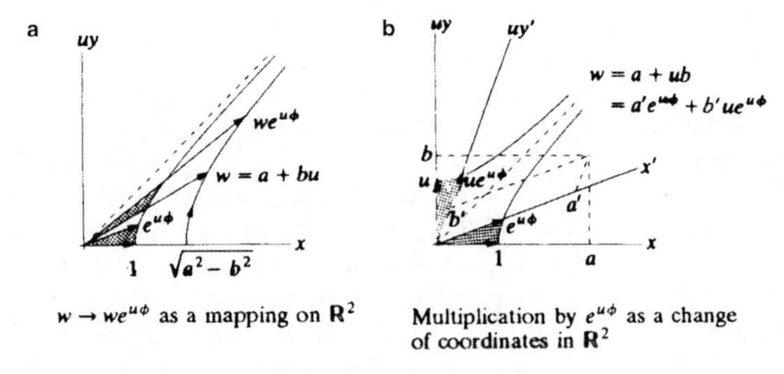


**Fig. 2.6** Hyperbolic rotation. The shaded areas each equal $\frac{1}{2}\phi$

called the *hyperbolic rotation* through the hyperbolic angle $\phi$. If $w = a + \mathrm{u}b$ is any vector, the coordinates $a', b'$ of $w$ with respect to the rotated basis satisfy

$$w = a'\exp(\phi\mathrm{u}) + b'\mathrm{u}\exp(\phi\mathrm{u}) = (a' + \mathrm{u}b')\exp(\phi\mathrm{u}),$$

so

$$a' + \mathrm{u}b' = (a + \mathrm{u}b)\exp(-\phi\mathrm{u}). \tag{2.10}$$

It follows at once that $|a' + \mathrm{u}b'|_h = |a + \mathrm{u}b|_h$, that is, the hyperbolic distance of a point from the origin is independent of which hyperbolic orthogonal basis is used to coordinatize the plane. Figures 2.5 and 2.6 show the geometric pictures of multiplication by $\exp(\mathrm{i}\theta)$ in the complex plane and multiplication by $\exp(\phi\mathrm{u})$ in the hyperbolic number plane, both as mappings sending each point to its image (alibi) and as a change of coordinates (alias). As a mapping, the hyperbolic rotation $\exp(\phi\mathrm{u})$ moves each point $a + \mathrm{u}b$ along the hyperbola $x^2 - y^2 = |a^2 - b^2|$, through the hyperbolic angle $\phi$, Fig. 2.6a. Equivalently, as a change of coordinates, the new coordinates of any point are found by the usual parallelogram construction after rotation of the axes of the original system through the hyperbolic angle $\phi$, as in Fig. 2.6b. (Note that the angles in Fig. 2.6 are hyperbolic angles, not Euclidean angles!)

Notice that multiplying by i sends the basis $\{1,\mathrm{i}\}$ into $\{\mathrm{i},-1\}$, a *counterclockwise* rotation about the origin taking the positive $x$-axis into the positive $y$-axis and the positive $y$-axis into the negative $x$-axis. Multiplying by $u$ in the hyperbolic plane sends $\{1,\mathrm{u}\}$ into $\{\mathrm{u},1\}$, a *hyperbolic* rotation about the line $y=x$ that interchanges the positive $x$- and $y$-axes.

## Exercises

1. Given the complex numbers $z_1 = 2+3\mathrm{i}$ and $z_2 = 3-5\mathrm{i}$, and the hyperbolic numbers $w_1 = 3+2\mathrm{u}$ and $w_2 = 5-3\mathrm{u}$,

    (a) Calculate the conjugate product $\overline{z}_1 z_2$ of $z_1$ with $z_2$. Calculate the inner product $< \overline{z}_1 z_2 >_0$ and the outer product $< \overline{z}_1 z_2 >_i$ between $z_1$ and $z_2$. Do the same calculations for the conjugate product $\overline{z}_2 z_1$ of $z_2$ with $z_1$. What is the relationship between the two conjugate products?
    (b) Calculate the conjugate product $w_1^{-} w_2$ of $w_1$ with $w_2$. Calculate the hyperbolic inner product $< w_1^{-} w_2 >_0$ and the outer product $< w_1^{-} w_2 >_u$ between $w_1$ and $w_2$. Do the same calculations for the conjugate product $w_2^{-} w_1$ of $w_2$ with $w_1$. What is the relationship between the two conjugate products?
    (c) What is the area of the parallelogram with the sides $z_1$ and $z_2$?
    (d) What is the area of the parallelogram with the sides $w_1$ and $w_2$?
    (e) What is the Euclidean angle between the vectors $z_1$ and $z_2$?
    (f) What is the hyperbolic angle between the vectors $w_1$ and $w_2$?

2. Given the complex numbers $z_1 = r_1 \mathrm{e}^{\mathrm{i}\theta_1}$ and $z_2 = r_2 \mathrm{e}^{\mathrm{i}\theta_2}$, and the hyperbolic numbers $w_1 = \rho_1 \mathrm{e}^{\mathrm{u}\phi_1}$ and $w_2 = \rho_2 \mathrm{e}^{\mathrm{u}\phi_2}$,

    (a) Calculate the conjugate product $\overline{z}_1 z_2$ of $z_1$ with $z_2$. Calculate the inner product $< \overline{z}_1 z_2 >_0$ and the outer product $< \overline{z}_1 z_2 >_\mathrm{i}$ between $z_1$ and $z_2$. Do the same calculations for the conjugate product $\overline{z}_2 z_1$ of $z_2$ with $z_1$. What is the relationship between the two conjugate products?
    (b) Calculate the conjugate product $w_1^{-} w_2$ of $w_1$ with $w_2$. Calculate the hyperbolic inner product $< w_1^{-} w_2 >_0$ and the outer product $< w_1^{-} w_2 >_\mathrm{u}$ between $w_1$ and $w_2$. Do the same calculations for the conjugate product $w_2^{-} w_1$ of $w_2$ with $w_1$. What is the relationship between the two conjugate products?
    (c) What is the area of the parallelogram with the sides $z_1$ and $z_2$?
    (d) What is the area of the parallelogram with the sides $w_1$ and $w_2$?
    (e) What is the Euclidean angle between the vectors $z_1$ and $z_2$?
    (f) What is the hyperbolic angle between the vectors $w_1$ and $w_2$?

## 2.4 Idempotent Basis

Besides the standard basis $\{1, \mathrm{u}\}$ in which every hyperbolic number can be expressed as $w = x + \mathrm{u}y$, the hyperbolic plane has another distinguished basis, associated with the two isotropic lines which separate the hyperbolic quadrants. The *idempotent basis* of the hyperbolic numbers is the pair $\{\mathrm{u}_+, \mathrm{u}_-\}$ where

$$\mathrm{u}_+ = \frac{1}{2}(1+\mathrm{u}) \quad \text{and} \quad \mathrm{u}_- = \frac{1}{2}(1-\mathrm{u}). \tag{2.11}$$

In terms of this basis, the expression for $w = x + \mathrm{u}y$ is $w = w_+\mathrm{u}_+ + w_-\mathrm{u}_-$, where

$$w_+ = x + y \quad \text{and} \quad w_- = x - y. \tag{2.12}$$

Conversely, given $w = w_+\mathrm{u}_+ + w_-\mathrm{u}_-$, we can recover the coordinates with respect to the standard basis by using the definitions of $w_+$ and $w_-$:

$$x = \frac{1}{2}(w_+ + w_-) \quad \text{and} \quad y = \frac{1}{2}(w_+ - w_-). \tag{2.13}$$

From the definitions of $\mathrm{u}_+$ and $\mathrm{u}_-$, it is clear that $\mathrm{u}_+ + \mathrm{u}_- = 1$ and $\mathrm{u}_+ - \mathrm{u}_- = \mathrm{u}$. We say that $\mathrm{u}_+$ and $\mathrm{u}_-$ are *idempotents* because $\mathrm{u}_+^2 = \mathrm{u}_+$ and $\mathrm{u}_-^2 = \mathrm{u}_-$, and they are *mutually annihilating* because $\mathrm{u}_+\mathrm{u}_- = 0$. Therefore, we have the *projective* properties

$$w\mathrm{u}_+ = w_+\mathrm{u}_+ \quad \text{and} \quad w\mathrm{u}_- = w_-\mathrm{u}_-. \tag{2.14}$$

(The decomposition $w = w_+\mathrm{u}_+ + w_-\mathrm{u}_-$ is the *spectral decomposition* of $w = x + \mathrm{u}y$. Under the identification of hyperbolic numbers with symmetric $2 \times 2$ matrices, it corresponds to the spectral decomposition [78] of the symmetric matrix $\begin{pmatrix} x & y \\ y & x \end{pmatrix}$.)

Because of its special properties, the idempotent basis is very nice for calculations. For example, from (2.12) we see that $w_+w_- = x^2 - y^2$. Since for any hyperbolic numbers $v$, $w$ we have

$$vw = (v_+\mathrm{u}_+ + v_-\mathrm{u}_-)(w_+\mathrm{u}_+ + w_-\mathrm{u}_-) = (v_+w_+)\mathrm{u}_+ + (v_-w_-)\mathrm{u}_-,$$

it follows that

$$|vw|_h = \sqrt{|(v_+w_+)(v_-w_-)|} = \sqrt{|v_+v_-|}\sqrt{|w_+w_-|} = |v|_h|w|_h.$$

In particular, if one of the factors in a product is isotropic, then the entire product is isotropic. Thus, although strictly speaking the polar decomposition (2.7) is defined only for nonisotropic vectors, we can extend it to the isotropic lines by assigning the isotropic vectors $\mathrm{u}_\pm$ to the exponentials of the hyperbolic arguments $\pm\infty$, respectively.

The binomial theorem takes a very simple form in the idempotent basis representation, so we can easily compute powers of hyperbolic numbers:

$$(w_+\mathrm{u}_+ + w_-\mathrm{u}_-)^k = (w_+)^k\mathrm{u}_+^k + (w_-)^k\mathrm{u}_-^k = (w_+)^k\mathrm{u}_+ + (w_-)^k\mathrm{u}_- \tag{2.15}$$

This formula is valid for *all* real numbers $k \in \mathbb{R}$ and not just the positive integers. For example, for $k=-1$, we find that

$$1/w = w^{-1} = (1/w_+)\mathrm{u}_+ + (1/w_-)\mathrm{u}_-,$$

a valid formula for the inverse (2) of $w \in \mathbb{H}$, provided that $|w|_h \neq 0$.

Indeed, the validity of (2.15) allows us to extend the definitions of all of the elementary functions to the elementary functions in the hyperbolic number plane. If $f(x)$ is such a function, for $w = w_+\mathrm{u}_+ + w_-\mathrm{u}_-$, we define

$$f(w) \equiv f(w_+)\mathrm{u}_+ + f(w_-)\mathrm{u}_- \tag{2.16}$$

provided that $f(w_+)$ and $f(w_-)$ are defined. It is natural to extend the hyperbolic numbers to the *complex hyperbolic numbers* or *unipodal numbers* by allowing $w_+, w_- \in \mathbb{C}$. In this extension, we assume that $\mathrm{i} = \sqrt{-1}$ *commutes* with the unipotent u, i.e., iu = ui. The unipodal numbers have been studied in [71]. They will be used in the next section to find the solutions of the cubic equation.

The extension of the real number system to include the unipodal numbers raises the question of the possibility of even further extensions. Shortly before his early death in 1879, William Kingdon Clifford *geometrically* extended the real number system to include the concept of direction, what is now called *geometric algebra* or *Clifford algebra*. In [43], geometric algebra provides a basis for a unified language for mathematics and physics. We begin our study of geometric algebra in the next chapter.

## Exercises

1. Given the hyperbolic numbers $w_1 = 3+2\mathrm{u}$ and $w_2 = 5-\mathrm{u}$,

   (a) Express $w_1$ and $w_2$ in the spectral basis $(\mathrm{u}_+, \mathrm{u}_-)$.
   (b) Calculate $w_1^3$, $w_1^{-1}$, $w_1w_2$, and $\sqrt{w_2}$.
   (c) Calculate $\mathrm{u}_+w_1$ and $\mathrm{u}_-w_1$ and show that $\mathrm{u}_+w_1 + \mathrm{u}_-w_1 = w_1$.
   (d) Calculate $\mathrm{e}^{w_1}$ and $\mathrm{e}^{w_2}$ by using the properties of the spectral basis.

2. Given the hyperbolic number $w = x + y\mathrm{u}$,

   (a) Calculate $\mathrm{e}^w$. (b) Calculate $\log w$.

3. Find $\sin(\frac{\pi}{4}\mathrm{u}_+ + \frac{\pi}{3}\mathrm{u}_-)$.

4. Find $\frac{1}{5-3\mathrm{u}}$.
5. Solve the following quadratic equation for the hyperbolic number $w$.

$$(5-4\mathrm{u})w^2+(5+3\mathrm{u})w-1=0.$$

## 2.5 The Cubic Equation

As mentioned in the introduction, the complex numbers were only grudgingly accepted because of their utility in solving the cubic equation. Following [42], we demonstrate here the usefulness of complex hyperbolic numbers by finding a formula for the solutions of the venerated reduced cubic equation

$$x^3+3ax+b=0. \tag{2.17}$$

(The general cubic $Ay^3+3By^2+Cy+D=0,\ A\neq 0$, can be transformed to the reduced form (2.17) by dividing through by the leading coefficient and then making the linear substitution $y=x-B/A$.) The basic unipodal equation $w^n=r$ can easily be solved using the idempotent basis, with the help of (2.15). Writing $w=w_+\mathrm{u}_++w_-\mathrm{u}_-$ and $r=r_+\mathrm{u}_++r_-\mathrm{u}_-$, we get

$$w^n=w_+^n\mathrm{u}_++w_-^n\mathrm{u}_-=r_+\mathrm{u}_++r_-\mathrm{u}_-, \tag{2.18}$$

so $w_+{}^n=r_+$ and $w_-^n=r_-$. It follows that $w_+=|r_+|^{\frac{1}{n}}\alpha^j$ and $w_-=|r_-|^{\frac{1}{n}}\alpha^k$ for some integers $0\leq j,k\leq n-1$, where $\alpha$ is a primitive $n$th root of unity http://en.wikipedia.org/wiki/Root_of_unity. This proves the following theorem.

**Theorem 2.5.1.** *For any positive integer $n$, the unipodal equation $w^n=r$ has $n^2$ solutions $w=\alpha^j r_+^{\frac{1}{n}}\mathrm{u}_++\alpha^k r_-^{\frac{1}{n}}\mathrm{u}_-$ for $j,k=0,1,\ldots,n-1$, where $\alpha\equiv\exp(2\pi\mathrm{i}/n)$.*

The number of roots to the equation $w^n=r$ can be reduced by adding constraints. The following corollary follows immediately from the theorem by noting that $w_+w_-=\rho\neq 0$ is equivalent to $w_-=\rho/w_+$.

**Corollary 2.5.2.** *The unipodal equation $w^n=r$, subject to the constraint $w_+w_-=\rho$, for a nonzero complex number $\rho$, has the $n$ solutions*

$$w=\alpha^j r_+^{\frac{1}{n}}\mathrm{u}_++\frac{\rho}{\alpha^j r_+^{\frac{1}{n}}}\mathrm{u}_-,$$

*for $j=0,1,\ldots,n-1$, where $\alpha\equiv\exp(2\pi\mathrm{i}/n)$ and $r_+^{\frac{1}{n}}$ denotes any $n$th root of the complex number $r_+$.*

It is interesting to note that Corollary 2.5.2 gives another way of solving the *dihedral equation*

$$z^n + \frac{1}{z^n} = \beta,$$

which is used by Felix Klein in his discussion of the solution of the cubic equation, [49, p.141]. We are now prepared to solve the reduced cubic equation (2.17).

**Theorem 2.5.3.** *The reduced cubic equation $x^3 + 3ax + b = 0$ has the solutions for $j = 0, 1, 2$,*

$$x = \frac{1}{2}\left(\alpha^j \sqrt[3]{s+t} + \frac{\rho}{\alpha^j \sqrt[3]{s+t}}\right), \tag{2.19}$$

*where $\alpha = \exp(2\pi\mathrm{i}/3) = -\frac{1}{2} + \frac{1}{2}\mathrm{i}\sqrt{3}$ is a primitive cube root of unity and $\rho = -4a$, $s = -4b$, and $t = \sqrt{s^2 - \rho^3} = 4\sqrt{b^2 + 4a^3}$.*

*Proof.* The unipodal equation $w^3 = r$, where $r = s + \mathrm{u}t$, is equivalent in the standard basis to $(x + y\mathrm{u})^3 = s + t\mathrm{u}$ or $(x^3 + 3xy^2) + \mathrm{u}(y^3 + 3x^2 y) = s + \mathrm{u}t$. Equating the complex scalar parts gives

$$x^3 + 3xy^2 - s = 0. \tag{2.20}$$

Making the additional constraint that $w_+ w_- = x^2 - y^2 = \rho$, we can eliminate $y^2$ from (2.20), getting the equivalent equation

$$x^3 - \frac{3}{4}\rho x - \frac{1}{4}s = 0. \tag{2.21}$$

The constraint $w_+ w_- = \rho$ further implies that

$$\rho^3 = (w_+ w_-)^3 = w_+{}^3 w_-^3 = r_+ r_- = s^2 - t^2,$$

which gives $t = \sqrt{s^2 - \rho^3}$. By letting $\rho = -4a$ and $s = -4b$, so $t = \sqrt{s^2 - \rho^3} = 4\sqrt{b^2 + 4a^3}$, (2.21) becomes the reduced cubic equation $x^3 + 3ax + b = 0$. Since $r_+ = s + t$, the desired solution (2.19) is then obtained by taking the complex scalar part of the solution given in Corollary 2.5.2. □

*Example.* Find the solutions of the reduced cubic $x^3 - 6x + 4 = 0$.

**Solution:** Here, $a = -2$, $b = 4$, so $\rho = 8$, $s = -16$, and we set $t = 16\mathrm{i}$. Then $s + t = 16(-1 + \mathrm{i}) = 2^{9/2}\exp(\mathrm{i}3\pi/4)$, so we may take $\sqrt[3]{s+t} = 2^{3/2}\exp(\mathrm{i}\pi/4)$. Thus,

$$\begin{aligned} x &= \frac{1}{2}\left(2^{3/2}\exp(\mathrm{i}\pi/4)\alpha^j + \frac{8}{2^{3/2}\exp(\mathrm{i}\pi/4)\alpha^j}\right) \\ &= 2^{1/2}\left(\exp(\mathrm{i}\pi/4)\alpha^j + \exp(-\mathrm{i}\pi/4)\alpha^{-j}\right), \end{aligned}$$

for $j = 0, 1, 2$. That is,

$$(i)\ x = 2^{1/2}\big(\exp(i\pi/4) + \exp(-i\pi/4)\big) = 2^{1/2}\big(2\cos(\pi/4)\big) = 2,$$

$$\begin{aligned}(ii)\ x &= 2^{1/2}\big(\exp(i\pi/4)\exp(2\pi i/3) + \exp(-i\pi/4)\exp(-2\pi i/3)\big)\\ &= 2^{1/2}\big(\exp(11\pi i/12) + \exp(-11\pi i/12)\big)\\ &= 2^{1/2}\big(2\cos(11\pi i/12)\big) = -(1+\sqrt{3}) \simeq -2.7321,\end{aligned}$$

or

$$\begin{aligned}(iii)\ x &= 2^{1/2}\big(\exp(i\pi/4)\exp(-2\pi i/3) + \exp(-i\pi/4)\exp(2\pi i/3)\big)\\ &= 2^{1/2}\big(\exp(-5\pi i/12) + \exp(5\pi i/12)\big)\\ &= 2^{1/2}\big(2\cos(5\pi i/12)\big) = (-1+\sqrt{3}) \simeq 0.7321.\end{aligned}$$

## Exercises

1. Find the solutions of the reduced cubic equation $x^3 - 3x - 1 = 0$.
2. Find the solutions of the reduced cubic equation $x^3 - 3x + 1 = 0$.
3. Find the solutions of the reduced cubic equation $x^3 - 4x + 2 = 0$.

## 2.6 Special Relativity and Lorentzian Geometry

In 1905, Albert Einstein, at that time only 26 years of age, published his special theory of relativity based on two postulates:

1. All coordinate systems (for measuring time and distance) moving with constant velocity relative to each other are equally valid for the formulation of the laws of physics.
2. Light propagates in every direction in a straight line and with the same speed $c$ in every valid coordinate system.

The formulas for transforming space and time coordinates between systems in uniform relative motion had been found somewhat earlier by the Dutch physicist H. A. Lorentz and were termed *Lorentz transformations* by H. Poincare. In 1907, Hermann Minkowski showed that Einstein's postulates imply a non-Euclidean geometry in four-dimensional space-time, called (flat) Minkowski space-time. Whereas more elaborate mathematical formalisms have been developed [3, 33, 68, 70], most of the features of the geometry can be understood by studying a two-dimensional subspace involving only one space dimension and time, which might naturally be called the Minkowski plane. But that term is already in use for another plane geometry that Minkowski developed for a quite different purpose, so this two-dimensional model of space-time is instead called the Lorentzian plane (even though Lorentz does not appear to have shared Minkowski's geometrical point of view). The hyperbolic numbers, which have also been called the "perplex numbers" [23], serve as coordinates in the Lorentzian plane in much the same way that the complex numbers serve as coordinates in the Euclidean plane.

An *event* $X$ that occurs at time $t$ and at the place $x$ is specified by its coordinates $t$ and $x$. If $c$ is the speed of light, the product $ct$ is a length, and the coordinates of the event $X$ can be combined into the hyperbolic number $X = ct + \mathrm{u}x$. By the *space-time distance* between two events $X_1$ and $X_2$, we mean the hyperbolic modulus $|X_1 - X_2|_h$ which is the hyperbolic distance of the point $X_1 - X_2$ to the origin. If $X_1 = ct + x_1\mathrm{u}$ and $X_2 = ct + x_2\mathrm{u}$ are two events occurring at the same time $t$, then by (2.1),

$$|X_1 - X_2|_h = \sqrt{|0^2 - (x_1 - x_2)^2|} = |x_1 - x_2|, \tag{2.22}$$

which is exactly the Euclidean distance between the points $x_1$ and $x_2$ on the $x$-axis.

If the coordinates of an event in two-dimensional space-time relative to one coordinate system are $X = ct + \mathrm{u}x$, what are its coordinates $X' = ct' + \mathrm{u}x'$ with respect to a second coordinate system that moves with uniform velocity $v < c$ with respect to the first system? We have already laid the groundwork for solving this problem. By (2.10), we have $X' = X\mathrm{e}^{-\phi\mathrm{u}}$. Calculating

$$X'X'^{-} = X\mathrm{e}^{-\phi\mathrm{u}}X^{-}\mathrm{e}^{\phi\mathrm{u}} = XX^{-},$$

we conclude that

$$X'X'^{-} = c^2t'^2 - (x')^2 = c^2t^2 - x^2 = XX^{-} \tag{2.23}$$

is invariant under the Lorentz transformation of coordinates $X' = X\mathrm{e}^{-\phi\mathrm{u}}$.

Let $\phi = \tanh^{-1}(v/c)$ so that $\tanh\phi = \frac{v}{c}$, which implies $\sinh(\phi) = \frac{v}{c}\cosh(\phi)$ and the identity

$$1 = \cosh^2\phi - \sinh^2\phi = \cosh^2\phi(1 - v^2/c^2).$$

It follows that $\gamma := \cosh\phi = \frac{1}{\sqrt{1-v^2/c^2}}$. The factor $\gamma$ often arises in special relativity which is why we denote it by a special symbol. Since $X\exp(-\phi\mathrm{u}) = X'$, we find

$$\begin{aligned}
&(ct + \mathrm{u}x)\big(\cosh(-\phi) + \mathrm{u}\sinh(-\phi)\big) \\
&= (ct\cosh\phi - x\sinh\phi) + \mathrm{u}(x\cosh\phi - ct\sinh\phi) \\
&= \frac{ct - xv/c}{\sqrt{1 - v^2/c^2}} + \mathrm{u}\frac{x - cvt/c}{\sqrt{1 - v^2/c^2}} \\
&= ct' + \mathrm{u}x'.
\end{aligned}$$

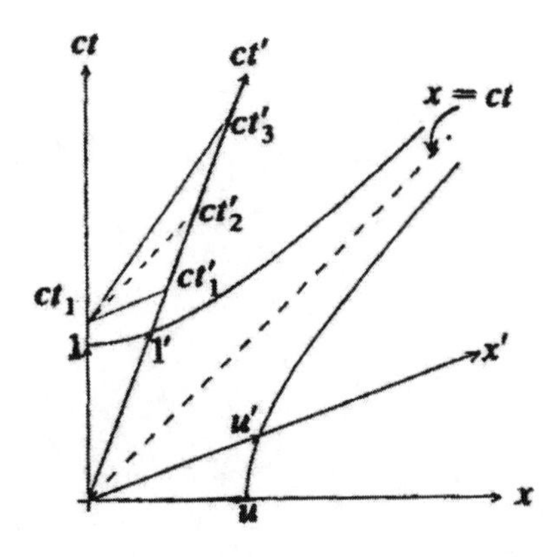


**Fig. 2.7** Minkowski diagram. Note that the unipotent u lies on the $x$-axis and that $ct$ lies on the real $y$-axis

Equating scalar and unipotent parts in the last equality gives the famous Lorentz equations:

$$t' = \frac{t - xv/c^2}{\sqrt{1 - v^2/c^2}}, \quad x' = \frac{x - vt}{\sqrt{1 - v^2/c^2}} \tag{2.24}$$

relating the times and positions of an event as measured by two observers in relative motion [9, p.236]. Diagrams such as Fig. 2.7, showing two coordinate systems in space-time in relative motion (using equal distance scales on the two axes), are called *Minkowski diagrams* [25].

In Fig. 2.7, we have adopted the convention that the unipotent u lies along the horizontal $x$-axis and 1 lies along the vertical time axis $ct$.

The histories of two particles in relative uniform motion are given by $X(t) = ct$ and $X'(t') = ct'$. Each observer has a *rest frame* or hyperbolic orthogonal coordinate system in which he or she measures the relative time $t$ and relative position $x$ of an event $X = ct + x\mathrm{u}$. The *history* or *worldline* of a particle is just the graph of its location as a function of time, $X(t) = ct + \mathrm{u}x(t)$. For example, in a coordinate system moving with the particle, so that the particle remains at $x(t) \equiv 0$ (the *rest frame* of the particle), the particle's history or worldline is just the straight line with equation $X = ct$, Fig. 2.7.

The division of the Lorentz plane into quadrants has an important physical interpretation. The interval between two events, say, the origin and $X$, is said to be *timelike* if there is a velocity $v < c$ such that in the coordinate system moving along the $x$-axis with velocity $v$, the history of the origin passes through $X$. Thus, to an observer in this coordinate system, the second event $X$ occurs at the same place $x = 0$ but at a later time $t > 0$. Events with a timelike separation are *causally* related in that the earlier event can influence the later one, since a decision at the later time might be made on the basis of information left at the given location at the earlier time.

In the Minkowski diagram, Fig. 2.7, each of the events $X_1 = ct_1$ and $X_2' = ct_2'$ have a timelike separation from the event $X_3' = ct_3'$. The events $X_1$ and $X_2'$ have a *lightlike* separation because the line segment connecting them, the history or worldline of a

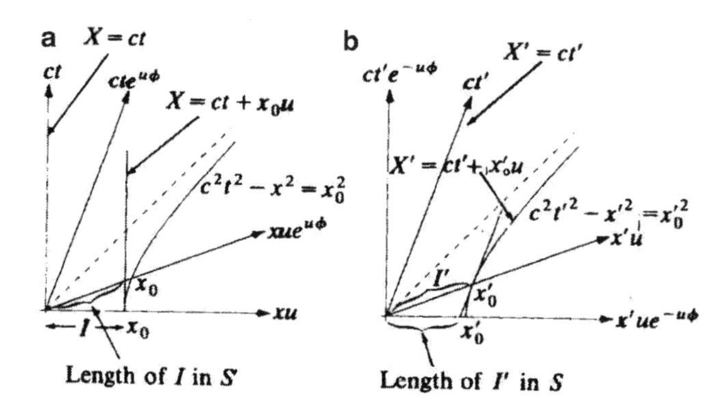


**Fig. 2.8** Fitzgerald–Lorentz contraction

*light signal*, is parallel to the isotropic line $x = ct$. (The isotropic lines $x = \pm ct$ are the worldlines of light signals moving at the velocity of light and passing through the origin $X = 0 = X'$.) The events $X_1$ and $X_1'$ have a *spacelike* separation because in the coordinate system $X' = ct'$ moving along the $x$-axis with velocity $v < c$, the two appear as simultaneous events at different locations, since the line segment joining them is parallel to the $x'$ axis.

Whereas the space-time distance reduces to the usual Euclidean distance (2.22) when an observer makes measurements of the distances between objects in her rest frame along her $x$-axis, the *Fitzgerald–Lorentz contraction* comes in to play when two coordinate systems are moving with respect to each other. An object at rest in either system appears *shorter* when measured from the other system than when measured in its rest system.

Suppose system $S'$ is moving with velocity $v < c$ with respect to system $S$. The histories of an interval $I$ at rest in $S$ with ends at the origin and point $X = x_0\mathrm{u}$ are the lines $X = ct$ and $X = ct + \mathrm{u}x_0$. Now recall that the points on the hyperbola $c^2t^2 - x^2 = \rho^2$, $\rho > 0$, all have hyperbolic distance $\rho$ from the origin. Thus, the point on the $x'$-axis at hyperbolic distance $x_0$ is the point where this axis intersects the hyperbola $c^2t^2 - x^2 = x_0^2$, and it is clear from the Minkowski diagram Fig. 2.8a that this is beyond the intersection of the $x'$-axis with the worldline $X = ct + \mathrm{u}x$. Thus, the hyperbolic length of the interval $I$ in the rest frame $S'$ is less than $x_0$, its hyperbolic length in $S$.

The history of a particle moving with constant velocity $v < c$ along the $x$-axis is $X(t) = ct + \mathrm{u}tv$. Differentiating $X(t)$ with respect to $t$ gives

$$V := \mathrm{d}X/\mathrm{d}t = c + \mathrm{u}v, \tag{2.25}$$

the *space-time velocity* of this particle: a vector parallel to the $ct'$-axis. Thus, the histories of the endpoints of an interval $I'$ of the $x'$-axis (i.e., an interval at rest in system $S'$) will be parallel to the $ct'$-axis, as in Fig. 2.8b, and again the Minkowski diagram shows that the length of this interval in the coordinate system S is less than its length in $S'$. The invariance of the hyperbolic distance under change of orthogonal coordinates in the Lorentz plane is seen to be a geometric expression of the relativity of physical length measurements. It follows that the length of an object is dependent upon the rest frame in which it is measured.

Experimentally, it has been determined in the coordinate system $S$ that the *momentum* of a particle with *rest mass m* moving with the velocity $v\mathrm{u} = \frac{dx}{dt}\mathrm{u}$ is $p = \gamma m v\mathrm{u}$ where $\gamma = \frac{1}{\sqrt{1-v^2/c^2}}$. The *space-time momentum* for a moving particle with the space-time velocity $V = \frac{dX}{dt}$ is defined to be

$$P = \gamma m V = \gamma m(c + v\mathrm{u}) = \left(\frac{E}{c} + p\right), \tag{2.26}$$

where $E$ is called the *total energy* of the particle with respect to the coordinate system $S$ at rest. Note that whereas we found in (2.23) the quantity $XX^-$ to be Lorentz invariant for the space-time event $X$, the corresponding quantity $VV^- = c^2 - v^2$ of the space-time velocity $V$ is *not* Lorentz invariant.

From the above relationships, we can easily derive Einstein's most famous formula for the total energy $E$ in the coordinate system $S$. We first calculate

$$PP^- = \frac{E^2}{c^2} - \gamma^2 m^2 v^2 = m^2 c^2$$

which is a Lorentz-invariant relation. Solving this relation for the total energy $E$ gives

$$E = \gamma m c^2 \tag{2.27}$$

relative to the coordinate system $S$ in which the particle has velocity $v\mathrm{u}$. The total energy (2.27) reduces to the energy $E_0 = mc^2$ of the particle when it is at rest.

The *kinetic energy K* of the particle with mass $m$ moving with the velocity $v\mathrm{u}$ can be thought of as the difference of its total energy $E$ and its energy $E_0$ when it is at rest. The kinetic energy is thus given by

$$K = E - E_0 = mc^2(\gamma - 1) = mc^2\left(\frac{1}{2}\frac{v^2}{c^2} + \frac{3}{8}\frac{v^4}{c^4} + \cdots\right),$$

where we are using a Taylor series expansion of the factor $\gamma$. It follows that for $v \ll c$, when the speed $v$ of the particle is much less than $c$, that $K \cong \frac{1}{2}mv^2$ which is in agreement with Newton's kinetic energy for a moving particle.

Many other fundamental results in special relativity theory can be derived using geometric properties of the 2-dimensional Lorentzian plane. As we have seen, the hyperbolic numbers play the same role for computations in the Lorentzian plane that the complex numbers have in the Euclidean plane. Whereas here we have limited our preliminary discussion of special relativity to the Lorentzian plane, we extend it to 4-dimensional space-time in Chap. 11.

## Exercises

1. Two twins are born on Earth on the same day. One twin travels with his father on a rocketship to the nearest star Alpha Centauri, a distance of 4.37 light-years from Earth. They then turn around and return to Earth just in time for his Earth twin's tenth birthday. Assuming that their outward and return journey was traveled roughly at the same speed,

   (a) Calculate the average velocity of the rocket (as a percentage of the speed of light).

   (b) Calculate the age of the returning twin in comparison to that of his sister who remained on Earth.

2. (a) Calculate the length of a meterstick in a direction parallel to its travel in a restframe in which it is moving at 95 % of the velocity of light. (The length of the same meterstick in a direction perpendicular to its travel is unchanged.)

3. Let $X = ct + \mathrm{u}vt$ be the history of a particle moving at a constant velocity $v\mathrm{u}$ in the coordinate system $S$, and let $X' = ct' = X\mathrm{e}^{-\phi\mathrm{u}}$ be the history of this same particle at rest in the coordinate system $S'$.

   (a) Show that

$$X'X'^{-} = c^2t'^2 = c^2t^2 - v^2t^2 = XX^{-},$$

   which implies that

$$t' = \frac{1}{\gamma}t \quad \Longleftrightarrow \quad \frac{\mathrm{d}t}{\mathrm{d}t'} = \gamma = \frac{1}{\sqrt{1-\frac{v^2}{c^2}}} = \cosh\phi.$$

   (b) Show that $V' = \frac{\mathrm{d}X'}{\mathrm{d}t'} = \gamma V\mathrm{e}^{-\phi\mathrm{u}}$.

   (c) Show that $V' = \gamma V\mathrm{e}^{-\phi\mathrm{u}}$ implies

$$V'V'^{-} = \gamma^2 VV^{-},$$

   so that the space-time velocity $V$ of a particle moving at constant velocity is not Lorentz invariant.

# Chapter 3
# Geometric Algebra

*That all our knowledge begins with experience, there is indeed no doubt ... but although our knowledge originates with experience, it does not all arise out of experience.*

–Immanuel Kant

The real number system $\mathbb{R}$ has a long and august history spanning a host of civilizations over a period of many centuries [17]. It may be considered the rock upon which many other mathematical systems are constructed and, at the same time, serves as a model of desirable properties that any extension of the real numbers should have. The real numbers $\mathbb{R}$ were extended to the *complex numbers* $\mathbb{C} = \mathbb{R}(\mathrm{i})$, where $\mathrm{i}^2 = -1$, principally because of the discovery of the solutions to the quadratic and cubic polynomials in terms of complex numbers during the Renaissance. The powerful *Euler formula* $z = r\exp(\mathrm{i}\theta)$ helps makes clear the geometric significance of the multiplication of complex numbers, as we have seen in Fig. 2.5 in the last chapter. Other extensions of the real and complex numbers have been considered but until recently have found only limited acceptance. For example, the extension of the complex numbers to Hamilton's quaternions was more divisive in its effects upon the mathematical community [14], one reason being the lack of universal commutativity and another the absence of a unique, clear geometric interpretation.

We have seen in the previous chapter that extending the real numbers $\mathbb{R}$ to include a new square root of $+1$ leads to the concept of the hyperbolic number plane $\mathbb{H}$, which in many ways is analogous to the complex number plane $\mathbb{C}$. Understanding the hyperbolic numbers is key to understanding even more general geometric extensions of the real numbers. Perhaps the extension of the real numbers to include a new square root $\mathrm{u} = \sqrt{+1} \notin \mathbb{R}$ only occurred much later because people were happy with the status quo that $\sqrt{1} = \pm 1$ and because such considerations were before the advent of Einstein's *theory of special relativity* and the study of *non-Euclidean* geometries.

Geometric algebra is the extension of the real number system to include new *anticommuting* square roots of $\pm 1$, which represent mutually orthogonal unit

G. Sobczyk, *New Foundations in Mathematics: The Geometric Concept of Number*, DOI 10.1007/978-0-8176-8385-6_3,

Hermann Gunther Grassmann (1809-1877) was a high-school teacher. His far reaching Ausdehnungslehre, "Theory of extension" lay the ground work for the development of the exterior or outer product of vectors. William Rowan Hamilton (1805-1865) was an Irish physicist, astronomer and mathematician. His invention of the quaternions as the natural generalization of the complex numbers of the plane to three dimensional space, together with the ideas of Grassmann, set the stage for William Kingdon Clifford's definition of geometric algebra. William Kingdon Clifford (1845-1879) was a professor of mathematics and mechanics at the University College of London. Tragically, he died at the early age of 33 before he could explore his profound ideas.

vectors in successively higher dimensions. The critical new insight is that by assuming that our new square roots of $\pm 1$ are anticommutative, we obtain a more general concept of number that will serve us well in the expression of geometrical ideas [72]. We begin with the extension of the real numbers to include two new anticommuting square roots of $+1$ which are given the interpretation of two *unit vectors* $\mathbf{e}_1$ and $\mathbf{e}_2$ lying along the $x$- and $y$-axes, respectively, as shown in Fig. 3.1. More generally, if we introduce $n$ anticommuting square roots of unity, together with their geometric sums and products, we can represent all the directions in an $n$-dimensional space.

Geometric algebra provides the framework for the rest of the material developed in this book, and it is for this reason that the book is entitled *New Foundations in Mathematics: The Geometric Concept of Number*. A brief history of the development of geometric algebra, also known as Clifford algebra, is given in the box.

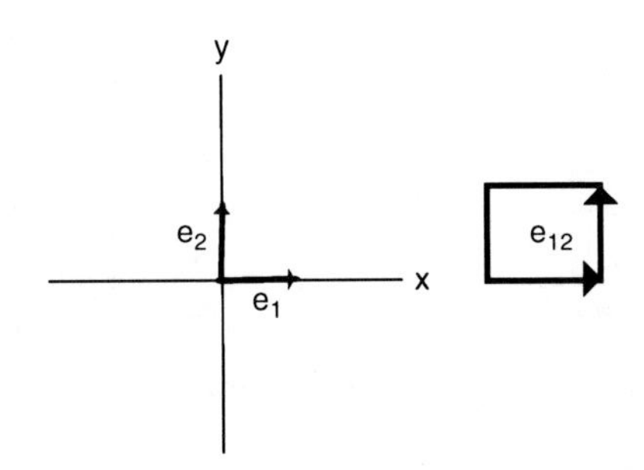


**Fig. 3.1** Anticommuting orthogonal unit vectors $\mathbf{e}_1$ and $\mathbf{e}_2$ along the $xy$-axes. Also pictured is the unit bivector $\mathbf{e}_{12} = \mathbf{e}_1\mathbf{e}_2$

## 3.1 Geometric Numbers of the Plane

Let $\mathbf{e}_1 = (1,0)$ and $\mathbf{e}_2 = (0,1)$ be orthonormal unit vectors along the $x$- and $y$-axes in $\mathbb{R}^2$ starting at the origin, and let $\mathbf{a} = (a_1,a_2) = a_1\mathbf{e}_1 + a_2\mathbf{e}_2$ and $\mathbf{b} = (b_1,b_2) = b_1\mathbf{e}_1 + b_2\mathbf{e}_2$, for $a_1,a_2,b_1,b_2 \in \mathbb{R}$, be arbitrary vectors in $\mathbb{R}^2$. The familiar *inner product* or "dot product" between the vectors $\mathbf{a}$ and $\mathbf{b}$ is given by

$$\mathbf{a}\cdot\mathbf{b} = |\mathbf{a}||\mathbf{b}|\cos\theta = a_1b_1 + a_2b_2, \tag{3.1}$$

where $|\mathbf{a}|$ and $|\mathbf{b}|$ are the *magnitudes* or *lengths* of the vectors $\mathbf{a}$ and $\mathbf{b}$ and $\theta$ is the angle between them.

The associative geometric algebra $\mathbb{G}_2 = \mathbb{G}(\mathbb{R}^2)$ of the plane $\mathbb{R}^2$ is generated by the *geometric multiplication* of the vectors in $\mathbb{R}^2$, subjected to the rule that given any vector $\mathbf{x} = (x_1,x_2) \in \mathbb{R}^2$,

$$\mathbf{x}^2 = \mathbf{x}\mathbf{x} = |\mathbf{x}|^2 = \mathbf{x}\cdot\mathbf{x} = x_1^2 + x_2^2. \tag{3.2}$$

This rule means that the square of any vector is equal to its magnitude squared. Geometric addition and multiplication of vectors satisfies all of the rules of an *associative algebra* with the *unity* 1. Indeed, geometric algebra satisfies all of the usual rules of addition and multiplication of real numbers, except that the geometric product of vectors is not universally commutative. The geometric algebra $\mathbb{G}_2$ is the *geometric extension* of the real numbers to include the two *anticommuting unit vectors* $\mathbf{e}_1$ and $\mathbf{e}_2$ along the $x$- and $y$-axes. See Fig. 3.1.

The fact that $\mathbf{e}_1$ and $\mathbf{e}_2$ are anticommuting can be considered to be a consequence of the rule (3.2). Since $\mathbf{e}_1$ and $\mathbf{e}_2$ are orthogonal unit vectors, by utilizing the *Pythagorean theorem*, we have

$$(\mathbf{e}_1+\mathbf{e}_2)^2 = \mathbf{e}_1^2 + \mathbf{e}_1\mathbf{e}_2 + \mathbf{e}_2\mathbf{e}_1 + \mathbf{e}_2^2 = 1 + \mathbf{e}_1\mathbf{e}_2 + \mathbf{e}_2\mathbf{e}_1 + 1 = 2,$$

from which it follows that $\mathbf{e}_1\mathbf{e}_2 + \mathbf{e}_2\mathbf{e}_1 = 0$, so $\mathbf{e}_1$ and $\mathbf{e}_2$ are anticommuting. We give the quantity $i = \mathbf{e}_{12} := \mathbf{e}_1\mathbf{e}_2$ the geometric interpretation of a *unit bivector* or

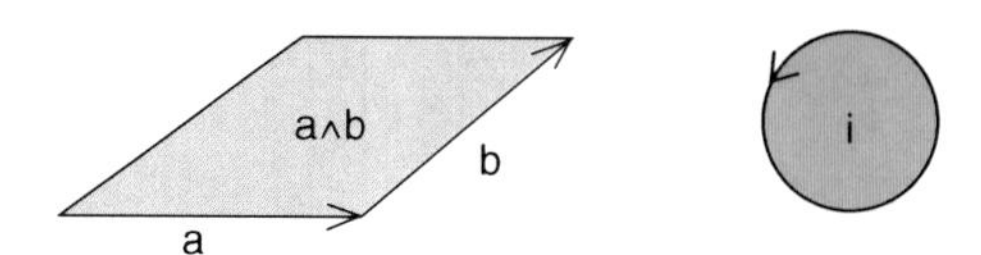


**Fig. 3.2** The bivector $\mathbf{a} \wedge \mathbf{b}$ is obtained by sweeping the vector $\mathbf{b}$ out along the vector $\mathbf{a}$. Also shown is the unit bivector $\mathbf{i}$ in the plane of $\mathbf{a}$ and $\mathbf{b}$

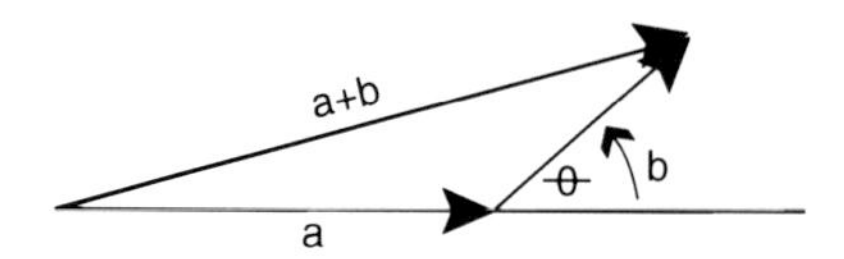


**Fig. 3.3** The law of cosines relates the lengths of the sides of the triangle to the cosine of the angle between the sides $\mathbf{a}$ and $\mathbf{b}$

*directed plane segment* in the $xy$-plane. Whereas by the rule (3.2) a unit vector has square $+1$, we find that

$$i^2 = (\mathbf{e}_1\mathbf{e}_2)(\mathbf{e}_1\mathbf{e}_2) = \mathbf{e}_1(\mathbf{e}_2\mathbf{e}_1)\mathbf{e}_2 = -\mathbf{e}_1(\mathbf{e}_1\mathbf{e}_2)\mathbf{e}_2 = -\mathbf{e}_1^2\mathbf{e}_2^2 = -1,$$

so that the unit bivector $i$ has square $-1$. The unit bivector $i$ is pictured in both Figs. 3.1 and 3.2. Notice that the *shape* of a bivector is unimportant; an oriented disk with area one in the $xy$-plane provides an equally valid picture of a unit bivector as does the oriented unit square in the $xy$-plane with the same orientation.

For any vectors $\mathbf{a}, \mathbf{b} \in \mathbb{R}^2$, we find by again using (3.2) that

$$(\mathbf{a}+\mathbf{b})^2 = \mathbf{a}^2 + (\mathbf{ab}+\mathbf{ba}) + \mathbf{b}^2, \tag{3.3}$$

or equivalently,

$$\mathbf{a}\cdot\mathbf{b} = \frac{1}{2}(\mathbf{ab}+\mathbf{ba}) = \frac{1}{2}\left(|\mathbf{a}+\mathbf{b}|^2 - |\mathbf{a}|^2 - |\mathbf{b}|^2\right) = |\mathbf{a}||\mathbf{b}|\cos\theta,$$

which is an expression of the *law of cosines*. See Fig. 3.3.

Directly calculating the geometric products $\mathbf{ab}$ and $\mathbf{ba}$ for the vectors $\mathbf{a} = a_1\mathbf{e}_1 + a_2\mathbf{e}_2$ and $\mathbf{b} = b_1\mathbf{e}_1 + b_2\mathbf{e}_2$, and simplifying, gives

$$\begin{aligned}\mathbf{ab} &= (a_1\mathbf{e}_1 + a_2\mathbf{e}_2)(b_1\mathbf{e}_1 + b_2\mathbf{e}_2) = a_1b_1\mathbf{e}_1^2 + a_2b_2\mathbf{e}_2^2 + a_1b_2\mathbf{e}_1\mathbf{e}_2 + a_2b_1\mathbf{e}_2\mathbf{e}_1 \\ &= (a_1b_1 + a_2b_2) + (a_1b_2 - a_2b_1)\mathbf{e}_{12}.\end{aligned} \tag{3.4}$$

and

$$\mathbf{ba} = (b_1\mathbf{e}_1 + b_2\mathbf{e}_2)(a_1\mathbf{e}_1 + a_2\mathbf{e}_2) = (a_1b_1 + a_2b_2) - (a_1b_2 - a_2b_1)\mathbf{e}_{12}. \tag{3.5}$$

Using (3.4) and (3.5), we decompose the geometric product $\mathbf{ab}$ into *symmetric* and *antisymmetric* parts, getting

$$\mathbf{ab} = \frac{1}{2}(\mathbf{ab}+\mathbf{ba}) + \frac{1}{2}(\mathbf{ab}-\mathbf{ba}) = \mathbf{a}\cdot\mathbf{b} + \mathbf{a}\wedge\mathbf{b}. \tag{3.6}$$

The symmetric part gives the inner product

$$\mathbf{a}\cdot\mathbf{b} = \frac{1}{2}(\mathbf{ab}+\mathbf{ba}) = \langle\mathbf{ab}\rangle_0 = a_1b_1 + a_2b_2,$$

where $\langle\mathbf{ab}\rangle_0$ denotes the 0-*vector part* or *scalar part* of the argument. The antisymmetric part, called the *outer product* or *Grassmann exterior product*, is given by

$$\mathbf{a}\wedge\mathbf{b} = \frac{1}{2}(\mathbf{ab}-\mathbf{ba}) = \langle\mathbf{ab}\rangle_2 = (a_1b_2 - a_2b_1)\mathbf{e}_1\mathbf{e}_2 = \mathrm{i}\det\begin{pmatrix} a_1 & a_2 \\ b_1 & b_2 \end{pmatrix},$$

where $\langle\mathbf{ab}\rangle_2$ denotes the 2-*vector part* or *bivector part* of the argument.

The outer product $\mathbf{a}\wedge\mathbf{b}$ defines the *direction* and *orientation* of an oriented or *directed plane segment*, just as a *vector* defines the *direction* and *orientation* of a directed line segment. Note that the bivector $\mathbf{a}\wedge\mathbf{b}$ has the direction of the unit bivector $\mathrm{i} = \mathbf{e}_1\mathbf{e}_2$ and the magnitude of the paralellogram defined by the vectors $\mathbf{a}$ and $\mathbf{b}$. See Fig. 3.2.

We can also express the outer product $\mathbf{a}\wedge\mathbf{b}$ in the form

$$\mathbf{a}\wedge\mathbf{b} = |\mathbf{a}||\mathbf{b}|\mathrm{i}\sin\theta \tag{3.7}$$

which is complimentary to (3.1) for the inner product. In the particular case that

$$0 = \mathbf{a}\cdot\mathbf{b} = \frac{1}{2}(\mathbf{ab}+\mathbf{ba}) \iff \mathbf{ab} = -\mathbf{ba},$$

we see once again that the geometric product of two nonzero orthogonal vectors is anticommutative. In this case, (3.3) reduces to the familiar Pythagorean theorem for a right triangle with the vectors $\mathbf{a}$ and $\mathbf{b}$ along its sides.

Putting together (3.1) and (3.7) into (3.6), and using (2.3), we find that the geometric product of the vectors $\mathbf{a}$ and $\mathbf{b}$ can be written in the *Euler form*

$$\mathbf{ab} = \mathbf{a}\cdot\mathbf{b} + \mathbf{a}\wedge\mathbf{b} = |\mathbf{a}||\mathbf{b}|(\cos\theta + \mathrm{i}\sin\theta) = |\mathbf{a}||\mathbf{b}|e^{\mathrm{i}\theta}. \tag{3.8}$$

We have made the fundamental discovery that the geometric product of the two vectors $\mathbf{a},\mathbf{b} \in \mathbb{R}^2$ is formally a complex number, most beautifully represented by the Euler formula above. Notice that whereas our derivation was in $\mathbb{R}^2$, the same derivation carries over for any two vectors in $\mathbb{R}^n$, since two linearly independent vectors in $\mathbb{R}^n$ will always determine a two-dimensional subspace of $\mathbb{R}^n$ which we can choose to be $\mathbb{R}^2$.

The *standard basis* of the geometric algebra $\mathbb{G}_2 = \mathbb{G}(\mathbb{R}^2)$ *over the real numbers* $\mathbb{R}$ is

$$\mathbb{G}_2 = \text{span}_{\mathbb{R}}\{1, \mathbf{e}_1, \mathbf{e}_2, \mathbf{e}_{12}\}, \tag{3.9}$$

where $\mathbf{e}_1^2 = \mathbf{e}_2^2 = 1$ and $\mathbf{e}_{12} := \mathbf{e}_1\mathbf{e}_2 = -\mathbf{e}_2\mathbf{e}_1$. We have seen that $\mathbf{e}_1$ and $\mathbf{e}_2$ have the interpretation of *orthonormal vectors* along the $x$- and $y$-axes of $\mathbb{R}^2$ and that the *imaginary unit* $\mathrm{i} = \mathbf{e}_{12}$ is the *unit bivector* of the plane spanned by $\mathbf{e}_1$ and $\mathbf{e}_2$. The most general geometric number $g \in \mathbb{G}_2$ has the form

$$g = (\alpha_1 + \alpha_2\mathbf{e}_{12}) + (v_1\mathbf{e}_1 + v_2\mathbf{e}_2), \tag{3.10}$$

where $\alpha_i, v_i \in \mathbb{R}$ for $i = 1, 2$, in the standard basis (3.9).

## Exercises

1. Let $\mathbf{x}_1 = (2,3)$, $\mathbf{x}_2 = (2,-3)$, and $\mathbf{x}_3 = (4,1)$. Calculate
   (a) $\mathbf{x}_1 \cdot \mathbf{x}_2$.
   (b) $\mathbf{x}_1 \wedge \mathbf{x}_2$. Graph this bivector.
   (c) $\mathbf{x}_2\mathbf{x}_3$. Find the Euler form (3.8) for this product.
   (d) Verify that $\mathbf{x}_1(\mathbf{x}_2 + \mathbf{x}_3) = \mathbf{x}_1\mathbf{x}_2 + \mathbf{x}_1\mathbf{x}_3$.
   (e) Graph $\mathbf{x}_1, \mathbf{x}_2$, and $\mathbf{x}_1 + \mathbf{x}_2$ in $\mathbb{R}^2$.
2. For $\mathbf{a} = (a_1, a_2)$, $\mathbf{b} = (b_1, b_2)$, $\mathbf{c} = (c_1, c_2)$ in $\mathbb{R}^2$, calculate
   (a) $\mathbf{a} \cdot \mathbf{b}$
   (b) $\mathbf{a} \wedge \mathbf{b}$. Graph this bivector.
   (c) Verify that $\mathbf{a}(\mathbf{b} + \mathbf{c}) = \mathbf{ab} + \mathbf{ac}$.
   (d) Verify that $(\mathbf{b} + \mathbf{c})\mathbf{a} = \mathbf{ba} + \mathbf{ca}$.
3. Let $\mathbf{x} = (x, y) = x\mathbf{e}_1 + y\mathbf{e}_2$.
   (a) Find the magnitude $|\mathbf{x}| = \sqrt{\mathbf{x}^2}$.
   (b) Find the unit vector $\hat{\mathbf{x}} := \frac{\mathbf{x}}{|\mathbf{x}|}$, and show that
   $$\mathbf{x}^{-1} = \frac{1}{\mathbf{x}} = \frac{\mathbf{x}}{|\mathbf{x}|^2} = \frac{\hat{\mathbf{x}}}{|\mathbf{x}|}$$
   where $\mathbf{x}^{-1}\mathbf{x} = 1 = \mathbf{x}\mathbf{x}^{-1}$.
   (c) Show that the equation of the **unit circle** in $\mathbb{R}^2$ with center at the point $\mathbf{a} \in \mathbb{R}^2$ is $(\mathbf{x} - \mathbf{a})^2 = 1$ .
4. Let $w_1 = 5 + 4\mathbf{e}_1$, $w_2 = 5 - 4\mathbf{e}_2$, and $z_3 = 2 + \mathbf{e}_1\mathbf{e}_2$ be geometric numbers in $\mathbb{G}_2$.
   (a) Show that $w_1w_2 - z_3 = 23 + 20\mathbf{e}_1 - 20\mathbf{e}_2 - 17\mathbf{e}_{12}$.

(b) Show that $w_1(w_2w_3) = (w_1w_2)w_3 = 66 + 60\mathbf{e}_1 - 20\mathbf{e}_2 - 7\mathbf{e}_{12}$ (geometric multiplication is associative).
(c) Show that $w_1(w_2 + w_3) = w_1w_2 + w_1w_3 = 35 + 28\mathbf{e}_1 - 16\mathbf{e}_2 - 11\mathbf{e}_{12}$ (distributive law).

5. Let $w = x + \mathbf{e}_1 y$ and $w^- = x - \mathbf{e}_1 y$. We define the magnitude $|w| = \sqrt{|ww^-|}$.

(a) Show that $|w| = \sqrt{|x^2 - y^2|}$.
(b) Show that the equation of the unit hyperbola in the hyperbolic number plane $\mathbb{H} = \text{span}_{\mathbb{R}}\{1, \mathbf{e}_1\}$ is $|w|^2 = |ww^-| = 1$ and has four branches.
(c) Hyperbolic Euler formula: Let $x > |y|$. Show that

$$w = x + \mathbf{e}_1 y = |w|\left(\frac{x}{|w|} + \mathbf{e}_1\frac{y}{|w|}\right) = \rho(\cosh\phi + \mathbf{e}_1 \sinh\phi) = \rho e^{\mathbf{e}_1\phi}$$

where $\rho = |w|$ is the hyperbolic magnitude of $w$ and $\phi$ is the hyperbolic angle that $w$ makes with the $x$-axis. The $(\rho, \phi)$ are also called the hyperbolic polar coordinates of the point $w = (x, y) = x + \mathbf{e}_1 y$. What happens in the case that $y > |x|$? Compare this with similar results in Chap. 2.
(d) Let $w_1 = \rho_1 \exp(\mathbf{e}_1\phi_1)$ and $w_2 = \rho_2 \exp(\mathbf{e}_1\phi_2)$. Show that

$$w_1w_2 = \rho_1\rho_2 \exp(\mathbf{e}_1(\phi_1 + \phi_2)).$$

What is the geometric interpretation of this result? Illustrate with a figure. Compare this with similar results in Chap. 2.
(f) Find the *square roots* of the geometric numbers $w = 5 + 4\mathbf{e}_1$ and $z = 2 + \mathbf{e}_{12}$. *Hint:* First express the numbers in Euler form.

6. Calculate

(a) $e^{i\theta}\mathbf{e}_1$ and $e^{i\theta}\mathbf{e}_2$, where $i = \mathbf{e}_1\mathbf{e}_2$, and graph the results on the unit circle in $\mathbb{R}^2$.
(b) Show that $e^{i\theta}\mathbf{e}_1 = \mathbf{e}_1 e^{-i\theta} = e^{\frac{i\theta}{2}}\mathbf{e}_1 e^{\frac{-i\theta}{2}}$.
(c) Show that $(e^{i\theta}\mathbf{e}_1) \wedge (e^{i\theta}\mathbf{e}_2) = \mathbf{e}_1 \wedge \mathbf{e}_2 = i$, and explain the geometric significance of this result.

7. Show that $e^{-i\theta}\mathbf{a}$ rotates the vector $\mathbf{a} = (a_1, a_2)$ counterclockwise in the $(x, y)$-plane through an angle of $\theta$.
8. Let the *geometric numbers* $A = 1 + 2\mathbf{e}_1 - \mathbf{e}_2 + 3i$ and $B = -2 - \mathbf{e}_1 + 2\mathbf{e}_2 - i$. Calculate the geometric product $AB$ and write it as the sum of its *scalar*, *vector*, and *bivector* parts.
9. (a) Show that $X = \mathbf{a}^{-1}(\mathbf{c} - \mathbf{b}) \in \mathbb{G}_2$ is the solution to the linear equation $\mathbf{a}X + \mathbf{b} = \mathbf{c}$ where $\mathbf{a}, \mathbf{b}, \mathbf{c} \in \mathbb{R}^2$.
(b) Show that $X = (\mathbf{c} - \mathbf{b})\mathbf{a}^{-1} \in \mathbb{G}_2$ is the solution to the linear equation $X\mathbf{a} + \mathbf{b} = \mathbf{c}$ where $\mathbf{a}, \mathbf{b}, \mathbf{c} \in \mathbb{R}^2$.
(c) Find the solution to the equation $\mathbf{a}X^2 + \mathbf{b} = \mathbf{c}$.

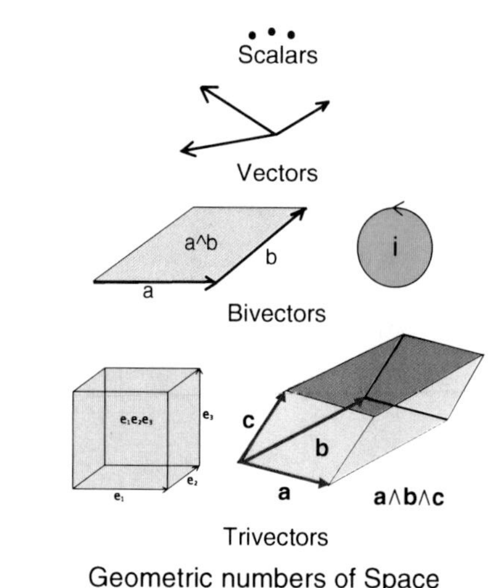


**Fig. 3.4** Geometric numbers of space

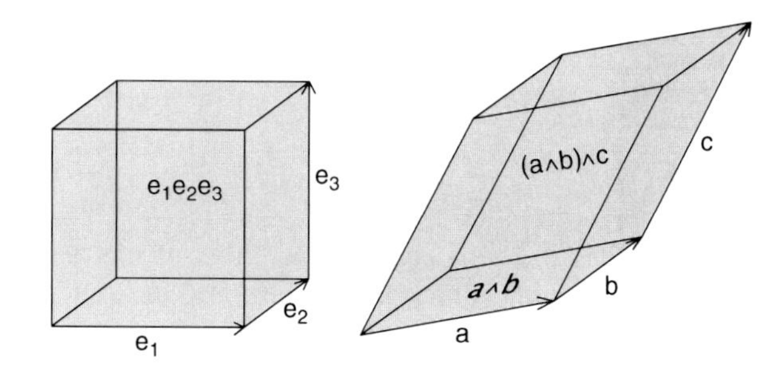


**Fig. 3.5** The trivector $\mathbf{a} \wedge \mathbf{b} \wedge \mathbf{c}$ is formed by sweeping the bivector $\mathbf{a} \wedge \mathbf{b}$ out along the vector $\mathbf{c}$. Also shown is the unit pseudoscalar $I = \mathbf{e}_{123}$

## 3.2 The Geometric Algebra $\mathbb{G}_3$ of Space

To arrive at the geometric algebra $\mathbb{G}_3 = \mathbb{G}(\mathbb{R}^3)$ of the 3-dimensional Euclidean space $\mathbb{R}^3$, we simply extend the geometric algebra $\mathbb{G}_2 = \mathbb{G}(\mathbb{R}^2)$ of the vector plane $\mathbb{R}^2$ to include the unit vector $\mathbf{e}_3$ which is orthogonal to the plane of the bivector $\mathbf{e}_{12}$. The *standard basis* of the geometric algebra $\mathbb{G}_3 = \mathbb{G}(\mathbb{R}^3)$ is

$$\mathbb{G}_3 = \text{span}_{\mathbb{R}}\{1, \mathbf{e}_1, \mathbf{e}_2, \mathbf{e}_3, \mathbf{e}_{23}, \mathbf{e}_{31}, \mathbf{e}_{12}, I\}, \tag{3.11}$$

where $\mathbf{e}_1^2 = \mathbf{e}_2^2 = \mathbf{e}_3^2 = 1$ and $\mathbf{e}_{ij} := \mathbf{e}_i\mathbf{e}_j = -\mathbf{e}_j\mathbf{e}_i$ for $i \neq j$ and $1 \leq i, j \leq 3$, and where the *trivector* $I := \mathbf{e}_{123} = \mathbf{e}_1\mathbf{e}_2\mathbf{e}_3$. The geometric numbers of 3-dimensional space are pictured in Fig. 3.4.

The $\mathbf{e}_1, \mathbf{e}_2, \mathbf{e}_3$ are orthonormal unit vectors along the $x, y, z$-axes, the $\mathbf{e}_{23}$, $\mathbf{e}_{31}$, and $\mathbf{e}_{12}$ are unit bivectors defining the directions of the $yz$-, $zx$-, and $xy$-planes, respectively, and the unit trivector $I = \mathbf{e}_{123}$ defines the *directed volume element* of space. The trivector $I$ is pictured in Fig. 3.5.

Calculating the square of the unit trivector or *pseudoscalar* of $\mathbb{G}_3$ gives

$$I^2 = \mathbf{e}_1\mathbf{e}_2\mathbf{e}_3\mathbf{e}_1\mathbf{e}_2\mathbf{e}_3 = -\mathbf{e}_1\mathbf{e}_2\mathbf{e}_1\mathbf{e}_3\mathbf{e}_2\mathbf{e}_3 = (\mathbf{e}_1\mathbf{e}_2)^2\mathbf{e}_3^2 = -1,$$

so the trivector $I$ is another geometric square root of minus one. The pseudoscalar $I$ also has the important property that it commutes with all the elements of $\mathbb{G}_3$ and is therefore in the *center* $Z(\mathbb{G}_3)$ of the algebra $\mathbb{G}_3$.

Noting that $\mathbf{e}_3 = -I\mathbf{e}_{12}, \mathbf{e}_{23} = I\mathbf{e}_1$ and $\mathbf{e}_{31} = I\mathbf{e}_2$, the standard basis of $\mathbb{G}_3$ takes the form

$$\mathbb{G}_3 = \text{span}_{\mathbb{R}}\{1, \mathbf{e}_1, \mathbf{e}_2, -I\mathbf{e}_{12}, I\mathbf{e}_1, I\mathbf{e}_2, \mathbf{e}_{12}, I\} = \text{span}_{\mathbb{C}}\{1, \mathbf{e}_1, \mathbf{e}_2, \mathbf{e}_{12}\} = \mathbb{G}_2(I) \quad (3.12)$$

The last two equalities in the above equation show that the geometric algebra $\mathbb{G}_3$ can be considered to be the geometric algebra $\mathbb{G}_2$ of the plane extended or *complexified* to include the pseudoscalar element $I = \mathbf{e}_{123}$. Any geometric number $g \in \mathbb{G}_3$ can be expressed in the form $g = A + IB$ where $A, B \in \mathbb{G}_2$. Later, it will be shown that there is an algebraic isomorphism between $\mathbb{G}_2$ and the $2 \times 2$ matrix algebra over the real numbers and that $\mathbb{G}_3$ is isomorphic to the $2 \times 2$ matrix algebra over the complex numbers.

Given the vectors $\mathbf{a}, \mathbf{b} \in \mathbb{R}^3$, where $\mathbf{a} = (a_1, a_2, a_3)$ and $\mathbf{b} = (b_1, b_2, b_3)$, a similar calculation of the geometric products (3.4) and (3.5) shows that

$$\mathbf{ab} = \frac{1}{2}(\mathbf{ab} + \mathbf{ba}) + \frac{1}{2}(\mathbf{ab} - \mathbf{ba}) = \mathbf{a} \cdot \mathbf{b} + \mathbf{a} \wedge \mathbf{b}, \quad (3.13)$$

where the inner product $\mathbf{a} \cdot \mathbf{b} = \langle \mathbf{ab} \rangle_0$ is given by

$$\mathbf{a} \cdot \mathbf{b} = \frac{1}{2}(\mathbf{ab} + \mathbf{ba}) = a_1 b_1 + a_2 b_2 + a_3 b_3 \quad (3.14)$$

and the corresponding outer product $\mathbf{a} \wedge \mathbf{b} = \langle \mathbf{ab} \rangle_2$ is given by

$$\mathbf{a} \wedge \mathbf{b} = \frac{1}{2}(\mathbf{ab} - \mathbf{ba}) = I \det \begin{pmatrix} \mathbf{e}_1 & \mathbf{e}_2 & \mathbf{e}_3 \\ a_1 & a_2 & a_3 \\ b_1 & b_2 & b_3 \end{pmatrix} = I(\mathbf{a} \times \mathbf{b}). \quad (3.15)$$

We have expressed the outer product $\mathbf{a} \wedge \mathbf{b}$ in terms of the familiar *cross product* $\mathbf{a} \times \mathbf{b}$ of the vectors $\mathbf{a}, \mathbf{b} \in \mathbb{R}^3$. The vector $\mathbf{a} \times \mathbf{b}$ is the right-handed normal to the plane of the bivector $\mathbf{a} \wedge \mathbf{b}$, and $|\mathbf{a} \times \mathbf{b}| = |\mathbf{a} \wedge \mathbf{b}|$. See Fig. 3.6.

Let us now explore the geometric product of the vector $\mathbf{a}$ with the bivector $\mathbf{b} \wedge \mathbf{c}$. We begin by decomposing the geometric product $\mathbf{a}(\mathbf{b} \wedge \mathbf{c})$ into antisymmetric and symmetric parts:

$$\mathbf{a}(\mathbf{b} \wedge \mathbf{c}) = \frac{1}{2}[\mathbf{a}(\mathbf{b} \wedge \mathbf{c}) - (\mathbf{b} \wedge \mathbf{c})\mathbf{a}] + \frac{1}{2}[\mathbf{a}(\mathbf{b} \wedge \mathbf{c}) + (\mathbf{b} \wedge \mathbf{c})\mathbf{a}].$$

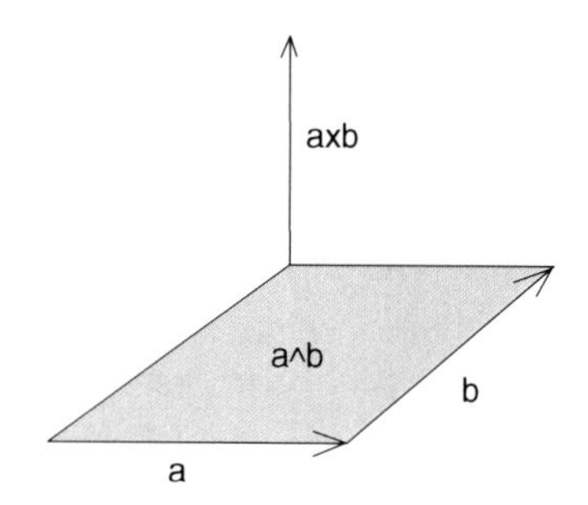


**Fig. 3.6** The vector **a** is swept out along the vector **b** to form the bivector $\mathbf{a}\wedge\mathbf{b}$. The vector $\mathbf{a}\times\mathbf{b}$ is the right-handed normal to this plane, whose length $|\mathbf{a}\times\mathbf{b}|$ is equal to the area or magnitude $|\mathbf{a}\wedge\mathbf{b}|$ of the bivector $\mathbf{a}\wedge\mathbf{b}$

For the antisymmetric part, we find that

$$\mathbf{a}\cdot(\mathbf{b}\wedge\mathbf{c}) := \frac{1}{2}[\mathbf{a}(\mathbf{b}\wedge\mathbf{c}) - (\mathbf{b}\wedge\mathbf{c})\mathbf{a}] = I\frac{1}{2}[\mathbf{a}(\mathbf{b}\times\mathbf{c}) - (\mathbf{b}\times\mathbf{c})\mathbf{a}] = -\mathbf{a}\times(\mathbf{b}\times\mathbf{c}), \tag{3.16}$$

and for the symmetric part, we find that

$$\begin{aligned}\mathbf{a}\wedge(\mathbf{b}\wedge\mathbf{c}) &:= \frac{1}{2}[\mathbf{a}(\mathbf{b}\wedge\mathbf{c}) + (\mathbf{b}\wedge\mathbf{c})\mathbf{a}] = I\frac{1}{2}[\mathbf{a}(\mathbf{b}\times\mathbf{c}) + (\mathbf{b}\times\mathbf{c})\mathbf{a}]\\ &= I[\mathbf{a}\cdot(\mathbf{b}\times\mathbf{c})].\end{aligned} \tag{3.17}$$

The quantity $\mathbf{a}\wedge(\mathbf{b}\wedge\mathbf{c}) = (\mathbf{a}\wedge\mathbf{b})\wedge\mathbf{c}$ is called the *trivector* obtained by sweeping the bivector $\mathbf{a}\wedge\mathbf{b}$ out along the vector **c**, as shown in Fig. 3.5.

Whereas we have carried out the above calculations for three vectors in $\mathbb{R}^3$, the calculation remains valid in $\mathbb{R}^n$ for the simple reason that three linearly independent vectors in $\mathbb{R}^n$, $n \geq 3$, define a three-dimensional subspace of $\mathbb{R}^n$. It is important, however, to give a direct argument for the identity

$$\mathbf{a}\cdot(\mathbf{b}\wedge\mathbf{c}) = (\mathbf{a}\cdot\mathbf{b})\mathbf{c} - (\mathbf{a}\cdot\mathbf{c})\mathbf{b}, \tag{3.18}$$

which is also valid in $\mathbb{R}^n$.

Decomposing the left side of this equation, using (3.16) and (3.15), we find

$$\mathbf{a}\cdot(\mathbf{b}\wedge\mathbf{c}) = \frac{1}{4}[\mathbf{abc} - \mathbf{acb} - \mathbf{bca} + \mathbf{cba}].$$

Decomposing the right hand side of the equation, using (3.14), we get

$$\begin{aligned}(\mathbf{a}\cdot\mathbf{b})\mathbf{c} - (\mathbf{a}\cdot\mathbf{c})\mathbf{b} &= \frac{1}{2}[(\mathbf{a}\cdot\mathbf{b})\mathbf{c} + \mathbf{c}(\mathbf{a}\cdot\mathbf{b}) - (\mathbf{a}\cdot\mathbf{c})\mathbf{b} - \mathbf{b}(\mathbf{a}\cdot\mathbf{c})]\\ &= \frac{1}{4}[(\mathbf{ab}+\mathbf{ba})\mathbf{c} + \mathbf{c}(\mathbf{ab}+\mathbf{ba}) - (\mathbf{ac}+\mathbf{ca})\mathbf{b} - \mathbf{b}(\mathbf{ac}+\mathbf{ca})].\end{aligned}$$

After cancellations, we see that the left and right sides of the equations are identical, so the identity (3.18) is proved.

We now give the definition of the inner product $(\mathbf{a}\wedge\mathbf{b})\cdot(\mathbf{c}\wedge\mathbf{d})$ of the two bivectors $\mathbf{a}\wedge\mathbf{b}$ and $\mathbf{c}\wedge\mathbf{d}$ where $\mathbf{a},\mathbf{b},\mathbf{c},\mathbf{d}\in\mathbb{R}^3$.

**Definition 3.2.1.** $(\mathbf{a}\wedge\mathbf{b})\cdot(\mathbf{c}\wedge\mathbf{d}) = \langle(\mathbf{a}\wedge\mathbf{b})(\mathbf{c}\wedge\mathbf{d})\rangle_0$ so that $(\mathbf{a}\wedge\mathbf{b})\cdot(\mathbf{c}\wedge\mathbf{d})$ is the *scalar part* of the geometric product $(\mathbf{a}\wedge\mathbf{b})(\mathbf{c}\wedge\mathbf{d})$ of the bivectors.

Using the above definition and (3.13), we can find a formula for calculating $(\mathbf{a}\wedge\mathbf{b})\cdot(\mathbf{c}\wedge\mathbf{d})$. We find that

$$(\mathbf{a}\wedge\mathbf{b})\cdot(\mathbf{c}\wedge\mathbf{d}) = <(\mathbf{a}\wedge\mathbf{b})(\mathbf{c}\wedge\mathbf{d})>_0 = <(\mathbf{ab}-\mathbf{a}\cdot\mathbf{b})(\mathbf{c}\wedge\mathbf{d})>_0$$
$$= <\mathbf{ab}(\mathbf{c}\wedge\mathbf{d})>_0 = <\mathbf{a}[\mathbf{b}\cdot(\mathbf{c}\wedge\mathbf{d})]>_0 = \mathbf{a}\cdot[\mathbf{b}\cdot(\mathbf{c}\wedge\mathbf{d})]. \tag{3.19}$$

As a special case of this identity, we define the *magnitude* $|\mathbf{a}\wedge\mathbf{b}|$ of the bivector $\mathbf{a}\wedge\mathbf{b}$ by

$$|\mathbf{a}\wedge\mathbf{b}| := \sqrt{(\mathbf{a}\wedge\mathbf{b})\cdot(\mathbf{b}\wedge\mathbf{a})} = \sqrt{|\mathbf{a}|^2|\mathbf{b}|^2-(\mathbf{a}\cdot\mathbf{b})^2}. \tag{3.20}$$

## Exercises

1. Verify the coordinate formulas (3.14) and (3.15) for the inner and outer products in $\mathbb{G}_3$.

   For problems 2–5, do the calculation for the vectors $\mathbf{a} = 2\mathbf{e}_1+3\mathbf{e}_2-\mathbf{e}_3$, $\mathbf{b} = -\mathbf{e}_1+2\mathbf{e}_2+\mathbf{e}_3, \mathbf{c} = 3\mathbf{e}_1-4\mathbf{e}_2+2\mathbf{e}_3\in\mathbb{R}^3$. The reader is strongly encouraged to check his hand calculations with the Clifford Algebra Calculator (CLICAL) developed by Pertti Lounesto [54]. The software for CLICAL can be downloaded from the site http://users.tkk.fi/ppuska/mirror/Lounesto/CLICAL.htm

2. (a) Calculate $\mathbf{a}\wedge\mathbf{b}$.
   (b) Calculate
   $$(\mathbf{a}\wedge\mathbf{b})^{-1} = \frac{1}{\mathbf{a}\wedge\mathbf{b}} = -\frac{\mathbf{a}\wedge\mathbf{b}}{|\mathbf{a}\wedge\mathbf{b}|^2}.$$
   Why is the minus sign necessary?
3. Calculate $\mathbf{a}\cdot(\mathbf{b}\wedge\mathbf{c}) = -\mathbf{a}\times(\mathbf{b}\times\mathbf{c})$ using formula (3.18)
4. Calculate $(\mathbf{a}\wedge\mathbf{b})\cdot(\mathbf{b}\wedge\mathbf{c}) := <(\mathbf{a}\wedge\mathbf{b})(\mathbf{b}\wedge\mathbf{c})>_0$ by using (3.19) and (3.18).
5. (a) Calculate $\mathbf{a}\wedge\mathbf{b}\wedge\mathbf{c}$ using the formula (3.17).
   (b) Using (3.15), calculate $(\mathbf{a}\wedge\mathbf{b})\boxtimes(\mathbf{b}\wedge\mathbf{c})$, where $A\boxtimes B := \frac{1}{2}(AB-BA)$. (The symbol $\boxtimes$ is used for the *antisymmetric product* to distinguish it from the symbol $\times$ used for the cross product of vectors in $\mathbb{R}^3$.)

For problems 6–16, let $\mathbf{a},\mathbf{b},\mathbf{c}$ be arbitrary vectors in $\mathbb{R}^3\subset\mathbb{G}_3$.

6. (a) Show that for any vectors $\mathbf{a},\mathbf{b},\mathbf{c} \in \mathbb{R}^3$,

$$\mathbf{a}\wedge\mathbf{b}\wedge\mathbf{c} = \frac{1}{4}(\mathbf{abc}-\mathbf{acb}+\mathbf{bca}-\mathbf{cba})$$
$$= \frac{1}{6}(\mathbf{abc}-\mathbf{acb}+\mathbf{bca}-\mathbf{cba}+\mathbf{cab}-\mathbf{bac})$$

   (b) Show that $(\mathbf{a}\wedge\mathbf{b})(\mathbf{b}\wedge\mathbf{c}) = (\mathbf{a}\wedge\mathbf{b})\cdot(\mathbf{b}\wedge\mathbf{c}) + (\mathbf{a}\wedge\mathbf{b}) \boxtimes (\mathbf{b}\wedge\mathbf{c})$.
7. Noting that $\mathbf{b} = (\mathbf{ba})\mathbf{a}^{-1} = (\mathbf{b}\cdot\mathbf{a})\mathbf{a}^{-1} + (\mathbf{b}\wedge\mathbf{a})\mathbf{a}^{-1}$, show that $\mathbf{b}_{\|}\mathbf{a} = \mathbf{b}_{\|}\cdot\mathbf{a}$ and $\mathbf{b}_{\perp}\mathbf{a} = \mathbf{b}_{\perp}\wedge\mathbf{a}$ where $\mathbf{b}_{\|} = (\mathbf{b}\cdot\mathbf{a})\mathbf{a}^{-1}$ and $\mathbf{b}_{\perp} = (\mathbf{b}\wedge\mathbf{a})\mathbf{a}^{-1} = \mathbf{b}-\mathbf{b}_{\|}$.
8. Find vectors $\mathbf{c}_{\|}$ and $\mathbf{c}_{\perp}$ such that $\mathbf{c} = \mathbf{c}_{\|}+\mathbf{c}_{\perp}$ and $\mathbf{c}_{\|}(\mathbf{a}\wedge\mathbf{b}) = \mathbf{c}\cdot(\mathbf{a}\wedge\mathbf{b})$ and $\mathbf{c}_{\perp}(\mathbf{a}\wedge\mathbf{b}) = \mathbf{c}\wedge(\mathbf{a}\wedge\mathbf{b})$. *Hint:* Use the fact that

$$\mathbf{c} = [\mathbf{c}(\mathbf{a}\wedge\mathbf{b})](\mathbf{a}\wedge\mathbf{b})^{-1} = [\mathbf{c}\cdot(\mathbf{a}\wedge\mathbf{b}) + \mathbf{c}\wedge(\mathbf{a}\wedge\mathbf{b})](\mathbf{a}\wedge\mathbf{b})^{-1}.$$

9. Using (3.15), show that $\mathbf{a}\times\mathbf{b} = -I(\mathbf{a}\wedge\mathbf{b})$ where $I = \mathbf{e}_{123}$.
10. Show that $\mathbf{a}\cdot[I(\mathbf{b}\wedge\mathbf{c})] = I(\mathbf{a}\wedge\mathbf{b}\wedge\mathbf{c})$ where $I = \mathbf{e}_{123}$.
11. Show that $\mathbf{a}\wedge[I(\mathbf{b}\wedge\mathbf{c})] = I[\mathbf{a}\cdot(\mathbf{b}\wedge\mathbf{c})]$ where $I = \mathbf{e}_{123}$.
12. Show that $\mathbf{a}(\mathbf{a}\wedge\mathbf{b}) = \mathbf{a}\cdot(\mathbf{a}\wedge\mathbf{b})$
13. Show that $(\mathbf{a}+\mathbf{b}\wedge\mathbf{c})^2 = \mathbf{a}^2 + 2\mathbf{a}\wedge\mathbf{b}\wedge\mathbf{c} + (\mathbf{b}\wedge\mathbf{c})^2$.
14. Show that $\mathbf{a}(\mathbf{a}\wedge\mathbf{b}) = -(\mathbf{a}\wedge\mathbf{b})\mathbf{a}$.
15. Show that $(\mathbf{a}+\mathbf{a}\wedge\mathbf{b})^2 = \mathbf{a}^2 + (\mathbf{a}\wedge\mathbf{b})^2$.
16. Show that $(\mathbf{a}\wedge\mathbf{b}+\mathbf{b}\wedge\mathbf{c})^2 = (\mathbf{a}\wedge\mathbf{b})^2 + 2(\mathbf{a}\wedge\mathbf{b})\cdot(\mathbf{b}\wedge\mathbf{c}) + (\mathbf{b}\wedge\mathbf{c})^2$.

## 3.3 Orthogonal Transformations

Much of the utility of geometric algebra stems from the simple and direct expression of orthogonal transformations. Specifically, given two vectors $\mathbf{a}$ and $\mathbf{b}$, such that $|\mathbf{a}| = |\mathbf{b}|$, let us find a *reflection* (mirror image) $L(\mathbf{x})$ and a rotation $R(\mathbf{x})$ with the property that $L(\mathbf{a}) = R(\mathbf{a}) = \mathbf{b}$.

Noting that

$$(\mathbf{a}+\mathbf{b})\cdot(\mathbf{a}-\mathbf{b}) = \mathbf{a}^2 - \mathbf{a}\cdot\mathbf{b} + \mathbf{b}\cdot\mathbf{a} - \mathbf{b}^2 = 0,$$

and assuming $\mathbf{a} \neq \mathbf{b}$, define the transformation

$$L(\mathbf{x}) = -(\mathbf{a}-\mathbf{b})\mathbf{x}(\mathbf{a}-\mathbf{b})^{-1}. \tag{3.21}$$

Writing $\mathbf{a} = \frac{1}{2}(\mathbf{a}+\mathbf{b}) + \frac{1}{2}(\mathbf{a}-\mathbf{b})$, it easily follows that

$$L(\mathbf{a}) = L\left(\frac{1}{2}(\mathbf{a}+\mathbf{b}) + \frac{1}{2}(\mathbf{a}-\mathbf{b})\right) = \frac{1}{2}(\mathbf{a}+\mathbf{b}) - \frac{1}{2}(\mathbf{a}-\mathbf{b}) = \mathbf{b}$$

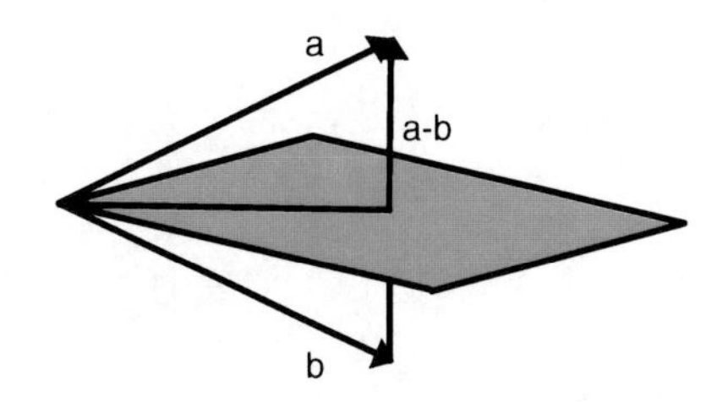


**Fig. 3.7** The vector **a** is reflected in the plane perpendicular to the vector **a** − **b** to give the vector **b**

as required. See Fig. 3.7. The transformation $L(\mathbf{x})$ represents a reflection through the plane normal to the vector $\mathbf{a}-\mathbf{b}$. In the case that $\mathbf{b}=\mathbf{a}$, the required reflection is given by

$$L(\mathbf{x}) = -\mathbf{cxc}^{-1},$$

where $\mathbf{c}$ is any nonzero vector orthogonal to $\mathbf{a}$. Another interesting case is when $\mathbf{b} = -\mathbf{a}$, in which case the $L_{\mathbf{a}}(\mathbf{x}) = -\mathbf{axa}^{-1}$ is a reflection with respect to the hyperplane that has $\mathbf{a}$ as its normal vector.

Let us now see how to define a rotation $R(\mathbf{x})$ with the property that $R(\mathbf{a}) = \mathbf{b}$ where $\mathbf{a}$ and $\mathbf{b}$ are any two vectors such that $|\mathbf{a}| = |\mathbf{b}|$. Recalling (3.8), we note that

$$\mathbf{b} = \mathbf{a}(\mathbf{a}^{-1}\mathbf{b}) = (\mathbf{ba}^{-1})^{\frac{1}{2}}\mathbf{a}(\mathbf{a}^{-1}\mathbf{b})^{\frac{1}{2}} = \mathbf{b}(\mathbf{a}^{-1}\mathbf{a}),$$

where $\mathbf{a}^{-1} = \frac{\mathbf{a}}{|\mathbf{a}|^2}$. The desired rotation is given by

$$R(\mathbf{x}) = (\mathbf{ba}^{-1})^{\frac{1}{2}}\mathbf{x}(\mathbf{a}^{-1}\mathbf{b})^{\frac{1}{2}}. \tag{3.22}$$

Since $|\mathbf{a}| = |\mathbf{b}|$, it follows that $\mathbf{a}^{-1}\mathbf{b} = \hat{\mathbf{a}}\hat{\mathbf{b}}$ where $\hat{\mathbf{a}} = \frac{\mathbf{a}}{|\mathbf{a}|}$ and $\hat{\mathbf{b}} = \frac{\mathbf{b}}{|\mathbf{b}|}$ are unit vectors.

There is a very simple formula that allows us to calculate $\sqrt{\hat{\mathbf{a}}\hat{\mathbf{b}}}$ algebraically. We have

$$\sqrt{\hat{\mathbf{a}}\hat{\mathbf{b}}} = \pm\hat{\mathbf{a}}\frac{\hat{\mathbf{a}}+\hat{\mathbf{b}}}{|\hat{\mathbf{a}}+\hat{\mathbf{b}}|}. \tag{3.23}$$

To verify (3.23), we simply calculate

$$\left(\hat{\mathbf{a}}\frac{\hat{\mathbf{a}}+\hat{\mathbf{b}}}{|\hat{\mathbf{a}}+\hat{\mathbf{b}}|}\right)^2 = \frac{\hat{\mathbf{a}}(\hat{\mathbf{a}}+\hat{\mathbf{b}})\hat{\mathbf{a}}(\hat{\mathbf{a}}+\hat{\mathbf{b}})\hat{\mathbf{b}}\hat{\mathbf{b}}}{(\hat{\mathbf{a}}+\hat{\mathbf{b}})^2}$$

$$= \frac{\hat{\mathbf{a}}(\hat{\mathbf{a}}+\hat{\mathbf{b}})(\hat{\mathbf{b}}+\hat{\mathbf{a}})\hat{\mathbf{b}}}{(\hat{\mathbf{a}}+\hat{\mathbf{b}})^2} = \hat{\mathbf{a}}\hat{\mathbf{b}}.$$

Equation (3.22) is called the *half angle* or *two-sided* representation of a rotation in the plane of the vectors $\mathbf{a}$ and $\mathbf{b}$. Writing the vector $\mathbf{x} = \mathbf{x}_{\|} + \mathbf{x}_{\perp}$ where

$$\mathbf{x}_{\|} = [\mathbf{x}\cdot(\mathbf{a}\wedge\mathbf{b})](\mathbf{a}\wedge\mathbf{b})^{-1} = \mathbf{x}\cdot(\mathbf{a}\wedge\mathbf{b})\frac{\mathbf{b}\wedge\mathbf{a}}{|\mathbf{a}\wedge\mathbf{b}|^2}$$

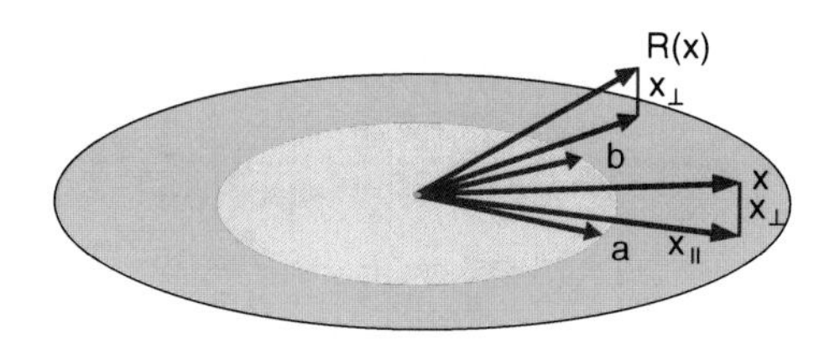


**Fig. 3.8** The vector $\mathbf{x} = \mathbf{x}_{\|} + \mathbf{x}_{\perp}$ is rotated in the plane of the bivector $\mathbf{a} \wedge \mathbf{b}$ into $R(\mathbf{x})$, leaving the perpendicular component unchanged. When $|\mathbf{a}| = |\mathbf{b}| = 1$, $R(\mathbf{x}) = e^{-i\frac{\theta}{2}}\mathbf{x}e^{i\frac{\theta}{2}}$, where $e^{i\frac{\theta}{2}} = \frac{\mathbf{a}(\mathbf{a}+\mathbf{b})}{|\mathbf{a}+\mathbf{b}|}$, $e^{-i\frac{\theta}{2}} = \frac{(\mathbf{a}+\mathbf{b})\mathbf{a}}{|\mathbf{a}+\mathbf{b}|}$ and $\theta$ is the angle between $\mathbf{a}$ and $\mathbf{b}$. If $\mathbf{a}$ and $\mathbf{b}$ lie in the $xy$-plane, then $i = \mathbf{e}_{12}$

so that $\mathbf{x}_{\|}$ is in the plane of $\mathbf{a} \wedge \mathbf{b}$, and $\mathbf{x}_{\perp} = \mathbf{x} - \mathbf{x}_{\|}$ is the component of $\mathbf{x}$ perpendicular to that plane, it follows that

$$R(\mathbf{x}) = R(\mathbf{x}_{\|} + \mathbf{x}_{\perp}) = R(\mathbf{x}_{\|}) + \mathbf{x}_{\perp}. \tag{3.24}$$

Thus, the two-sided representation of the rotation leaves the perpendicular part of the vector $\mathbf{x}$ to the plane of the rotation $\mathbf{a} \wedge \mathbf{b}$ invariant. See Fig. 3.8.

The rotation found in (3.22) can equally well be considered to be the composition of the two reflections

$$L_{\mathbf{a}}(x) = -\mathbf{a}\mathbf{x}\mathbf{a}^{-1} \quad \text{and} \quad L_{\mathbf{a}+\mathbf{b}}(x) = -(\mathbf{a}+\mathbf{b})\mathbf{x}(\mathbf{a}+\mathbf{b})^{-1}.$$

Using (3.23), we find that

$$R(\mathbf{x}) = (\mathbf{b}\mathbf{a}^{-1})^{\frac{1}{2}}\mathbf{x}(\mathbf{a}^{-1}\mathbf{b})^{\frac{1}{2}} = L_{\mathbf{a}+\mathbf{b}} \circ L_{\mathbf{a}}(\mathbf{x}). \tag{3.25}$$

## Exercises

Given the vectors $\mathbf{a} = (1,2,2)$, $\mathbf{b} = (1,2,-2)$, and $\mathbf{c} = (2,1,2)$ in $\mathbb{R}^3$. The reader is strongly encouraged to check his hand calculations with Pertti Lounesto's CLICAL [54].

1. Calculate
   (a) $\mathbf{a} \cdot \mathbf{b}$ and $\mathbf{a} \wedge \mathbf{b}$
   (b) $\mathbf{ab}$
   (c) $(\mathbf{ab})\mathbf{c}$
   (d) $\mathbf{a}(\mathbf{bc})$
   (e) $\mathbf{a}(\mathbf{b}+\mathbf{c})$
   (f) $\mathbf{ab}+\mathbf{ac}$.
2. (a) Find the magnitude $|\mathbf{a} \wedge \mathbf{b}|$ of the bivector $\mathbf{a} \wedge \mathbf{b}$.
   (b) Graph the bivector $\mathbf{a} \wedge \mathbf{b}$.

(c) Find the Euler form of **ac**.
(d) Find $\sqrt{\mathbf{ab}}$ and $\sqrt{\mathbf{ac}}$.

3. (a) Find a reflection $L(\mathbf{x})$ and a rotation $R(\mathbf{x})$ which takes **a** into **b**.
(b) Find a reflection $L(\mathbf{x})$ and a rotation $R(\mathbf{x})$ which takes **b** into **c**.
(c) Find a reflection $L(\mathbf{x})$ and a rotation $R(\mathbf{x})$ which takes **a** into **c**.
4. Verify (3.24) for the rotation (3.22).
5. Verify (3.25) for the rotation (3.22).
6. Hamilton's quaternions: In the geometric algebra $\mathbb{G}_3 = \mathbb{G}(\mathbb{R}^3)$, define

$$i = \mathbf{e}_{23}, \quad j = \mathbf{e}_{31}, \quad k = \mathbf{e}_{21}.$$

(a) Show that $i^2 = j^2 = k^2 = ijk = -1$. These are the famous relationships that Hamilton carved into stone on the Bougham Bridge on 16 October 1843 in Dublin.

http://en.wikipedia.org/wiki/Quaternion

(b) Show that

$$ij = -ji = k, \quad jk = -kj = i, \quad ik = -ki = -j.$$

We see that the quaternions $\{1, i, j, k\}$ can be identified as the *even subalgebra* of scalars and bivectors of the geometric algebra $\mathbb{G}_3$. For a further discussion of these issues, see [55, p.68,190].

## 3.4 Geometric Algebra of $\mathbb{R}^n$

Geometric algebra was introduced by William Kingdon Clifford in 1878 as a generalization and unification of the ideas of Hermann Grassmann and William Hamilton, who came before him [12–14]. Whereas Hamilton's quaternions are known and Grassmann algebras appear in the guise of differential forms and antisymmetric tensors, the utility of Clifford's geometric algebras is just beginning to be widely appreciated [6, 20, 24, 41].

http://en.wikipedia.org/wiki/Differential_form

Let

$$(\mathbf{e})_{(n)} := (\mathbf{e}_i)_{i=1}^n = \begin{pmatrix}\mathbf{e}_1 & \mathbf{e}_2 & \dots & \mathbf{e}_n\end{pmatrix} \tag{3.26}$$

be the standard orthonormal basis of the $n$-dimensional Euclidean space $\mathbb{R}^n$. In anticipation of our work with matrices whose elements are vectors, or even more general geometric numbers, we are deliberately expressing the basis (3.26) as a *row matrix* of the orthonormal basis vectors $\mathbf{e}_i \in \mathbb{R}^n$ for $i = 1, \dots, n$. We will have much more to say about this later.

If the vectors $\mathbf{a},\mathbf{b} \in \mathbb{R}^n$ are represented by *column vectors* of their components,

$$\mathbf{a} = \begin{pmatrix} a^1 \\ a^2 \\ \cdot \\ \cdot \\ a_n \end{pmatrix} = \sum_i a^i \mathbf{e}_i \text{ and } \mathbf{b} = \begin{pmatrix} b^1 \\ b^2 \\ \cdot \\ \cdot \\ b_n \end{pmatrix} = \sum_i b^i \mathbf{e}_i,$$

then their *inner product* is given by

$$\mathbf{a}\cdot\mathbf{b} = |\mathbf{a}||\mathbf{b}|\cos\theta = \sum_i a^i b^i$$

where $\theta$ is the angle between them.

Just as we write the standard orthonormal basis of $\mathbb{R}^n$ as the row of vectors $(\mathbf{e})_{(n)}$, we write the *standard orthonormal reciprocal basis* as the *column* of vectors:

$$(\mathbf{e})^{(n)} := \begin{pmatrix} \mathbf{e}^1 \\ \mathbf{e}^2 \\ \cdot \\ \cdot \\ \mathbf{e}^n \end{pmatrix}. \tag{3.27}$$

However, since we are dealing with an orthonormal basis, the reciprocal basis vectors $\mathbf{e}^i$ can be identified with the basis vectors $\mathbf{e}_i$, i.e., $\mathbf{e}^i = \mathbf{e}_i$ for $i = 1,\ldots,n$. We can also represent a vector $\mathbf{a} \in \mathbb{R}^n$ in the *row vector form*

$$\mathbf{a} = \begin{pmatrix} a_1 & a_2 & \cdots & a_n \end{pmatrix} = \sum_i a_i \mathbf{e}^i.$$

More generally, given any basis $(\mathbf{v})_{(n)}$ of $\mathbb{R}^n$, it is always possible to construct a reciprocal basis $(\mathbf{v})^{(n)}$, as will be discussed in Chap. 7.

The *associative geometric algebra* $\mathbb{G}_n = \mathbb{G}(\mathbb{R}^n)$ of the Euclidean space $\mathbb{R}^n$ is the *geometric extension* of the real number system $\mathbb{R}$ to include $n$ new *anticommuting square roots* of unity, which we identify with the orthonormal basis vectors $\mathbf{e}_i$ of $\mathbb{R}^n$, and their geometric sums and products. Each vector $\mathbf{a} \in \mathbb{R}^n$ has the property that

$$\mathbf{a}^2 = \mathbf{a}\mathbf{a} = \mathbf{a}\cdot\mathbf{a} = |\mathbf{a}|^2, \tag{3.28}$$

where $|\mathbf{a}|$ is the Euclidean length or *magnitude* of the vector $\mathbf{a}$.

The fundamental *geometric product* $\mathbf{ab}$ of the vectors $\mathbf{a},\mathbf{b} \in \mathbb{R}^n \subset \mathbb{G}_n$ can be decomposed into the symmetric inner product and the antisymmetric outer product

$$\mathbf{ab} = \frac{1}{2}(\mathbf{ab}+\mathbf{ba}) + \frac{1}{2}(\mathbf{ab}-\mathbf{ba}) = \mathbf{a}\cdot\mathbf{b} + \mathbf{a}\wedge\mathbf{b} = |\mathbf{a}||\mathbf{b}|e^{\mathrm{i}\theta}, \tag{3.29}$$

where

$$\mathbf{a}\cdot\mathbf{b}=\frac{1}{2}(\mathbf{ab}+\mathbf{ba})=|\mathbf{a}||\mathbf{b}|\cos\theta$$

is the inner product and

$$\mathbf{a}\wedge\mathbf{b}=\frac{1}{2}(\mathbf{ab}-\mathbf{ba})=|\mathbf{a}||\mathbf{b}|\mathrm{i}\sin\theta$$

is the outer product of the vectors $\mathbf{a}$ and $\mathbf{b}$. The outer product $\mathbf{a}\wedge\mathbf{b}$ is given the geometric interpretation of a *directed plane segment* or *bivector* and characterizes the direction of the plane of the subspace of $\mathbb{R}^n$ spanned by the vectors $\mathbf{a}$ and $\mathbf{b}$, recall Fig. 3.2. The unit bivector $i$ orients the plane of the vectors $\mathbf{a}$ and $\mathbf{b}$, and, as we have seen in Sect. 3.2, has the property that $\mathrm{i}^2=-1$.

The real $2^n$-dimensional geometric algebra $\mathbb{G}_n=\mathbb{G}(\mathbb{R}^n)$ has the *standard basis*

$$\mathbb{G}_n=\mathrm{span}\{\mathbf{e}_{\lambda_{(k)}}\}_{k=0}^{n}, \tag{3.30}$$

where the $\binom{n}{k}$ $k$-vector basis elements of the form $\mathbf{e}_{\lambda_{(k)}}$ are defined by

$$\mathbf{e}_{\lambda_{(k)}}=e_{\lambda_1,\ldots,\lambda_k}=e_{\lambda_1}\cdots e_{\lambda_k}$$

for each $\lambda_{(k)}=\lambda_1,\ldots,\lambda_k$ where $1\le\lambda_1<\cdots<\lambda_k\le n$. When $k=0$, $\lambda_0=0$ and $\mathbf{e}_0=1$. Thus, as we have already seen in (3.11), the $2^3$-dimensional geometric algebra of $\mathbb{R}^3$ has the standard basis

$$\mathbb{G}_3=\mathrm{span}\{\mathbf{e}_{\lambda_{(k)}}\}_{k=0}^{3}=\mathrm{span}\{1,\mathbf{e}_1,\mathbf{e}_2,\mathbf{e}_3,\mathbf{e}_{12},\mathbf{e}_{13},\mathbf{e}_{23},\mathbf{e}_{123}\}$$

over the real numbers $\mathbb{R}$.

In geometric algebra, deep geometrical relationships are expressed directly in terms of the multivectors of the algebra without having to constantly refer to a basis. On the other hand, the language gives geometric meaning to the powerful matrix formalism that has developed over the last 150 years. As a real associative algebra, each geometric algebra is isomorphic to a corresponding algebra or subalgebra of real matrices, and we have advocated elsewhere the need for a uniform approach to both of these structures [65, 86]. Matrices are invaluable for systematizing calculations, but geometric algebra provides deep geometrical insight and powerful algebraic tools for the efficient expression of geometrical properties. Clifford algebras and their relationships to matrix algebra and the classical groups have been thoroughly studied in [55, 64] and elsewhere and will be examined in many of the succeeding chapters in this book.

We denote the *pseudoscalar* of $\mathbb{R}^n$ by the special symbol $I=\mathbf{e}_{1\ldots n}$. The pseudoscalar gives a unique orientation to $\mathbb{R}^n$ and $I^{-1}=\mathbf{e}_{n\ldots 1}$. For an $n$-vector $\mathbf{A}_n\in\mathbb{G}_n$, the *determinant function* is defined by $\mathbf{A}_n=\det(\mathbf{A}_n)I$ or $\det(\mathbf{A}_n)=\mathbf{A}_nI^{-1}$. The determinant function will be studied in Chap. 5.

In dealing with geometric numbers, three different types of *conjugation operations* are important. Let $A \in \mathbb{G}_n$ be a geometric number. By the *reverse* $A^\dagger$ of $A$, we mean the geometric number obtained by reversing the order of all the geometric products of vectors in $A$. For example, if $A = 2\mathbf{e}_1 + 3\mathbf{e}_{12} - 2\mathbf{e}_{123} + \mathbf{e}_{1234}$, then

$$A^\dagger = 2\mathbf{e}_1 + 3\mathbf{e}_{21} - 2\mathbf{e}_{321} + \mathbf{e}_{4321} = 2\mathbf{e}_1 - 3\mathbf{e}_{12} + 2\mathbf{e}_{123} + \mathbf{e}_{1234}. \tag{3.31}$$

By the *grade inversion* $B^-$ of a geometric number $B \in \mathbb{G}_n$, we mean the geometric number obtained by replacing each vector that occurs in $B$ by the negative of that vector. For the geometric number $A$ given above, we find that

$$A^- = -2\mathbf{e}_1 + 3\mathbf{e}_{12} + 2\mathbf{e}_{123} + \mathbf{e}_{1234}. \tag{3.32}$$

The third conjugation, called the *Clifford conjugation* $\widetilde{C} = (C^\dagger)^-$ of $C \in \mathbb{G}_n$, is the composition of the operations of reversal together with grade inversion. For the geometric number $A$, we find that

$$\widetilde{A} = (A^\dagger)^- = (2\mathbf{e}_1 - 3\mathbf{e}_{12} + 2\mathbf{e}_{123} + \mathbf{e}_{1234})^- = -2\mathbf{e}_1 - 3\mathbf{e}_{12} - 2\mathbf{e}_{123} + \mathbf{e}_{1234}. \tag{3.33}$$

For $A, B, C \in \mathbb{G}_n$, the following properties are easily verified. For the operation of reverse, we have

$$(A+B)^\dagger = A^\dagger + B^\dagger, \quad \text{and} \quad (AB)^\dagger = B^\dagger A^\dagger. \tag{3.34}$$

For the grade involution, we have

$$(A+B)^- = A^- + B^-, \quad \text{and} \quad (AB)^- = A^- B^-. \tag{3.35}$$

Clifford conjugation, like the conjugation of reverse, satisfies

$$\widetilde{(A+B)} = \widetilde{A} + \widetilde{B}, \quad \text{and} \quad \widetilde{AB} = \widetilde{B}\widetilde{A}. \tag{3.36}$$

In general, the geometric product of a vector $\mathbf{a}$ and a $k$-vector $\mathbf{B}_k$ for $k \geq 1$ can always be decomposed into the sum of an *inner product* and an *outer product*:

$$\mathbf{a}\mathbf{B}_k = \mathbf{a} \cdot \mathbf{B}_k + \mathbf{a} \wedge \mathbf{B}_k. \tag{3.37}$$

The inner product $\mathbf{a} \cdot \mathbf{B}_k$ is the $(k-1)$-vector

$$\mathbf{a} \cdot \mathbf{B}_k := \frac{1}{2}(\mathbf{a}\mathbf{B}_k + (-1)^{k+1}\mathbf{B}_k\mathbf{a}) = \langle \mathbf{a}\mathbf{B}_k \rangle_{k-1}, \tag{3.38}$$

and the outer product $\mathbf{a} \wedge \mathbf{B}_k$ is the $(k+1)$-vector

$$\mathbf{a} \wedge \mathbf{B}_k := \frac{1}{2}(\mathbf{a}\mathbf{B}_k - (-1)^{k+1}\mathbf{B}_k\mathbf{a}) = \langle \mathbf{a}\mathbf{B}_k \rangle_{k+1}. \tag{3.39}$$

More generally, if $\mathbf{A}_r$ and $\mathbf{B}_s$ are $r$- and $s$-vectors of $\mathbb{G}_n$, where $r>0, s>0$, we define $\mathbf{A}_r\cdot\mathbf{B}_s=\langle\mathbf{A}_r\mathbf{B}_s\rangle_{|r-s|}$ and $\mathbf{A}_r\wedge\mathbf{B}_s=\langle\mathbf{A}_r\mathbf{B}_s\rangle_{r+s}$. In the exceptional cases when $r=0$ or $s=0$, we define $\mathbf{A}_r\cdot\mathbf{B}_s=0$ and $\mathbf{A}_r\wedge\mathbf{B}_s=\mathbf{A}_r\mathbf{B}_s$.

In a similar way that we established the identity (3.18), we now show that for the vectors $\mathbf{a}$ and $\mathbf{b}$ and the $k$-vector $\mathbf{C}_k$, where $k\geq 1$,

$$\mathbf{a}\cdot(\mathbf{b}\wedge\mathbf{C}_k)=(\mathbf{a}\cdot\mathbf{b})\mathbf{C}_k-\mathbf{b}\wedge(\mathbf{a}\cdot\mathbf{C}_k). \tag{3.40}$$

Decomposing the left side of this equation, with the help of (3.38) and (3.39), gives

$$\mathbf{a}\cdot(\mathbf{b}\wedge\mathbf{C}_k)=\frac{1}{4}[\mathbf{abC}_k+(-1)^k\mathbf{aC}_k\mathbf{b}+(-1)^k\mathbf{bC}_k\mathbf{a}+\mathbf{C}_k\mathbf{ba}].$$

Decomposing the right hand side of the equation, with the help of (3.38) and (3.39), we get

$$(\mathbf{a}\cdot\mathbf{b})\mathbf{C}_k-\mathbf{b}\wedge(\mathbf{a}\cdot\mathbf{C}_k)=\frac{1}{2}[(\mathbf{a}\cdot\mathbf{b})\mathbf{C}_k+\mathbf{C}_k(\mathbf{a}\cdot\mathbf{b})-\mathbf{b}(\mathbf{a}\cdot\mathbf{C}_k)+(-1)^k(\mathbf{a}\cdot\mathbf{C}_k)\mathbf{b}]$$

$$=\frac{1}{4}[(\mathbf{ab}+\mathbf{ba})\mathbf{C}_k+\mathbf{C}_k(\mathbf{ab}+\mathbf{ba})-\mathbf{baC}_k+(-1)^k\mathbf{bC}_k\mathbf{a}+(-1)^k\mathbf{aC}_k\mathbf{b}-\mathbf{C}_k\mathbf{ab}].$$

After cancellations, we see that the left and right sides of the equations are the same, so the identity (3.40) is proved.

In the equations above, three different products have been used, the geometric product $\mathbf{ab}$, the outer or wedge product $\mathbf{a}\wedge\mathbf{b}$, and the inner or dot product $\mathbf{a}\cdot\mathbf{b}$, often more than one such product in the same equation. In order to simplify the writing of such equations and to eliminate the use of a host of parentheses, we establish a convention regarding the order in which these operations are to be carried out. *Outer*, *inner*, and then *geometric* products are to be carried out in that order in evaluating any expression. For example, for the vectors $\mathbf{a},\mathbf{b},\mathbf{c},\mathbf{d}$,

$$\mathbf{a}\cdot\mathbf{b}\wedge\mathbf{c}\,\mathbf{d}=\mathbf{a}\cdot(\mathbf{b}\wedge\mathbf{c})\mathbf{d}=[\mathbf{a}\cdot(\mathbf{b}\wedge\mathbf{c})]\mathbf{d}. \tag{3.41}$$

One very useful identity, which follows from (3.40), is

$$\mathbf{a}\cdot(\mathbf{A}_r\wedge\mathbf{B}_s)=(\mathbf{a}\cdot\mathbf{A}_r)\wedge\mathbf{B}_s+(-1)^r\mathbf{A}_r\wedge(\mathbf{a}\cdot\mathbf{B}_s)=(-1)^{r+s+1}(\mathbf{A}_r\wedge\mathbf{B}_s)\cdot\mathbf{a} \tag{3.42}$$

and gives the *distributive law* for the inner product of a vector over the outer product of an $r$- and an $s$-vector. In Chap. 5, we study the *outer product* $\mathbf{a}_1\wedge\cdots\wedge\mathbf{a}_k$ of $k$ vectors $\mathbf{a}_i\in\mathbb{R}^n$. This outer product can be directly expressed as the completely antisymmetric geometric product of those vectors. We have

$$\mathbf{a}_1\wedge\cdots\wedge\mathbf{a}_k=\frac{1}{k!}\sum_{\pi\in\Pi}\operatorname{sgn}(\pi)\mathbf{a}_{\pi(1)}\cdots\mathbf{a}_{\pi(k)}, \tag{3.43}$$

where $\pi$ is a permutation on the indices $(1,2,\ldots,k)$, and $\text{sgn}(\pi) = \pm 1$ according to whether $\pi$ is an even or odd permutation, respectively.

A *simple k-vector* or $k$-blade is any geometric number which can be expressed as the outer product of $k$ vectors. For example, the bivector $\mathbf{e}_{12} + \mathbf{e}_{23}$ is a 2-blade because $\mathbf{e}_{12} + \mathbf{e}_{23} = (\mathbf{e}_1 - \mathbf{e}_3) \wedge \mathbf{e}_2$. On the other hand, the bivector $\mathbf{e}_{12} + \mathbf{e}_{34}$ is not a simple bivector. The *magnitude* of a $k$-vector $\mathbf{A}_k$ is defined by

$$|\mathbf{A}_k| = \sqrt{|\mathbf{A}_k \cdot \mathbf{A}_k^\dagger|}, \tag{3.44}$$

where $\mathbf{A}_k^\dagger = (-1)^{\frac{k(k-1)}{2}} \mathbf{A}_k$. If $\mathbf{A}_k$ is a $k$-blade for $k \geq 1$, then $\mathbf{A}_k^{-1} = (-1)^{\frac{k(k-1)}{2}} \frac{\mathbf{A}_k}{|\mathbf{A}_k|^2}$.

We have given here only a few of the most basic algebraic identities, others are given in the exercises below. For further discussion of basic identities, we suggest the references [1, 36, 43, 55]. In addition, the following links provide good online references:

http://users.tkk.fi/ppuska/mirror/Lounesto/

http://geocalc.clas.asu.edu/

http://www.mrao.cam.ac.uk/~clifford/pages/introduction.htm

http://www.science.uva.nl/ga/index.html

http://en.wikipedia.org/wiki/Geometric_algebra

## Exercises

1. Let $\mathbf{a},\mathbf{b},\mathbf{c},\mathbf{d} \in \mathbb{R}^n$. Show that

$$\mathbf{abc} = (\mathbf{b}\cdot\mathbf{c})\mathbf{a} - (\mathbf{a}\cdot\mathbf{c})\mathbf{b} + (\mathbf{a}\cdot\mathbf{b})\mathbf{c} + \mathbf{a}\wedge\mathbf{b}\wedge\mathbf{c}.$$

2. Show that

$$\mathbf{a}\cdot(\mathbf{b}\wedge\mathbf{c}\wedge\mathbf{d}) = (\mathbf{a}\cdot\mathbf{b})\mathbf{c}\wedge\mathbf{d} - \mathbf{b}\wedge[\mathbf{a}\cdot(\mathbf{c}\wedge\mathbf{d})].$$

3. Show that

$$\mathbf{a}\cdot(\mathbf{b}\wedge\mathbf{c}\wedge\mathbf{d}) = (\mathbf{a}\cdot\mathbf{b})\mathbf{c}\wedge\mathbf{d} - (\mathbf{a}\cdot\mathbf{c})\mathbf{b}\wedge\mathbf{d} + (\mathbf{a}\cdot\mathbf{d})\mathbf{b}\wedge\mathbf{c}.$$

4. Show that

$$(\mathbf{a}\wedge\mathbf{b})\boxtimes(\mathbf{cd}) = [(\mathbf{a}\wedge\mathbf{b})\cdot\mathbf{c}]\mathbf{d} + \mathbf{c}[(\mathbf{a}\wedge\mathbf{b})\cdot\mathbf{d}],$$

   where $A\boxtimes B = \frac{1}{2}(AB - BA)$ is the *anti-symmetric part* of the geometric product of $A$ and $B$.

5. Show that

$$(\mathbf{a}\wedge\mathbf{b})\boxtimes(\mathbf{c}\wedge\mathbf{d}) = [(\mathbf{a}\wedge\mathbf{b})\cdot\mathbf{c}]\wedge\mathbf{d} + \mathbf{c}\wedge[(\mathbf{a}\wedge\mathbf{b})\cdot\mathbf{d}].$$

6. Show that

$$(\mathbf{a} \wedge \mathbf{b})(\mathbf{c} \wedge \mathbf{d}) = (\mathbf{a} \wedge \mathbf{b}) \cdot (\mathbf{c} \wedge \mathbf{d}) + (\mathbf{a} \wedge \mathbf{b}) \boxtimes (\mathbf{c} \wedge \mathbf{d}) + (\mathbf{a} \wedge \mathbf{b}) \wedge (\mathbf{c} \wedge \mathbf{d}).$$

7. Prove the identity (3.42) by using (3.40).
8. Show that if $\mathbf{A}_k \in \mathbb{G}_n^k$, then $\mathbf{A}_k^\dagger = (-1)^{\frac{k(k-1)}{2}} \mathbf{A}_k$. Also find a formula for $\mathbf{A}_k^-$ and $\widetilde{\mathbf{A}_k}$.
9. Using (3.44), find the magnitudes of the bivectors $\mathbf{e}_{12} + 2\mathbf{e}_{34}$ and $\mathbf{e}_{12} + 2\mathbf{e}_{23}$.
10. Let $\mathbf{A}_k$ be a $k$-blade, and define $\hat{\mathbf{A}}_k = \frac{\mathbf{A}_k}{|\mathbf{A}_k|}$. Show that $\hat{\mathbf{A}}_k^2 = \pm 1$. For what values of $k$ is $\hat{\mathbf{A}}_k^2 = 1$, and for what values of $k$ is $\hat{\mathbf{A}}_k^2 = -1$? We say that $\hat{\mathbf{A}}_k$ is a unit $k$-vector.
11. Let $\mathbf{F} = \mathbf{F}_k$ be a $k$-vector where $k \geq 1$, $\mathbf{x}$ a vector, and $\mathbf{B} = \mathbf{B}_r$ an $r$-vector where $r \geq k$. Prove the identity,

$$(\mathbf{B} \wedge \mathbf{x}) \cdot \mathbf{F} = \mathbf{B} \cdot (\mathbf{x} \cdot \mathbf{F}) + (-1)^k (\mathbf{B} \cdot \mathbf{F}) \wedge \mathbf{x}.$$

   This identity will be needed in Chap. 16 when studying Lie algebras.
12. Let $\mathbf{A}_k, \mathbf{B}_k \in \mathbb{G}_n^k$ be $k$-vectors and $\mathbf{a}, \mathbf{b} \in \mathbb{G}_n^1$ be 1-vectors.

   (a) Show that

$$\mathbf{A}_k \cdot (\mathbf{B}_k \wedge \mathbf{b}) = (\mathbf{A}_k \cdot \mathbf{B}_k)\mathbf{b} - (\mathbf{b} \cdot \mathbf{A}_k) \cdot \mathbf{B}_k.$$

   (b) Show that

$$(\mathbf{a} \wedge \mathbf{A}_k) \cdot (\mathbf{B}_k \wedge \mathbf{b}) = (\mathbf{a} \cdot \mathbf{b})(\mathbf{A}_k \cdot \mathbf{B}_k) - (\mathbf{b} \cdot \mathbf{A}_k)(\mathbf{B}_k \cdot \mathbf{a}).$$

13. (a) Let $\mathbf{A}, \mathbf{B}, \mathbf{C}$ be bivectors. Show that

$$\mathbf{A} \boxtimes (\mathbf{B} \boxtimes \mathbf{C}) = (\mathbf{A} \cdot \mathbf{B})\mathbf{C} - (\mathbf{A} \cdot \mathbf{C})\mathbf{B} + (\mathbf{A} \wedge \mathbf{B}) \cdot \mathbf{C} - (\mathbf{A} \wedge \mathbf{C}) \cdot \mathbf{B}.$$

   (b) Show that $\mathbf{A} \circ (\mathbf{B} \boxtimes \mathbf{C}) = (\mathbf{A} \boxtimes \mathbf{B}) \circ \mathbf{C}$, where $\mathbf{A} \circ \mathbf{B} = \frac{1}{2}(\mathbf{AB} + \mathbf{BA})$ is the symmetric part of the geometric product of $\mathbf{A}$ and $\mathbf{B}$.

## 3.5 Vector Derivative in $\mathbb{R}^n$

A powerful *vector deriviative* $\partial_{\mathbf{x}}$ at a point $\mathbf{x} \in \mathbb{R}^n$ can be easily defined in terms of its basic properties. Given a direction $\mathbf{a} \in \mathbb{R}^n$ at the point $\mathbf{x}$, we first define the $\mathbf{a}$-derivative $\mathbf{a} \cdot \partial_{\mathbf{x}}$ in the direction $\mathbf{a}$. Let $F : \mathscr{D} \longrightarrow \mathbb{G}_n$ be any continuous geometric algebra-valued function defined at all points $\mathbf{x} \in \mathscr{D}$ where $\mathscr{D} \subset \mathbb{R}^n$ is an open domain.[1] Recall that an *open r-ball* around a point $\mathbf{x}_0 \in \mathbb{R}^n$ is defined by

[1] An open domain is an open connected subset of $\mathbb{R}^n$. The topological propererties of $\mathbb{R}^n$ are rigorously defined and discussed in Michael Spivak's book, "Calculus on Manifolds" [92].

$$B_r(\mathbf{x}_0) := \{\mathbf{x} \in \mathbb{R}^n |\ |\mathbf{x} - \mathbf{x}_0| < r\}.$$

We say that $F(\mathbf{x})$ has an **a**-*derivative* $F_{\mathbf{a}}(\mathbf{x}) = \mathbf{a} \cdot \partial_{\mathbf{x}} F(\mathbf{x})$ at the point $\mathbf{x} \in \mathscr{D}$ provided the limit

$$F_{\mathbf{a}}(\mathbf{x}) = \lim_{h \to 0} \frac{F(\mathbf{x} + h\mathbf{a}) - F(\mathbf{x})}{h} \tag{3.45}$$

exists, where $h \in \mathbb{R}$. We say that $F$ is $C^1$ *differentiable* on the open domain $\mathscr{D}$ if $F_{\mathbf{a}}(\mathbf{x})$ exists and is continuous for all directions $\mathbf{a} \in \mathbb{R}^n$ and at all points $\mathbf{x} \in \mathscr{D}$. The definition of the **a**-derivative is equivalent to the *directional derivative* on $\mathbb{R}^n$, but the range of the function $F$ lies in the much larger geometric algebra $\mathbb{G}_n$ of $\mathbb{R}^n$. In Chap. 15, when we study the differential geometry of a vector manifold, we will refine this definition.

To complete the definition of the vector derivative $\partial_{\mathbf{x}} F(\mathbf{x})$ of the geometric algebra-valued function $F(\mathbf{x})$, we require that the vector derivative $\partial_{\mathbf{x}}$ has the algebraic properties of a vector in $\mathbb{R}^n$. This means that for the orthonormal basis $(\mathbf{e})_{(n)}$ of $\mathbb{R}^n$, we have

$$\partial_{\mathbf{x}} = \sum_{i=1}^{n} \mathbf{e}_i\, (\mathbf{e}_i \cdot \partial_{\mathbf{x}}) = \sum_{i=1}^{n} \mathbf{e}_i\, \frac{\partial}{\partial x^i}, \tag{3.46}$$

where $\mathbf{e}_i \cdot \partial_{\mathbf{x}} = \frac{\partial}{\partial x^i}$ is the ordinary partial derivatives in the directions of the coordinate axis at the point $\mathbf{x} = \begin{pmatrix} x^1 & x^2 & \cdots & x^n \end{pmatrix} \in \mathbb{R}^n$. The geometric function $F(\mathbf{x})$ is said to have *order* $C^p$ at the point $\mathbf{x} \in \mathbb{R}^n$ for a positive integer $p \in \mathbb{N}$, if all $p^{th}$ order partial derivatives of $F(\mathbf{x})$ exist and are continuous at the point $\mathbf{x}$. If $F(\mathbf{x})$ has continuous partial derivatives of all orders, we say that $F(\mathbf{x})$ is a $C^\infty$ function [92, p.26].

As our first example, consider the identity function $f(\mathbf{x}) = \mathbf{x}$ for all $\mathbf{x} \in \mathbb{R}^n$. We calculate

$$f_{\mathbf{a}}(\mathbf{x}) = \mathbf{a} \cdot \partial_{\mathbf{x}} \mathbf{x} = \lim_{h \to 0} \frac{f(\mathbf{x} + h\mathbf{a}) - f(\mathbf{x})}{h} = \lim_{h \to 0} \frac{\mathbf{x} + h\mathbf{a} - \mathbf{x}}{h} = \mathbf{a} \tag{3.47}$$

for all $\mathbf{x} \in \mathbb{R}^n$ and all $\mathbf{a} \in \mathbb{G}^1$. It follows that the identity function is differentiable on $\mathbb{R}^n$. From (3.46) and (3.47), we calculate the vector derivative of the identity function

$$\partial_{\mathbf{x}} \mathbf{x} = \sum_{i=1}^{n} \mathbf{e}_i (\mathbf{e}_i \cdot \partial_{\mathbf{x}} \mathbf{x}) = \sum_{i=1}^{n} \mathbf{e}_i \mathbf{e}_i = n, \tag{3.48}$$

which gives the number of *degrees of freedom* at the point $\mathbf{x} \in \mathbb{R}^n$. It follows immediately from (3.48) that

$$\partial_{\mathbf{x}} \cdot \mathbf{x} = n, \text{ and } \partial_{\mathbf{x}} \wedge \mathbf{x} = 0. \tag{3.49}$$

From the basic identities (3.47) and (3.49), together with Leibniz product rule for directional or partial deriviatives, other identities easily follow. For example, since $\partial_{\mathbf{x}} \wedge \mathbf{x} = 0$, it follows that

$$0 = \mathbf{a} \cdot (\partial_{\mathbf{x}} \wedge \mathbf{x}) = \mathbf{a} \cdot \partial_{\mathbf{x}} \mathbf{x} - \partial_{\mathbf{x}} \mathbf{x} \cdot \mathbf{a},$$

so that

$$\partial_{\mathbf{x}} \mathbf{x} \cdot \mathbf{a} = \mathbf{a} \cdot \partial_{\mathbf{x}} \mathbf{x} = \mathbf{a}. \tag{3.50}$$

We also have

$$\mathbf{a} \cdot \partial_{\mathbf{x}} |\mathbf{x}|^2 = \mathbf{a} \cdot \partial_{\mathbf{x}} (\mathbf{x} \cdot \mathbf{x}) = 2\mathbf{a} \cdot \mathbf{x}, \tag{3.51}$$

$$\partial_{\mathbf{x}} \mathbf{x}^2 = \dot{\partial}_{\mathbf{x}}\, \dot{\mathbf{x}} \cdot \mathbf{x} + \dot{\partial}_{\mathbf{x}}\, \mathbf{x} \cdot \dot{\mathbf{x}} = 2\mathbf{x}, \tag{3.52}$$

and consequently,

$$\partial_{\mathbf{x}} |\mathbf{x}| = \partial_{\mathbf{x}} (\mathbf{x}^2)^{1/2} = \frac{1}{2} (\mathbf{x}^2)^{-1/2} 2\mathbf{x} = \hat{\mathbf{x}}. \tag{3.53}$$

The dots over the arguments denote which argument is being differentiated.

There is one further property of the vector derivative $\partial_{\mathbf{x}}$ in $\mathbb{R}^n$ that needs to be discussed, and that is the *integrability condition*

$$\partial_{\mathbf{x}} \wedge \partial_{\mathbf{x}} = \sum_{i,j} \mathbf{e}_{ij} \frac{\partial}{\partial x^i} \frac{\partial}{\partial x^j} = \sum_{i<j} \mathbf{e}_{ij} \Big( \frac{\partial}{\partial x^i} \frac{\partial}{\partial x^j} - \frac{\partial}{\partial x^j} \frac{\partial}{\partial x^i} \Big) = 0, \tag{3.54}$$

since partial derivatives commute for differentiable functions. This property for the vector derivative in $\mathbb{R}^n$ depends heavily upon the fact that in a flat space a constant orthonormal basis, in this case the standard orthonormal basis $(\mathbf{e})_{(n)}$, can be chosen at all points $\mathbf{x} \in \mathbb{R}^n$, thus making the derivatives $\frac{\partial \mathbf{e}_j}{\partial x^j} = 0$ for all $1 \le i, j \le n$. When we discuss the vector derivative on a curved space, say, a cylinder or a sphere, this property is no longer true as we will later see.

The above differentiation formulas will be used in the study of the structure of a linear operator on $\mathbb{R}^n$. In Chap. 13, and in later chapters, we generalize the concept of both the **a**-derivative and the vector derivative in $\mathbb{R}^n$ to the concept of the corresponding derivatives on a $k$-dimensional surface in $\mathbb{R}^n$ or in an even more general *pseudo-Euclidean space* $\mathbb{R}^{p,q}$. Because the **a**-derivative and the vector derivative are defined ultimately in terms of partial derivatives, the usual rules of differentiation remain valid. However, care must be taken when applying these rules since we no longer have universal commutativity in the geometric algebra $\mathbb{G}_n$. Other differentiation formulas, which follow easily from the basic formulas above, are given as exercises below. All of these formulas remain valid when we generalize to the vector derivative for a curved surface, except formulas using the property (3.54) which only apply in flat spaces.

## Exercises

Calculate or verify the following vector derivatives:

1. $\partial_{\mathbf{x}}|\mathbf{x}|^k = k|\mathbf{x}|^{k-2}\mathbf{x}$,
2. $\mathbf{a}\cdot\partial_{\mathbf{x}}|\mathbf{x}|^k = k|\mathbf{x}|^{k-2}\mathbf{a}\cdot\mathbf{x}$,
3. $\partial_{\mathbf{x}}\frac{\mathbf{x}}{|\mathbf{x}|^k} = \frac{n-k}{|\mathbf{x}|^k}$,
4. $\mathbf{a}\cdot\partial_{\mathbf{x}}\frac{\mathbf{x}}{|\mathbf{x}|^k} = \frac{\mathbf{a}-k(\mathbf{a}\cdot\hat{\mathbf{x}})\hat{\mathbf{x}}}{|\mathbf{x}|^k}$,
5. $\partial_{\mathbf{x}}\log|\mathbf{x}| = \frac{\mathbf{x}}{|\mathbf{x}|^2} = \mathbf{x}^{-1}$.
6. $\mathbf{a}\cdot\partial_{\mathbf{x}}\log|\mathbf{x}| = \frac{\mathbf{a}\cdot\mathbf{x}}{|\mathbf{x}|^2}$.
7. $\mathbf{a}\cdot\partial_{\mathbf{x}}\hat{\mathbf{x}} = \frac{(\mathbf{a}\wedge\hat{\mathbf{x}})\hat{\mathbf{x}}}{|\mathbf{x}|}$.
8. $\mathbf{a}\cdot\partial_{\mathbf{x}}\sin\mathbf{x} = \mathbf{a}\cdot\partial_{\mathbf{x}}\hat{\mathbf{x}}\sin|\mathbf{x}| = \frac{(\mathbf{a}\wedge\hat{\mathbf{x}})\hat{\mathbf{x}}}{|\mathbf{x}|}\sin|\mathbf{x}| + (\mathbf{a}\cdot\hat{\mathbf{x}})\hat{\mathbf{x}}\cos|\mathbf{x}|$.
9. $\mathbf{a}\cdot\partial_{\mathbf{x}}\sinh\mathbf{x} = \mathbf{a}\cdot\partial_{\mathbf{x}}\hat{\mathbf{x}}\sinh|\mathbf{x}| = \frac{(\mathbf{a}\wedge\hat{\mathbf{x}})\hat{\mathbf{x}}}{|\mathbf{x}|}\sinh|\mathbf{x}| + (\mathbf{a}\cdot\hat{\mathbf{x}})\hat{\mathbf{x}}\cosh|\mathbf{x}|$.
10. $\partial_{\mathbf{x}}\sin\mathbf{x} = \frac{n-1}{|\mathbf{x}|}\sin|\mathbf{x}| + \cos|\mathbf{x}|$.
11. $\partial_{\mathbf{x}}\sinh\mathbf{x} = \frac{n-1}{|\mathbf{x}|}\sinh|\mathbf{x}| + \cosh|\mathbf{x}|$.
12. $\mathbf{a}\cdot\partial_{\mathbf{x}}\exp\mathbf{x} = (\mathbf{a}\cdot\hat{\mathbf{x}})\hat{\mathbf{x}}\mathrm{e}^{\mathbf{x}} + \frac{(\mathbf{a}\wedge\hat{\mathbf{x}})\hat{\mathbf{x}}}{|\mathbf{x}|}\sinh|\mathbf{x}|$.
13. $\partial_{\mathbf{x}}\exp\mathbf{x} = \mathrm{e}^{\mathbf{x}} + \frac{n-1}{|\mathbf{x}|}\sinh|\mathbf{x}| = (1 + \frac{n-1}{|\mathbf{x}|}\mathrm{e}^{-\mathbf{x}}\sinh|\mathbf{x}|)\mathrm{e}^{\mathbf{x}}$.
14. $\mathbf{a}\cdot\partial_{\mathbf{x}}(\mathbf{x}\wedge\mathbf{b})^2 = 2(\mathbf{a}\wedge\mathbf{b})\cdot(\mathbf{x}\wedge\mathbf{b})$.
15. $\partial_{\mathbf{x}}(\mathbf{x}\wedge\mathbf{b})^2$.
16. $\mathbf{a}\cdot\partial_{\mathbf{x}}|\mathbf{x}\wedge\mathbf{b}| = -\frac{(\mathbf{a}\wedge\mathbf{b})\cdot(\mathbf{x}\wedge\mathbf{b})}{|\mathbf{x}\wedge\mathbf{b}|}$.
17. $\partial_{\mathbf{x}}|\mathbf{x}\wedge\mathbf{b}|$.
18. $\mathbf{a}\cdot\partial_{\mathbf{x}}\frac{\mathbf{x}\wedge\mathbf{b}}{|\mathbf{x}\wedge\mathbf{b}|} = \frac{\mathbf{a}\wedge\mathbf{b}+(\mathbf{a}\wedge\mathbf{b})\cdot(\widehat{\mathbf{x}\wedge\mathbf{b}})\widehat{\mathbf{x}\wedge\mathbf{b}}}{|\mathbf{x}\wedge\mathbf{b}|}$.
19. $\partial_{\mathbf{x}}\frac{\mathbf{x}\wedge\mathbf{b}}{|\mathbf{x}\wedge\mathbf{b}|}$.
20. $\mathbf{a}\cdot\partial_{\mathbf{x}}\exp(\mathbf{x}\wedge\mathbf{b})$.
21. $\partial_{\mathbf{x}}\exp(\mathbf{x}\wedge\mathbf{b})$.
22. Using the property (3.54), show that the vector derivative in $\mathbb{R}^n$ satisfies the property

$$(\mathbf{a}\wedge\mathbf{b})\cdot(\partial_{\mathbf{x}}\wedge\partial_{\mathbf{x}}) = [\mathbf{b}\cdot\partial_{\mathbf{x}},\mathbf{a}\cdot\partial_{\mathbf{x}}] - [\mathbf{b},\mathbf{a}]\cdot\partial_{\mathbf{x}} = 0,$$

where the brackets are defined by

$$[\mathbf{b}\cdot\partial_{\mathbf{x}},\mathbf{a}\cdot\partial_{\mathbf{x}}] = \mathbf{b}\cdot\partial_{\mathbf{x}}\,\mathbf{a}\cdot\partial_{\mathbf{x}} - \mathbf{a}\cdot\partial_{\mathbf{x}}\,\mathbf{b}\cdot\partial_{\mathbf{x}}, \tag{3.55}$$

and

$$[\mathbf{b},\mathbf{a}] = \mathbf{b}\cdot\partial_{\mathbf{x}}\mathbf{a} - \mathbf{a}\cdot\partial_{\mathbf{x}}\mathbf{b}, \tag{3.56}$$

and where $\mathbf{a}=\mathbf{a}(\mathbf{x})$ and $\mathbf{b}=\mathbf{b}(\mathbf{x})$ are any differentiable vector-valued functions at the point $\mathbf{x}\in\mathbb{R}^n$.

# Chapter 4
# Vector Spaces and Matrices

*Pure mathematics is, in its way, the poetry of logical ideas.*
–Albert Einstein

We begin this chapter with the formal definition of a vector space. Multiplication of matrices is defined in terms of the product of a row vector and a column vector. Since the rules of matrix algebra over the real and complex numbers are identical to the rules of the addition and multiplication of geometric numbers, it makes sense to consider matrices whose entries are geometric numbers alongside the more traditional matrices of real and complex numbers. Whereas geometric algebra provides geometric insight into what otherwise is just a table of numbers with an unintuitive rule of multiplication, the use of matrices augments the computational power of geometric algebra. The geometric algebras of the plane and 3-dimensional space can be represented by square $2 \times 2$ matrices of real or complex numbers, respectively.

## 4.1 Definitions

In studying modular number systems $\mathbb{Z}_h$ and modular polynomial number systems $\mathbb{K}[x]_{h(x)}$, we have met with the concept of a basis and the fact that elements in either $\mathbb{Z}_h$ or $\mathbb{K}[x]_h$ are linear combinations of the respective basis elements. We now formalize these properties in the definition of a *vector space* over a field $\mathbb{K}$. As before, we are mainly interested in the cases when the scalar field $\mathbb{K}$ is either the real numbers $\mathbb{R}$ or complex numbers $\mathbb{C}$.

**Definition 4.1.1.** By a vector space over a field $\mathbb{K}$, we mean a set of elements $\mathbb{V}$ on which is defined the two binary operations of addition and scalar multiplication.

G. Sobczyk, *New Foundations in Mathematics: The Geometric Concept of Number*, DOI 10.1007/978-0-8176-8385-6_4,

More precisely, let $\mathbf{w}, \mathbf{x}, \mathbf{y} \in \mathbb{V}$ and $\alpha, \beta, \gamma \in \mathbb{K}$. Then the following properties are satisfied:

1. (a) $\mathbf{x}+\mathbf{y} \in \mathbb{V}$. *Addition of vectors is a vector.*
   (b) There exists a *zero vector* with the property that $\mathbf{x}+0 = \mathbf{x}$ for all vectors $\mathbf{x} \in \mathbb{V}$. (Note: We use the same symbol for the zero vector and the zero scalar.)
   (c) $\mathbf{x}+\mathbf{y} = \mathbf{y}+\mathbf{x}$. *Commutative law of addition of vectors.*
   (d) $\mathbf{w}+(\mathbf{x}+\mathbf{y}) = (\mathbf{w}+\mathbf{x})+\mathbf{y}$. *Associative law of addition of vectors.*
   (e) For every $\mathbf{x} \in \mathbb{V}$, there exists a unique $-\mathbf{x} \in \mathbb{V}$ such that $\mathbf{x}+(-\mathbf{x}) = 0$. *Additive inverse.*
2. (a) $\alpha\mathbf{x} = \mathbf{x}\alpha \in \mathbb{V}$. *Multiplication of a vector by a scalar.*
   (b) $(\alpha\beta)\mathbf{x} = \alpha(\beta\mathbf{x})$. *Associative law of scalar multiplication.*
3. $\alpha(\mathbf{x}+\mathbf{y}) = \alpha\mathbf{x}+\alpha\mathbf{y}$ and $(\alpha+\beta)\mathbf{x} = \alpha\mathbf{x}+\beta\mathbf{x}$. *Distributive properties.*

It is often expedient to replace the concept of a vector space over a field by the more general concept of a *module* over a *ring*. For example, we will often consider matrices whose elements are geometric numbers. The addition and multiplication of matrices are fully compatible with the addition and multiplication of geometric numbers.

## Examples

1. The set of all polynomials $\mathbb{R}[x]$ over the real numbers $\mathbb{R}$ satisfies Definition 4.1.1 and is therefore a vector space, as is the set of polynomials $\mathbb{C}[z]$ over the complex numbers $\mathbb{C}$. We can also consider the set of polynomials over the geometric numbers $\mathbb{G}_3$, which make up a module.
2. The set of all polynomials $\mathbb{Z}[x]$ over the intergers $\mathbb{Z}$ is not a vector space because the integers $\mathbb{Z}$ do not form a field.
3. The set of all polynomials $\mathbb{Q}[x]$ over the rational numbers $\mathbb{Q}$ is a vector space. Because the polynomials in $\mathbb{Q}[x]$ can also be multiplied together, we say that $\mathbb{Q}[x]$ has the structure of a *ring*.
4. The set of all polynomials $\mathbb{Z}_p[x]$ over the finite field $\mathbb{Z}_p$ for a prime number $p \in \mathbb{N}$ makes up a vector space under the addition of polynomials. Because polynomials in $\mathbb{Z}_p[x]$ can be *multiplied* as well as added, $\mathbb{Z}_p[x]$ is also a *ring*.
5. Let $\mathbb{R}^n$ denote the set of all *column n-tuples* of the form

$$\mathbb{R}^n = \{\mathbf{x} | \ \mathbf{x} = \begin{pmatrix} x^1 \\ x^2 \\ \cdot \\ \cdot \\ x^n \end{pmatrix}, \text{ where } x^1, x^2, \ldots, x^n \in \mathbb{R}\}.$$

With the usual addition of components and multiplication by scalars, $\mathbb{R}^n$ is called the *n-dimensional vector space of real column vectors.*

6. Let $\mathbb{R}^n$ denote the set of all *row n-tuples* of the form

$$\mathbb{R}^n = \{\mathbf{x} |\ \mathbf{x} = \begin{pmatrix} x_1 & x_2 & \dots & x_n \end{pmatrix},\ \text{where}\ x_1, x_2, \dots, x_n \in \mathbb{R}\}.$$

With the usual addition of components and multiplication by scalars, $\mathbb{R}^n$ is called the *n-dimensional vector space of real row vectors.* No distinction is made between row and column vectors with respect to the *standard orthonormal basis* $\{\mathbf{e}_i\}_{i=1}^n$ of $\mathbb{R}^n$. However, the distinction between a row vector and a column vector becomes important in the definition of matrix multiplication. The identification of both row and column *n-tuples* as names for points in the $n$-dimensional coordinate plane is very common, as is the geometric interpretation of addition and scalar multiplication of vectors in $\mathbb{R}^n$ in terms of directed line segments.

There are many, many other important examples of finite and infinite dimensional vector spaces, which are exactly the reason for defining them in the first place! There is one more important example which we will give here:

7. Let $\mathbb{M}_{\mathbb{K}}(m,n)$ denote the set of all $(m,n)$-dimensional *matrices* of the form

$$A = \begin{pmatrix} a_{11} & a_{12} & \dots & a_{1n} \\ a_{21} & a_{22} & \dots & a_{2n} \\ \cdot & \cdot & \dots & \cdot \\ \cdot & \cdot & \dots & \cdot \\ a_{m1} & a_{m2} & \dots & a_{mn} \end{pmatrix}$$

where the scalars $a_{ij} \in \mathbb{K}$ for all $i = 1, \dots, m$ and $j = 1, \dots, n$. The set $M_{\mathbb{K}}(m,n)$ of all matrices makes up a vector space where addition is component-wise and scalar multiplication is as usual. By allowing the entries in the matrices to be geometric numbers, say in $\mathbb{G}_3$, the matrices $M_{\mathbb{G}_3}(m,n)$ become a module over the geometric algebra $\mathbb{G}_3$. Left and right scalar multiplication by geometric numbers is well defined but no longer universally commutative.

## Exercises

1. Show that 6 of the 7 examples given above are vector spaces.
2. Give vector diagrams to represent the addition and scalar multiplication of vectors in the plane. Represent each of the properties of a vector space in Definition 1, in terms of a corresponding vector diagram.
3. Show that $\mathbf{e}_1 \begin{pmatrix} 1 & \mathbf{e}_2 \\ \mathbf{e}_1 & \mathbf{e}_{12} \end{pmatrix} \neq \begin{pmatrix} 1 & \mathbf{e}_2 \\ \mathbf{e}_1 & \mathbf{e}_{12} \end{pmatrix} \mathbf{e}_1$, where $\mathbf{e}_1, \mathbf{e}_2, \mathbf{e}_{12} \in \mathbb{G}_2$.

## 4.2 Matrix Algebra

The reader may be wondering why, in the definition of a vector space, the multiplication of vectors with themselves is not defined as it was in the definition of geometric algebra in the last chapter. This restriction is removed when we use, instead, the structure of a module and allow matrices to be multiplied by the elements of a geometric algebra instead of just by elements of a field.

Whereas in Chap. 3, we constructed the geometric algebra $\mathbb{G}_n(\mathbb{R}^n)$ of the Euclidean space $\mathbb{R}^n$ by taking the geometric sums and geometric products of all vectors in $\mathbb{R}^n$, in this chapter, we define the quite different algebra of matrices which we construct from the row and column vectors of $\mathbb{R}^n$. Whereas the addition and multiplication of matrices satisfy the same algebraic rules as the addition and multiplication of geometric numbers, care must be taken to distinguish between these two different algebras. Although the algebra of matrices and the algebra of geometric numbers are very different objects, they are completely compatible structures. For this reason, it makes sense to consider matrices whose elements are geometric numbers and to multiply them by geometric numbers. We shall have much to say about matrix algebras of geometric numbers, and we will show how they provide us with powerful new tools.

The basic building blocks of a matrix algebra are *rows* and *columns* of elements of a field or more generally of a geometric algebra. We have already considered the row and column vectors of the Euclidean space $\mathbb{R}^n$, given in Examples 5 and 6 above.

For an $m$-row matrix of vectors from a geometric algebra $\mathbb{G}$, we write

$$(\mathbf{a})_{(m)} = \begin{pmatrix}\mathbf{a}_1 & \mathbf{a}_2 & \dots & \mathbf{a}_m\end{pmatrix}, \tag{4.1}$$

where each $\mathbf{a}_i \in \mathbb{G}$ for $i = 1, \dots, m$. The *transpose* $(\mathbf{a})^{\mathrm{T}}_{(m)}$ of the row of vectors $(\mathbf{a})_{(m)}$ gives the $m$-column

$$(\mathbf{a})^{\mathrm{T}}_{(m)} = \begin{pmatrix}\mathbf{a}_1 \\ \mathbf{a}_2 \\ \cdot \\ \cdot \\ \cdot \\ \mathbf{a}_m\end{pmatrix} \tag{4.2}$$

of these same vectors.

Consider now the $(m,n)$-matrix $A \in \mathbb{M}_{\mathbb{K}}(m,n)$, given in Example 7. The matrix $A$ can be expressed in terms of either the *column vectors* or *row vectors* of $\mathbb{K}^m$ or $\mathbb{K}^n$, respectively, given in Examples 5 and 6. When considering row or column vectors of $\mathbb{K}^m$ or $\mathbb{K}^n$ where matrix multiplication applies, we must be careful to distinguish them from the vectors from a geometric algebra $\mathbb{G}$ where geometric multiplication applies.

We write

$$A = (\mathbf{a})_{(n)} \equiv \left(\mathbf{a}_1 \; \mathbf{a}_2 \; \ldots \; \mathbf{a}_n\right)$$

when we think about $A$ as consisiting of the $n$ column vectors,

$$\mathbf{a}_j \equiv \begin{pmatrix} a_{1j} \\ a_{2j} \\ \cdot \\ \cdot \\ a_{mj} \end{pmatrix},$$

where $a_{ij} \in \mathbb{K}$ for all $i = 1, \ldots, m$ and $j = 1, 2, \ldots, n$. But, we may equally well think about the matrix $A$ as consisting of $m$ row vectors,

$$A = (\mathbf{a})^{(m)} \equiv \begin{pmatrix} \mathbf{a}^1 \\ \mathbf{a}^2 \\ \cdot \\ \cdot \\ \mathbf{a}^m \end{pmatrix},$$

where the row vectors $\mathbf{a}^i$, for $i = 1, \ldots, m$ are defined by

$$\mathbf{a}^i \equiv \left(a_{i1} \; a_{i2} \; \ldots \; a_{in}\right).$$

The use of lower and upper indicies is an artifice often used to distinguish between column and row vectors, respectively.

We also write the matrix $A$ in the form

$$A = (a)_{(m,n)} = \begin{pmatrix} a_{11} & \cdots & a_{1n} \\ \cdot & \cdots & \cdot \\ \cdot & \cdots & \cdot \\ \cdot & \cdots & \cdot \\ a_{m1} & \cdots & a_{mn} \end{pmatrix} \tag{4.3}$$

where the scalars $a_{ij} \in \mathbb{K}$ for $1 \leq i \leq m$ and $1 \leq j \leq n$. The *transpose* $A^{\mathrm{T}}$ of the matrix $A$ is obtained by interchanging the rows and columns of the matrix $A$,

$$A^{\mathrm{T}} = (a)^{\mathrm{T}}_{m,n} = \begin{pmatrix} a_{11} & \cdots & a_{m1} \\ \cdot & \cdots & \cdot \\ \cdot & \cdots & \cdot \\ \cdot & \cdots & \cdot \\ a_{1n} & \cdots & a_{mn} \end{pmatrix}. \tag{4.4}$$

For example, the $(3,2)$-matrix

$$A = \begin{pmatrix} 1 & 2 \\ 3 & 4 \\ 5 & 6 \end{pmatrix}$$

can be considered to consist of the column vectors

$$(\mathbf{a})_{(2)} = \begin{pmatrix} \mathbf{a}_1 & \mathbf{a}_2 \end{pmatrix} \text{ for } \mathbf{a}_1 = \begin{pmatrix} 1 \\ 3 \\ 5 \end{pmatrix} \text{ and } \mathbf{a}_2 = \begin{pmatrix} 2 \\ 4 \\ 6 \end{pmatrix}$$

or of the row vectors

$$(\mathbf{a})^{(3)} = \begin{pmatrix} \mathbf{a}^1 \\ \mathbf{a}^2 \\ \mathbf{a}^3 \end{pmatrix} \text{ for}$$

$$\mathbf{a}^1 = \begin{pmatrix} 1 & 2 \end{pmatrix}, \ \mathbf{a}^2 = \begin{pmatrix} 3 & 4 \end{pmatrix}, \text{ and } \mathbf{a}^3 = \begin{pmatrix} 5 & 6 \end{pmatrix}.$$

The transpose of the matrix $A$ is

$$A^{\mathrm{T}} = \begin{pmatrix} 1 & 3 & 5 \\ 2 & 4 & 6 \end{pmatrix} = \begin{pmatrix} \mathbf{a}_1 & \mathbf{a}_2 \end{pmatrix}^{\mathrm{T}} = \begin{pmatrix} \mathbf{a}_1^{\mathrm{T}} \\ \mathbf{a}_2^{\mathrm{T}} \end{pmatrix}.$$

Any $(m,n)$-matrix (4.3) can be multiplied by a scalar $\alpha \in \mathbb{K}$ simply by multiplying all its components by the scalar $\alpha$,

$$\alpha A = (\alpha a)_{(m,n)} = \begin{pmatrix} \alpha a_{11} & \cdots & \alpha a_{1n} \\ \cdot & \cdots & \cdot \\ \cdot & \cdots & \cdot \\ \cdot & \cdots & \cdot \\ \alpha a_{m1} & \cdots & \alpha a_{mn} \end{pmatrix} = A\alpha.$$

The addition $A+B$ of two $(m,n)$-matrices $A = (a_{ij})$ and $B = (b_{ij})$ is defined by

$$\begin{pmatrix} a_{11} & \cdots & a_{1n} \\ \cdot & \cdots & \cdot \\ \cdot & \cdots & \cdot \\ \cdot & \cdots & \cdot \\ a_{m1} & \cdots & a_{mn} \end{pmatrix} + \begin{pmatrix} b_{11} & \cdots & b_{1n} \\ \cdot & \cdots & \cdot \\ \cdot & \cdots & \cdot \\ \cdot & \cdots & \cdot \\ b_{m1} & \cdots & b_{mn} \end{pmatrix} = \begin{pmatrix} a_{11}+b_{11} & \cdots & a_{1n}+b_{1n} \\ \cdot & \cdots & \cdot \\ \cdot & \cdots & \cdot \\ \cdot & \cdots & \cdot \\ a_{m1}+b_{m1} & \cdots & a_{mn}+b_{mn} \end{pmatrix}. \tag{4.5}$$

The definition of matrix multiplication is given in the next section.

## Exercises

1. Let $(\mathbf{a})_{(3)} = \begin{pmatrix}\mathbf{e}_1 & 2\mathbf{e}_2 & \mathbf{e}_{12}\end{pmatrix}$ and $(\mathbf{b})_{(3)} = \begin{pmatrix}\mathbf{e}_2 & 2\mathbf{e}_{12} & \mathbf{e}_1\end{pmatrix}$ be row matrices over the geometric algebra $\mathbb{G}_2$. Calculate:

   (a) $2(\mathbf{a})_{(3)} - 3(\mathbf{b})_{(3)}$.
   (b) $\mathbf{e}_1(\mathbf{a})_{(3)} - \mathbf{e}_{12}(\mathbf{b})_{(3)}$.

2. Let $A = \begin{pmatrix}1 & 2 & 3\\ 2 & -4 & 0\end{pmatrix}$ and $B = \begin{pmatrix}-1 & 3 & -2\\ 1 & -5 & 2\end{pmatrix}$. Calculate:

   (a) $A - 3B$.
   (b) Show that $A^{\mathrm{T}} - 3B^{\mathrm{T}} = (A - 3B)^{\mathrm{T}}$.

3. Show that $\mathbf{e}_{123}\begin{pmatrix}1 & \mathbf{e}_2\\ \mathbf{e}_1 & \mathbf{e}_{12}\end{pmatrix} = \begin{pmatrix}1 & \mathbf{e}_2\\ \mathbf{e}_1 & \mathbf{e}_{12}\end{pmatrix}\mathbf{e}_{123}$ over the geometric algebra $\mathbb{G}_3$.

4. Show that $\mathbf{e}_1\begin{pmatrix}1\\ \mathbf{e}_1\end{pmatrix}\begin{pmatrix}1 & \mathbf{e}_1\end{pmatrix} = \begin{pmatrix}\mathbf{e}_1 & 1\\ 1 & \mathbf{e}_1\end{pmatrix}$ over the geometric algebra $\mathbb{G}_1$.

5. Let $u_+ = \frac{1}{2}(1+\mathbf{e}_2)$. Show that $u_+\begin{pmatrix}1\\ \mathbf{e}_2\end{pmatrix}\begin{pmatrix}1 & \mathbf{e}_2\end{pmatrix} = u_+\begin{pmatrix}1 & 1\\ 1 & 1\end{pmatrix}$ over the geometric algebra $\mathbb{G}_2$.

## 4.3 Matrix Multiplication

Matrix multiplication can be most easily defined in terms of the multiplication of row and column vectors. The multiplication of the row vector of real numbers $\begin{pmatrix}a_1 & a_2 & \dots & a_m\end{pmatrix}$ by the column vector of real numbers $\begin{pmatrix}b^1\\ b^2\\ \cdot\\ \cdot\\ b^m\end{pmatrix}$ is defined to be the real number

$$\begin{pmatrix}a_1 & a_2 & \dots & a_m\end{pmatrix}\begin{pmatrix}b^1\\ b^2\\ \cdot\\ \cdot\\ b^m\end{pmatrix} \equiv a_1b^1 + a_2b^2 + \cdots + a_mb^m.$$

Note that the matrix multiplication of a row vector times a column vector is only defined when they each have the same number $m$ of components.

On the other hand, the matrix product of a column vector of real numbers $\begin{pmatrix} b^1 \\ b^2 \\ \cdot \\ \cdot \\ b^m \end{pmatrix}$ with the row vector $\begin{pmatrix} a_1 & a_2 & \dots & a_n \end{pmatrix}$ of real numbers is the $(m,n)$-matrix

$$\begin{pmatrix} b^1 \\ b^2 \\ \cdot \\ \cdot \\ b^m \end{pmatrix} \begin{pmatrix} a_1 & a_2 & \dots & a_n \end{pmatrix} \equiv \begin{pmatrix} b^1 a_1 & b^1 a_2 & \dots & b^1 a_n \\ b^2 a_1 & b^2 a_2 & \dots & b^2 a_n \\ \cdot & \cdot & \dots & \cdot \\ \cdot & \cdot & \dots & \cdot \\ b^m a_1 & b^m a_2 & \dots & b^m a_n \end{pmatrix}.$$

In terms of the previous definitions of the product of row and column vectors, we can now easily define the general product of a real $(m,n)$-matrix

$$A = \begin{pmatrix} \mathbf{a}^1 \\ \mathbf{a}^2 \\ \cdot \\ \cdot \\ \mathbf{a}^m \end{pmatrix} = \begin{pmatrix} a_{11} & \cdots & a_{1n} \\ \cdot & \cdots & \cdot \\ \cdot & \cdots & \cdot \\ \cdot & \cdots & \cdot \\ a_{m1} & \cdots & a_{mn} \end{pmatrix}$$

with a real $(n,s)$-matrix

$$B = \begin{pmatrix} \mathbf{b}_1 & \mathbf{b}_2 & \dots & \mathbf{b}_s \end{pmatrix} = \begin{pmatrix} b_{11} & \cdots & b_{1s} \\ \cdot & \cdots & \cdot \\ \cdot & \cdots & \cdot \\ \cdot & \cdots & \cdot \\ b_{n1} & \cdots & a_{ns} \end{pmatrix}$$

to be the real $(m,s)$-matrix $C$ given by

$$C = \begin{pmatrix} \mathbf{a}^1\mathbf{b}_1 & \mathbf{a}^1\mathbf{b}_2 & \dots & \mathbf{a}^1\mathbf{b}_s \\ \mathbf{a}^2\mathbf{b}_1 & \mathbf{a}^2\mathbf{b}_2 & \dots & \mathbf{a}^2\mathbf{b}_s \\ \cdot & \cdot & \dots & \cdot \\ \cdot & \cdot & \dots & \cdot \\ \mathbf{a}^m\mathbf{b}_1 & \mathbf{a}^m\mathbf{b}_2 & \dots & \mathbf{a}^m\mathbf{b}_s \end{pmatrix}.$$

Note that multiplying a *row* of real column vectors times a *column* of real row vectors is a generalization of the definition of a real row vector times a real column vector. We must warn the reader that here we are *not* employing the geometric product of vectors. We give simple examples below in order to minimize possible confusion.

## 4.4 Examples of Matrix Multiplication

1. For $\mathbf{a} = \begin{pmatrix} 1 & 2 & 3 \end{pmatrix} \in \mathbb{R}^3$ and $\mathbf{b} = \begin{pmatrix} 2 \\ 5 \\ 6 \end{pmatrix} \in \mathbb{R}^3$,

$$\mathbf{ab} = \begin{pmatrix} 1 & 2 & 3 \end{pmatrix} \begin{pmatrix} 2 \\ 5 \\ 6 \end{pmatrix} = 1 \cdot 2 + 2 \cdot 5 + 3 \cdot 6 = 30 \in \mathbb{R}.$$

   We are not using the geometric product here since we are considering **a** and **b** to be strictly a real row vector and a real column vector, respectively.

2. For $\mathbf{b} = \begin{pmatrix} 2 \\ 5 \\ 6 \end{pmatrix}$ and $\mathbf{a} = \begin{pmatrix} 1 & 3 \end{pmatrix}$, we get

$$\mathbf{ba} = \begin{pmatrix} 2 \\ 5 \\ 6 \end{pmatrix} \begin{pmatrix} 1 & 3 \end{pmatrix} = \begin{pmatrix} 2 \cdot 1 & 2 \cdot 3 \\ 5 \cdot 1 & 5 \cdot 3 \\ 6 \cdot 1 & 6 \cdot 3 \end{pmatrix} = \begin{pmatrix} 2 & 6 \\ 5 & 15 \\ 6 & 18 \end{pmatrix}.$$

   We are not using the geometric product.

3. For $A = \begin{pmatrix} 1 & 2 & 3 \\ 4 & 5 & 6 \end{pmatrix}$ and $B = \begin{pmatrix} 2 & -2 \\ 2 & 0 \\ 1 & -1 \end{pmatrix}$, we have

$$AB = \begin{pmatrix} 1 & 2 & 3 \\ 4 & 5 & 6 \end{pmatrix} \begin{pmatrix} 2 & -2 \\ 2 & 0 \\ 1 & -1 \end{pmatrix} = \begin{pmatrix} 9 & -5 \\ 24 & -14 \end{pmatrix}$$

   and

$$BA = \begin{pmatrix} 2 & -2 \\ 2 & 0 \\ 1 & -1 \end{pmatrix} \begin{pmatrix} 1 & 2 & 3 \\ 4 & 5 & 6 \end{pmatrix} = \begin{pmatrix} -6 & -6 & -6 \\ 2 & 4 & 6 \\ -3 & -3 & -3 \end{pmatrix}.$$

   No geometric product is used.

4. For $(\mathbf{e})_{(3)} = \begin{pmatrix} \mathbf{e}_1 & \mathbf{e}_2 & \mathbf{e}_3 \end{pmatrix}$ and $(\mathbf{e})^{\mathrm{T}}_{(3)} = \begin{pmatrix} \mathbf{e}_1 \\ \mathbf{e}_2 \\ \mathbf{e}_3 \end{pmatrix}$ over $\mathbb{G}_3$, we calculate:

   (a)

$$(\mathbf{e})_{(3)}(\mathbf{e})^{\mathrm{T}}_{(3)} = (\mathbf{e})_{(3)} \cdot (\mathbf{e})^{\mathrm{T}}_{(3)} + (\mathbf{e})_{(3)} \wedge (\mathbf{e})^{\mathrm{T}}_{(3)} = \sum_{i=1}^{3} \mathbf{e}_i \cdot \mathbf{e}_i + \sum_{i=1}^{3} \mathbf{e}_i \wedge \mathbf{e}_i = 3$$

(b)

$$(\mathbf{e})^{\mathrm{T}}_{(3)}(\mathbf{e})_{(3)} = (\mathbf{e})^{\mathrm{T}}_{(3)} \cdot (\mathbf{e})_{(3)} + (\mathbf{e})^{\mathrm{T}}_{(3)} \wedge (\mathbf{e})_{(3)}$$

$$= \begin{pmatrix} \mathbf{e}_1\cdot\mathbf{e}_1 & \mathbf{e}_1\cdot\mathbf{e}_2 & \mathbf{e}_1\cdot\mathbf{e}_3 \\ \mathbf{e}_2\cdot\mathbf{e}_1 & \mathbf{e}_2\cdot\mathbf{e}_2 & \mathbf{e}_2\cdot\mathbf{e}_3 \\ \mathbf{e}_3\cdot\mathbf{e}_1 & \mathbf{e}_3\cdot\mathbf{e}_2 & \mathbf{e}_3\cdot\mathbf{e}_3 \end{pmatrix} + \begin{pmatrix} \mathbf{e}_1\wedge\mathbf{e}_1 & \mathbf{e}_1\wedge\mathbf{e}_2 & \mathbf{e}_1\wedge\mathbf{e}_3 \\ \mathbf{e}_2\wedge\mathbf{e}_1 & \mathbf{e}_2\wedge\mathbf{e}_2 & \mathbf{e}_2\wedge\mathbf{e}_3 \\ \mathbf{e}_3\wedge\mathbf{e}_1 & \mathbf{e}_3\wedge\mathbf{e}_2 & \mathbf{e}_3\wedge\mathbf{e}_3 \end{pmatrix} = \begin{pmatrix} 1 & \mathbf{e}_{12} & \mathbf{e}_{13} \\ \mathbf{e}_{21} & 1 & \mathbf{e}_{23} \\ \mathbf{e}_{31} & \mathbf{e}_{32} & 1 \end{pmatrix}.$$

Here, *both* the geometric product and matrix product are being used.

## Exercises

1. Compute $AB$ and $BA$ for

$$A = \begin{pmatrix} 1 & -2 & -1 \\ 4 & 2 & -1 \end{pmatrix} \text{ and } B = \begin{pmatrix} 2 & -1 \\ 2 & 1 \\ 1 & -1 \end{pmatrix}.$$

2. Show that the associative law $A(BC) = (AB)C$ is valid for matrix multiplication for

$$A = \begin{pmatrix} 1 & 2 \\ -3 & 2 \end{pmatrix},\ B = \begin{pmatrix} 2 & -2 \\ 1 & 2 \end{pmatrix},\text{and } C = \begin{pmatrix} 0 & 1 \\ 0 & 2 \end{pmatrix}.$$

3. Show that the system of linear equations

$$\begin{pmatrix} x_1 + 2x_2 + 3x_3 = 3 \\ 4x_1 + 5x_2 + 6x_3 = 4 \end{pmatrix}$$

can be expressed in the matrix form $AX = B$ for the matrices

$$A = \begin{pmatrix} 1 & 2 & 3 \\ 4 & 5 & 6 \end{pmatrix},\ X = \begin{pmatrix} x_1 \\ x_2 \\ x_3 \end{pmatrix},\text{and } B = \begin{pmatrix} 3 \\ 4 \end{pmatrix}.$$

4. Show that the matrix $P = \begin{pmatrix} 1 & 2 & 3 \\ 0 & 0 & 0 \\ 0 & 0 & 0 \end{pmatrix}$ has the property that $P^2 = P$. Such a matrix is called an *idempotent* or a *projection*.

5. Show that the matrix $Q = \begin{pmatrix} 0 & 1 & 2 & 3 \\ 0 & 0 & 0 & 0 \\ 0 & 0 & 0 & 4 \\ 0 & 0 & 0 & 0 \end{pmatrix}$ has the property that $Q^3 = 0$, but $Q^2 \neq 0$.

   Such a matrix is called a *nilpotent* with the *index of nilpotency* of $m = 3$.

6. Let $P_1 = \begin{pmatrix} 1 & 1 & 1 \\ 0 & 0 & 0 \\ 0 & 0 & 0 \end{pmatrix}$ and $P_2 = \begin{pmatrix} 0 & -1 & -1 \\ 0 & 1 & 0 \\ 0 & 0 & 1 \end{pmatrix}$.

   (a) Calculate $A = 2P_1 + 3P_2$.
   (b) Show that $A^2 = 4P_1 + 9P_2$.
   (c) Show that $P_1^2 = P_1$, $P_2^2 = P_2$ and $P_1P_2 = P_2P_1 = 0$. The matrices $P_1$ and $P_2$ are called *mutually annihilating idempotents*.

7. Express the system of linear equations

$$\begin{pmatrix} 2x - 5y + z = 5 \\ x + 2y - z = 3 \\ 4x - y - z = 11 \end{pmatrix}$$

   as a matrix equation in the form $AX = B$.

8. Let $A = \begin{pmatrix} 1 & 2 & 0 \\ 0 & 1 & 1 \\ 1 & 0 & 2 \end{pmatrix}$. The matrix $A^{-1} = \frac{1}{4}\begin{pmatrix} 2 & -4 & 2 \\ 1 & 2 & -1 \\ -1 & 2 & 1 \end{pmatrix}$ is called the *inverse* of $A$ because it satisfies $AA^{-1} = I = A^{-1}A$ where $I = \begin{pmatrix} 1 & 0 & 0 \\ 0 & 1 & 0 \\ 0 & 0 & 1 \end{pmatrix}$ is the *identity* matrix. Verify this relationship.

9. Show that if $A \in M_{\mathbb{R}}(3)$ is any $(3,3)$-matrix, then $IA = A = AI$ for $I = \begin{pmatrix} 1 & 0 & 0 \\ 0 & 1 & 0 \\ 0 & 0 & 1 \end{pmatrix}$.

10. Let

$$D = \frac{1}{2\sqrt{21}} \begin{pmatrix} \sqrt{7} - 3\sqrt{3} & \sqrt{7} + 3\sqrt{3} \\ \sqrt{3} - \sqrt{7} & -\sqrt{3} - \sqrt{7} \\ -2\sqrt{3} - 2\sqrt{7} & 2\sqrt{3} - 2\sqrt{7} \end{pmatrix}.$$

   Using the computer software package MATHEMATICA, MATLAB, or MAPLE, show that

$$D^{\mathrm{T}} D = \begin{pmatrix} 1 & 0 \\ 0 & 1 \end{pmatrix} \quad \text{and} \quad DD^{\mathrm{T}} = P$$

   where $P^2 = P$.

11. Let $(\mathbf{a})_{(2)} = \begin{pmatrix}\mathbf{a}_1 & \mathbf{a}_2\end{pmatrix}$ where $\mathbf{a}_1 = 2\mathbf{e}_1 + 3\mathbf{e}_2 - \mathbf{e}_3$ and $\mathbf{a}_2 = -\mathbf{e}_1 + \mathbf{e}_2 + \mathbf{e}_3$ are vectors of $\mathbb{R}^3$ *considered as vectors in* $\mathbb{G}_3$. Let $(\mathbf{e})_{(3)} = \begin{pmatrix}\mathbf{e}_1 & \mathbf{e}_2 & \mathbf{e}_3\end{pmatrix}$ be the standard basis of $\mathbb{R}^3$ *considered as vectors in* $\mathbb{G}_3$. Show that:

(a) $(\mathbf{e})_{(3)}^{\mathrm{T}} \cdot (\mathbf{a})_{(2)} = \begin{pmatrix}\mathbf{e}_1\\ \mathbf{e}_2\\ \mathbf{e}_3\end{pmatrix} \cdot \begin{pmatrix}\mathbf{a}_1 & \mathbf{a}_2\end{pmatrix} = \begin{pmatrix}\mathbf{e}_1\cdot\mathbf{a}_1 & \mathbf{e}_1\cdot\mathbf{a}_2\\ \mathbf{e}_2\cdot\mathbf{a}_1 & \mathbf{e}_2\cdot\mathbf{a}_2\\ \mathbf{e}_3\cdot\mathbf{a}_1 & \mathbf{e}_3\cdot\mathbf{a}_2\end{pmatrix} = \begin{pmatrix}2 & -1\\ 3 & 1\\ -1 & 1\end{pmatrix}$

(b) $(\mathbf{a})_{(2)}^{\mathrm{T}} \cdot (\mathbf{e})_{(3)} = \begin{pmatrix}\mathbf{a}_1\\ \mathbf{a}_2\end{pmatrix} \cdot \begin{pmatrix}\mathbf{e}_1 & \mathbf{e}_2 & \mathbf{e}_3\end{pmatrix} = \begin{pmatrix}2 & 3 & -1\\ -1 & 1 & 1\end{pmatrix}$

so

$$\left((\mathbf{e})_{(3)}^{\mathrm{T}} \cdot (\mathbf{a})_{(2)}\right)^{\mathrm{T}} = (\mathbf{a})_{(2)}^{\mathrm{T}} \cdot (\mathbf{e})_{(3)}.$$

(c) Show that

$$(\mathbf{e})_{(3)}^{\mathrm{T}} \wedge (\mathbf{a})_{(2)} = \begin{pmatrix}\mathbf{e}_1\\ \mathbf{e}_2\\ \mathbf{e}_3\end{pmatrix} \wedge \begin{pmatrix}\mathbf{a}_1 & \mathbf{a}_2\end{pmatrix} = \begin{pmatrix}\mathbf{e}_1\wedge\mathbf{a}_1 & \mathbf{e}_1\wedge\mathbf{a}_2\\ \mathbf{e}_2\wedge\mathbf{a}_1 & \mathbf{e}_2\wedge\mathbf{a}_2\\ \mathbf{e}_3\wedge\mathbf{a}_1 & \mathbf{e}_3\wedge\mathbf{a}_2\end{pmatrix}$$

$$= \begin{pmatrix}3\mathbf{e}_{12} - \mathbf{e}_{13} & \mathbf{e}_{12} + \mathbf{e}_{13}\\ 2\mathbf{e}_{12} - \mathbf{e}_{23} & -\mathbf{e}_{12} + \mathbf{e}_{23}\\ 2\mathbf{e}_{31} + 3\mathbf{e}_{32} & -\mathbf{e}_{31} + \mathbf{e}_{32}\end{pmatrix}$$

and

$$(\mathbf{a})_{(2)}^{\mathrm{T}} \wedge (\mathbf{e})_{(3)} = \begin{pmatrix}\mathbf{a}_1\\ \mathbf{a}_2\end{pmatrix} \wedge \begin{pmatrix}\mathbf{e}_1 & \mathbf{e}_2 & \mathbf{e}_3\end{pmatrix}$$

$$= \begin{pmatrix}3\mathbf{e}_{21} - \mathbf{e}_{31} & 2\mathbf{e}_{12} - \mathbf{e}_{32} & 2\mathbf{e}_{13} + 3\mathbf{e}_{23}\\ \mathbf{e}_{21} + \mathbf{e}_{31} & -\mathbf{e}_{12} + \mathbf{e}_{32} & -\mathbf{e}_{13} + \mathbf{e}_{23}\end{pmatrix},$$

so

$$\left((\mathbf{e})_{(3)}^{\mathrm{T}} \wedge (\mathbf{a})_{(2)}\right)^{\mathrm{T}} = -(\mathbf{a})_{(2)}^{\mathrm{T}} \cdot (\mathbf{e})_{(3)}.$$

## 4.5 Rules of Matrix Algebra

Let $A, B, C \in M_{\mathbb{R}}(n)$ where $M_{\mathbb{R}}(n)$ denotes the set of all $(n,n)$-matrices over the real numbers $\mathbb{R}$. Addition and multiplication of matrices satisfy the rules:

1. $A+B$ and $AB$ are in $M_{\mathbb{R}}(n)$. (Closure of addition and multiplication.)
2. (a) $(A+B)+C = A+(B+C)$. (Associative law of addition.)
   (b) $(AB)C = A(BC)$. (Associative law of multiplication.)
3. $A+B = B+A$. (Commutative law of addition.)
   Note that in general, the multiplication of matrices is **not** commutative.
4. (a) $\alpha(A+B) = \alpha A + \alpha B$. (Distributive law of scalar multiplication over matrix addition.)

(b) $\alpha A = A\alpha$. (Scalars commute with matrices.)
(c) $\alpha(AB) = A(\alpha B)$.
(d) $A(B+C) = AB + AC$. (Distributive law of matrix multiplication over addition.)

The addition and multiplication of matrices satisfy the same rules as the addition and multiplication of real numbers, with the exception of the commutative law of multiplication and the existence of zero divisors in $M_{\mathbb{R}}(n)$. The matrices $M_{\mathbb{G}}(n)$ over the geometric algebra $\mathbb{G}$ make up a *module* and obey the same rules as the matrix algebra $M_{\mathbb{R}}(n)$, except that the rules 4(b) and 4(c) for scalar multiplication by elements of $\mathbb{G}$ no longer remain universally valid.

## 4.6 The Matrices of $\mathbb{G}_2$ and $\mathbb{G}_3$

Recall (3.10) that the most general geometric number $g \in \mathbb{G}_2$ has the form

$$g = (\alpha_1 + \alpha_2 \mathbf{e}_{12}) + (v_1 \mathbf{e}_1 + v_2 \mathbf{e}_2), \tag{4.6}$$

where $\alpha_i, v_i \in \mathbb{R}$ for $i = 1, 2$, in the *standard basis* $\{1, \mathbf{e}_1, \mathbf{e}_2, \mathbf{e}_{12}\}$ of $\mathbb{G}_2$. The geometric algebra $\mathbb{G}_2$ obeys all the algebraic rules of the real numbers $\mathbb{R}$, except that $\mathbb{G}_2$ has zero divisors and is not universally commutative.

The algebraic rules satified by elements of $\mathbb{G}_2$ are identical to the rules of matrix algebra and are therefore completely compatible with them [71]. Indeed, the elements of the geometric algebra $\mathbb{G}_2$ provide a natural *geometric basis* for matrices. By the *spectral basis* of $\mathbb{G}_2$, we mean

$$\begin{pmatrix} 1 \\ \mathbf{e}_1 \end{pmatrix} \mathrm{u}_+ \begin{pmatrix} 1 & \mathbf{e}_1 \end{pmatrix} = \begin{pmatrix} \mathrm{u}_+ & \mathrm{u}_+\mathbf{e}_1 \\ \mathbf{e}_1\mathrm{u}_+ & \mathbf{e}_1\mathrm{u}_+\mathbf{e}_1 \end{pmatrix} = \begin{pmatrix} \mathrm{u}_+ & \mathbf{e}_1\mathrm{u}_- \\ \mathbf{e}_1\mathrm{u}_+ & \mathrm{u}_- \end{pmatrix}, \tag{4.7}$$

where $\mathrm{u}_\pm = \frac{1}{2}(1 \pm \mathbf{e}_2)$ are mutually annihilating idempotents. The spectral basis of $\mathbb{G}_2$ has the same number of elements as the standard basis (3.9) but different algebraic properties. The *position* of the spectral basis elements in (4.7) determines the matrix which represents that element. For example, if

$$[g] = \begin{pmatrix} a & b \\ c & d \end{pmatrix},$$

for $a, b, c, d \in \mathbb{R}$, then the corresponding element $g \in \mathbb{G}_2$ is specified by

$$g = \begin{pmatrix} 1 & \mathbf{e}_1 \end{pmatrix} \mathrm{u}_+ [g] \begin{pmatrix} 1 \\ \mathbf{e}_1 \end{pmatrix} = a\mathrm{u}_+ + b\mathbf{e}_1\mathrm{u}_- + c\mathbf{e}_1\mathrm{u}_+ + d\mathrm{u}_- \tag{4.8}$$

as is demonstrated below.

Noting that $\mathbf{e}_1 \mathrm{u}_+ = \mathrm{u}_- \mathbf{e}_1$ and $\mathbf{e}_1^2 = 1$, it follows that

$$(1\ \mathbf{e}_1)\,\mathrm{u}_+ \begin{pmatrix} 1 \\ \mathbf{e}_1 \end{pmatrix} = \mathrm{u}_+ + \mathbf{e}_1 \mathrm{u}_+ \mathbf{e}_1 = \mathrm{u}_+ + \mathrm{u}_- = 1.$$

Then for $g \in \mathbb{G}_2$, given by (4.6), we find that

$$\begin{aligned} g &= (1\ \mathbf{e}_1)\,\mathrm{u}_+ \begin{pmatrix} 1 \\ \mathbf{e}_1 \end{pmatrix} g\,(1\ \mathbf{e}_1)\,\mathrm{u}_+ \begin{pmatrix} 1 \\ \mathbf{e}_1 \end{pmatrix} \\ &= (1\ \mathbf{e}_1)\,\mathrm{u}_+ \begin{pmatrix} g & g\mathbf{e}_1 \\ \mathbf{e}_1 g & \mathbf{e}_1 g \mathbf{e}_1 \end{pmatrix} \mathrm{u}_+ \begin{pmatrix} 1 \\ \mathbf{e}_1 \end{pmatrix} \\ &= (1\ \mathbf{e}_1)\,\mathrm{u}_+ \begin{pmatrix} \alpha_1 + v_2 & v_1 - \alpha_2 \\ v_1 + \alpha_2 & \alpha_1 - v_2 \end{pmatrix} \begin{pmatrix} 1 \\ \mathbf{e}_1 \end{pmatrix}. \end{aligned}$$

The last equality in the steps above follows by noting that

$$\begin{aligned} \mathrm{u}_+ g \mathrm{u}_+ &= \mathrm{u}_+ \Big( (\alpha_1 + \alpha_2 \mathbf{e}_{12}) + (v_1 \mathbf{e}_1 + v_2 \mathbf{e}_2) \Big) \mathrm{u}_+ = \mathrm{u}_+ (\alpha_1 + v_2), \\ \mathrm{u}_+ g \mathbf{e}_1 \mathrm{u}_+ &= \mathrm{u}_+ \Big( (\alpha_1 + \alpha_2 \mathbf{e}_{12}) + (v_1 \mathbf{e}_1 + v_2 \mathbf{e}_2) \Big) \mathbf{e}_1 \mathrm{u}_+ = \mathrm{u}_+ (v_1 - \alpha_2), \\ \mathrm{u}_+ \mathbf{e}_1 g \mathrm{u}_+ &= \mathrm{u}_+ \mathbf{e}_1 \Big( (\alpha_1 + \alpha_2 \mathbf{e}_{12}) + (v_1 \mathbf{e}_1 + v_2 \mathbf{e}_2) \Big) \mathrm{u}_+ = \mathrm{u}_+ (v_1 + \alpha_2), \end{aligned}$$

and

$$\mathrm{u}_+ \mathbf{e}_1 g \mathbf{e}_1 \mathrm{u}_+ = \mathrm{u}_+ \mathbf{e}_1 \Big( (\alpha_1 + \alpha_2 \mathbf{e}_{12}) + (v_1 \mathbf{e}_1 + v_2 \mathbf{e}_2) \Big) \mathbf{e}_1 \mathrm{u}_+ = \mathrm{u}_+ (\alpha_1 - v_2).$$

The real matrix

$$[g] := \begin{pmatrix} \alpha_1 + v_2 & v_1 - \alpha_2 \\ v_1 + \alpha_2 & \alpha_1 - v_2 \end{pmatrix} \tag{4.9}$$

is called the *matrix* of $g$ with respect to the spectral basis (4.7).

By the *inner automorphism* or $\mathbf{e}_1$-*conjugate* $g^{\mathbf{e}_1}$ of $g \in \mathbb{G}_2$ with respect to the unit vector $\mathbf{e}_1$, we mean

$$g^{\mathbf{e}_1} := \mathbf{e}_1 g \mathbf{e}_1. \tag{4.10}$$

Consider now the equation

$$g = (1\ \mathbf{e}_1)\,\mathrm{u}_+ [g] \begin{pmatrix} 1 \\ \mathbf{e}_1 \end{pmatrix} \tag{4.11}$$

relating a geometric number $g \in \mathbb{G}_2$ to the corresponding real matrix $[g] \in \mathbb{M}_{\mathbb{R}}(2)$.

We can explicitly solve (4.11) for the matrix $[g]$ of $g$ as follows: Multiplying each side on the left and right by $\begin{pmatrix} 1 \\ \mathbf{e}_1 \end{pmatrix}$ and $(1\ \mathbf{e}_1)$, respectively, gives

$$\begin{pmatrix}1\\ \mathbf{e}_1\end{pmatrix} g \,(1\ \mathbf{e}_1) = \begin{pmatrix}1 & \mathbf{e}_1\\ \mathbf{e}_1 & 1\end{pmatrix} \mathrm{u}_+[g] \begin{pmatrix}1 & \mathbf{e}_1\\ \mathbf{e}_1 & 1\end{pmatrix}.$$

Then, multiplying on the left and right sides of this equation by $\mathrm{u}_+$, we get

$$\mathrm{u}_+ \begin{pmatrix}1\\ \mathbf{e}_1\end{pmatrix} g \,(1\ \mathbf{e}_1)\,\mathrm{u}_+ = \mathrm{u}_+ \begin{pmatrix}1 & \mathbf{e}_1\\ \mathbf{e}_1 & 1\end{pmatrix} \mathrm{u}_+[g] \begin{pmatrix}1 & \mathbf{e}_1\\ \mathbf{e}_1 & 1\end{pmatrix} \mathrm{u}_+ = \mathrm{u}_+[g],$$

and taking the e-conjugate of this equation gives

$$\mathrm{u}_- \begin{pmatrix}1\\ \mathbf{e}_1\end{pmatrix} g^{\mathbf{e}_1} \,(1\ \mathbf{e}_1)\,\mathrm{u}_- = \mathrm{u}_-[g].$$

Adding the last two expressions gives the desired result that

$$[g] = \mathrm{u}_+ \begin{pmatrix} g & g\mathbf{e}_1\\ \mathbf{e}_1 g & \mathbf{e}_1 g \mathbf{e}_1\end{pmatrix} \mathrm{u}_+ + \mathrm{u}_- \begin{pmatrix} \mathbf{e}_1 g \mathbf{e}_1 & \mathbf{e}_1 g\\ g\mathbf{e}_1 & g\end{pmatrix} \mathrm{u}_-. \tag{4.12}$$

We now introduce the *spectral basis* for the geometric numbers $\mathbb{G}_3$ of the 3-dimensional Euclidean space $\mathbb{R}^3$. Recall that the standard basis of $\mathbb{G}_3$, given in (3.11), is

$$\mathbb{G}_3 := \mathrm{span}_{\mathbb{R}}\{1, \mathbf{e}_1, \mathbf{e}_2, \mathbf{e}_3, \mathbf{e}_2\mathbf{e}_3, \mathbf{e}_3\mathbf{e}_1, \mathbf{e}_1\mathbf{e}_2, \mathrm{i} = \mathbf{e}_{123}\}.$$

In the complexified standard basis (3.12), over the center $Z(\mathbb{G}_3)$ of $\mathbb{G}_3$,

$$\mathbb{G}_3 = \mathrm{span}\{1, \mathbf{e}_1, \mathbf{e}_2, \mathbf{e}_3\}_{Z(\mathbb{G}_3)},$$

an element has the form $g = \alpha + x\mathbf{e}_1 + y\mathbf{e}_2 + z\mathbf{e}_3$, where $\alpha, x, y, z \in Z(\mathbb{G}_3)$. Recall that $Z(\mathbb{G}_3) = \mathrm{span}\{1, \mathrm{i}\}$, where $\mathrm{i} = \mathbf{e}_{123}$. Making the substitutions

$$\alpha_1 \to \alpha,\ v_1 \to x,\ v_2 \to y,\ \alpha_2 \to -\mathrm{i}z$$

in (4.9) or (4.12), we find that the matrix representing $g \in \mathbb{G}_3$ is

$$[g] = \begin{pmatrix} \alpha + y & x + \mathrm{i}z\\ x - \mathrm{i}z & \alpha - y\end{pmatrix} = \begin{pmatrix} a & b\\ c & d\end{pmatrix} \tag{4.13}$$

where $a, b, c, d \in Z(\mathbb{G}_3)$. The spectral basis of $\mathbb{G}_3$ is is identical to the spectral basis (4.7) of $\mathbb{G}_2$, except that the spectral basis of $\mathbb{G}_3$ is over the center $Z(\mathbb{G}_3) = \mathrm{span}\{1, \mathrm{i}\}$.

Recall that the geometric algebra $\mathbb{G}_3$ has three involutions which are related to complex conjugation. The *main involution* (3.32) is obtained by changing the sign of all vectors,

$$g^- := \overline{\alpha} - \overline{x}\mathbf{e}_1 - \overline{y}\mathbf{e}_2 - \overline{z}\mathbf{e}_3. \tag{4.14}$$

*Reversion* (3.31) is obtained by reversing the order of the products of vectors,

$$g^{\dagger} := \overline{\alpha} + \overline{x}\mathbf{e}_1 + \overline{y}\mathbf{e}_2 + \overline{z}\mathbf{e}_3, \tag{4.15}$$

and *Clifford conjugation* (3.33) is obtained by combining the above two operations,

$$\widetilde{g} := (g^{-})^{\dagger} = \alpha - x\mathbf{e}_1 - y\mathbf{e}_2 - z\mathbf{e}_3. \tag{4.16}$$

We can now calculate the corresponding operations on the matrix (4.13) of $g$. We find

$$[g]^{-} := [g^{-}] = \begin{pmatrix} \overline{\alpha} - \overline{y} & -\overline{x} - \mathrm{i}\overline{z} \\ -\overline{x} + \mathrm{i}\overline{z} & \overline{\alpha} + \overline{y} \end{pmatrix} = \begin{pmatrix} \overline{d} & -\overline{c} \\ -\overline{b} & \overline{a} \end{pmatrix},$$

$$[g]^{\dagger} := [g^{\dagger}] = \begin{pmatrix} \overline{\alpha} + \overline{y} & \overline{x} + \mathrm{i}\overline{z} \\ \overline{x} - \mathrm{i}\overline{z} & \overline{\alpha} - \overline{y} \end{pmatrix} = \begin{pmatrix} \overline{a} & \overline{c} \\ \overline{b} & \overline{d} \end{pmatrix}, \text{ and}$$

$$\widetilde{[g]} := [\widetilde{g}] = \begin{pmatrix} \alpha - y & -x - \mathrm{i}z \\ -x + \mathrm{i}z & \alpha + y \end{pmatrix} = \begin{pmatrix} d & -b \\ -c & a \end{pmatrix}.$$

Since $g\widetilde{g} = \det[g]$, it follows that $g^{-1} = \frac{\widetilde{g}}{\det[g]}$.

The geometric algebra $\mathbb{G}_3$ is algebraically closed, with $\mathrm{i} = \mathbf{e}_{123} \in Z(\mathbb{G}_3)$, and enjoys the same algebraic rules as the complex number system $\mathbb{C}$, except that $\mathbb{G}_3$ is not universally commutative and has zero divisors.

## Exercises

1. (a) Find the matrix $[\mathbf{e}_k]$ of the unit vectors $\mathbf{e}_k$ for $k = 1,2,3$.
   (b) Show that the matrix $[1]$ that represents the identity element $g = 1$ is the identity matrix $[1] = \begin{pmatrix} 1 & 0 \\ 0 & 1 \end{pmatrix}$.
   (c) Show that $[\mathbf{e}_1][\mathbf{e}_2][\mathbf{e}_3] = \mathrm{i}[1]$.
   (d) More generally, show that $[g_1 + g_2] = [g_1] + [g_2]$ and $[g_1 g_2] = [g_1][g_2]$ for any elements $g_1, g_2 \in \mathbb{G}_3$.
2. Find the matrix $[\mathbf{e}_1 + \mathrm{i}\mathbf{e}_2]$ which represents the *nilpotent* $\mathbf{e}_1 + \mathrm{i}\mathbf{e}_2$.
3. Find the inverse $(5 + 3\mathbf{e}_1)^{-1}$ of the element $5 + 3\mathbf{e}_1$, and show that

$$[(5 + 3\mathbf{e}_1)^{-1}] = [5 + 3\mathbf{e}_1]^{-1}.$$

4. Solve the *characteristic equation* $(x - 5 - 3\mathbf{e}_1)(x - 5 + 3\mathbf{e}_1) = 0$ for the *eigenvalues* $x = x_1, x_2$ of the element $5 + 3\mathbf{e}_1 \in \mathbb{G}_3$. Show that $\text{trace}[5 + 3\mathbf{e}_1] = x_1 + x_2$ and $\det[5 + 3\mathbf{e}_1] = x_1 x_2$.

5. More generally, let $a = \alpha + \mathbf{A} \in \mathbb{G}_3$ where $\alpha \in Z(\mathbb{G}_3)$ and $\mathbf{A}$ is a complex vector. The characteristic equation of the element $a$ is

$$(a-x)(\widetilde{a}-x) = (\alpha - x + \mathbf{A})(\alpha - x - \mathbf{A}) = 0.$$

   (a) Show that the eigenvalues of $a$ (solutions of the characteristic equations) are $x = \alpha \pm \sqrt{\mathbf{A}^2}$.
   (b) Show that $\operatorname{trace}[a] = x_1 + x_2$ and $\det[a] = x_1 x_2$.

6. (a) Calculate the multiplication table for the standard basis of $\mathbb{G}_2$ by calculating the matrix product $\mathscr{B}^{\mathrm{T}}\mathscr{B}$, where

$$\mathscr{B} = \begin{pmatrix} 1 & \mathbf{e}_1 & \mathbf{e}_2 & \mathbf{e}_{12} \end{pmatrix}.$$

   (b) Calculate the multiplication table for the spectral basis of $\mathbb{G}_2$ by calculating the matrix product $\mathscr{S}^{\mathrm{T}}\mathscr{S}$, where

$$\mathscr{S} = \begin{pmatrix} \mathrm{u}_+ & \mathbf{e}_1\mathrm{u}_- & \mathbf{e}_1\mathrm{u}_+ & \mathrm{u}_- \end{pmatrix}.$$

   (c) Show that $A\mathscr{S}^{\mathrm{T}} = \mathscr{B}^{\mathrm{T}}$ and $\mathscr{S}^{\mathrm{T}} = \frac{1}{2}A^{\mathrm{T}}\mathscr{B}^{\mathrm{T}}$, where

$$A = \begin{pmatrix} 1 & 0 & 0 & 1 \\ 0 & 1 & 1 & 0 \\ 1 & 0 & 0 & -1 \\ 0 & 1 & -1 & 0 \end{pmatrix}.$$

7. Show that the $\mathbf{e}_1$ conjugation, defined in (4.10), satisfies the properties

$$(g_1 + g_2)^{\mathbf{e}_1} = g_1^{\mathbf{e}_1} + g_2^{\mathbf{e}_1} \text{ and } (g_1 g_2)^{\mathbf{e}_1} = g_1^{\mathbf{e}_1} g_2^{\mathbf{e}_1}$$

where $g_1, g_2 \in \mathbb{G}_2$. What is the most general element $g \in \mathbb{G}_3$ for which $g^{\mathbf{e}_1} = g$?

# Chapter 5
# Outer Product and Determinants

*... A good notation has a subtlety and suggestiveness which at times make it seem almost like a live teacher.*

—Bertrand Russell

The outer product of vectors is closely related to the concept of a determinant. The outer products of vectors and determinants are very important tools in the study of the structure of a linear transformation. A set of vectors is *linearly independent* if and only if the outer product of those vectors is not zero. Whereas a determinant is scalar valued, the outer product characterizes the *oriented direction* of the subspace spanned by the set of vectors.

## 5.1 The Outer Product

As noted in Chap. 3, the geometric product of two vectors, consisting of the sum of its symmetric and antisymmetric parts, was discovered in 1878 by William Kingdon Clifford. The geometric significance of the antisymmetric outer product was discovered separately and considerably earlier by the German mathematician Hermann Grassmann in his famous *Ausdehnungslehre (1844)* (which is translated as *the Theory of Extension*). If $(\mathbf{v})_{(k)} = (\mathbf{v}_1, \mathbf{v}_2, \ldots, \mathbf{v}_k)$ are any $k$ vectors in the geometric algebra $\mathbb{G}_n$, then

$$\mathbf{v}_{(k)} := \wedge(\mathbf{v})_{(k)} = \mathbf{v}_1 \wedge \mathbf{v}_2 \wedge \cdots \wedge \mathbf{v}_k \tag{5.1}$$

is *antisymmetric* (changes sign) over the interchange of any two of its vectors in (5.1) and has the geometric interpretation of a *simple k-vector* or *k-blade* as given in Chap. 3. If the $k$-vector $\mathbf{v}_{(k)} = 0$, the vectors $\mathbf{v}_1, \ldots, \mathbf{v}_k$ are *linearly dependent*, and if $\mathbf{v}_{(k)} \neq 0$, the vectors are *linearly independent.*

G. Sobczyk, *New Foundations in Mathematics: The Geometric Concept of Number*, DOI 10.1007/978-0-8176-8385-6_5,

The outer product of $k$ vectors is the completely antisymmetric part of the geometric products of those vectors (3.43). We now explore the properties of the outer product of vectors and its relationship to the determinant in more detail. The most simple case of (5.1) is when $k = 2$,

$$\mathbf{v}_{(2)} = \mathbf{v}_1 \wedge \mathbf{v}_2 = -\mathbf{v}_2 \wedge \mathbf{v}_1,$$

which has the geometric interpretation of a simple *bivector*, or 2-*blade*, as we have seen in Fig. 3.1 of Chap. 3. Bivectors in the geometric algebra $\mathbb{G}_n$, just like vectors, are added and multiplied by scalars and make up the subspace $\mathbb{G}_n^2 \subset \mathbb{G}_n$.

The *standard basis* $\mathbb{G}_n^2$ of the bivectors of the geometric algebra $\mathbb{G}_n$ is given by

$$\mathbb{G}_n^2 = \text{span}\{\mathbf{e}_{(2)}^{(n)}\} = \{\mathbf{e}_{ij} | \text{ for } 1 \le i < j \le n\},$$

where

$$\mathbf{e}_{ij} = \mathbf{e}_i \wedge \mathbf{e}_j.$$

It follows that $\dim(\mathbb{G}_n^2) = \binom{n}{2}$, as we have seen earlier in (3.30) of Chap. 3.

Letting

$$\mathbf{a} = (\mathbf{e})_{(n)}(a)^{(n)} = \sum_{i=1}^{n} a^i \mathbf{e}_i,$$

$$\mathbf{b} = (\mathbf{e})_{(n)}(b)^{(n)} = \sum_{i=1}^{n} b^i \mathbf{e}_i,$$

we find, by using the fact that the outer product is antisymmetric, that

$$\mathbf{a} \wedge \mathbf{b} = (\sum_{i=1}^{n} a^i \mathbf{e}_i) \wedge (\sum_{j=1}^{n} b^j \mathbf{e}_j) = \mathbf{e}_{(2)}^{(n)} \det \left(a^{(2)}\ b^{(2)}\right)_{(n)},$$

where

$$\mathbf{e}_{(2)}^{(n)} \det \left(a^{(2)}\ b^{(2)}\right)_{(n)} \equiv \sum_{i<j} \mathbf{e}_{ij} \det \begin{pmatrix} a^i & b^i \\ a^j & b^j \end{pmatrix}.$$

Note that

$$\det \begin{pmatrix} a^i & b^i \\ a^j & b^j \end{pmatrix} = a^i b^j - a^j b^i$$

is the usual determinant function.

It is worthwhile to go over these calculations in detail for the simple cases of bivectors of $\mathbb{G}_n^2$ when $n = 2$ and $n = 3$. This will give insight into the notation being used and the exact relationship between a bivector and a determinant of order 2.

For $n = 2$, let $\mathbf{a} = (\mathbf{e})_{(2)}(a)^{(2)} = a^1\mathbf{e}_1 + a^2\mathbf{e}_2$ and $\mathbf{b} = (\mathbf{e})_{(2)}(b)^{(2)} = b^1\mathbf{e}_1 + b^2\mathbf{e}_2$. Then

$$\mathbf{a} \wedge \mathbf{b} = \mathbf{e}^{(2)}_{(2)} \det\left(a^{(2)}\ b^{(2)}\right)_{(2)} = \mathbf{e}_{12} \det\begin{pmatrix} a^1 & b^1 \\ a^2 & b^2 \end{pmatrix} = (a^1b^2 - a^2b^1)\mathbf{e}_{12} \tag{5.2}$$

We see that $\det\begin{pmatrix} a^1 & b^1 \\ a^2 & b^2 \end{pmatrix}$ relates the bivector $\mathbf{a} \wedge \mathbf{b}$ to the size and orientation of the bivector $\mathbf{e}_{12}$.

Similarly, for $n = 3$, let $\mathbf{a} = (\mathbf{e})_{(3)}(a)^{(3)}$ and $\mathbf{b} = (\mathbf{e})_{(3)}(b)^{(3)}$. Then

$$\begin{aligned} \mathbf{a} \wedge \mathbf{b} &= \mathbf{e}^{(3)}_{(2)} \det\left(a^{(2)}\ b^{(2)}\right)_{(3)} \\ &= \mathbf{e}_{12} \det\begin{pmatrix} a^1 & b^1 \\ a^2 & b^2 \end{pmatrix} + \mathbf{e}_{13} \det\begin{pmatrix} a^1 & b^1 \\ a^3 & b^3 \end{pmatrix} + \mathbf{e}_{23} \det\begin{pmatrix} a^2 & b^2 \\ a^3 & b^3 \end{pmatrix} \\ &= (a^1b^2 - a^2b^1)\mathbf{e}_{12} + (a^1b^3 - a^3b^1)\mathbf{e}_{13} + (a^2b^3 - a^2b^3)\mathbf{e}_{23} \end{aligned} \tag{5.3}$$

We see that the determinants $\det\begin{pmatrix} a^i & b^i \\ a^j & b^j \end{pmatrix}$ relate the bivector $\mathbf{a} \wedge \mathbf{b}$ to the size and orientation of the bivectors $\mathbf{e}_{12}$, $\mathbf{e}_{13}$, and $\mathbf{e}_{23}$.

Comparing the bivectors in the standard basis for $n = 2$ and $n = 3$, we see that $\mathbf{a} \wedge \mathbf{b}$ in (5.2) has only one term, whereas in (5.3), there are three distinct components for a general bivector in a three-dimensional space. However, if we consider a *trivector* (or 3-vector) in three-dimensional space, we once again get a simple relationship. Specifically, for $n = 3$ and the vectors

$$\mathbf{a} = (\mathbf{e})_{(3)}(a)^{(3)},\ \mathbf{b} = (\mathbf{e})_{(3)}(b)^{(3)},\ \mathbf{c} = (\mathbf{e})_{(3)}(c)^{(3)},$$

in terms of the basis vectors $(\mathbf{e})_{(3)} = (\mathbf{e}_1, \mathbf{e}_2, \mathbf{e}_3)$, we calculate

$$\mathbf{a} \wedge \mathbf{b} \wedge \mathbf{c} = \mathbf{e}_{123} \det\left((a)^{(3)}\ (b)^{(3)}\ (c)^{(3)}\right)_{(3)},$$

where

$$\det\left((a)^{(3)}\ (b)^{(3)}\ (c)^{(3)}\right)_{(3)} = \det\begin{pmatrix} a^1 & b^1 & c^1 \\ a^2 & b^2 & c^2 \\ a^3 & b^3 & c^3 \end{pmatrix}$$

is the usual 3-dimensional determinant function.

We have introduced the concept of the standard basis $(\mathbf{e})_{(n)}$ of $\mathbb{R}^n$. More generally we make the following:

**Definition 5.1.1.** A row $(\mathbf{a})_{(n)} = \left(\mathbf{a}_1\ \ldots\ \mathbf{a}_n\right)$ of vectors of $\mathbb{R}^n$ make up a basis of $\mathbb{R}^n$ iff

$$\mathbf{a}_{(n)} = \wedge(\mathbf{a})_{(n)} = \mathbf{a}_1 \wedge \cdots \wedge \mathbf{a}_n \neq 0.$$

Suppose that we are given two sets of $n$ vectors of $\mathbb{R}^n$

$$(\mathbf{a})_{(n)} = \begin{pmatrix}\mathbf{a}_1 & \mathbf{a}_2 & \dots & \mathbf{a}_n\end{pmatrix}, \quad (\mathbf{b})_{(n)} = \begin{pmatrix}\mathbf{b}_1 & \mathbf{b}_2 & \dots & \mathbf{b}_n\end{pmatrix},$$

related by a *matrix of transition* $C = (c_{ij})$, for $c_{ij} \in \mathbb{R}$, where $(\mathbf{a})_{(n)} = (\mathbf{b})_{(n)}C$, then

$$\mathbf{a}_{(n)} = \wedge(\mathbf{a})_{(n)} = \wedge\big((\mathbf{b})_{(n)}C\big) = \mathbf{b}_{(n)} \det C. \tag{5.4}$$

This fundamental relationship follows immediately from the antisymmetric properties of the outer product and the definition of the determinant function. In view of the above definition, $(\mathbf{a})_{(n)}$ is a basis of $\mathbb{R}^n$ iff $(\mathbf{b})_{(n)}$ is a basis of $\mathbb{R}^n$ and $\det C \neq 0$. Further relationships between the outer product and the determinant function are explored in the exercises.

The elements of the geometric algebra $\mathbb{G}_n = \mathbb{G}(\mathbb{R}^n)$ of the Euclidean space $\mathbb{R}^n$ have a natural *grading* into the sum of Grassmann subspaces $\mathbb{G}_n^k$ of $k$-vectors over the real numbers $\mathbb{R}$. We identify $\mathbb{G}_n^1 = \mathbb{R}^n$ and write

$$\mathbb{G}_n = \mathbb{R} + \mathbb{G}_n^1 + \cdots + \mathbb{G}_n^k + \cdots + \mathbb{G}_n^n, \tag{5.5}$$

where each $\mathbb{G}_n^k$ represents the linear space of all $k$-vectors generated by taking all outer products of $k$ vectors in $\mathbb{R}^n$. A basis for the space of $k$-vectors $\mathbb{G}_n^k$ is constructed by taking all distinct outer products of $k$ basis vectors from $(\mathbf{e})_{(n)}$ of $\mathbb{R}^n$ to get

$$\mathbb{G}_n^k = \text{span}\{\mathbf{e}_{\lambda_k} | \ \lambda_k = i_1 i_2 \cdots i_k \ \text{ where } \ 1 \le i_1 < i_2 < \cdots < i_k \le n\}. \tag{5.6}$$

A general multivector $M \in \mathbb{G}_n$ has the form

$$M = <M>_0 + <M>_1 + <M>_2 + \cdots + <M>_n \tag{5.7}$$

where $<M>_k$ denotes the $k$-vector part of $M$. We denote the Grassmann algebra of the $n$-dimensional vector space $\mathbb{R}^n$ with the same symbol as used for the geometric algebra $\mathbb{G}_n$, although in the Grassmann algebra only the antisymmetric outer product is used.

## Exercises

1. In $\mathbb{R}^3$, let $\mathbf{a} = (\mathbf{e})_{(3)}(a)^{(3)} = \mathbf{e}_1 + 2\mathbf{e}_2 - \mathbf{e}_3$ and $\mathbf{b} = (\mathbf{e})_{(3)}(b)^{(3)} = 2\mathbf{e}_1 - 2\mathbf{e}_2 + 3\mathbf{e}_3$.

   (a) Calculate $\mathbf{a} \wedge \mathbf{b}$, and draw a picture of this bivector in the $xy$-plane. Also draw a picture of the bivector $\mathbf{b} \wedge \mathbf{a}$.

   (b) Calculate the trivectors $\mathbf{e}_1 \wedge \mathbf{a} \wedge \mathbf{b}$, $\mathbf{e}_2 \wedge \mathbf{a} \wedge \mathbf{b}$, and $\mathbf{e}_3 \wedge \mathbf{a} \wedge \mathbf{b}$. What is the relationship between these trivectors and the bivector $\mathbf{a} \wedge \mathbf{b}$?

(c) Calculate the bivectors $\mathbf{e}_1 \wedge \mathbf{a}$, $\mathbf{e}_2 \wedge \mathbf{a}$, and $\mathbf{e}_3 \wedge \mathbf{a}$. What is the relationship between these bivectors and the vector $\mathbf{a}$?

(d) What is the relationship between the bivectors $\mathbf{a} \wedge \mathbf{b}$ and $\mathbf{a} \wedge (2\mathbf{a} + \mathbf{b})$?

2. Let $\mathbf{a} = (\mathbf{e})_{(4)}(a)^{(4)}$ and $\mathbf{b} = (\mathbf{e})_{(4)}(b)^{(4)}$ be vectors in $\mathbb{R}^4$ expressed in the basis $(\mathbf{e})_{(4)} = (\mathbf{e}_1, \mathbf{e}_2, \mathbf{e}_3, \mathbf{e}_4)$.

   (a) Find the formula for the bivector $\mathbf{a} \wedge \mathbf{b}$ in terms of the 6 components in the standard basis $\{\mathbf{e}^{(4)}_{(2)}\}$ of bivectors in $\mathbb{G}_4^2$ of $\mathbb{G}_4$.

   (b) Find the 4-blade $\mathbf{e}_1 \wedge \mathbf{e}_2 \wedge \mathbf{a} \wedge \mathbf{b}$. What is the relationship between this 4-blade and the bivector $\mathbf{a} \wedge \mathbf{b}$?

3. Let $\mathbf{a}, \mathbf{b}, \mathbf{c}$ be 3 vectors expressed in the standard basis $(\mathbf{e})_{(4)} = (\mathbf{e}_1, \mathbf{e}_2, \mathbf{e}_3, \mathbf{e}_4)$ of $\mathbb{R}^4$.

   (a) Find the explicit formula for the 4 components of the trivector $\mathbf{a} \wedge \mathbf{b} \wedge \mathbf{c}$ in terms of the standard basis $\{\mathbf{e}^{(4)}_{(3)}\}$ of the space of trivectors $\mathbb{G}_4^3$ in $\mathbb{G}_4$.

   (b) Find the 4-blade $\mathbf{e}_1 \wedge \mathbf{a} \wedge \mathbf{b} \wedge \mathbf{c}$. What is the relationship between this 4-blade and the trivector $\mathbf{a} \wedge \mathbf{b} \wedge \mathbf{c}$?

4. Let $\mathbf{a}, \mathbf{b}, \mathbf{c}, \mathbf{d}$ be 4 vectors expressed in the basis $(\mathbf{e})_{(4)} = (\mathbf{e}_1, \mathbf{e}_2, \mathbf{e}_3, \mathbf{e}_4)$ of $\mathbb{R}^4$.

   (a) Find the explicit formula for the the 4-blade $\mathbf{a} \wedge \mathbf{b} \wedge \mathbf{c} \wedge \mathbf{d}$ in terms of the standard basis $\{\mathbf{e}_{(4)}\} = \{\mathbf{e}_{1234}\}$ of the space of 4 vectors $\mathbb{G}_4^4$ of $\mathbb{G}_4$.

   (b) What is the relationship between the 4-blade $\mathbf{e}_{1234}$ and the 4-blade $\mathbf{a} \wedge \mathbf{b} \wedge \mathbf{c} \wedge \mathbf{d}$?

   (c) Express the trivector $\mathbf{a} \wedge \mathbf{b} \wedge \mathbf{c}$ in terms of the basis $\{\mathbf{e}^{(4)}_{(3)}\}$ of trivectors of the vector space $\mathbb{R}^4$. Show that there are exactly $\binom{4}{3} = 4$ standard basis elements of the trivector space $\mathbb{G}_4^3$ of $\mathbb{G}_4$.

5. Let $(\mathbf{e})_{(n)} = \begin{pmatrix}\mathbf{e}_1 & \mathbf{e}_2 & \dots & \mathbf{e}_n\end{pmatrix}$ be the standard basis of $\mathbb{R}^n$, and let $\mathbf{v}_1, \mathbf{v}_2, \dots, \mathbf{v}_n \in \mathbb{R}^n$.

   (a) Show that

$$\mathbf{v}_{(n)} = \mathbf{v}_1 \wedge \mathbf{v}_2 \wedge \cdots \wedge \mathbf{v}_n = \mathbf{e}_{12\cdots n} \det \begin{pmatrix}\mathbf{v}_1 & \mathbf{v}_2 & \dots & \mathbf{v}_n\end{pmatrix},$$

   where $\mathbf{v}_i$ are the column vectors $\mathbf{v}_i = \begin{pmatrix} v_{1i} \\ v_{2i} \\ \cdot \\ \cdot \\ v_{ni} \end{pmatrix}$.

   (b) Show that the vectors $(\mathbf{v})_{(n)}$ are linearly independent if and only if

$$\sum_{i=1}^{n} v^i \mathbf{v}_i = 0 \quad \Longleftrightarrow \quad v^i = 0 \text{ for } i = 1, \dots, n.$$

(c) Show that the vectors $\mathbf{v}_1,\ldots,\mathbf{v}_n$ are linearly dependent iff

$$\wedge(\mathbf{v})_{(n)} = 0 \quad \Longleftrightarrow \quad \sum_{i=1}^{n} v^i \mathbf{v}_i = 0$$

and not all $v_i = 0$ for $i = 1,\ldots,n$.

6. In $\mathbb{R}^3$, suppose that

$$(\mathbf{a})_{(3)} = (\mathbf{a}_1,\mathbf{a}_2,\mathbf{a}_3) = (\mathbf{e})_{(3)} \begin{pmatrix} 1 & 2 & 3 \\ -1 & 0 & 1 \\ 2 & 3 & 0 \end{pmatrix}.$$

Show that

$$\mathbf{a}_{(3)} = \mathbf{a}_1 \wedge \mathbf{a}_2 \wedge \mathbf{a}_3 = \mathbf{e}_{123} \det \begin{pmatrix} 1 & 2 & 3 \\ -1 & 0 & 1 \\ 2 & 3 & 0 \end{pmatrix}.$$

Conclude that the vectors $\mathbf{a}_1,\mathbf{a}_2,\mathbf{a}_3$ form a basis of $V^3$ if and only if

$$\det \mathbf{a}_{(3)} \neq 0.$$

7. In $\mathbb{R}^n$, suppose that

$$(\mathbf{a})_{(n)} = (\mathbf{a}_1,\mathbf{a}_2,\ldots,\mathbf{a}_n) = (\mathbf{e})_{(n)}A, \text{ and } \mathbf{a}_{(n)} = \wedge(\mathbf{a})_{(n)},$$
$$(\mathbf{b})_{(n)} = (\mathbf{b}_1,\mathbf{b}_2,\ldots,\mathbf{b}_n) = (\mathbf{e})_{(n)}B, \text{ and } \mathbf{b}_{(n)} = \wedge(\mathbf{b})_{(n)},$$
$$(\mathbf{c})_{(n)} = (\mathbf{c}_1,\mathbf{c}_2,\ldots,\mathbf{c}_n) = (\mathbf{e})_{(n)}C, \text{ and } \mathbf{c}_{(n)} = \wedge(\mathbf{c})_{(n)},$$

and that the matrices $A,B,C$ satisfy $C = AB$. Show, by using (5.4), that

$$\mathbf{c}_{(n)} = \mathbf{a}_{(n)} \det B = \mathbf{e}_{(n)} \det A \det B,$$

and conclude that $\det(C) = \det(A)\det(B)$.

8. Show that the Grassmann algebra $\mathbb{G}_n$, considered as a linear (vector) space, has $2^n$ dimensions just like the geometric algebra $\mathbb{G}(\mathbb{R}^n)$.
9. Let $(\mathbf{a})_{(n)} = (\mathbf{e})_{(n)}A$, where $A$ is the $n \times n$ matrix

$$A = \begin{pmatrix} a_{11} & a_{12} & \ldots & a_{1n} \\ a_{21} & a_{22} & \ldots & a_{2n} \\ \cdot & \cdot & \cdot & \cdot \\ \cdot & \cdot & \cdot & \cdot \\ a_{n1} & a_{12} & \ldots & a_{nn} \end{pmatrix}$$

which defines the $n$ vectors $\mathbf{a}_i$ in $\mathbb{R}^n$. Show that

$$\mathbf{a}_1 \wedge (\mathbf{a}_2 \wedge \mathbf{a}_3 \wedge \cdots \wedge \mathbf{a}_n) = \sum_{i=1}^{n} a_{i1} \mathbf{e}_i \wedge \mathbf{a}_2 \wedge \mathbf{a}_3 \wedge \cdots \wedge \mathbf{a}_n$$

gives the formula for the expansion of a determinant by cofactors of elements in the first column.

10. Let $(\mathbf{a})_{(n)} = (\mathbf{e})_{(n)}A$ as in Problem 9. We know that

$$\mathbf{a}_{(n)} \equiv \mathbf{a}_1 \wedge \cdots \wedge \mathbf{a}_n = \mathbf{e}_{12\ldots n} \det A.$$

Now define

$$(\mathbf{a})_{(n)_{ij}} = \{\mathbf{a}_1, \ldots, \mathbf{a}_{j-1}, \mathbf{e}_i, \mathbf{a}_{j+1}, \ldots, \mathbf{a}_n\} = (\mathbf{e})_{(n)}A_{(ij)}.$$

The matrix $A_{(ij)}$ is called the *(ij)-minor* of the matrix $A$ and is formed by replacing the $j$th column of the matrix $A$ with the column vector $\mathbf{e}_i$. Taking the determinant of this matrix gives

$$A_{ij} = \det A_{(ij)},$$

called the *cofactor* of the element $a_{ij}$ of the matrix $A$.

(a) For $A = \begin{pmatrix} 1 & 2 & 3 \\ 0 & -1 & 1 \\ 1 & 1 & 1 \end{pmatrix}$, explicitly calculate $A_{(12)}$ and $A_{(23)}$ and their corresponding determinants $A_{12}$ and $A_{23}$, respectively. The *adjoint matrix* of the matrix $A$ is defined in terms of its cofactors $A_{ij}$ by

$$\operatorname{adj}(A) = \begin{pmatrix} A_{11} & A_{12} & \ldots & A_{1n} \\ A_{21} & A_{22} & \ldots & A_{2n} \\ \cdot & \cdot & \cdot & \cdot \\ \cdot & \cdot & \cdot & \cdot \\ A_{n1} & A_{12} & \ldots & A_{nn} \end{pmatrix}^{\mathrm{T}}.$$

(b) Show that $\sum_{i=1}^{n} a_{ij}A_{ij} = \det A$ for $j = 1, 2, \ldots, n$.
(c) Show that $\sum_{i=1}^{n} a_{ij}A_{ik} = 0$ for $j \neq k$.
(d) Use parts (b) and (c) above to show that

$$A \operatorname{adj}(A) = \det(A)I = \operatorname{adj}(A)A.$$

Using this result, show that

$$A^{-1} = \frac{\operatorname{adj}(A)}{\det A}.$$

It follows that a square matrix $A$ has an inverse $A^{-1}$ if and only if $\det A \neq 0$.
(e) Use the formula given in part (d) to find the inverse of the matrix given in part (a).

## 5.2 Applications to Matrices

Suppose that $AB = C$ for the three $(n,n)$-matrices $A, B, C$. Multiplying this matrix equation on the left by the standard basis $(\mathbf{e})_{(n)}$ of $\mathbb{R}^n$ gives the vector equation

$$(\mathbf{e})_{(n)}AB = (\mathbf{a})_{(n)}B = (\mathbf{e})_{(n)}C = (\mathbf{c})_{(n)},$$

which means that the vectors $\mathbf{c}_i$ are linear combinations of the vectors $(\mathbf{a})_{(n)}$. Taking the outer products of both sides of this last equation, and using (5.4), gives

$$\mathbf{c}_{(n)} = \mathbf{a}_{(n)} \det B \tag{5.8}$$

and defines the *determinant function* of the matrix $B$ relating the $n$-vectors $\mathbf{c}_{(n)}$ and $\mathbf{a}_{(n)}$. But $(\mathbf{c})_{(n)} = (\mathbf{e})_{(n)}C$ and $(\mathbf{a})_{(n)} = (\mathbf{e})_{(n)}A$ imply that $\mathbf{c}_{(n)} = \mathbf{e}_{(n)} \det C$ and $\mathbf{a}_{(n)} = \mathbf{e}_{(n)} \det A$. Using these relationships in (5.8) gives the result

$$\mathbf{e}_{(n)} \det C = \mathbf{e}_{(n)} \det A \det B,$$

or $\det(AB) = \det(A)\det(B)$.

We have noted that a system of linear equations can be written in the matrix form

$$A \begin{pmatrix} x^1 \\ x^2 \\ \cdot \\ \cdot \\ x^n \end{pmatrix} = \begin{pmatrix} b^1 \\ b^2 \\ \cdot \\ \cdot \\ b^n \end{pmatrix},$$

where $A = (a_{ij})$ is an $(n,n)$-matrix. Multiplying both sides of this equation on the left by $(\mathbf{e})_{(n)}$ gives the equivalent *vector equation*

$$(\mathbf{a})_{(n)}(x)^{(n)} = \mathbf{b}$$

of this system of equations.

We can easily solve this system of equations for $x^i$, for $1 \le i \le n$, by taking the outer product on the left and right sides of the last equation, with $\mathbf{a}_1 \wedge \cdots \wedge \mathbf{a}_{i-1}$ and $\mathbf{a}_{i+1} \wedge \cdots \wedge \mathbf{a}_n$, respectively, to get

$$x^i \det(\mathbf{a})_{(n)}\mathbf{e}_{12\cdots n} = \det(\mathbf{a}_1 \wedge \cdots \wedge \mathbf{a}_{i-1} \wedge \mathbf{b} \wedge \mathbf{a}_{i+1} \wedge \cdots \wedge \mathbf{a}_n)\mathbf{e}_{12\cdots n}$$

or

$$x^i = \frac{\det(\mathbf{a}_1 \wedge \cdots \wedge \mathbf{a}_{i-1} \wedge \mathbf{b} \wedge \mathbf{a}_{i+1} \wedge \cdots \wedge \mathbf{a}_n)}{\det(\mathbf{a})_{(n)}}. \tag{5.9}$$

The formula (5.9) is known as *Cramer's rule* and is well defined if $\det(\mathbf{a})_{(n)} \neq 0$.

## Exercises

1. Let $A = \begin{pmatrix} 1 & 2 \\ 3 & 4 \end{pmatrix}$, $B = \begin{pmatrix} 1 & -1 \\ 1 & 1 \end{pmatrix}$. Calculate $(\mathbf{a})_{(2)} = (\mathbf{e})_{(2)}A$ and $(\mathbf{c})_{(2)} = (\mathbf{e})_{(2)}(AB)$. Verify that $\mathbf{a}_1 \wedge \mathbf{a}_2 \det B = \mathbf{c}_1 \wedge \mathbf{c}_2$ and $\det(A)\det(B) = \det C$.
2. (a) Write the system of equation $\begin{pmatrix} a_1x + b_1y = c_1 \\ a_2x + b_2y = c_2 \end{pmatrix}$ in matrix form.
   (b) Write the system of equations in equivalent vector form and solve by using Cramer's rule.
3. (a) Write the system of equation $\begin{pmatrix} x_1 + 2x_2 - x_3 = 5 \\ x_1 + x_2 + x_3 = 1 \\ 2x_1 - 2x_2 + x_3 = 4 \end{pmatrix}$ in matrix form.
   (b) Write the system of equations in equivalent vector form and solve by using Cramer's rule.

# Chapter 6
# Systems of Linear Equations

*Even stranger things have happened; and perhaps the strangest of all is the marvel that mathematics should be possible to a race akin to the apes.*

—Eric T. Bell

We give a rather conventional treatment of the study of a system of linear equations by using the *augmented matrix* of the system. The so-called LU decomposition of a matrix is introduced, together with the concept of elementary row and column operations and their corresponding matrices. All vectors are to be considered *row vectors* or *column vectors* of $\mathbb{R}^n$. Geometric algebra is not used in this chapter.

## 6.1 Elementary Operations and Matrices

The theory of matrices arose primarily from techniques for solving systems of linear equations. A system of $m$ linear equations in $n$ unknowns has the form

$$\begin{pmatrix} a_{11}x_1 + a_{12}x_2 + \cdots + a_{1n}x_n = b_1 \\ a_{21}x_1 + a_{22}x_2 + \cdots + a_{2n}x_n = b_2 \\ \cdot \\ \cdot \\ \cdot \\ a_{m1}x_1 + a_{m2}x_2 + \cdots + a_{mn}x_n = b_m \end{pmatrix} \tag{6.1}$$

We have already noted that this system of linear equations can be expressed by the matrix equation $AX = B$ where



G. Sobczyk, *New Foundations in Mathematics: The Geometric Concept of Number*, DOI 10.1007/978-0-8176-8385-6_6,

$$A = \begin{pmatrix} a_{11} & a_{12} & \dots & a_{1n} \\ a_{21} & a_{22} & \dots & a_{2n} \\ \cdot & \cdot & \dots & \cdot \\ \cdot & \cdot & \dots & \cdot \\ \cdot & \cdot & \dots & \cdot \\ a_{m1} & a_{m2} & \dots & a_{mn} \end{pmatrix}, \quad X = \begin{pmatrix} x_1 \\ x_2 \\ \cdot \\ \cdot \\ \cdot \\ x_n \end{pmatrix}, \quad B = \begin{pmatrix} b_1 \\ b_2 \\ \cdot \\ \cdot \\ \cdot \\ b_m \end{pmatrix}.$$

The *augmented matrix* $M$ of the coefficients of the equations is defined by

$$M = \begin{pmatrix} a_{11} & a_{12} & \dots & a_{1n} & b_1 \\ a_{21} & a_{22} & \dots & a_{2n} & b_2 \\ \cdot & \cdot & \dots & \cdot & \cdot \\ \cdot & \cdot & \dots & \cdot & \cdot \\ \cdot & \cdot & \dots & \cdot & \cdot \\ a_{m1} & a_{m2} & \dots & a_{mn} & b_m \end{pmatrix}. \tag{6.2}$$

The following *elementary (row) operations* can be performed on the system of linear equations (6.1) or on the corresponding augmented matrix (6.2), giving an equivalent system of linear equations with exactly the same solutions as (6.1):

1. Multiply any equation (row) by a nonzero scalar.
2. Multiply the $i$th equation (row) by a nonzero scalar and add it to the $j$th equation (row).
3. Interchange the $i$th equation (row) with the $j$th equation (row).

For example, given the system of linear equations

$$\begin{pmatrix} x_1 + 2x_2 + 3x_3 = 3 \\ 4x_1 + 5x_2 + 6x_3 = 6 \end{pmatrix}, \tag{6.3}$$

the augmented matrix is

$$M = \begin{pmatrix} 1 & 2 & 3 & 3 \\ 4 & 5 & 6 & 6 \end{pmatrix}.$$

If we multiply the first equation by 2, we get the equivalent set of equations

$$\begin{pmatrix} 2x_1 + 4x_2 + 6x_3 = 6 \\ 4x_1 + 5x_2 + 6x_3 = 6 \end{pmatrix},$$

and the corresponding augmented matrix of the system becomes

$$\begin{pmatrix} 2 & 4 & 6 & 6 \\ 4 & 5 & 6 & 6 \end{pmatrix}.$$

This is an example of the first elementary (row) operation. On the other hand, if we multiply the 1st equation by $-4$ and add it to the 2nd equation of (6.3), we get

$$\begin{pmatrix} x_1 + 2x_2 + 3x_3 = 3 \\ -3x_2 - 6x_3 = -6 \end{pmatrix}.$$

The corresponding row operation on the matrix $M$ is

$$M = \begin{pmatrix} 1 & 2 & 3 & 3 \\ 0 & -3 & -6 & -6 \end{pmatrix}.$$

The third elementary operation of interchanging the first and second equations (rows) is easily written down and is omitted. We see that elementary operations on a system of linear equations correspond exactly to the same row operation on its augmented matrix. This makes it more economical to work with the augmented matrix of the system. We now observe that the same elementary row operations on the augmented matrix $M$ can be accomplished by multiplying the matrix $M$ on the left by the corresponding *elementary row matrix*. The elementary row matrices are constructed from the standard $m$ row vectors

$$\mathbf{e}^i = \begin{pmatrix} \delta_1^i & \delta_2^i & \cdots & \delta_m^i \end{pmatrix}$$

where $\delta_i^j = 0$ for $i \neq j$ and $\delta_i^j = 1$ when $i = j$.

For the matrix

$$A = (a)_{(m,n)} = (a_{ij}),$$

where $1 \leq i \leq m$ and $1 \leq j \leq n$, we define the following three kinds of elementary matrices:

1. The elementary square $(m,m)$-matrix

$$E^i(\alpha) = \begin{pmatrix} \mathbf{e}^1 \\ \mathbf{e}^2 \\ \cdot \\ \alpha\mathbf{e}^i \\ \cdot \\ \mathbf{e}^m \end{pmatrix},$$

in the product $E^i(\alpha)A$ multiplies the $i$th row of the matrix $A$ by the nonzero constant $\alpha \in \mathbb{R}$.

2. The elementary square matrix

$$E^{ij}(\alpha) = \begin{pmatrix} \mathbf{e}^1 \\ \mathbf{e}^2 \\ \cdot \\ \alpha\mathbf{e}^i + \mathbf{e}^j \\ \cdot \\ \mathbf{e}^m \end{pmatrix}$$

in the product $E^{ij}(\alpha)A$ multiplies the $i$th row of the matrix $A$ by $\alpha$ and adds it to the $j$th row of $A$.

3. The elementary square matrix

$$E^{ij} = \begin{pmatrix} \mathbf{e}^1 \\ \cdot \\ \mathbf{e}^j \\ \cdot \\ \mathbf{e}^i \\ \cdot \\ \mathbf{e}^m \end{pmatrix}$$

in the product $E^{ij}A$ interchanges the $i$th and $j$th rows of the matrix $A$, leaving the other rows unaffected. This 3rd type of elementary matrix is a *permutation matrix* and, indeed, is just the permutation of the $i$th and $j$th rows of the identity matrix.

Notice that the $(m,m)$ identity matrix $[1]_m$ can be expressed in terms of the row vectors $\mathbf{e}^i$,

$$[1]_m = \begin{pmatrix} 1 & 0 & 0 \dots 0 \\ 0 & 1 & 0 \dots 0 \\ \dots & & \dots \\ \dots & & \dots \\ 0 & \dots & 0 \quad 1 \end{pmatrix} = \begin{pmatrix} \mathbf{e}^1 \\ \mathbf{e}^2 \\ \cdot \\ \mathbf{e}^m \end{pmatrix}$$

The elementary (row) matrices are formed by simply performing the desired row operation on the identity matrix $[1]_m$.

## Exercises

1. Let $A = \begin{pmatrix} 1 & 2 & 3 & 4 \\ 0 & -1 & 1 & 0 \\ 1 & 3 & 0 & -1 \end{pmatrix}$

(a) Show that

$$E^2(3)A = \begin{pmatrix} 1 & 2 & 3 & 4 \\ 0 & -3 & 3 & 0 \\ 1 & 3 & 0 & -1 \end{pmatrix}$$

where the elementary matrix

$$E^2(3) = \begin{pmatrix} \mathbf{e}^1 \\ 3\mathbf{e}^2 \\ \mathbf{e}^3 \end{pmatrix} = \begin{pmatrix} 1 & 0 & 0 \\ 0 & 3 & 0 \\ 0 & 0 & 1 \end{pmatrix}$$

(b) Show that

$$E^{13}(-2)A = \begin{pmatrix} 1 & 2 & 3 & 4 \\ 0 & -1 & 1 & 0 \\ -1 & -1 & -6 & -9 \end{pmatrix}$$

where the elementary matrix

$$E^{13}(-2) = \begin{pmatrix} \mathbf{e}^1 \\ \mathbf{e}^2 \\ -2\mathbf{e}^1 + \mathbf{e}^3 \end{pmatrix} = \begin{pmatrix} 1 & 0 & 0 \\ 0 & 1 & 0 \\ -2 & 0 & 1 \end{pmatrix}$$

(c) Show that

$$E^{12}A = \begin{pmatrix} 0 & -1 & 1 & 0 \\ 1 & 2 & 3 & 4 \\ 1 & 3 & 0 & -1 \end{pmatrix}$$

interchanges the first two rows of the matrix $A$ where the elementary matrix

$$E^{12} = \begin{pmatrix} \mathbf{e}^2 \\ \mathbf{e}^1 \\ \mathbf{e}^3 \end{pmatrix} = \begin{pmatrix} 0 & 1 & 0 \\ 1 & 0 & 0 \\ 0 & 0 & 1 \end{pmatrix}$$

2. For the matrix $A = \begin{pmatrix} 1 & 2 & 3 & 4 \\ 0 & -1 & 1 & 0 \\ 1 & 3 & 0 & -1 \end{pmatrix}$, show that $E^{23}(-2)A$ multiplies the second row of $A$ by $-2$ and adds it to the third row of $A$, where the elementary matrix

$$E^{23}(-2) = \begin{pmatrix} \mathbf{e}^1 \\ \mathbf{e}^2 \\ -2\mathbf{e}^2 + \mathbf{e}^3 \end{pmatrix} = \begin{pmatrix} 1 & 0 & 0 \\ 0 & 1 & 0 \\ 0 & -2 & 1 \end{pmatrix}$$

3. Show that $E^{23}(2)E^{23}(-2) = [1]_3$ where $[1]_3$ is the identity matrix; we see that $E^{23}(-2)$ is the inverse elementary matrix to $E^{23}(2)$.

4. Show that $E^2(1/2)$ is the inverse of the elementary matrix $E^2(2)$.
5. Show that $E^{12}$ is the inverse of the elementary matrix $E^{12}$; thus, $E^{12}$ is its own inverse.
6. We have seen how the elementary (row) matrices perform elementary row operations on a matrix $A$ by multiplication on the left. Similarly, elementary (column) matrices can be defined, which perform elementary column operations on a matrix $A$ by multiplication on the right. For the Problems 1–5 above, find the elementary (column) matrices which perform the corresponding elementary column operations.

## 6.2 Gauss–Jordan Elimination

The method of *Gauss–Jordan elimination* is used to find the solutions of a system of linear equations. We have already noted that, given the augmented matrix $M$ of a system of linear equations, we are free to perform an arbitrary number of elementary row operations without fear of altering its solutions. Thus, the idea of Gauss–Jordan elimination is to simplify the matrix $M$ as much as possible by performing elementary row operations on $M$.

For example,

$$M = \begin{pmatrix} -2 & 2 & -4 & -6 & 2 \\ -3 & 6 & 3 & -15 & 3 \\ 5 & -8 & -1 & 17 & 5 \\ 2 & -2 & 2 & 2 & 8 \end{pmatrix} \tag{6.4}$$

is the augmented matrix of the system of linear equations $AX = B$, where

$$A = \begin{pmatrix} -2 & 2 & -4 & -6 \\ -3 & 6 & 3 & -15 \\ 5 & -8 & -1 & 17 \\ 2 & -2 & 2 & 2 \end{pmatrix}, \quad X = \begin{pmatrix} x_1 \\ x_2 \\ x_3 \\ x_4 \end{pmatrix}, \quad B = \begin{pmatrix} 2 \\ 3 \\ 5 \\ 8 \end{pmatrix}$$

Multiplying $M$ on the left by successive elementary (row) matrices gives

$$E^1(-1/2)M = \begin{pmatrix} 1 & -1 & 2 & 3 & -1 \\ -3 & 6 & 3 & -15 & 3 \\ 5 & -8 & -1 & 17 & 5 \\ 2 & -2 & 2 & 2 & 8 \end{pmatrix}$$

$$E^{14}(-2)E^{13}(-5)E^{12}(3)E^1(-1/2)M = \begin{pmatrix} 1 & -1 & 2 & 3 & -1 \\ 0 & 3 & 9 & -6 & 0 \\ 0 & -3 & -11 & 2 & 10 \\ 0 & 0 & -2 & -4 & 10 \end{pmatrix}.$$

Continuing this process ultimately gives

$$\begin{pmatrix} 1 & 0 & 0 & -9 & 24 \\ 0 & 1 & 0 & -8 & 15 \\ 0 & 0 & 1 & 2 & -5 \\ 0 & 0 & 0 & 0 & 0 \end{pmatrix}, \tag{6.5}$$

Using this final row-reduced form of $M$, the solution to the original system of equations is easily calculated to be

$$\begin{pmatrix} x_1 = 9x_4 + 24 \\ x_2 = 8x_4 + 15 \\ x_3 = -2x_4 - 5 \\ x_4 = x_4 \end{pmatrix} \iff \begin{pmatrix} x_1 \\ x_2 \\ x_3 \\ x_4 \end{pmatrix} = x_4 \begin{pmatrix} 9 \\ 8 \\ -2 \\ 1 \end{pmatrix} + \begin{pmatrix} 24 \\ 15 \\ -5 \\ 0 \end{pmatrix}.$$

Since $x_4 \in \mathbb{R}$ is arbitrary, we see that there are infinitely many solutions parameterized by $x_4$. Geometrically, the solutions lie on a straight line in $\mathbb{R}^4$, passing through the point $\begin{pmatrix} 24 & 15 & -5 & 0 \end{pmatrix}^{\mathrm{T}}$ in the direction of the vector $\begin{pmatrix} 9 & 8 & -2 & 1 \end{pmatrix}^{\mathrm{T}}$.

The Gauss–Jordan elimination method can also be used to find the inverse $A^{-1}$ of a matrix $A$, provided that $\det A \neq 0$. We will illustrate the method by using the first 3 rows and columns of the matrix (6.4). To find the inverse of the matrix

$$A = \begin{pmatrix} -2 & 2 & -4 \\ -3 & 6 & 3 \\ 5 & -8 & -1 \end{pmatrix},$$

where $\det A = 12$, we apply Gauss–Jordan elimination to the augmented matrix

$$M = \begin{pmatrix} -2 & 2 & -4 & 1 & 0 & 0 \\ -3 & 6 & 3 & 0 & 1 & 0 \\ 5 & -8 & -1 & 0 & 0 & 1 \end{pmatrix}$$

to get

$$E^{13}(-5)E^{12}(3)E^{1}(-1/2)M = \begin{pmatrix} 1\,0\,0 & \\ 0\,1\,0 & E^{13}(-5)E^{12}(3)E^{1}(-1/2) \\ 0\,0\,1 & \end{pmatrix}. \tag{6.6}$$

It follows that

$$A^{-1} = E^{13}(-5)E^{12}(3)E^{1}(-1/2) = \frac{1}{6} \begin{pmatrix} 9 & 17 & 15 \\ 6 & 11 & 9 \\ -3 & -3 & -3 \end{pmatrix}.$$

## Exercises

Use the Gauss–Jordan elimination method to find the solutions (if they exist) of the following systems of linear equations for which the augmented matrices are the following:

1. $A = \begin{pmatrix} 0 & 0 & -4 & 3 \\ -2 & -1 & -1 & 4 \\ 3 & 1 & -3 & 0 \end{pmatrix}$
2. $B = \begin{pmatrix} -2 & -4 & -4 & 0 \\ 0 & 0 & -4 & 3 \\ 4 & 0 & 0 & 3 \end{pmatrix}$
3. $C = \begin{pmatrix} 2 & 4 & -4 & 3 \\ 2 & 3 & 1 & 3 \\ 4 & 0 & -1 & 2 \end{pmatrix}$
4. $D = \begin{pmatrix} 0 & 2 & -1 & 4 \\ -1 & -2 & 0 & -4 \\ 4 & 1 & -3 & -4 \end{pmatrix}$
5. $E = \begin{pmatrix} 4 & -1 & -4 & 3 & 0 \\ -3 & 1 & 8 & -6 & -2 \\ 2 & 4 & 15 & -13 & -4 \end{pmatrix}$
6. $F = \begin{pmatrix} -4 & 0 & -1 & 2 & 3 \\ 3 & 0 & -4 & 3 & 4 \\ 1 & -3 & -3 & 0 & -1 \end{pmatrix}$
7. $G = \begin{pmatrix} -1 & -2 & -3 & 2 & 0 \\ -3 & 1 & 0 & -2 & 1 \\ -3 & 1 & -1 & 3 & -4 \\ -4 & -4 & 4 & -3 & 2 \end{pmatrix}$
8. Show that

$$\begin{pmatrix} a & b \\ c & d \end{pmatrix}^{-1} = \frac{1}{ad - bc} \begin{pmatrix} d & -b \\ -c & a \end{pmatrix}$$

provided that $\det \begin{pmatrix} a & b \\ c & d \end{pmatrix} = ad - bc \neq 0$.

9. Show that

$$\begin{pmatrix} 1 & 2 & 3 \\ 2 & -1 & 0 \\ -2 & 2 & 1 \end{pmatrix}^{-1} = \begin{pmatrix} -1 & 4 & 3 \\ -2 & 7 & 6 \\ 2 & -6 & -5 \end{pmatrix}.$$

## 6.3 LU Decomposition

A square matrix $L$ is called a *lower triangular matrix* if all of its elements above its main diagonal are zero. Similarly, a square matrix $U$ is called an *upper triangular matrix* if all of its elements below its main diagonal are zero. A matrix which is both lower triangular and upper triangular is called a *diagonal matrix.*

It is not hard to verify that the inverse of a *lower triangular* elementary row matrix is also a lower triangular elementary matrix and that the product of lower triangular matrices is lower triangular. Given a square matrix $M$, if we can perform the Gauss–Jordan elimination process by only using the first 2 kinds of elementary row operations on $M$, then we can express $M$ in the form

$$M = E_k^{-1} \cdots E_1^{-1} E_1 \cdots E_k M = LU \tag{6.7}$$

where $L = E_k^{-1} \cdots E_1^{-1}$ is a lower triangular matrix and $U = E_1 \cdots E_k M$ is the resulting reduced upper triangular matrix form of $M$.

The $LU$ decomposition (6.7) of a matrix $M$ is possible only when the first 2 kinds of elementary matrices are used in the Gauss–Jordan elimination process. When the 3rd kind of elementary matrix is used, we can only find an $LU$ decomposition up to the inverse of a permutation matrix. This allows us to write $PM = LU$, where $P$ is a permutation matrix. We give a simple example of this modified $LU$ decomposition below.

Let

$$M = \begin{pmatrix} 0 & 0 & 4 \\ 2 & 4 & 6 \\ -4 & -8 & -10 \end{pmatrix},$$

and let $P = E^{12}$ so that

$$PM = E^{12} \begin{pmatrix} 0 & 0 & 4 \\ 2 & 4 & 6 \\ -4 & -8 & -10 \end{pmatrix}.$$

We then successively calculate

$$PM = \begin{pmatrix} 2 & 4 & 6 \\ 0 & 0 & 4 \\ -4 & -8 & -10 \end{pmatrix} = E^1(2) \begin{pmatrix} 1 & 2 & 3 \\ 0 & 0 & 4 \\ -4 & -8 & -10 \end{pmatrix}$$

$$= E^1(2)E^{13}(-4) \begin{pmatrix} 1 & 2 & 3 \\ 0 & 0 & 4 \\ 0 & 0 & 2 \end{pmatrix} = E^1(2)E^{13}(-4)E^2(4) \begin{pmatrix} 1 & 2 & 3 \\ 0 & 0 & 1 \\ 0 & 0 & 2 \end{pmatrix}$$

$$= E^1(2)E^{13}(-4)E^2(4)E^{23}(2) \begin{pmatrix} 1 & 2 & 3 \\ 0 & 0 & 1 \\ 0 & 0 & 0 \end{pmatrix} = LU,$$

where

$$L = E^1(2)E^{13}(-4)E^2(4)E^{23}(2)$$
$$= \begin{pmatrix} 2 & 0 & 0 \\ 0 & 1 & 0 \\ 0 & 0 & 1 \end{pmatrix} \begin{pmatrix} 1 & 0 & 0 \\ 0 & 1 & 0 \\ -4 & 0 & 1 \end{pmatrix} \begin{pmatrix} 1 & 0 & 0 \\ 0 & 4 & 0 \\ 0 & 0 & 1 \end{pmatrix} \begin{pmatrix} 1 & 0 & 0 \\ 0 & 1 & 0 \\ 0 & 2 & 1 \end{pmatrix} = \begin{pmatrix} 2 & 0 & 0 \\ 0 & 4 & 0 \\ -4 & 2 & 1 \end{pmatrix}$$

and

$$U = \begin{pmatrix} 1 & 2 & 3 \\ 0 & 0 & 1 \\ 0 & 0 & 0 \end{pmatrix}.$$

## Exercises

1. Give the LU decomposition for the following matrices:
   (a) $A = \begin{pmatrix} 1 & 2 \\ 3 & 4 \end{pmatrix}$
   (b) $B = \begin{pmatrix} 2 & 1 \\ 8 & 7 \end{pmatrix}$
   (c) $C = \begin{pmatrix} 1 & 0 \\ 8 & 1 \end{pmatrix}$
2. Give the LU decomposition for A and solve the upper triangular system $UX = L^{-1}C$ for $X$ where
$$AX = \begin{pmatrix} 2 & 3 & 3 \\ 0 & 5 & 7 \\ 6 & 9 & 8 \end{pmatrix} \begin{pmatrix} x_1 \\ x_2 \\ x_3 \end{pmatrix} = \begin{pmatrix} 2 \\ 2 \\ 5 \end{pmatrix}$$
   for
$$A = \begin{pmatrix} 2 & 3 & 3 \\ 0 & 5 & 7 \\ 6 & 9 & 8 \end{pmatrix}, \quad X = \begin{pmatrix} x_1 \\ x_2 \\ x_3 \end{pmatrix} \text{ and } C = \begin{pmatrix} 2 \\ 2 \\ 5 \end{pmatrix}.$$
3. Find the LU decomposition for $A = \begin{pmatrix} 0 & 0 & -4 \\ -2 & -1 & -1 \\ 3 & 1 & -3 \end{pmatrix}$. If this is not possible, then find the permutation matrix $P$ such that $PA = LU$.
4. Find the LU decomposition for $B = \begin{pmatrix} -2 & -4 & -4 \\ 0 & 0 & -4 \\ 4 & 0 & 0 \end{pmatrix}$. If this is not possible, then find the permutation matrix $P$ such that $PB = LU$.

5. Find the LU decomposition for $C = \begin{pmatrix} 2 & 4 & -4 \\ 2 & 3 & 1 \\ 4 & 0 & -1 \end{pmatrix}$. If this is not possible, then find the permutation matrix $P$ such that $PC = LU$.
6. Find the LU decomposition for $D = \begin{pmatrix} 0 & 2 & -1 \\ -1 & -2 & 0 \\ 4 & 1 & -3 \end{pmatrix}$. If this is not possible, then find the permutation matrix $P$ such that $PD = LU$.
7. Find the LU decomposition for $E = \begin{pmatrix} -1 & -2 & -3 & 2 \\ -3 & 1 & 0 & -2 \\ -3 & 1 & -1 & 3 \\ -4 & -4 & 4 & -3 \end{pmatrix}$. If this is not possible, then find the permutation matrix $P$ such that $PE = LU$.

# Chapter 7
# Linear Transformations on $\mathbb{R}^n$

*Life is good for only two things, discovering mathematics and teaching mathematics.*

—*Simón Poisson*

The definition of a linear transformation on $\mathbb{R}^n$, and its natural extension to an outermorphism on all of the geometric algebra $\mathbb{G}_n$, is given. The tools of geometric algebra, such as the **a**-derivative and the simplicial $k$-derivative, are used to study its basic properties. We introduce the adjoint linear transformation and use it to derive the inverse of a nonsingular transformation.

## 7.1 Definition of a Linear Transformation

Let $\mathbb{K}$ be the real number field $\mathbb{R}$ or the complex number field $\mathbb{C}$, and let $f : \mathscr{A}^n \to \mathscr{B}^m$, where $\mathscr{A}^n$ is an $n$-dimensional vector space over $\mathbb{K}$ and $\mathscr{B}^m$ is an $m$-dimensional vector space over $\mathbb{K}$. We say that $f$ is a *linear transformation* over $\mathbb{K}$ if for all $\mathbf{x}, \mathbf{y} \in \mathscr{A}^n$, and all $\alpha, \beta \in \mathbb{K}$,

$$f(\alpha\mathbf{x} + \beta\mathbf{y}) = \alpha f(\mathbf{x}) + \beta f(\mathbf{y}) \in \mathscr{B}^m. \tag{7.1}$$

It is easily checked that this is equivalent to saying that

$$f(\mathbf{x} + \mathbf{y}) = f(\mathbf{x}) + f(\mathbf{y}) \in \mathscr{B}^m \text{ and } f(\alpha\mathbf{x}) = \alpha f(\mathbf{x}) \in \mathscr{B}^m,$$

for all $\mathbf{x}, \mathbf{y} \in \mathscr{A}^n$ and $\alpha \in \mathbb{K}$. A linear transformation $f$ is also called a *linear mapping* or a *linear operator*. In this chapter, we are primarily interested in studying the basic properties of linear transformations $f : \mathbb{R}^n \to \mathbb{R}^n$, using the rich algebraic structure of the geometric algebra $\mathbb{G}_n = \mathbb{G}(\mathbb{R}^n)$.

G. Sobczyk, *New Foundations in Mathematics: The Geometric Concept of Number*, DOI 10.1007/978-0-8176-8385-6_7,

In order to be able to use the full power of the geometric algebra $\mathbb{G}_n$ in the study of the linear transformation $f$, we extend $f$ to an *outermorphism* $\underline{f}$ on the geometric algebra $\mathbb{G}_n$. We extend $f$ to $\underline{f}$ by the following.

**Outermorphism Rule 7.1.1.** *There are three parts to the rule:*

1. $\underline{f}(\alpha) = \alpha$ for all $\alpha \in \mathbb{R}$.
2. $\underline{f}(\mathbf{v}) = f(\mathbf{v})$ for all $\mathbf{v} \in \mathbb{R}^n$.
3. $\underline{f}(\mathbf{v}_1 \wedge \cdots \wedge \mathbf{v}_k) = f(\mathbf{v}_1) \wedge \cdots \wedge f(\mathbf{v}_k)$.

Note that $\underline{f}$ inherits the property of linearity from the linearity of $f$ so that

$$\underline{f}(\alpha A + \beta B) = \alpha \underline{f}(A) + \beta \underline{f}(B) \text{ for all } A, B \in \mathbb{G}_n.$$

Recalling the definition (3.45) of the $\mathbf{a}$-derivative of a function on $\mathbb{R}^n$, it immediately follows that if $f$ is a linear transformation on $\mathbb{R}^n$, then its $\mathbf{a}$-derivative

$$\mathbf{a} \cdot \partial_{\mathbf{x}} f(\mathbf{x}) = \lim_{h \to 0} \frac{f(\mathbf{x} + h\mathbf{a}) - f(\mathbf{x})}{h} = f(\mathbf{a}) \tag{7.2}$$

for all $\mathbf{a} \in \mathbb{R}^n$. Indeed, the $\mathbf{a}$-derivative of a function $f$, defined on $\mathbb{R}^n$, evaluates to $f(\mathbf{a})$ for all $\mathbf{a} \in \mathbb{R}^n$ iff $f(\mathbf{x})$ is a linear transformation on $\mathbb{R}^n$. Introducing the *simplicial k-derivative*

$$\partial_{(k)} = \frac{1}{k!} \partial_{\mathbf{x}_k} \wedge \partial_{\mathbf{x}_{k-1}} \wedge \cdots \wedge \partial_{\mathbf{x}_1}, \tag{7.3}$$

for $k = 1, \ldots, n$, then for each $k$-vector $A_k \in \mathbb{G}_n^k$,

$$A_k \cdot \partial_{(k)} f_{(k)} = \underline{f}(A_k) \tag{7.4}$$

where $f_{(k)} := \underline{f}(\mathbf{x}_{(k)}) = f(\mathbf{x}_1) \wedge \cdots \wedge f(\mathbf{x}_k)$. Using the simplicial derivative, we can express the determinant of a linear function $f$ by

$$\det f = \mathbf{e}_{(n)}^{-1} \underline{f}(\mathbf{e}_{(n)}) = \partial_{(n)} \cdot f_{(n)}, \tag{7.5}$$

which will be very useful in the next chapter.

For example, for $A_2 = \mathbf{a}_1 \wedge \mathbf{a}_2$, applying (7.2) and (7.3), we have

$$\begin{aligned}(\mathbf{a}_1 \wedge \mathbf{a}_2) \cdot \partial_{(2)} f_{(2)} &= \frac{1}{2}(\mathbf{a}_2 \cdot \partial_{\mathbf{x}_2} \mathbf{a}_1 \cdot \partial_{\mathbf{x}_1} - \mathbf{a}_1 \cdot \partial_{\mathbf{x}_2} \mathbf{a}_2 \cdot \partial_{\mathbf{x}_1}) f(\mathbf{x}_1) \wedge f(\mathbf{x}_2) \\ &= \frac{1}{2}\big(f(\mathbf{a}_1) \wedge f(\mathbf{a}_2) - f(\mathbf{a}_2) \wedge f(\mathbf{a}_1)\big) = \underline{f}(\mathbf{a}_1 \wedge \mathbf{a}_2),\end{aligned}$$

and the more general (7.4) follows by a similar argument. If we apply the operator $(\mathbf{a}_1 \wedge \mathbf{a}_2) \cdot \partial_{(2)}$ to the geometric product $f(\mathbf{x}_1) g(\mathbf{x}_2)$ of two linear operators $f$ and $g$, we get the antisymmetric operator

$$H(\mathbf{a}_1 \wedge \mathbf{a}_2) = (\mathbf{a}_1 \wedge \mathbf{a}_2) \cdot \partial_{(2)} f(\mathbf{x}_1) g(\mathbf{x}_2) = \frac{f(\mathbf{a}_1)g(\mathbf{a}_2) - f(\mathbf{a}_2)g(\mathbf{a}_1)}{2} \tag{7.6}$$

defined in terms of $f$ and $g$.

Let $(\mathbf{e})_{(n)} = (\mathbf{e}_1 \, \ldots \, \mathbf{e}_n)$ be the standard orthonormal (row) basis vectors of $\mathbb{R}^n$. Then the *matrix* $[f]$ of $f$ with respect to the standard row basis is defined by

$$f(\mathbf{e})_{(n)} := (f\mathbf{e})_{(n)} = (\mathbf{f}_1 \cdots \mathbf{f}_n) = (\mathbf{e})_{(n)}[f], \tag{7.7}$$

where

$$[f] = [f_{ij}] = (\mathbf{e})^{\mathrm{T}}_{(n)} \cdot (f\mathbf{e})_{(n)} = \begin{pmatrix} \mathbf{e}_1 \\ \cdot \\ \cdot \\ \mathbf{e}_n \end{pmatrix} \cdot (\mathbf{f}_1 \, \ldots \, \mathbf{f}_n)$$

$$= \begin{pmatrix} f_{11} & \ldots & f_{1n} \\ \cdot & \cdot & \cdot \\ \cdot & \cdot & \cdot \\ \cdot & \cdot & \cdot \\ f_{m1} & \ldots & f_{mn} \end{pmatrix},$$

$\mathbf{f}_i = f(\mathbf{e}_i)$, and $f_{ij} = \mathbf{e}_i \cdot \mathbf{f}_j$. By taking the transpose of (7.7), we get

$$f(\mathbf{e})^{\mathrm{T}}_{(n)} = (f\mathbf{e})^{\mathrm{T}}_{(n)} = \begin{pmatrix} \mathbf{f}_1 \\ \cdot \\ \cdot \\ \mathbf{f}_n \end{pmatrix} = [f]^{\mathrm{T}} (\mathbf{e})^{\mathrm{T}}_{(n)}, \tag{7.8}$$

the relationship of $f$ to its transpose matrix $[f]^{\mathrm{T}}$.

Note that $\mathbf{e}_i = \mathbf{e}^i$ for $1 \le i \le n$, since the reciprocal basis of a given orthonormal basis of $\mathbb{R}^n$ is identical to the given orthonormal basis. We can thus write the *reciprocal basis* $(\mathbf{e})^{(n)}$ as a column matrix of the basis vectors $\mathbf{e}_i$,

$$(\mathbf{e})^{(n)} = \begin{pmatrix} \mathbf{e}^1 \\ \mathbf{e}^2 \\ \cdot \\ \cdot \\ \mathbf{e}^n \end{pmatrix} = (\mathbf{e})^{\mathrm{T}}_{(n)}. \tag{7.9}$$

Since $(\mathbf{e})^{(n)} \cdot (\mathbf{e})_{(n)} = [1]_n$, we sometimes write $(\mathbf{e})^{(n)} = (\mathbf{e})^{-1}_{(n)}$, where $[1]_n$ is the $(n,n)$-identity matrix. Because of the noncommutative nature of matrix multiplication,

$$[1]_n = (\mathbf{e})^{(n)} \cdot (\mathbf{e})_{(n)} \neq (\mathbf{e})_{(n)} \cdot (\mathbf{e})^{(n)} = n, \tag{7.10}$$

so these operations must be used carefully.

Suppose now that $(\mathbf{a})_{(n)} = (\mathbf{e})_{(n)}A$ is an arbitrary basis of $\mathbb{R}^n$, where $A$ is the matrix of transition. We wish to find the unique reciprocal basis $(\mathbf{a})^{(n)} = (\mathbf{a})^{-1}_{(n)}$ such that

$$(\mathbf{a})^{(n)} \cdot (\mathbf{a})_{(n)} = [1]_n.$$

We find immediately that

$$(\mathbf{a})^{(n)} = (\mathbf{a})^{-1}_{(n)} = A^{-1}(\mathbf{e})^{(n)} \tag{7.11}$$

as can be easily verified.

We can find the reciprocal basis $(\mathbf{a})^{(n)}$ in another way by finding the linear mapping $g$ with the property that $g(\mathbf{a}^i) = \mathbf{a}_i$ for $1 \leq i \leq n$. With the help of (7.8), we see that this is equivalent to the requirement that

$$g(\mathbf{a})^{(n)} = (g\mathbf{a})^{(n)} = (\mathbf{a})^{\mathrm{T}}_{(n)} = [g]^{\mathrm{T}}(\mathbf{a})^{(n)}.$$

Dotting each side of this equation on the right with $(\mathbf{a})_{(n)}$ gives

$$[g]^{\mathrm{T}} = (\mathbf{a})^{\mathrm{T}}_{(n)} \cdot (\mathbf{a})_{(n)} = [g],$$

which is called the *Gramian matrix* of the basis $(\mathbf{a})_{(n)}$. It follows that

$$(\mathbf{a})^{(n)} = [g]^{-1}(\mathbf{a})^{\mathrm{T}}_{(n)}. \tag{7.12}$$

Comparing (7.11) with (7.12), we find that $[g]^{-1}A^{\mathrm{T}} = A^{-1}$ which is equivalent to $A^{\mathrm{T}}A = [g]$.

Given a basis $(\mathbf{a})_{(n)}$ of $\mathbb{R}^n$, it is always possible to construct an orthogonal basis $(\mathbf{b})_{(n)}$ with the property that

$$\mathbf{b}_1 = \mathbf{a}_1 \quad \text{and} \quad \mathbf{b}_{(k)} = \mathbf{a}_{(k)}, \tag{7.13}$$

for $k = 2, \ldots, n$. The construction proceeds recursively as follows:

$$\mathbf{b}_1 = \mathbf{a}_1, \ \ \mathbf{b}_k = \hat{\mathbf{b}}^{\dagger}_{(k-1)}\left(\hat{\mathbf{b}}_{(k-1)} \wedge \mathbf{a}_k\right) = \mathbf{a}_k - \hat{\mathbf{b}}^{\dagger}_{(k-1)}\left(\hat{\mathbf{b}}_{(k-1)} \cdot \mathbf{a}_k\right) \tag{7.14}$$

for $k = 2, \ldots, n$, where

$$\hat{\mathbf{b}}^{\dagger}_{(k-1)} = \hat{\mathbf{b}}^{-1}_{(k-1)} = \frac{\mathbf{b}^{\dagger}_{(k-1)}}{|\mathbf{b}^{\dagger}_{(k-1)}|} = \hat{\mathbf{b}}_{k-1} \wedge \cdots \wedge \hat{\mathbf{b}}_1 = \hat{\mathbf{b}}_{k-1} \cdots \hat{\mathbf{b}}_1$$

since $\mathbf{b}_i$'s, by construction, are orthogonal. This construction is known as the *Gram-Schmidt orthogonalization process*, [15, p.160]. Later, in Chap. 10, we will generalize this process to apply in more general pseudo-Euclidean spaces $\mathbb{R}^{p,q}$.

As an example, we will construct an orthogonal basis for $\mathbb{R}^4$ from the basis of column vectors

$$A = (\mathbf{a})_{(4)} = \begin{pmatrix} 1 & 0 & 1 & 1 \\ 0 & 1 & 2 & 0 \\ 1 & 1 & 0 & -1 \\ 0 & 0 & 1 & 2 \end{pmatrix}$$

Thus,

$$\mathbf{b}_1 = \mathbf{a}_1 = (1,0,1,0)^{\mathrm{T}}$$

$$\mathbf{b}_2 = \mathbf{a}_2 - \frac{(\mathbf{a}_2 \cdot \mathbf{b}_1)\mathbf{b}_1}{\mathbf{b}_1 \cdot \mathbf{b}_1} = \left(-\frac{1}{2}, 1, \frac{1}{2}, 0\right)^{\mathrm{T}}$$

$$\mathbf{b}_3 = \mathbf{a}_3 - \frac{(\mathbf{a}_3 \cdot \mathbf{b}_1)\mathbf{b}_1}{\mathbf{b}_1 \cdot \mathbf{b}_1} - \frac{(\mathbf{a}_3 \cdot \mathbf{b}_2)\mathbf{b}_2}{\mathbf{b}_2 \cdot \mathbf{b}_2} = (1,1,-1,1)^{\mathrm{T}}$$

$$\mathbf{b}_4 = \mathbf{a}_4 - \frac{(\mathbf{a}_4 \cdot \mathbf{b}_1)\mathbf{b}_1}{\mathbf{b}_1 \cdot \mathbf{b}_1} - \frac{(\mathbf{a}_4 \cdot \mathbf{b}_2)\mathbf{b}_2}{\mathbf{b}_2 \cdot \mathbf{b}_2} - \frac{(\mathbf{a}_4 \cdot \mathbf{b}_3)\mathbf{b}_3}{\mathbf{b}_3 \cdot \mathbf{b}_3} = \left(-\frac{1}{3}, -\frac{1}{3}, \frac{1}{3}, 1\right)^{\mathrm{T}}.$$

We can also write $(\mathbf{b})_{(4)} = (\mathbf{a})_{(4)}\triangle$ where

$$\triangle = (\mathbf{a}_1, \mathbf{a}_2, \mathbf{a}_3, \mathbf{a}_4)^{-1}(\mathbf{b}_1, \mathbf{b}_2, \mathbf{b}_3, \mathbf{b}_4) = \begin{pmatrix} 1 & -\frac{1}{2} & 0 & -\frac{1}{3} \\ 0 & 1 & -1 & \frac{5}{3} \\ 0 & 0 & 1 & -1 \\ 0 & 0 & 0 & 1 \end{pmatrix}$$

We can now use the orthogonal basis (7.14) defined by the basis $(\mathbf{a})_{(n)}$ to demonstrate the *Schwarz inequality*. Using (7.13), we have

$$(\mathbf{a}_1 \wedge \mathbf{a}_2) \cdot (\mathbf{a}_2 \wedge \mathbf{a}_1) = |\mathbf{a}_1|^2 |\mathbf{a}_2|^2 - (\mathbf{a}_1 \cdot \mathbf{a}_2)^2 = |\mathbf{b}_1|^2 |\mathbf{b}_2|^2 \geq 0. \tag{7.15}$$

In the case of an orthonormal basis, indices can be raised or lowered at will. For example, for any vector $\mathbf{x} \in \mathbb{R}^n$, then

$$\mathbf{x} = (x)_{(n)}(\mathbf{e})^{(n)} = \mathbf{x} \cdot (\mathbf{e})_{(n)}(\mathbf{e})^{(n)} = (\mathbf{e})_{(n)}(x)^{(n)} = (\mathbf{e})_{(n)}(\mathbf{e})^{(n)} \cdot \mathbf{x}.$$

Since $\mathbf{e}_i = \mathbf{e}^i$ for $1 \leq i \leq n$, it follows that $(x)^{(n)} = (x)_{(n)}^{\mathrm{T}}$. Suppose now that $(\mathbf{a})_{(n)} = (\mathbf{e})_{(n)}A$ is any other basis of $\mathbb{R}^n$. Then

$$\mathbf{x} = ((\mathbf{e})_{(n)}A)A^{-1}(x)^{(n)} = (x)_{(n)}A^{-1}(A(\mathbf{e})^{(n)}), \tag{7.16}$$

which gives the transformation rule for the vector $\mathbf{x}$. In words, the components $(x)^{(n)}$ with respect to the standard basis $(\mathbf{e})_{(n)}$ transform into the components $(x)_{\mathbf{a}}^{(n)} = A^{-1}(x)^{(n)}$ with respect to the basis $(\mathbf{a})_{(n)}$.

We can now find the transformation rule for the matrix $[f]$ of a linear transformation $f$. We have

$$f((\mathbf{e})_{(n)}) = (\mathbf{e})_{(n)}[f] = (\mathbf{e})_{(n)}AA^{-1}[f] = (\mathbf{a})_{(n)}A^{-1}[f],$$

which implies that

$$f(\mathbf{x}) = f((\mathbf{e})_{(n)})(x)^{(n)} = (\mathbf{e})_{(n)}[f](x)^{(n)} = (\mathbf{e})_{(n)}AA^{-1}[f]AA^{-1}(x)^{(n)}. \qquad (7.17)$$

It follows that the matrix of the transformation $f$ with respect to the bases $(\mathbf{a})_{(n)}$ is $[f]_a = A^{-1}[f]A$.

Let us see how the formulas (7.16) and (7.17) work for the following simple example: Given the basis

$$(\mathbf{a})_{(2)} = (\mathbf{e})_{(2)}A = \begin{pmatrix}\mathbf{e}_1 & \mathbf{e}_2\end{pmatrix}\begin{pmatrix}3 & 2\\ 7 & 5\end{pmatrix} = (3\mathbf{e}_1 + 7\mathbf{e}_2, 2\mathbf{e}_1 + 5\mathbf{e}_2)$$

for the matrix $A = \begin{pmatrix}3 & 2\\ 7 & 5\end{pmatrix}$, whose inverse is $A^{-1} = \begin{pmatrix}5 & -2\\ -7 & 3\end{pmatrix}$, a vector

$$\mathbf{x} = (\mathbf{e})_{(2)}(x)^{(2)} = \begin{pmatrix}\mathbf{e}_1 & \mathbf{e}_2\end{pmatrix}\begin{pmatrix}x^1\\ x_2\end{pmatrix} = x^1\mathbf{e}_1 + x^2\mathbf{e}_2$$

in the standard basis $(\mathbf{e})_{(2)}$ becomes

$$\mathbf{x} = (\mathbf{e})_{(2)}(x)^{(2)} = (\mathbf{e})_{(2)}AA^{-1}(x)^{(2)} = (\mathbf{a})_{(2)}(x)^{(2)}_a = x^1_a\mathbf{a}_1 + x^2_a\mathbf{a}_2$$

where

$$(x)^{(2)}_a = A^{-1}(x)^{(2)} = \begin{pmatrix}5 & -2\\ -7 & 3\end{pmatrix}\begin{pmatrix}x^1\\ x^2\end{pmatrix} = \begin{pmatrix}5x^1 - 2x^2\\ -7x^1 + 3x^2\end{pmatrix}_a,$$

when expressed in the basis $(\mathbf{a})_{(2)}$. For a linear transformation

$$f(\mathbf{x}) = (\mathbf{e})_{(2)}[f](x)^{(2)} = (\mathbf{e})_{(n)}AA^{-1}[f]AA^{-1}(x)^{(2)} = (\mathbf{a})_{(2)}[f]_a(x)^{(2)}_a,$$

the matrix $[f]_{\mathbf{a}}$ of the transformation $f$ with respect to the basis $(\mathbf{a})_{(2)}$ is

$$[f]_a = A^{-1}[f]A = \begin{pmatrix}5 & -2\\ -7 & 3\end{pmatrix}\begin{pmatrix}f_{11} & f_{12}\\ f_{21} & f_{22}\end{pmatrix}\begin{pmatrix}3 & 2\\ 7 & 5\end{pmatrix}.$$

## Exercises

In the following exercises, let $(\mathbf{a})_{(3)} = (\mathbf{e})_{(3)}A$ for the matrix $A = \begin{pmatrix} 1 & 2 & 3 \\ 2 & -1 & 0 \\ -2 & 2 & 1 \end{pmatrix}$,

given $A^{-1} = \begin{pmatrix} -1 & 4 & 3 \\ -2 & 7 & 6 \\ 2 & -6 & -5 \end{pmatrix}$.

1. Using (7.16), find the components of $\mathbf{x} = (\mathbf{e})_{(3)} \begin{pmatrix} 3 \\ 2 \\ -1 \end{pmatrix}$ with respect to the basis $(\mathbf{a})_{(3)}$.
2. Find the Gramian matrix $[g]$ of the basis $(\mathbf{a})_{(3)}$.
3. Find the reciprocal basis $(\mathbf{a})^{(3)}$ of the basis $(\mathbf{a})_{(3)}$.
4. Given the linear transformation $f$ defined by

$$f(\mathbf{e})_{(3)} = (\mathbf{e})_{(3)}[f] = (\mathbf{e})_{(3)} \begin{pmatrix} 1 & -1 & 2 \\ 2 & 2 & 1 \\ 0 & 1 & 3 \end{pmatrix},$$

   using (7.17), find the matrix $[f]_{\mathbf{a}}$ of $f$ with respect to the basis $(\mathbf{a})_{(3)}$.
5. (a) Using the Gram-Schmidt orthogonalization process (7.14), find the orthogonal basis $(\mathbf{b})_{(3)}$ for the given basis $(\mathbf{a})_{(3)}$.
   (b) Show that $\mathbf{b}_1\mathbf{b}_2 = \mathbf{a}_1 \wedge \mathbf{a}_2$ and $\mathbf{b}_1\mathbf{b}_2\mathbf{b}_3 = \mathbf{a}_1 \wedge \mathbf{a}_2 \wedge \mathbf{a}_3$.
   (c) Verify the Schwarz inequality (7.15) that $|\mathbf{a}_1|^2|\mathbf{a}_2|^2 - (\mathbf{a}_1 \cdot \mathbf{a}_2)^2 \geq 0$.
6. For the linear transformation given in problem 4, find

   (a) $f(\mathbf{e}_1)$, (b) $f(\mathbf{e}_2)$, (c) $f(\mathbf{e}_3)$, (d) $\underline{f}(\mathbf{e}_{12})$, (e) $\underline{f}(\mathbf{e}_{13})$, (f) $\underline{f}(\mathbf{e}_{23})$.

   (g) Show that $\underline{f}(\mathbf{e}_{123}) = \det[f]\mathbf{e}_{123}$.
7. Find the antisymmetric operator defined by (7.6) for the two linear operators $f$ and $g$ specified by

$$f(\mathbf{e})_{(2)} = (\mathbf{e})_{(2)} \begin{pmatrix} 1 & 2 \\ 0 & 1 \end{pmatrix} \quad \text{and} \quad g(\mathbf{e})_{(2)} = (\mathbf{e})_{(2)} \begin{pmatrix} 1 & 0 \\ 0 & 2 \end{pmatrix}.$$

   (Note that this operator has a scalar and a bivector part.)

## 7.2 The Adjoint Transformation

A useful auxiliary tool in the study of a linear transformation $f$ and its outermorphism $\underline{f}$ is the *adjoint transformation* and its outermorphism, both denoted by the same symbol $\overline{f}$. The adjoint $\overline{f}$ is defined by the condition that for all $\mathbf{x}, \mathbf{y} \in \mathbb{R}^n$,

$$\overline{f}(\mathbf{x}) \cdot \mathbf{y} = \mathbf{x} \cdot \underline{f}(\mathbf{y}). \tag{7.18}$$

We can solve for an explicit expression for $\overline{f}(\mathbf{x})$, getting

$$\overline{f}(\mathbf{x}) = \overline{f}(\mathbf{x}) \cdot (\mathbf{e})_{(n)}(\mathbf{e})^{(n)} = \mathbf{x} \cdot (f\mathbf{e})_{(n)}(\mathbf{e})^{(n)} = \partial_{\mathbf{y}} \underline{f}(\mathbf{y}) \cdot \mathbf{x}$$

The last equality expresses the adjoint transformation in terms of the vector derivative, defined in (3.46).

There is a simple relationship between the matrix $[f]$ and the matrix $[\overline{f}]$ of the adjoint transformation. Using (7.8), we find that

$$[\underline{f}]^{\mathrm{T}} = \underline{f}(\mathbf{e})^{(n)} \cdot (\mathbf{e})_{(n)} = (\mathbf{e})^{(n)} \cdot \overline{f}(\mathbf{e})_{(n)} = [\overline{f}].$$

Note that $[f] = [\underline{f}]$, since the outermorphism $\underline{f}$ is identical to the linear transformation $f$ when restricted to a vector argument.

The following two basic relationships between the outermorphism $\underline{f}$ and its adjoint outermorphism $\overline{f}$ are important and easily established. Let $A_r, B_s \in \mathbb{G}_n$ be an $r$-vector and an $s$-vector, respectively, where $r \geq s$. Then

$$\underline{f}(A_r) \cdot B_s = \underline{f}[A_r \cdot \overline{f}(B_s)], \quad \text{and} \quad \overline{f}(A_r) \cdot B_s = \overline{f}[A_r \cdot \underline{f}(B_s)]. \tag{7.19}$$

Using (7.19) and the fact that $\det f = \det \overline{f}$, we can establish the classical formula for the inverse $[f]^{-1}$ of the matrix $[f]$ of an invertible linear transformation $f$. In the second relationship, let $A_r = I_n = \mathbf{e}_{12\ldots n}$ the unit pseudoscalar element of $\mathbb{G}_n$, $B_s = \mathbf{x}$, and $\mathbf{y} = f(\mathbf{x})$. Then

$$\det(f) I_n \mathbf{x} = \overline{f}(I_n \mathbf{y}) \quad \iff \quad \det(f) \mathbf{x} I_n = \overline{f}(\mathbf{y} I_n),$$

or

$$\mathbf{x} = \frac{\overline{f}(\mathbf{y} I_n) I_n^{-1}}{\det f} = \frac{\overline{f}(\mathbf{y}\mathbf{e}_{1\ldots n})\mathbf{e}_{n\ldots 1}}{\det f} = f^{-1}(\mathbf{y}), \tag{7.20}$$

which gives $\mathbf{x} = f^{-1}(\mathbf{y})$ for all $\mathbf{y} \in \mathbb{R}^n$. Expressing both sides of (7.20) in terms of the standard basis gives

$$(\mathbf{e})_{(n)}(x)^{(n)} = \frac{\overline{f}[(\mathbf{e})_{(n)}(y)^{(n)}\mathbf{e}_{1\ldots n}]\mathbf{e}_{n\ldots 1}}{\det f} = \frac{\overline{f}[(\mathbf{e})_{(n)}\mathbf{e}_{1\ldots n}]\mathbf{e}_{n\ldots 1}(y)^{(n)}}{\det f}$$
$$= (\mathbf{e})_{(n)}[f^{-1}](y)^{(n)}.$$

Dotting both sides of this equation on the left by $(\mathbf{e})^{(n)}$, and simplifying, gives

$$[f^{-1}](y)^{(n)} = [f_{ij}^{-1}](y)^{(n)} = \frac{\left((\mathbf{e}\mathbf{e}_{1\ldots n})^{(n)} \cdot \underline{f}[(\mathbf{e}_{n\ldots 1}\mathbf{e})_{(n)}]\right)^{\mathrm{T}}}{\det f}(y)^{(n)}$$

or

$$f_{ij}^{-1} = \frac{F_{ji}}{\det f} \tag{7.21}$$

where

$$F_{ji} = (\mathbf{e}_j \mathbf{e}_{1\ldots n}) \cdot \underline{f}[(\mathbf{e}_{n\ldots 1}\mathbf{e}_i)]$$

is the *algebraic cofactor* of the element $f_{ji}$ in the determinant $\det[f]$ of $f$ [29, p. 15]. See also Problem 10 of the exercises in Sect. 5.1. The formula (7.21) is the classical formula for the inverse of a square matrix in terms of its algebraic cofactors.

For example, if

$$[f] = (\mathbf{f})_{(n)} = (\mathbf{f}_1, \mathbf{f}_2, \mathbf{f}_3) = \begin{pmatrix} f_{11} & f_{12} & f_{13} \\ f_{21} & f_{22} & f_{23} \\ f_{31} & f_{32} & f_{33} \end{pmatrix},$$

then

$$[f^{-1}] = [f_{ij}^{-1}] = \frac{\left((\mathbf{e}\mathbf{e}_{123})^{(3)} \cdot \underline{f}[(\mathbf{e}_{321}\mathbf{e})_{(3)}]\right)^{\mathrm{T}}}{\det f}$$

$$= \frac{1}{\det f} \begin{pmatrix} f_{22}f_{33} - f_{32}f_{23} & f_{13}f_{32} - f_{33}f_{12} & f_{12}f_{23} - f_{22}f_{13} \\ f_{23}f_{31} - f_{33}f_{21} & f_{11}f_{33} - f_{31}f_{13} & f_{13}f_{21} - f_{23}f_{11} \\ f_{21}f_{32} - f_{31}f_{22} & f_{12}f_{31} - f_{32}f_{11} & f_{11}f_{22} - f_{21}f_{12} \end{pmatrix}.$$

## Exercises

1. Using the formula (7.21), find the inverse of the matrix $A = \begin{pmatrix} 1 & 2 & 3 \\ 2 & -1 & 0 \\ -2 & 2 & 1 \end{pmatrix}$.

2. Show that the inverse of the matrix $\begin{pmatrix} a & b \\ c & d \end{pmatrix}$ is

$$\begin{pmatrix} a & b \\ c & d \end{pmatrix}^{-1} = \frac{1}{ad - bc} \begin{pmatrix} d & -b \\ -c & a \end{pmatrix}.$$

3. Given the linear transformation defined by

$$f(\mathbf{x}) = (\mathbf{e})_{(3)} \begin{pmatrix} 1 & 2 & 3 \\ 2 & -1 & 0 \\ -2 & 2 & 1 \end{pmatrix} (x)^{(3)},$$

where $\mathbf{x} = (\mathbf{e})_{(3)}(x)^{(3)}$, find

(a) $f^{-1}(\mathbf{e}_1)$, (b) $f^{-1}(\mathbf{e}_2)$, (c) $f^{-1}(\mathbf{e}_3)$, (d) $\underline{f}^{-1}(\mathbf{e}_{12})$, (e) $\underline{f}^{-1}(\mathbf{e}_{13})$, (f) $\underline{f}^{-1}(\mathbf{e}_{23})$.

(g) Show that $\underline{f}^{-1}(\mathbf{e}_{123}) = \det[f^{-1}]\mathbf{e}_{123}$.

4. Establish the basic relationship (7.19). It may be helpful to first show $\underline{f}(A_r) \cdot B_r = A_r \cdot \overline{f}(B_r)$ for all $r$-vectors $A_r, B_r \in \mathbb{G}_n^r$.
5. Show that $\det \underline{f} = \det \overline{f}$ by using (7.5).

# Chapter 8
# Structure of a Linear Operator

*Mathematicians are like Frenchmen: whatever you say to them they translate into their own language, and forthwith it is something entirely different.*

–Goethe

We show how the basic structure of a linear operator follows from its minimal polynomial. The *spectral decomposition* of a linear operator follows immediately from the spectral basis of idempotents and nilpotents which was developed in Chap. 1 for modular polynomials. The *Jordan form* of a linear operator, while technically more difficult, is just a refinement of its more fundamental spectral decomposition.

## 8.1 Rank of a Linear Operator

Before we can talk about the structure of a linear operator, we must understand what is meant by the *rank* of a linear operator. Let $g$ be a linear operator, $g : \mathbb{R}^n \rightarrow \mathbb{R}^m$, where $\mathbb{R}^n$ and $\mathbb{R}^m$ are $n$ and $m$-dimensional Euclidean spaces over the real numbers $\mathbb{R}$. With respect to the standard basis $(\mathbf{e})_{(n)} = \begin{pmatrix}\mathbf{e}_1 & \mathbf{e}_2 & \ldots & \mathbf{e}_n\end{pmatrix}$ of $\mathbb{R}^n$ and the standard basis $(\mathbf{f})_{(m)} = \begin{pmatrix}\mathbf{f}_1 & \mathbf{f}_2 & \ldots & \mathbf{f}_m\end{pmatrix}$ of $\mathbb{R}^m$, we have

$$g(\mathbf{x}) = g(\mathbf{e})_{(n)}(x)^{(n)} = (\mathbf{f})_{(m)}[g]_{(m,n)}(x)^{(n)}$$

where $[g]_{(m,n)}$ is the $m \times n$ matrix of $g$ with respect to the bases $(\mathbf{e})_{(n)}$ and $(\mathbf{f})_{(m)}$. Note, if $(\mathbf{a})_{(n)} = (\mathbf{e})_{(n)}A$ and $(\mathbf{b})_{(m)} = (\mathbf{f})_{(m)}B$ are any other bases of $\mathbb{R}^n$ and $\mathbb{R}^m$, then

$$g(\mathbf{x}) = (\mathbf{f})_{(m)}[g]_{(m,n)}(x)^{(n)} = (\mathbf{f})_{(m)}BB^{-1}[g]_{(m,n)}AA^{-1}(x)^{(n)} = (\mathbf{b})_{(m)}G_{ba}(x)_a^{(n)},$$

G. Sobczyk, *New Foundations in Mathematics: The Geometric Concept of Number*, DOI 10.1007/978-0-8176-8385-6_8,

where $G_{ba} = B^{-1}[g]_{(m,n)}A$ is the matrix of $g$ with respect to the new bases $(\mathbf{a})_{(n)}$ and $(\mathbf{b})_{(m)}$, and $(x)_a^{(n)} = A^{-1}(x)^{(n)}$.

The *rank* of the linear operator $g$ is the rank of either of the matrices $[g]_{(m,n)}$ or $G_{ba}$, or it can be defined independent of a basis by

**Definition 8.1.1.** An operator $g$ is said to have rank($g$)$= k$ if there is a set of $k$ vectors $\{\mathbf{a}_1, \mathbf{a}_2, \ldots, \mathbf{a}_k\}$ in $\mathbb{R}^n$ such that

$$\underline{g}(\mathbf{a}_{(k)}) = g(\mathbf{a}_1) \wedge g(\mathbf{a}_2) \wedge \cdots \wedge g(\mathbf{a}_k) \neq 0,$$

and there is no larger set of vectors from $\mathbb{R}^n$ with this property.

Suppose that rank($g$)$= k$, and $\underline{g}(\mathbf{a}_{(k)}) \neq 0$. (Recall that the vectors $\mathbf{a}_1, \mathbf{a}_2, \ldots, \mathbf{a}_k$ are linearly independent if and only if $\mathbf{a}_{(k)} = \mathbf{a}_1 \wedge \mathbf{a}_2 \wedge \cdots \wedge \mathbf{a}_k \neq 0$.) We can then find vectors $\mathbf{a}_{k+1}, \ldots, \mathbf{a}_n \in \mathbb{R}^n$, such that $(\mathbf{a})_{(n)} = \begin{pmatrix}\mathbf{a}_1 & \mathbf{a}_2 & \ldots & \mathbf{a}_n\end{pmatrix} = (\mathbf{e})_{(n)}A$ is a basis of $\mathbb{R}^n$, where $A$ is the matrix of transition from the basis $(\mathbf{e})_{(n)}$ to the basis $(\mathbf{a})_{(n)}$.

Indeed, we can do even better by requiring the vectors $\mathbf{a}_{k+1}, \mathbf{a}_{k+2}, \ldots, \mathbf{a}_n$ to be in the *kernel* of $g$; that is, $g(\mathbf{a}_j) = 0$ for $j = k+1, \ldots, n$. For suppose that for some such $j$, $g(\mathbf{a}_j) \neq 0$. Since we know that

$$g(\mathbf{a}_1) \wedge \cdots \wedge g(\mathbf{a}_k) \wedge g(\mathbf{a}_j) = 0,$$

or else rank$(g) > k$, it follows that

$$g(\mathbf{a}_j) - \sum_{i=1}^{k} \alpha_i g(\mathbf{a}_i) = 0,$$

or that

$$g(\mathbf{a}_j - \sum_{i=1}^{k} \alpha_i \mathbf{a}_i) = 0.$$

So if $g(\mathbf{a}_j) \neq 0$ for some $j > k$, we can redefine the vector $\mathbf{a}_j$ to be the vector

$$(\mathbf{a}_j)_{\text{new}} = (\mathbf{a}_j)_{\text{old}} - \sum_{i=1}^{k} \alpha_i \mathbf{a}_i$$

which is in the kernel of $g$. We have proved

**Theorem 8.1.2.** *Given a linear operator $g : \mathbb{R}^n \rightarrow \mathbb{R}^m$ with* rank$(g) = k$, *there exists a basis $(\mathbf{a})_{(n)} = (\mathbf{e})_{(n)}A$ such that*

$$\underline{g}(\mathbf{a}_{(k)}) = g(\mathbf{a}_1) \wedge g(\mathbf{a}_2) \wedge \cdots \wedge g(\mathbf{a}_k) \neq 0$$

*and $g(\mathbf{a}_j) = 0$ for $j = k+1, \ldots, n$.*

Theorem 8.1.2 has two important corollaries which easily follow.

**Corollary 8.1.3.** *For every linear operator $g$ on $\mathbb{R}^n$,*

$$\text{rank}[g] + \dim(\ker[g]) = n.$$

**Corollary 8.1.4.** *There exist bases* $(\mathbf{a})_{(n)}$ *and*

$$(\mathbf{b})_{(m)} = \left(g(\mathbf{a}_1) \ldots g(\mathbf{a}_k)\ \mathbf{b}_{k+1} \ldots \mathbf{b}_m\right),$$

*such that the matrix* $G_{ba} = [g_{ij}^{ba}]$ *is diagonal in the sense that* $g_{ij}^{ba} = 0$ *for* $i \neq j$.

## Exercises

1. Give a proof of Corollary 8.1.3 of Theorem 8.1.2.
2. Give a proof of Corollary 8.1.4 of Theorem 8.1.2.
3. Let $g : \mathbb{R}^3 \to \mathbb{R}^3$, be defined by

$$g(\mathbf{x}) = (\mathbf{e})_{(3)} \begin{pmatrix} 2 & 3 & 3 \\ 0 & 5 & 7 \\ 6 & 9 & 8 \end{pmatrix} \begin{pmatrix} x_1 \\ x_2 \\ x_3 \end{pmatrix}$$

   (a) What is the rank of $g$?
   (b) Find a basis $(\mathbf{a})_{(3)}$ of vectors such that the matrix $[g]_a$ is diagonal.

4. Let $g : \mathbb{R}^4 \to \mathbb{R}^3$ be a linear operator, and let

$$g(\mathbf{e})_{(4)} = (\mathbf{f})_{(3)}[g]_{(3,4)} = (\mathbf{f})_{(3)} \begin{pmatrix} 0 & 0 & -4 & 3 \\ -2 & -1 & -1 & 4 \\ 3 & 1 & -3 & 0 \end{pmatrix}$$

   where $[g]_{3,4}$ is the matrix of $g$ with respect to the standard basis $(\mathbf{e})_{(4)}$ of $\mathbb{R}^4$ and $(\mathbf{f})_{(3)}$ of $\mathbb{R}^3$.

   (a) What is the rank of the matrix $[g]_{3,4}$, and what is the rank of $g$?
   (b) Find the bases $(\mathbf{a})_{(4)}$ and $(\mathbf{b})_{(3)}$ of vectors of $\mathbb{R}^4$ and $\mathbb{R}^3$, such that the matrix $[g]_{ba}$ is diagonal.

5. Let $g : \mathbb{R}^4 \to \mathbb{R}^3$ be a linear operator, and let

$$g(\mathbf{e})_{(4)} = (\mathbf{f})_{(3)}[g]_{3,4} = (\mathbf{f})_{(3)} \begin{pmatrix} -2 & -4 & -4 & 0 \\ 0 & 0 & -4 & 3 \\ 4 & 0 & 0 & 3 \end{pmatrix}$$

   where $[g]_{(3,4)}$ is the matrix of $g$ with respect to the standard basis $(\mathbf{e})_{(4)}$ of $\mathbb{R}^4$ and $(\mathbf{f})_{(3)}$ is the standard basis of $\mathbb{R}^3$.

   (a) What is the rank of the matrix $[g]_{(3,4)}$, and what is the rank of $g$?
   (b) Find a bases $(\mathbf{a})_{(4)}$ and $(\mathbf{f})_{(3)}$ of vectors of $\mathbb{R}^4$ and $\mathbb{R}^3$, such that the matrix $[g]_{ba}$ is diagonal.

## 8.2 Characteristic Polynomial

Let $f : \mathbb{R}^n \rightarrow \mathbb{R}^n$ be a linear transformation. The set of all such operators $\mathrm{End}(\mathbb{R}^n)$ are called *linear endomorphisms*. We define a new auxiliary operator $h \in \mathrm{End}(\mathbb{R}^n)$ by $h(\mathbf{x}) = f(\mathbf{x}) - \lambda \mathbf{x}$ where $\lambda \in \mathbb{R}$. Recall (7.5) that $\det f = \partial_{(n)} \cdot f_{(n)}$. The *characteristic polynomial* $\varphi_f(\lambda)$ of $f$ is defined by

$$\varphi_f(\lambda) = \det h = \partial_{(n)} \cdot h_{(n)} \tag{8.1}$$

We directly calculate

$$\begin{aligned}\varphi_f(\lambda) = \det h &= \partial_{(n)} \cdot \left[(f(\mathbf{x}_1) - \lambda \mathbf{x}_1) \wedge \cdots \wedge (f(\mathbf{x}_n) - \lambda \mathbf{x}_n)\right] \\ &= \partial_{(n)} \cdot \left[f_{(n)} - \lambda\, f_{(n-1)} \wedge \mathbf{x}_n + \lambda^2\, f_{(n-2)} \wedge \mathbf{x}_{n-1} \wedge \mathbf{x}_n + \cdots + (-1)^n \lambda^n\, \mathbf{x}_{(n)}\right] \\ &= \partial_{(n)} \cdot f_{(n)} - \lambda\, \partial_{(n-1)} \cdot f_{(n-1)} + \lambda^2\, \partial_{(n-2)} \cdot f_{(n-2)} + \cdots + (-1)^n \lambda^n. \end{aligned} \tag{8.2}$$

The scalars $\partial_{(k)} \cdot f_{(k)}$ are called the *scalar invariants* of the operator $f$. Most widely known are the *trace* $\partial \cdot f = \partial_{\mathbf{x}} \cdot f(\mathbf{x})$ and the *determinant* $\det f = \partial_{(n)} \cdot f_{(n)}$.

The characteristic polynomial is a polynomial with real coefficients. Such a polynomial will not necessarily have real roots. However, using Gauss' fundamental theorem of algebra, http://en.wikipedia.org/wiki/Fundamental_theorem_of_algebra we may always factor $\varphi_f(\lambda)$ into a unique product of its distinct roots $\lambda_i$, called the *eigenvalues* of $f$. Thus,

$$\varphi_f(\lambda) = (-1)^n \prod_{i=1}^{r} (\lambda - \lambda_i)^{n_i}. \tag{8.3}$$

Each distinct root $\lambda_i \in \mathbb{C}$ has *algebraic multiplicity* $n_i$, and since $\varphi_f(\lambda)$ is a real polynomial, complex eigenvalues will always occur in conjugate pairs; if $\lambda_i = x_i + \mathrm{i} y_i$ with $y_i \neq 0$ is an eigenvalue with multiplicity $n_i$, then $\overline{\lambda}_i = x_i - \mathrm{i} y_i$ will also be an eigenvalue with the same multiplicity $n_i$ as $\lambda_i$.

For each eigenvalue $\lambda_i$, $\det(f - \lambda_i) = 0$ guarantees the existence of a vector $\mathbf{v}_i \in \mathbb{C}^n$, called a *complex eigenvector* of the operator $f$, with the property that

$$(f - \lambda_i)\mathbf{v}_i = f(\mathbf{v}_i) - \lambda_i \mathbf{v}_i = 0, \tag{8.4}$$

or equivalently, $f(\mathbf{v}_i) = \lambda_i \mathbf{v}_i$. Since we began our discussion assuming that $f \in \mathrm{End}(\mathbb{R}^n)$, we are in the awkward situation of having to explain our acceptance of complex eigenvalues and complex eigenvectors. The defensive explanation is that we have no choice but to consider complex eigenvalues and their corresponding complex eigenvectors because of the algebraic fact that polynomial equations over the real numbers will not always have real roots, but will always have a complete set of roots over the complex numbers.

It is common practice to simply replace the real scalar field $\mathbb{R}$ with the complex scalar field $\mathbb{C}$ and the real vector space $\mathbb{R}^n$ with the complex vector space $\mathbb{C}^n$.

In a like manner, the real geometric algebra $\mathbb{G}(\mathbb{R}^n)$ can be "complexified" to give the complex geometric algebra $\mathbb{G}(\mathbb{C}^n)$, where it is assumed that the complex scalars commute with vectors and therefore do not affect the linearity of the inner and outer product [10, 11]. We will leave the question about a geometric interpretation of i as unresolved at this point, although one possibility is to assume that it is a bivector from a larger geometric algebra [74]. In Chap. 10, i will be given an unambiguous geometric interpretation in the appropriate geometric algebra.

To clarify the issue, consider the following simple example. Let $f \in \text{End}(\mathbb{R}^3)$ be defined by $f(\mathbf{x}) = \mathbf{x} \cdot \mathbf{e}_{12}$ so that $f(\mathbf{e}_1) = \mathbf{e}_2$, $f(\mathbf{e}_2) = -\mathbf{e}_1$, and $f(\mathbf{e}_3) = 0$. The matrix $[f]$ of $f$ is given by

$$f(\mathbf{e})_{(3)} = (\mathbf{e})_{(3)} \begin{pmatrix} 0 & -1 & 0 \\ 1 & 0 & 0 \\ 0 & 0 & 0 \end{pmatrix}.$$

Calculating the characteristic polynomial $\varphi_f(\lambda)$ of $f$, we find

$$\varphi_f(\lambda) = \det(f - \lambda) = \det \begin{pmatrix} -\lambda & -1 & 0 \\ 1 & -\lambda & 0 \\ 0 & 0 & -\lambda \end{pmatrix} = -\lambda(\lambda - i)(\lambda + i),$$

so the eigenvalues are $\lambda_1 = i, \lambda_2 = -i$, and $\lambda_3 = 0$. The corresponding complex eigenvectors are $\mathbf{v}_1 = (i, 1, 0) = i\mathbf{e}_1 + \mathbf{e}_2$, $\mathbf{v}_2 = (-i, 1, 0) = -i\mathbf{e}_1 + \mathbf{e}_2$, and $\mathbf{v}_3 = \mathbf{e}_3$, respectively. It is interesting to note that $f(\mathbf{e}_1) = \mathbf{e}_1\mathbf{e}_{12}$ and $f(\mathbf{e}_2) = -\mathbf{e}_2\mathbf{e}_{12}$, opening up the possibility of giving $i$ the geometric interpretation of the bivector $\mathbf{e}_{12}$ the generator of rotations in the plane of $\mathbf{e}_{12}$. This interpretation is not entirely satisfactory because $\mathbf{e}_{12}$ anticommutes with all of the vectors in the plane of $\mathbf{e}_{12}$, but commutes with $\mathbf{e}_3$. We defer further consideration of these issues until we resolve them in Chap. 10.

We now come to one of the most beautiful, famous, and important theorems in linear algebra, known as the *Cayley–Hamilton theorem* after its discoverers. When the author proved this theorem using geometric algebra as a Ph.D. student at Arizona State University, he realized that geometric algebra has precisely the rich algebraic structure necessary for the efficient expressions and proofs of the key theorems of linear algebra [67]. The Cayley–Hamilton theorem tells us that if $\lambda$ is replaced by $f$ in (8.2), then we get the valid operator equation $\varphi_f(f)(\mathbf{x}) = 0$ for all $\mathbf{x} \in \mathbb{R}^n$. In this equation, 1 is treated as the identity operator on $\mathbb{R}^n$, and powers $f^k$ of $f$ are interpreted as functional composition.

The proof of the Cayley–Hamilton theorem follows directly from the identity

$$\partial_{(k)} \cdot (f_{(k)} \wedge \mathbf{x}) = \partial_{(k)} \cdot f_{(k)}\, \mathbf{x} - \partial_{(k-1)} \cdot (f_{(k-1)} \wedge f(\mathbf{x})), \tag{8.5}$$

for all $\mathbf{x} \in \mathbb{R}^n$. Choosing $k = n$ in the equation above, and calculating recursively, we get

$$
\begin{aligned}
0 &= \partial_{(n)} \cdot (f_{(n)} \wedge \mathbf{x}) = \partial_{(n)} \cdot f_{(n)}\, \mathbf{x} - \partial_{(n-1)} \cdot (f_{(n-1)} \wedge f(\mathbf{x})) \\
&= \partial_{(n)} \cdot f_{(n)}\, \mathbf{x} - \partial_{(n-1)} \cdot f_{(n-1)}\, f(\mathbf{x}) + \partial_{(n-2)} \cdot (f_{(n-2)} \wedge f^2(\mathbf{x})) = \dots \\
&= \partial_{(n)} \cdot f_{(n)}\, \mathbf{x} - \partial_{(n-1)} \cdot f_{(n-1)}\, f(\mathbf{x}) + \partial_{(n-2)} \cdot f_{(n-2)}\, f^2(\mathbf{x}) + \dots + (-1)^n f^n(\mathbf{x})
\end{aligned}
$$

for all $\mathbf{x} \in \mathbb{R}^n$.

## Exercises

1. Given the linear transformation defined by $f(\mathbf{e})_{(2)} = (\mathbf{e})_{(2)} \begin{pmatrix} 1 & 1 \\ 4 & 1 \end{pmatrix}$.
   (a) Find the characteristic polynomial of $f$. What are its eigenvalues?
   (b) Find the eigenvectors of $f$.
   (c) Verify that $f$ satisfies its characteristic polynomial. What are the trace and determinant of $f$?
2. Given the linear transformation defined by $f(\mathbf{e})_{(3)} = (\mathbf{e})_{(3)} \begin{pmatrix} 0 & -2 & -3 \\ -1 & 1 & -1 \\ 2 & 2 & 5 \end{pmatrix}$.
   (a) Find the characteristic polynomial of $f$. What are its eigenvalues?
   (b) Find the eigenvectors of $f$.
   (c) Verify that $f$ satisfies its characteristic polynomial. What are the trace and determinant of $f$?
3. (a) For a linear transformation $f : \mathbb{R}^2 \to R^2$ and $h(\mathbf{x}) = f(\mathbf{x}) - \lambda \mathbf{x}$, carry out the calculation (8.2),
   $$\varphi_f(\lambda) = \partial_{(2)} \cdot h_{(2)} = \partial_{(2)} \cdot f_{(2)} - \lambda \partial \cdot f + \lambda^2,$$
   for the characteristic polynomial of $f$.
   (b) Prove the Cayley–Hamilton theorem for $f$, that is,
   $$0 = \partial_{(2)} \cdot (f_{(2)} \wedge \mathbf{x}) = \partial_{(2)} \cdot f_{(2)}\, \mathbf{x} - (\partial \cdot f)\, f(\mathbf{x}) + f^2(\mathbf{x}).$$

## 8.3 Minimal Polynomial of *f*

Let $f \in \mathrm{End}(\mathbb{R}^n)$, $\mathbf{v} \in \mathbb{R}^n$, and $\mathbf{v} \neq 0$. Then for some integer $s > 1$,

$$\mathbf{v} \wedge f(\mathbf{v}) \wedge \dots \wedge f^s(\mathbf{v}) = 0, \tag{8.6}$$

but $\mathbf{v} \wedge f(\mathbf{v}) \wedge \cdots \wedge f^{s-1}(\mathbf{v}) \neq 0$. Equation (8.6) implies that there exists a *unique monic polynomial* $\psi_{\mathbf{v}}(\lambda)$ of smallest degree $s \leq n$, called the *minimal polynomial* of $\mathbf{v}$ with respect to $f$, with the property that $\psi_{\mathbf{v}}(f)(\mathbf{v}) = 0$. Whereas the polynomial $\psi_{\mathbf{v}}(\lambda)$ is a real polynomial, its roots may well be complex,

$$\psi_{\mathbf{v}}(\lambda) = \prod_{i=1}^{k}(\lambda - \beta_i)^{b_i} \tag{8.7}$$

where $\beta_i \in \mathbb{C}$ and $b_i \in \mathbb{N}$. Since the minimal polynomial $\psi_{\mathbf{v}}(\lambda)$ is of smallest degree and since the characteristic polynomial $\varphi_f(\lambda)$ given by (8.1) also annihilates $\mathbf{v}$, i.e., $\varphi_f(f)\mathbf{v} = 0$ for each $\mathbf{v} \in \mathbb{R}^n$, it follows that $\psi_{\mathbf{v}}(\lambda)|\varphi_f(\lambda)$ for each $\mathbf{v} \in \mathbb{R}^n$, meaning that the polynomial $\psi_{\mathbf{v}}(\lambda)$ divides the polynomial $\varphi_f(\lambda)$ with 0 remainder. This implies that $\beta_i = \lambda_i$ and $b_i \leq n_i$. We also know that for each root $\lambda_i$ of the characteristic polynomial, there is a complex eigenvector (8.4) such that $(f - \lambda_i)\mathbf{e}_i = 0$. Putting (8.7) and (8.4) together, it follows that the *minimal polynomial* $\psi_f(\lambda)$ of $f \in \text{End}(\mathbb{R}^n)$ is the *unique* monic polynomial of the form

$$\psi_f(\lambda) = \prod_{i=1}^{r}(\lambda - \lambda_i)^{m_i},$$

where $1 \leq m_i \leq n_i$ for each $1 \leq i \leq r$.

We now give the formal definition of the minimal polynomial $\psi_f(\lambda)$ of $f$.

**Definition 8.3.1.** The minimal polynomial of $f \in \text{End}(\mathbb{R}^n)$ is the unique monic polynomial of smallest degree of the form

$$\psi_f(\lambda) = \prod_{i=1}^{r}(\lambda - \lambda_i)^{m_i},$$

where $1 \leq m_i \leq n_i$ for each $1 \leq i \leq r$, which has the property that $\psi_f(f)(\mathbf{x}) = 0$ for all $\mathbf{x} \in \mathbb{R}^n$.

The minimal polynomial of $f \in \text{End}(\mathbb{R}^n)$ can always be found by first calculating the characteristic polynomial and then reducing the multiplicities $n_i$ of its eigenvalues until the minimal values $m_i$ are found with the property that $\psi_f(f)(\mathbf{x}) = 0$ for all $\mathbf{x} \in \mathbb{R}^n$.

## Exercises

1. Find the minimal polynomial for the following matrices:

(a) $T = \begin{pmatrix} 3 & 2 & 2 \\ 1 & 4 & 1 \\ -2 & -4 & -1 \end{pmatrix}$.

(b) $T = \begin{pmatrix} 5 & -6 & -6 \\ -1 & 4 & 2 \\ 3 & -6 & -4 \end{pmatrix}$.

2. Let $\mathbb{V}$ be a vector space of finite dimension. Find the minimal polynomial for the identity and the zero operators on $\mathbb{V}$.
3. In each case, find the minimal polynomial:

(a) $T = \begin{pmatrix} 0 & 0 & c \\ 1 & 0 & b \\ 0 & 1 & a \end{pmatrix}$.

(b) $T = \begin{pmatrix} 1 & 1 & 0 & 0 \\ -1 & -1 & 0 & 0 \\ -2 & -2 & 2 & 1 \\ 1 & 1 & -1 & 0 \end{pmatrix}$.

(c) $T = \begin{pmatrix} 0 & 1 & 0 & 1 \\ 1 & 0 & 1 & 0 \\ 0 & 1 & 0 & 1 \\ 1 & 0 & 1 & 0 \end{pmatrix}$.

4. Find a $3 \times 3$ matrix such that its minimal polynomial is $x^2$. How about for $x^3$?
5. Show that

$$\psi(x) = x^n + \alpha_{n-1}x^{n-1} + \alpha_{n-2}x^{n-2} + \cdots + \alpha_1 x + a_0,$$

for $\alpha_0, \alpha_1, \ldots, \alpha_{n-1} \in \mathbb{R}$, is the minimal polynomial for the matrix

$$A_\psi = \det \begin{pmatrix} 0 & 1 & \cdot & \ldots & 0 \\ 0 & 0 & 1 & \ldots & 0 \\ \cdot & \cdot & \cdot & \cdot & \cdot \\ \cdot & \cdot & \cdot & \cdot & 1 \\ -\alpha_0 & -\alpha_1 & \ldots & -\alpha_{n-2} & -\alpha_{n-1} \end{pmatrix}.$$

The matrix $A_\psi$ is called the *companion matrix* of the polynomial $\psi(x)$.
6. Let P be an operator on $R^n$ such that

$$P \begin{pmatrix} x_1 \\ x_2 \\ \cdot \\ \cdot \\ \cdot \\ \cdot \\ x_n \end{pmatrix} = \begin{pmatrix} x_1 \\ x_2 \\ \cdot \\ \cdot \\ x_k \\ 0 \\ \cdot \\ 0 \end{pmatrix}.$$

Thus, $P$ is the projection onto the first $k$ components of the vector $\mathbf{x}$.

(a) What is the characteristic polynomial for $P$?
(b) What is the minimal polynomial for $P$?
(c) Find the matrix $P$.

## 8.4 Spectral Decomposition

Let $f \in End(\mathbb{R}^n)$, and suppose that the minimal polynomial $\psi(x)$ has been found,

$$\psi(x) = \prod_{i=1}^{r}(x - x_i)^{m_i} \tag{8.8}$$

where the $x_i \in \mathbb{C}$ are distinct, and that the algebraic multiplicities have been ordered to satisfy

$$1 \le m_1 \le m_2 \le \cdots \le m_r.$$

In Chapter 1, for each polynomial of the form (8.8), we learned how to find a *spectral basis* of the form

$$\{s_1, q_1, \ldots, q_1^{m_i-1}, \ldots, s_r, q_r, \ldots, q_r^{m_r-1}\},$$

where $q_i^0 = s_i(x)$ for $i = 1, \ldots, r$ and $q_i^j \overset{\text{h}}{=} (x - x_i)^j s_i(x) \neq 0$ for $j = 1, \ldots, m_i - 1$. We can immediately apply this spectral basis to the linear operator $f$ to find the *spectral basis* of $f$ by simply making the replacement $x \to f$ in the spectral basis of $\psi(x)$. Since $\psi(\mathbf{x})$ is the minimal polynomial of $f$, the resulting spectral basis of operators

$$\{p_1, q_1, \ldots, q_1^{m_i-1}, \ldots, p_r, q_r, \ldots, q_r^{m_r-1}\},$$

where $p_i = s_i(f), q_i = q_i(f)$ for $i = 1, \cdots, r$, obeys exactly the same rules as the algebra of modular polynomials modulo $\psi(x)$.

Applying the spectral decomposition formula (1.18) to the operator $f$ gives the corresponding *spectral decomposition* of $f$,

$$f = \sum_{i=1}^{r}(f - x_i + x_i)p_i = \sum_{i=1}^{r}(x_i + q_i)p_i = \sum_{i=1}^{r} x_i p_i + q_i. \tag{8.9}$$

The reader may compare the spectral decomposition given here with that given by Nering in [60, p.270].

The following theorem summarizes the basic properties of a linear operator that follow from its spectral decomposition (8.9).

**Theorem 8.4.1.** *The spectral decomposition*

$$f = \sum_{i=1}^{r}(x_i + q_i)p_i,$$

*of* $f \in \mathrm{End}(\mathbb{R}^n)$ *implies:*

*(i)* *For each* $i = 1,2,\ldots,r$, $m_i \leq \mathrm{rank}(q_i) + 1 \leq n_i$, *where* $n_i \equiv \mathrm{rank}(p_i)$. *Also,* $\sum_{i=1}^{r} n_i = n$.

*(ii)* *The distinct eigenvalues of* $f$ *are* $x_1, x_2, \ldots, x_r$ *and* $\deg(\psi(x)) \leq n$. *The characteristic polynomial* $\varphi(x)$ *of* $f$ *is given by*

$$\varphi(x) = \prod_{i=1}^{r} (x - x_i)^{n_i}$$

*where* $n_i = \mathrm{rank}(p_i)$.

*(iii)* $f$ *is diagonalizable iff* $m_1 = m_2 = \cdots = m_r = 1$.

*(iv)* *The minimal polynomial* $\psi(x)$ *divides* $\varphi(x)$ *(Cayley- Hamilton theorem).*

*Proof.* (i) Let $n_i \equiv \mathrm{rank}(p_i)$. Then $n - n_i = \mathrm{kernel}(p_i)$ for each $i = 1,2,\ldots,r$. Since $p_i p_j = 0$ for $i \neq j$, it follows that

$$n_1 + n_2 + \cdots + n_r = n.$$

On the other hand, since $p_i q_i = q_i$, it follows that $\mathrm{rank}(q_i) \leq \mathrm{rank}(p_i)$. Since $q_i^{m_i} = 0$, we can find a vector $p_i(\mathbf{a}) = \mathbf{a}$ with the property that

$$\mathbf{a} \wedge q_i(\mathbf{a}) \wedge \cdots \wedge q_i^{m_i - 1}(\mathbf{a}) \neq 0.$$

It follows from this that

$$m_i \leq \mathrm{rank}(q_i) + 1 \leq \mathrm{rank}(p_i) = n_i.$$

(ii) Note that for each $i$, $f(q_i^{m_i-1}(\mathbf{v}_i)) = x_i q_i^{m_i-1}(\mathbf{v}_i)$ for some $p_i(\mathbf{v}_i) = \mathbf{v}_i$. It follows that $q_i^{m_i-1}(\mathbf{v}_i)$ is an eigenvector of $f$ with eigenvalue $x_i$ for $i = 1,2,\ldots,r$. Let us now calculate the characteristic polynomial $\varphi(x) \equiv \det(x - f)$. We find that

$$\varphi(x) = \prod_{i=1}^{r} (x - x_i)^{n_i}.$$

(iii) If $m_i = 1$, then $q_i^1 = 0$, for $i = 1,\ldots,r$ and

$$f = \sum_{i=1}^{r} x_i p_i,$$

so that $f$ is diagonalizable.

(iv) We have seen from part (i) that $m_i \leq n_i$ for $i = 1,2,\ldots,r$. It follows from part (ii) that

$$\psi(x) | \varphi(x).$$

□

Because the minimal polynomial, by part (iv) of the spectral decomposition theorem, always divides the characteristic polynomial, one way of finding the minimal polynomial is to first calculate the spectral decomposition (8.9) of $f$ using the characteristic polynomial and then by trial and error calculate the minimal values $m_i \le n_i$ for which $q_i^{m_i} = 0$, for $i = 1, \ldots, r$. Once again, we defer discussion of the meaning of complex eigenvalues until Chap. 10.

Let us see the spectral decomposition Theorem 8.4.1 in action. Suppose that the linear operator $f$ is specified by $f(\mathbf{e})_{(4)} = (\mathbf{e})_{(4)}[f]$ for the matrix

$$[f] = \begin{pmatrix} 1 & 0 & 2 & 3 \\ 0 & 1 & -1 & 2 \\ 0 & 0 & 1 & 1 \\ 0 & 0 & 0 & -1 \end{pmatrix}.$$

The characteristic polynomial of $f$ is $\varphi(\lambda) = (\lambda+1)(\lambda-1)^3$, and the minimal polynomial of $f$ is $\psi(\lambda) = (\lambda+1)(\lambda-1)^2$. Using the spectral basis for the minimal polynomial calculated in (1.19) of Chap. 1, we find that the spectral basis for $f$ is

$$[p_1] = \frac{1}{4}([f]-[1]_4)^2 = \begin{pmatrix} 0 & 0 & 0 & -1 \\ 0 & 0 & 0 & -\frac{5}{4} \\ 0 & 0 & 0 & -\frac{1}{2} \\ 0 & 0 & 0 & 1 \end{pmatrix},$$

$$[p_2] = -\frac{1}{4}([f]-3[1]_4)([f]+[1]_4) = \begin{pmatrix} 1 & 0 & 0 & 1 \\ 0 & 1 & 0 & \frac{5}{4} \\ 0 & 0 & 1 & \frac{1}{2} \\ 0 & 0 & 0 & 0 \end{pmatrix},$$

and

$$[q_2] = \frac{1}{2}([f]-[1]_4)([f]+[1]_4) = \begin{pmatrix} 0 & 0 & 2 & 1 \\ 0 & 0 & -1 & -\frac{1}{2} \\ 0 & 0 & 0 & 0 \\ 0 & 0 & 0 & 0 \end{pmatrix}.$$

In terms of the matrix $[f]$ of $f$, the spectral decomposition of $f$ is

$$[f] = -[p_1] + [p_2] + [q_2] \tag{8.10}$$

as can be easily checked.

From the spectral decomposition (8.10) of $[f]$, we can easily find a complete set of eigenvectors for $f$. For the eigenvalue $\lambda_1 = -1$, we have the column eigenvector

$$\mathbf{v}_1 = [p_1]\begin{pmatrix} 0 \\ 0 \\ 0 \\ 1 \end{pmatrix} = \begin{pmatrix} -1 \\ -\frac{5}{4} \\ -\frac{1}{2} \\ 1 \end{pmatrix}.$$

For the eigenvalue $\lambda_2 = 1$, we have the column eigenvectors

$$\mathbf{v}_2 = [p_2] \begin{pmatrix} 1 \\ 0 \\ 0 \\ 0 \end{pmatrix} = \begin{pmatrix} 1 \\ 0 \\ 0 \\ 0 \end{pmatrix}, \quad \mathbf{v}_3 = [q_2] \begin{pmatrix} 0 \\ 0 \\ 1 \\ 0 \end{pmatrix} = \begin{pmatrix} 2 \\ -1 \\ 0 \\ 0 \end{pmatrix}.$$

In addition, we have the *generalized column eigenvector*

$$\mathbf{v}_4 = [p_2] \begin{pmatrix} 0 \\ 0 \\ 1 \\ 0 \end{pmatrix} = \begin{pmatrix} 0 \\ 0 \\ 1 \\ 0 \end{pmatrix}.$$

In the basis of column eigenvectors $T = [\mathbf{v}]_{(4)}$ of $\mathbb{R}^4$, we find that

$$[f]_v = T^{-1}[f]T = \begin{pmatrix} -1 & 0 & 0 & 0 \\ 0 & 1 & 0 & 0 \\ 0 & 0 & 1 & 1 \\ 0 & 0 & 0 & 1 \end{pmatrix}.$$

## Exercises

1. Let f be a linear operator on $\mathbb{R}^3$, specified by $f(\mathbf{e})_{(3)} = (\mathbf{e})_{(3)}[f]$, where

$$[f] = \begin{pmatrix} 1 & -1 & 0 \\ -1 & 2 & 1 \\ 0 & -1 & 1 \end{pmatrix}.$$

   (a) Show that $\varphi(x) = (x-2)(x-1)^2$ is the characteristic polynomial $f$.
   (b) What is the minimal polynomial $\psi(x)$ of $f$?
   (c) Find the spectral basis and the spectral decomposition of $f$.
   (d) What are the eigenvectors and eigenvalues of $f$?
   (e) Can $f$ be diagonalized?

2. Let f be a linear operator on $\mathbb{R}^3$ specified by

$$[f] = \begin{pmatrix} 1 & 1 & 1 \\ 1 & 1 & 1 \\ 1 & 1 & 3 \end{pmatrix}.$$

(a) Show that $\varphi(x) = x(x-1)(x-4)$ is the characteristic polynomial $f$.
(b) What is the minimal polynomial $\psi(x)$ of $f$?
(c) Find the spectral basis and the spectral decomposition of $f$.
(d) What are the eigenvectors and eigenvalues of $f$?
(e) Can $f$ be diagonalized?

3. (a) Find the companion matrix $A_{h(x)}$, defined in Problem 5 of Section 8.3, for the polynomial
$$h(x) = (x-1)^2(x+1)^2.$$
(b) Find the spectral basis and the spectral decomposition of $A_{h(x)}$.
(c) What are the eigenvectors and eigenvalues of $A_{h(x)}$?
(d) Can $A_{h(x)}$ be diagonalized?
(e) Show that the characteristic polynomial and minimal are the same for $A_{h(x)}$.

4. Given the characteristic polynomials for the following matrices, find the minimal polynomials and spectral decompositions of these matrices.

(a) For $F = \begin{pmatrix} 3 & 2 & 2 \\ 1 & 4 & 1 \\ -2 & -4 & 1 \end{pmatrix}$, the characteristic polynomial is $\varphi(x) = (x-2)(x-3-2i)(x-3+i)$.

(b) For $F = \begin{pmatrix} -1 & 1 & -1 \\ 2 & 0 & -1 \\ 2 & 0 & 0 \end{pmatrix}$, the characteristic polynomial is $\varphi(x) = (x-1)(x-1+i)(x-1-i)$.

(c) For $F = \begin{pmatrix} 0 & 0 & -1 & 1 \\ 1 & 0 & 0 & 1 \\ 0 & 0 & -1 & 1 \\ -1 & 0 & 0 & 1 \end{pmatrix}$, the characteristic polynomial is $\varphi(x) = x^4$.

5. Write down the minimal polynomial and spectral basis for the following matrices:

(a) $T = \begin{pmatrix} 1 & 0 & -1 & 1 \\ 0 & 2 & 0 & 1 \\ 0 & 0 & 1 & 1 \\ 0 & 0 & 0 & 1 \end{pmatrix}$.

(b) $F = \begin{pmatrix} 3 & 1 & 2 \\ 0 & 1 & 1 \\ 0 & 0 & 1 \end{pmatrix}$,

(c) $S = \begin{pmatrix} 0 & 0 & -1 & 1 \\ 0 & 2 & 0 & 1 \\ 0 & 0 & 2 & 3 \\ 0 & 0 & 0 & 2 \end{pmatrix}$.

6. Let $f \in \mathrm{End}(\mathbb{R}^3)$ be specified by $[f] = \begin{pmatrix} 7 & 4 & -4 \\ 4 & -8 & -1 \\ -4 & -1 & -8 \end{pmatrix}$.

   (a) Find the characteristic polynomial of $f$.
   (b) Find the spectral basis of $f$ using the characteristic polynomial.
   (c) Using the spectral decomposition of $f$ based on the characteristic polynomial, find the minimal polynomial of $f$.

## *8.5 Jordan Normal Form

We show in this section how the *Jordan normal form* of a linear operator follows from the spectral decomposition (8.9) of $f$ [78]. Jordan form has been discussed by many authors [15, p.197] [46]. The key idea is to understand the structure of a nilpotent $q$ with $\mathrm{index}(q) = m$ so that $q^m = 0$, but $q^{m-1} \neq 0$. Our discussion is similar to that given in [15, p.189]. See also [75,76].

We can assume that $q^0 = p = 1$, where $p$ is the identity operator on $\mathbb{R}^n$. Since $q^{m-1} \neq 0$, there exists a vector $p(\mathbf{a}) = \mathbf{a}$ in $\mathbb{R}^n$ such that $q^{m-1}(\mathbf{a}) \neq 0$, for otherwise $\mathrm{index}(q) < m$ which would be a contradiction. This implies that

$$q^{(m-1)}(\mathbf{a}) \equiv \mathbf{a} \wedge q(\mathbf{a}) \wedge \cdots \wedge q^{m-2}(\mathbf{a}) \wedge q^{m-1}(\mathbf{a}) \neq 0. \tag{8.11}$$

For if $q^{(m-1)}(\mathbf{a}) = 0$, there would exists scalars $\alpha_i$ not all zero such that

$$\sum_{i=0}^{m-1} \alpha_i q^i(\mathbf{a}) = 0.$$

Multiplying this sum successively by $q^k$, for $k = m-1, m-2, \ldots, 1$ , shows that each $\alpha_i = 0$, a contradiction. We say that the vectors

$$\mathbf{a}, q(\mathbf{a}), \ldots, q^{m-2}(\mathbf{a}), q^{m-1}(\mathbf{a})$$

make up an *ascending Jordan chain* with height $m$.

More generally, we have the following:

**Lemma 8.5.1.** *If $q^{k_1-1}(\mathbf{a}_1) \wedge \cdots \wedge q^{k_r-1}(\mathbf{a}_r) \neq 0$ and $q^{k_1}(\mathbf{a}_1) = \cdots = q^{k_r}(\mathbf{a}_r) = 0$, then*

$$q^{(k_1-1)}(\mathbf{a}_1) \wedge \cdots \wedge q^{(k_r-1)}(\mathbf{a}_r) \neq 0.$$

*Proof.* Without loss of generality, we can assume that $k_1 \geq \cdots \geq k_r$. Suppose now that

$$\sum_{i=0}^{k_1-1} \alpha_{1i} q^i(\mathbf{a}_1) + \cdots + \sum_{j=0}^{k_r-1} \alpha_{rj} q^j(\mathbf{a}_r) = 0.$$

Applying $q^{k_1-1}$ to both sides of this equation gives

$$\alpha_{10}q^{k_1-1}(\mathbf{a}_1)+\cdots+\alpha_{r0}q^{k_r-1}(\mathbf{a}_r)=0,$$

which implies that $\alpha_{s0}=0$ whenever $k_s=k_1$. Successively applying

$$q^{k_1-2},\ldots,q^0=1,$$

to each side of the equation further reduces the equation until all $\alpha_{ij}$ are shown $=0$, proving the linear independence of all the $q^i(\mathbf{a}_j)$ and establishing the lemma. □

Armed with the above lemma, we finish the construction by finding a set of Jordan chains $q^{(k_1-1)}(\mathbf{a}_1),\ldots,q^{(k_r-1)}(\mathbf{a}_r)$ for the nilpotent $q$ of index $m=k_1$ which make up a basis for the block $p(\mathbb{R}^n)=\mathbb{R}^n$. This is equivalent to finding a set of ascending Jordan chains, as given in the lemma, which satisfies the additional restraint that $n=\sum_{i=1}^r k_i$. To see that this additional constraint can be satisfied, we systematically calculate of the ranks

$$n_i=\mathrm{rank}(q^{m-i})=n-\dim(\ker(q^{m-i}))$$

for $i=1,\ldots,m$. The successive ranks $n_i$, in turn, determine the number of *mountain tops* $r_i$ of ascending Jordan chains of height $m-i$.

The following recursive relationships between ranks $n_i$ and the corresponding $r_i$ easily follow:

$$n_1=\mathrm{rank}(q^{m-1})=\#(\mathbf{a}_1,\ldots,\mathbf{a}_{r_1})=r_1,$$

$$n_2=\mathrm{rank}(q^{m-2})=2r_1+r_2$$

$$n_3=\mathrm{rank}(q^{m-3})=3r_1+2r_2+r_3,$$

and more generally,

$$n_s=\mathrm{rank}(q^{m-s})=sr_1+(s-1)r_2+\cdots+r_s.$$

For $s=m$, we get the relationship

$$n_m=\mathrm{rank}(q^0)=mr_1+(m-1)r_2+\cdots+r_m=n.$$

Solving this system of equations for the numbers $r_i$ in terms of the ranks $n_i$ gives the recursive *mountain top formulas*

$$r_1=n_1,\ r_2=-2n_1+n_2,\ r_3=n_1-2n_2+n_3 \tag{8.12}$$

and more generally, $r_{3+i}=n_{1+i}-2n_{i+2}+n_{i+3}$ for $i\geq 1$.

The construction is illustrated in Fig. 8.1

Since the problem of finding the Jordan normal form of a linear operator reduces to finding the Jordan normal form of the nilpotents in its Jordan blocks, it is worthwhile to consider an example of a nilpotent. Suppose that we are given the nilpotent $q(\mathbf{e})_{(5)}=(\mathbf{e})_{(5)}[q]$, where

$$\begin{array}{llll}
q^{k_1-1} & \mathbf{a}_1 & \mathbf{a}_{r_1} & \\
q^{k_1-2} & \mathbf{a}_1\ q\mathbf{a}_1 & \mathbf{a}_{r_1}\ q\mathbf{a}_{r_1} & \cdots \\
\vdots & \cdots\quad\cdots & \cdots & \\
p & \mathbf{a}_1\ q\mathbf{a}_1\ \ldots\ q^{k_1-1}\mathbf{a}_1 & \mathbf{a}_{r_1}\ q\mathbf{a}_{r_1}\ \ldots\ q^{k_1-1}\mathbf{a}_r\ \ldots\ \mathbf{a}_{h_1}\ \cdots\ \mathbf{a}_{h_2} &
\end{array}$$

where $h_1 = r_1 + \cdots + r_{m-1} + 1$ and $h_2 = r_1 + \cdots + r_m$.

**Fig. 8.1** The operator $q^{k_1-j}$ on the left side of the figure operates on the vectors in the pyramids at the same level to form a basis of the subspace $q^{k_1-j}(R^n)$ and determines its rank. The first level of the pyramids is $q^0 \equiv p$. The rank of $q^{k_1-1}$ is the number $r_1$ of highest pyramids or mountain tops. If the next highest mountain top is more than one unit lower, then $\operatorname{rank}(q^{k_1-2}) = 2r_1$. If there are $r_2$ mountain tops which are one unit lower, then rank $q^{k_1-2} = 2r_1 + r_2$

$$[q] = \begin{pmatrix} 12 & 19 & 50 & -28 & 2 \\ 1 & 2 & 5 & -2 & 0 \\ 4 & 5 & 14 & -8 & 0 \\ 14 & 20 & 54 & -30 & 1 \\ 13 & 20 & 53 & -30 & 2 \end{pmatrix}.$$

We calculate

$$[q]^2 = \begin{pmatrix} -3 & -4 & -11 & 6 & 0 \\ 6 & 8 & 22 & -12 & 0 \\ -3 & -4 & -11 & 6 & 0 \\ -3 & -4 & -11 & 6 & 0 \\ -6 & -8 & -22 & 12 & 0 \end{pmatrix}$$

and $[q]^3 = 0$ so that $\operatorname{index}[q] = 3$. We then calculate the ranks $n_1 = \operatorname{rank}[q^2] = 1 = r_1$, $n_2 = \operatorname{rank}[q] = 2r_1 + r_2$ so that $r_2 = 1$, and $n_3 = \operatorname{rank}[q^0] = 3r_1 + 2r_2 = 5$. Choosing $\mathbf{a}_1 = \mathbf{e}_1$ and $\mathbf{a}_2 = \mathbf{e}_5$, then

$$A = \begin{pmatrix} \mathbf{a}_1 & q\mathbf{a}_1 & q^2\mathbf{a}_1 & \mathbf{a}_2 & q\mathbf{a}_2 \end{pmatrix} = \begin{pmatrix} 1 & 12 & -3 & 0 & 2 \\ 0 & 1 & 6 & 0 & 0 \\ 0 & 4 & -3 & 0 & 0 \\ 0 & 14 & -3 & 0 & 1 \\ 0 & 13 & -6 & 1 & 2 \end{pmatrix}$$

is a lower Jordan chain of generalized eigenvectors for which

$$[q]_{\mathbf{a}} = A^{-1}[q]A = \begin{pmatrix} 0 & 0 & 0 & 0 & 0 \\ 1 & 0 & 0 & 0 & 0 \\ 0 & 1 & 0 & 0 & 0 \\ 0 & 0 & 0 & 0 & 0 \\ 0 & 0 & 0 & 1 & 0 \end{pmatrix}$$

is the (lower) Jordan normal form of the nilpotent $q$. If, instead, we choose the generalized eigenvectors in reverse order,

$$B = \begin{pmatrix} q^2\mathbf{a}_1 & q\mathbf{a}_1 & \mathbf{a}_1 & q\mathbf{a}_2 & \mathbf{a}_2 \end{pmatrix} = \begin{pmatrix} -3 & 12 & 1 & 2 & 0 \\ 6 & 1 & 0 & 0 & 0 \\ -3 & 4 & 0 & 0 & 0 \\ -3 & 14 & 0 & 1 & 0 \\ -6 & 13 & 0 & 2 & 1 \end{pmatrix},$$

we obtain a (upper) Jordan chain of generalized eigenvectors for which

$$[q]_{\mathbf{b}} = B^{-1}[q]B = \begin{pmatrix} 0 & 1 & 0 & 0 & 0 \\ 0 & 0 & 1 & 0 & 0 \\ 0 & 0 & 0 & 0 & 0 \\ 0 & 0 & 0 & 0 & 1 \\ 0 & 0 & 0 & 0 & 0 \end{pmatrix}$$

is in (upper) Jordan normal form.

We can now finish the construction of the Jordan normal form of a linear operator.

**Theorem 8.5.2 (Jordan normal form).** *Given a linear operator $f(\mathbf{e})_{(n)} = (\mathbf{e})_{(n)}[f]$ on $\mathbb{R}^n$, there exists a basis $(\mathbf{a})_{(n)} = (\mathbf{e})_{(n)}A$ of Jordan chains for which $f(\mathbf{a})_{(n)} = (\mathbf{a})_{(n)}[f]_{\mathbf{a}}$, where $[f]_{\mathbf{a}} = A^{-1}[f]A$ is in Jordan normal form.*

*Proof.* From the spectral decomposition of $f$, we know that

$$f = \sum_{i=1}^{r}(x_i + q_i)p_i.$$

Let $(\mathbf{a})_{(n)}$ be a basis for $\mathbb{R}^n$ in which the operator

$$g = \sum_{i=1}^{r} x_i p_i$$

has the property that $g(\mathbf{a})_{(n)} = (\mathbf{a})_{(n)}[g]_{\mathbf{a}}$ where $D = [g]_{\mathbf{a}}$ is the diagonal matrix whose elements are

$$\operatorname{diag}(D) = \begin{pmatrix} x_1 & \dots & x_1 & x_2 & \dots & x_2 & \dots & x_r & \dots x_r \end{pmatrix}.$$

We now use the construction given in the lemma to construct an ascending Jordan basis of Jordan chains for each of the Jordan blocks $p_i, q_i$ in the spectral decomposition.

□

## Exercises

For the matrices $[f]$ given in Exercises 1–7, find the characteristic polynomial, the minimal polynomial, and the spectral form of $f$. Find a Jordan basis $B$ of generalized eigenvectors of $[f]$ for which $[f]_B = B^{-1}TB$ is in upper Jordan normal form.

1. $T = \begin{pmatrix} 1 & 2 \\ -2 & -1 \end{pmatrix}$.
2. $T = \begin{pmatrix} 1 & 1 \\ 1 & 1 \end{pmatrix}$.
3. $T = \begin{pmatrix} 1 & 1 \\ -1 & -1 \end{pmatrix}$.
4. $T = \begin{pmatrix} -1 & 1 & 0 \\ 0 & -1 & 0 \\ 0 & 0 & -1 \end{pmatrix}$.
5. $T = \begin{pmatrix} 0 & 1 & 2 \\ 0 & 0 & 0 \\ 0 & 0 & 0 \end{pmatrix}$.
6. $T = \begin{pmatrix} 1 & 2 & 3 \\ 0 & 4 & 5 \\ 0 & 0 & 6 \end{pmatrix}$.
7. $T = \begin{pmatrix} 0 & 0 & 1 & 0 & 0 \\ 0 & 0 & 0 & 1 & 0 \\ 0 & 0 & 0 & 0 & 1 \\ 0 & 0 & 0 & 0 & 0 \\ 0 & 0 & 0 & 0 & 0 \end{pmatrix}$.
8. Let $\mathrm{Poly}_4[x]$ be the space of polynomials of degree $\leq 3$,

$$\mathrm{Poly}_4[x] = \{a_1 + a_2x + a_3x^2 + a_4x^3 \mid a_1, a_2, a_3, a_4 \in \mathbb{R}.\}$$

We can represent a polynomial $f(x) = a_1 + a_2x + a_3x^2 + a_4x^3$ by the column vector $(a)^{\mathrm{T}}_{(4)} = \begin{pmatrix} a_1 & a_2 & a_3 & a_4 \end{pmatrix}^{\mathrm{T}}$ so that $f(x) = \begin{pmatrix} 1 & x & x^2 & x^3 \end{pmatrix} (a)^{\mathrm{T}}_{(4)}$. The derivative operator is represented with respect to the standard basis $\mathrm{Poly}_4[x] = \begin{pmatrix} 1 & x & x^2 & x^3 \end{pmatrix}$ by the matrix

$$D = \begin{pmatrix} 0 & 1 & 0 & 0 \\ 0 & 0 & 2 & 0 \\ 0 & 0 & 0 & 3 \\ 0 & 0 & 0 & 0 \end{pmatrix},$$

so that $f'(x) = \begin{pmatrix} 1 & x & x^2 & x^3 \end{pmatrix} D(a)^{\mathrm{T}}_{(4)}$. What is the Jordan form for the matrix $D$?

9. Let $A$ be a $5 \times 5$ complex matrix with characteristic polynomial $\varphi(x) = (x-2)^3(x+7)^2$ and minimal polynomial $p = (x-2)^2(x+7)$. What is the Jordan form for A ?
10. Given the matrices

$$R = \begin{pmatrix} 1 & -1 \\ -12 & 2 \end{pmatrix}, \quad S = \begin{pmatrix} 1 & -1 \\ -12 & 2 \end{pmatrix}, \quad T = \begin{pmatrix} 1 & -1 \\ -12 & 2 \end{pmatrix}$$

(a) Find the spectral basis of each of these matrices.
(b) Find a Jordan basis $C$ of column vectors for each of these matrices.
(c) Find the Jordan form for each of these matrices.

11. Given the matrices

$$R = \begin{pmatrix} 1 & -1 & 1 \\ 0 & -1 & 0 \\ 0 & -1 & 0 \end{pmatrix}, \quad S = \begin{pmatrix} 6 & -2 & 2 \\ 1 & 2 & 0 \\ -1 & -2 & 0 \end{pmatrix}, \quad T = \begin{pmatrix} 3 & -1 & 1 \\ 1 & 1 & 0 \\ -1 & -1 & 0 \end{pmatrix}$$

(a) Find the spectral basis of each of these matrices.
(b) Find a Jordan basis $C$ of column vectors for each of these matrices.
(c) Find the Jordan form for each of these matrices.

12. Given the minimal polynomial $\psi(x) = x(x-2)(x-1)^2$ of $T = \begin{pmatrix} 1 & -1 & 1 & 1 \\ 0 & 1 & 1 & 0 \\ 0 & 0 & 2 & 1 \\ 0 & 0 & 0 & 0 \end{pmatrix}$.

(a) Find the spectral basis for this matrix.
(b) Find a Jordan basis $C$ of column vectors this matrix.
(c) Find the Jordan form of this matrix.

13. Given the minimal polynomial $\psi(x) = x(x-2)(x-1)$ of $T = \begin{pmatrix} 1 & 0 & 0 & -1 \\ 0 & 1 & 0 & -1 \\ 0 & 0 & 2 & 1 \\ 0 & 0 & 0 & 0 \end{pmatrix}$.

(a) Find the spectral basis of this matrix.
(b) Find a Jordan basis $C$ of column vectors for this matrix.
(c) Find the Jordan form for this matrix.

14. Given the minimal polynomial $\psi(x) = (x-2)(x-1)^2$ of $T = \begin{pmatrix} 1 & 0 & 1 & 0 \\ 0 & 1 & 1 & 0 \\ 0 & 0 & 2 & 1 \\ 0 & 0 & 0 & 1 \end{pmatrix}$.

(a) Find the spectral basis for this matrix.
(b) Find a Jordan basis $C$ of column vectors for this matrix.
(c) Find the Jordan form of this matrix.

15. Given the minimal polynomial $\psi(x) = (x-1)^3$ of $T = \begin{pmatrix} 1 & 0 & 1 & 0 \\ 0 & 1 & 1 & 0 \\ 0 & 0 & 1 & 1 \\ 0 & 0 & 0 & 1 \end{pmatrix}$.

(a) Find the spectral basis for this matrix.
(b) Find a Jordan basis $C$ of column vectors for this matrix.
(c) Find the Jordan form of this matrix.

# Chapter 9
# Linear and Bilinear Forms

*I am not for imposing any sense on your words: you are at liberty to explain them as you please. Only, I beseech you, make me understand something by them.*

–Bishop Berkeley

Geometric algebra is not used in this chapter. The material presented is closely related to the material in Sect. 7.1 but represents a change of viewpoint. Instead of talking about the *reciprocal basis* of a given basis, we introduce the concept of a *dual basis*. The relationship between a bilinear and a quadratic form is discussed, and Sylvester's famous law of inertia is proven. The material lays the foundation for studying geometric algebras of arbitrary signatures in later chapters.

## 9.1 The Dual Space

Let $\mathbb{V} = \mathbb{K}^n$ be an $n$-dimensional vector space over the field $\mathbb{K}$ where $\mathbb{K} = \mathbb{R}$ or $\mathbb{K} = \mathbb{C}$ the real or complex numbers.

**Definition 9.1.1.** A linear form is a function $f : \mathbb{V} \to \mathbb{K}$ which satisfies the property that

$$f(\alpha\mathbf{x} + \beta\mathbf{y}) = \alpha f(\mathbf{x}) + \beta f(\mathbf{y})$$

for all $\mathbf{x}, \mathbf{y} \in \mathbb{V}$ and $\alpha, \beta \in \mathbb{K}$.

The set $\mathbb{V}^*$ of all linear forms,

$$\mathbb{V}^* \equiv \{f |\ f : \mathbb{V} \to \mathbb{K}\},$$

makes up an $n$-dimensional vector space, called the *dual space* of the vector space $\mathbb{V}$, with the operations of *addition* and *scalar multiplication* of forms $f, g \in \mathbb{V}^*$ defined by

G. Sobczyk, *New Foundations in Mathematics: The Geometric Concept of Number*, DOI 10.1007/978-0-8176-8385-6_9,

$$(f+g)(\mathbf{x}) = f(\mathbf{x}) + g(\mathbf{x})$$

for all $\mathbf{x} \in \mathbb{V}$ and

$$(\alpha f)(\mathbf{x}) = \alpha f(\mathbf{x})$$

for all $\mathbf{x} \in \mathbb{V}$ and all $\alpha \in \mathbb{K}$. Once a basis $(\mathbf{e})_{(n)}$ of $\mathbb{V}$ is chosen, a corresponding *dual basis* $(\mathbf{e})^{(n)}$ of $\mathbb{V}^*$ is defined by the requirement that

$$\mathbf{e}^i(\mathbf{e}_j) = \delta^i_j \tag{9.1}$$

for all $i, j = 1, 2, \ldots, n$, where $\delta^i_j$ is the Kronecker delta ($\delta^i_i = 1$ and $\delta^i_j = 0$ for $i \neq j$).

Recall that a vector $\mathbf{x} \in \mathbb{V}$ is a column vector

$$\mathbf{x} = (x)^{\mathrm{T}}_{(n)} = \begin{pmatrix} x^1 \\ x^2 \\ \cdot \\ \cdot \\ x^n \end{pmatrix} = (\mathbf{e})_{(n)}(x)^{(n)},$$

where $x^i \in \mathbb{K}$ for $i = 1, 2, \ldots, n$, and where the *standard basis* of column vectors $\mathbf{e}_i$ in $(\mathbf{e})_{(n)}$ is defined by

$$\mathbf{e}_1 = \begin{pmatrix} 1 \\ 0 \\ 0 \\ \cdot \\ \cdot \\ 0 \end{pmatrix}, \ \mathbf{e}_2 = \begin{pmatrix} 0 \\ 1 \\ 0 \\ \cdot \\ \cdot \\ 0 \end{pmatrix}, \ \ldots, \ \mathbf{e}_n = \begin{pmatrix} 0 \\ 0 \\ \cdot \\ \cdot \\ 0 \\ 1 \end{pmatrix}.$$

Analogously, by the *standard row basis* $(\mathbf{e})^{(n)}$ of $\mathbb{V}^*$, we mean

$$(\mathbf{e})^{(n)} = \begin{pmatrix} \mathbf{e}^1 \\ \mathbf{e}^2 \\ \cdot \\ \cdot \\ \mathbf{e}^n \end{pmatrix} = (\mathbf{e})^{\mathrm{T}}_{(n)},$$

where

$$\mathbf{e}^1 = (1 \quad 0 \quad 0 \quad \ldots \quad 0),\ \mathbf{e}^2 = (0 \quad 1 \quad 0 \quad \ldots \quad 0), \ldots,$$
$$\mathbf{e}^n = (0 \quad 0 \quad \ldots \quad 0 \quad 1),$$

A row vector $\mathbf{y} \in \mathbb{V}^*$ can now be expressed as the linear combination of the standard row basis vectors,

$$\mathbf{y} = \begin{pmatrix} y_1 & y_2 & \ldots & y_n \end{pmatrix} = (y)_{(n)}(\mathbf{e})^{(n)}.$$

Since the dimension $n$ of the column space $\mathbb{V}$ is equal to the dimension $n$ of the row space $\mathbb{V}^*$, the matrix multiplication of a row vector and a column vector can be used to define the form $f \in \mathbb{V}^*$. Thus, we can think about the elements of the dual space $\mathbb{V}^*$ as row vectors,

$$\mathbf{e}^i = \mathbf{e}^i(\mathbf{e})_{(n)} = (\delta_1^i, \ldots, \delta_n^i).$$

Similarly, we can write

$$\mathbf{e}_i = (\mathbf{e})^{(n)}\mathbf{e}_i$$

where $\mathbf{e}_i \in \mathbb{V}$ is the corresponding column vector in $\mathbb{V}$.

We have the following theorem:

**Theorem 9.1.2.** *In terms of the dual basis* $(\mathbf{e})^{(n)}$ *of* $\mathbb{V}^*$, *the form* $f \in \mathbb{V}^*$ *can be written* $f = \sum f_i \mathbf{e}^i = (f)_{(n)}(\mathbf{e})^{(n)}$ *where* $\mathbf{e}^i(\mathbf{e}_j) = \delta_j^i$ *and* $f_i \in \mathbb{K}$, *and*

$$(f)_{(n)} = (f_1, f_2, \ldots, f_n)$$

*is the row vector of components of $f$ with respect to* $(\mathbf{e})^{(n)}$.

*Proof.* Define $f_i \equiv f(\mathbf{e}_i)$. Then

$$f(\mathbf{e})_{(n)} = (f\mathbf{e})_{(n)} = \begin{pmatrix} f_1 & f_2 & \ldots & f_n \end{pmatrix}.$$

By using the linearity of $f$, we then find for $\mathbf{x} = (\mathbf{e})_{(n)}(x)^{(n)} \in \mathbb{V}$,

$$f(\mathbf{x}) = f(\mathbf{e})_{(n)}(x)^{(n)} = (f)_{(n)}(x)^{(n)} = \sum_{i=1}^{n} x_i f_i$$

as required. □

From the last equation of the above proof, we see that the equation of an $(n-1)$-dimensional hyperplane passing through the origin of $\mathbb{V}$, defined by the linear form $f \in \mathbb{V}^*$, is given by

$$f(\mathbf{x}) = (f)_{(n)}(x)^{(n)} = \sum_{i=1}^{n} x_i f_i = 0. \tag{9.2}$$

Taking advantage of matrix notation, notice that the relationship between the standard basis $(\mathbf{e})_{(n)}$ of $\mathbb{V}$ and the dual standard basis of $\mathbb{V}^*$ can be expressed in the form

$$(\mathbf{e})^{(n)}(\mathbf{e})_{(n)} = [1]_n$$

where $[1]_n$ is the $n \times n$ identity matrix.

*Example:* Let $\mathbb{V}=\mathbb{R}^3$ be a 3 dimensional vector space over $\mathbb{R}$ with the standard bases $(\mathbf{e})_{(3)}$, and let $(\mathbf{a})_{(3)}=(\mathbf{e})_{(3)}A$ be a second basis of $\mathbb{V}$ defined by

$$A=\begin{pmatrix}1 & 1 & 0\\ 1 & -1 & 2\\ 2 & 0 & -1\end{pmatrix}.$$

Find the corresponding dual bases $(\mathbf{a})^{(3)}$ of $\mathbb{V}^*$ in terms of the dual basis $(\mathbf{e})^{(3)}$ of $\mathbb{V}^*$.

**Solution.** The dual basis $(\mathbf{a})^{(3)}$ must satisfy the defining relationship

$$(\mathbf{a})^{(3)}(\mathbf{a})_{(3)}=A^{-1}(\mathbf{e})^{(3)}(\mathbf{e})_{(3)}A=[1]_3.$$

Thus, the problem has been reduced to finding the inverse of the matrix $A$,

$$A^{-1}=\frac{1}{6}\begin{pmatrix}1 & 1 & 2\\ 5 & -1 & -2\\ 2 & 2 & -2\end{pmatrix}.$$

Whereas the column vectors of $A$ define the vectors $(\mathbf{a})_{(3)}$, the row vectors of $A^{-1}$ define the corresponding dual vectors $(\mathbf{a})^{(3)}=(\mathbf{a})^{(3)}(\mathbf{e})^{(3)}$.

The above example shows how to change basis in the dual vector space $\mathbb{V}^*$. In general, since

$$(\mathbf{a})^{(n)}(\mathbf{a})_{(n)}=[1]_n=A^{-1}(\mathbf{e})^{(n)}(\mathbf{e})_{(n)}A,$$

it follows that

$$(\mathbf{a})^{(n)}=A^{-1}(\mathbf{e})^{(n)}.$$

Given $\mathbf{x}\in\mathbb{V}$,

$$\mathbf{x}=(\mathbf{e})_{(n)}(x)^{(n)}=(\mathbf{e})_{(n)}AA^{-1}(x)^{(n)}=(\mathbf{a})_{(n)}A^{-1}(x)^{(n)}.$$

To express the linear form $f\in\mathbb{V}^*$ in the dual basis $(\mathbf{a})^{(n)}$, we write

$$f(\mathbf{x})=f(\mathbf{e})_{(n)}(x)^{(n)}=f(\mathbf{e})_{(n)}AA^{-1}(x)^{(n)}=f(\mathbf{a})_{(n)}A^{-1}(x)^{(n)}$$

or

$$f(\mathbf{x})=f(\mathbf{a})_{(n)}(x)_{\mathbf{a}}^{(n)},$$

where $(\mathbf{x})_{\mathbf{a}}^{(n)}=A^{-1}(x)^{(n)}$.

## Exercises

1. Let $\mathbb{V} = \mathbb{V}^3$ be a 3-dimensional vector space over $\mathbb{R}$, and let $(\mathbf{a})_{(3)} = (\mathbf{e})_{(3)}A$ be a basis of $\mathbb{V}$ where the matrix $A = \begin{pmatrix} 1 & 1 & 0 \\ 1 & -1 & 2 \\ 2 & 0 & -1 \end{pmatrix}$ is the same as in the above example.

   (a) Let $f = 5\mathbf{e}^1 - 3\mathbf{e}^2 + \mathbf{e}^3$. Find $f$ in the dual basis $(\mathbf{a})^{(3)}$.
   (b) Find a vector $\mathbf{x} \in \mathbb{V}$, $\mathbf{x} \neq 0$, such that $f(\mathbf{x}) = 0$.
   (c) Find $f(2\mathbf{e}_1 - \mathbf{e}_2 + 3\mathbf{e}_3)$ and $f(2\mathbf{a}_1 - \mathbf{a}_2 + 3\mathbf{a}_3)$.

2. Let $f \in \mathbb{V}^*$. Show that if $f(\mathbf{x}) = 0$ for all $\mathbf{x} \in \mathbb{V}$, then $f = 0$.
3. For each $\mathbf{x} \in \mathbb{V}$, define $f(\mathbf{x}) = 2$. Is $f \in \mathbb{V}^*$? Justify your answer.
4. For the basis $(\mathbf{a})_{(3)} = \begin{pmatrix} 1 & 0 & 0 \\ 1 & 1 & 0 \\ 1 & 1 & 1 \end{pmatrix}$ of $\mathbb{V}^3$, find the corresponding dual basis $(\mathbf{a})^{(3)}$ of $\mathbb{V}^*$.
5. For the basis $(\mathbf{a})_{(3)} = \begin{pmatrix} 1 & 0 & -1 \\ -1 & 1 & 0 \\ 0 & 1 & 1 \end{pmatrix}$ of $\mathbb{V}^3$, find the corresponding dual basis $(\mathbf{a})^{(3)}$ of $\mathbb{V}^*$.
6. Let $\mathbb{R}[x]$ denote the vector space of all polynomials over $\mathbb{R}$. Determine which of the following are linear forms in the dual space $\mathbb{R}[x]^*$:

   (a) For $g(x) \in \mathbb{R}[x]$, define $f(g(x)) = g(0) + g(1)$.
   (b) For $g(x) \in \mathbb{R}[x]$, define $f(g(x)) = g(0)$.
   (c) For $g(x) \in \mathbb{R}[x]$, define $f(g(x)) = \int_0^1 g(x)\mathrm{d}x$.

7. Let $f \in \mathbb{V}^*$ the dual space of $\mathbb{V} = \mathbb{R}^3$. Show that the set of all points $\mathbf{x} \in \mathbb{R}^3$ such that $f(\mathbf{x}) = 0$ is a plane passing through the origin. Show this for the general case for $f \in \mathbb{V}^*$ the dual space of $\mathbb{V} = \mathbb{R}^n$ as stated in (9.2).
8. Let $\mathbb{R}[x]_{h(x)}$ for $h(x) = (x-1)x^2$.

   (a) Show that $\{D_0\} \equiv \{D_0^0, D_0^1, 1/2D_0^2\}$ is a dual basis of $\mathbb{R}[x]_h^*$ corresponding to the standard basis $\{x\} = \{1, x, x^2\}$ of $\mathbb{R}[x]_h$, where $D_0^i g(x) = g^{(i)}(0)$ is the $i$th derivative of $g(x) \in \mathbb{R}[x]_h$ evaluated at $x = 0$.
   (b) Find the spectral basis $\{p_1, p_2, q_2\}$ of $\mathbb{R}[x]_h$ and the corresponding dual spectral basis of $\mathbb{R}[x]_h^*$. Finding the Lagrange–Sylvester interpolation polynomial of $f(x) \in \mathbb{R}[x]$ is equivalent to representing $f(x)$ in terms of the spectral basis of $\mathbb{R}[x]_h$.
   (c) Find the matrix of transition between the standard basis and the spectral basis of $\mathbb{R}[x]_h$ and the corresponding matrix of transition for the dual spectral basis.
   (d) Show that $\mathbb{R}[x]_h{}^{**} \equiv \mathbb{R}[x]_h$.

## 9.2 Bilinear Forms

Let $\mathbb{V} = \mathbb{R}^m$ and $\mathbb{W} = \mathbb{R}^n$ be $m$- and $n$-dimensional vector spaces over the real numbers $\mathbb{R}$.

**Definition 9.2.1.** A bilinear form is a function $b : \mathbb{V} \times \mathbb{W} \to \mathbb{R}$ satisfying the properties

(i) $b(\alpha\mathbf{x}+\beta\mathbf{y},\mathbf{z}) = \alpha b(\mathbf{x},\mathbf{z})+\beta b(\mathbf{y},\mathbf{z})$
(ii) $b(\mathbf{x},\alpha\mathbf{w}+\beta\mathbf{z}) = \alpha b(\mathbf{x},\mathbf{w})+\beta b(\mathbf{y},\mathbf{z})$

for all scalars $\alpha,\beta \in \mathbb{R}$ and vectors $\mathbf{x},\mathbf{y} \in \mathbb{V}$ and $\mathbf{w},\mathbf{z} \in \mathbb{W}$.

We begin the study of a bilinear form $b$ by representing it in terms of a given basis $(\mathbf{e})_{(m)}$ of $\mathbb{V}$ and $(\mathbf{f})_{(n)}$ of $\mathbb{W}$. Letting $\mathbf{x} = (x)_{(m)}(\mathbf{e})^{\mathrm{T}}_{(m)}$ and $\mathbf{y} = (\mathbf{f})_{(n)}(y)^{\mathrm{T}}_{(n)}$ be vectors in $\mathbb{V}$ and $\mathbb{W}$ expressed in the bases $(\mathbf{e})_{(m)}$ and $(\mathbf{f})_{(n)}$, respectively, we find

$$b(\mathbf{x},\mathbf{y}) = (x)_{(m)}b((\mathbf{e})^{\mathrm{T}}_{(m)},(\mathbf{f})_{(n)})(y)^{\mathrm{T}}_{(n)} = (x)_{(m)}(b)_{(m,n)}(y)^{\mathrm{T}}_{(n)}$$

where

$$(b)_{(m,n)} = \begin{pmatrix} b(\mathbf{e}_1,\mathbf{f}_1) & b(\mathbf{e}_1,\mathbf{f}_2) & \dots & b(\mathbf{e}_1,\mathbf{f}_n) \\ b(\mathbf{e}_2,\mathbf{f}_1) & b(\mathbf{e}_2,\mathbf{f}_2) & \dots & b(\mathbf{e}_2,\mathbf{f}_n) \\ \dots & \dots & \dots & \dots \\ b(\mathbf{e}_m,\mathbf{f}_1) & b(\mathbf{e}_m,\mathbf{f}_2) & \dots & b(\mathbf{e}_m,\mathbf{f}_n) \end{pmatrix}$$

is the *matrix* of the bilinear form $b$ with respect to the basis $(\mathbf{e})_{(m)}$ of $\mathbb{V}$ and $(\mathbf{f})_{(n)}$ of $\mathbb{W}$.

We now restrict attention to the case when $\mathbb{V} = \mathbb{W} = \mathbb{R}^n$ and $(\mathbf{e})_{(n)} = (\mathbf{f})_{(n)}$. A bilinear form $b$, $b : \mathbb{V} \times \mathbb{V} \to \mathbb{R}$, is said to be *symmetric* if $b(\mathbf{x},\mathbf{y}) = b(\mathbf{y},\mathbf{x})$ for all vectors $\mathbf{x},\mathbf{y} \in \mathbb{V}$. Clearly, if $b$ is a symmetric bilinear form, then its matrix $B = (b)_{(n,n)}$ is a symmetric matrix, i.e., $B^{\mathrm{T}} = B$. Let us now see how the matrix of a bilinear form transforms when we change from the basis $(\mathbf{e})_{(n)}$ to a new basis $(\mathbf{a})_{(n)} = (\mathbf{e})_{(n)}A$.

Recall that

$$\mathbf{x} = (\mathbf{e})_{(n)}(x)^{\mathrm{T}}_{(n)} = (\mathbf{e})_{(n)}AA^{-1}(x)^{\mathrm{T}}_{(n)} = (\mathbf{a})_{(n)}(x_a)^{\mathrm{T}}_{(n)}$$

where $(x_a)^{\mathrm{T}}_{(n)} = A^{-1}(x)^{\mathrm{T}}_{(n)}$. Since real numbers commute with vectors, we can take the transpose of this equation, getting

$$\mathbf{x} = \mathbf{x}^{\mathrm{T}} = (x)_{(n)}(\mathbf{e})^{\mathrm{T}}_{(n)} = (x_a)_{(n)}(\mathbf{a})^{\mathrm{T}}_{(n)},$$

where

$$(x_a)_{(n)} = (x)_{(n)}(A^{-1})^{\mathrm{T}} = (x)_{(n)}(A^{\mathrm{T}})^{-1}.$$

For the bilinear form $b(\mathbf{x},\mathbf{y})$, we find

$$b(\mathbf{x},\mathbf{y}) = (x)_{(n)}B(y)^{\mathrm{T}}_{(n)} = (x)_{(n)}(A^{\mathrm{T}})^{-1}A^{\mathrm{T}}BAA^{-1}(y)^{\mathrm{T}}_{(n)} = (x_a)_{(n)}B_a(y_a)^{\mathrm{T}}_{(n)} \quad (9.3)$$

where $B_a = A^{\mathrm{T}}BA$. We have thus found the transformation rule $B_a = A^{\mathrm{T}}BA$ for a bilinear form. It is important to note that this rule is quite different from the rule $T_a = A^{-1}TA$ which we previously derived for the transformation of the matrix $T$ of a linear transformation. However, in the special case that $A^{\mathrm{T}} = A^{-1}$, the rules become the same.

A matrix $A$ is called *symmetric* if $A^{\mathrm{T}} = A$ where $A^{\mathrm{T}}$ is the transpose of $A$. A matrix $S$ is said to be *skew-symmetric* if $S^{\mathrm{T}} = -S$. Two matrices $S$ and $T$ are said to be *similar* if $T = A^{-1}SA$, and $S$ and $T$ are said to be *congruent* if $T = A^{\mathrm{T}}SA$ for some nonsingular matrix of transition $A$.

The *rank* of a bilinear form is defined to be the rank of it's matrix: $\mathrm{rank}(b) \equiv \mathrm{rank}(B)$. Note that the rank of a bilinear form is independent of the choice of a basis. A bilinear form $b$ is *nonsingular* if $\mathrm{rank}(b) = \dim(\mathbb{V})$, or, equivalently, if $\det(B) \neq 0$. Suppose that in some basis $(\mathbf{a})_{(n)}$ the matrix $B$ has the diagonal form

$$B_a = \begin{pmatrix} \beta_1 & 0 & \dots & 0 \\ 0 & \beta_2 & \dots & 0 \\ \cdot & \cdot & \dots & \cdot \\ 0 & \dots & 0 & \beta_n \end{pmatrix}.$$

Then in the basis $(\mathbf{a})_{(n)} = \begin{pmatrix} \mathbf{a}_1 & \mathbf{a}_2 & \dots & \mathbf{a}_n \end{pmatrix}$

$$b(\mathbf{x}, \mathbf{y}) = (x_a)_{(n)} B_a (y_a)^{\mathrm{T}}_{(n)} = \beta_1 x_{a1} y_{a1} + \beta_2 x_{a2} y_{a2} + \dots + \beta_n x_{an} y_{an},$$

called the *diagonal form* of $b(\mathbf{x}, \mathbf{y})$. We will soon give a method for finding a basis which diagonalizes a given bilinear form.

## Exercises

1. Show that it makes sense to take the *transpose of a vector* by verifying for $\mathbf{x} \in \mathbb{R}^n$,

$$\mathbf{x} = (\mathbf{e})_{(n)} (x)^{\mathrm{T}}_{(n)} = \sum_{i=1}^{n} \mathbf{e}_i x_i = \sum_{i=1}^{n} x_i \mathbf{e}_i = (x)_{(n)} (\mathbf{e})^{\mathrm{T}}_{(n)} = \mathbf{x}^{\mathrm{T}}.$$

2. Let the bilinear form $b : \mathbb{R}^2 \times \mathbb{R}^3 \to \mathbb{R}$ be defined by

$$b(\mathbf{x}, \mathbf{y}) = x_1 y_1 + 2x_1 y_2 - x_2 y_1 - x_2 y_2 + 6x_1 y_3$$

for

$$\mathbf{x} = (\mathbf{e})_{(2)} (x)^{\mathrm{T}}_{(2)} = (\mathbf{e})_{(2)} \begin{pmatrix} x_1 \\ x_2 \end{pmatrix}, \quad \text{and} \quad \mathbf{y} = (\mathbf{e})_{(3)} (y)^{\mathrm{T}}_{(3)} = (\mathbf{e})_{(3)} \begin{pmatrix} y_1 \\ y_2 \\ y_3 \end{pmatrix}.$$

Determine the $2 \times 3$ matrix $B$ such that $b(\mathbf{x}, \mathbf{y}) = (x)_{(2)} B (y)^{\mathrm{T}}_{(3)}$.

3. Show that if $B^{\mathrm{T}} = B$, then $A^{\mathrm{T}}BA$ is symmetric for each matrix $A$. Also show that if $B$ is skew-symmetric, then $A^{\mathrm{T}}BA$ is also skew-symmetric.
4. Express the matrix

$$B = \begin{pmatrix} 1 & 2 & 3 \\ -1 & 2 & 5 \\ 2 & -1 & -2 \end{pmatrix}$$

as the sum of a symmetric matrix and a skew-symmetric matrix.
5. Show that if $A = A_{(m,n)}$ is an $m \times n$ matrix, then both $A^{\mathrm{T}}A$ and $AA^{\mathrm{T}}$ are symmetric matrices.
6. Show that a skew-symmetric matrix $A$ of odd order must be singular, i.e., $\det A = 0$.
7. Let $b : \mathbb{V} \times \mathbb{W} \to \mathbb{R}$ be a bilinear form. Show that for each fixed $\mathbf{a} \in \mathbb{V}$, $f_a(\mathbf{y}) = b(\mathbf{a}, \mathbf{y})$ defines a linear form on $\mathbb{W}$ and thus $f_a \in \mathbb{V}^*$.
8. Show by an example that if $A$ and $B$ are symmetric $2 \times 2$ matrices, then $AB$ may not be symmetric.
9. Show that if $B^{\mathrm{T}} = -B$, then $(x)_{(n)}B(x)^{\mathrm{T}}_{(n)} = 0$ for all $\mathbf{x} = (\mathbf{e})_{(n)}(x)^{\mathrm{T}}_{(n)} \in \mathbb{R}^n$.
10. A square matrix $P$ is said to be an *idempotent* or a *projection* if $P^2 = P$. If in addition $P = P^{\mathrm{T}}$, then $P$ is an *orthogonal projection*.

    (a) Give an example of a $2 \times 2$ matrix which is an idempotent.
    (b) Give an example of a $2 \times 2$ matrix which is an orthogonal projection.

## 9.3 Quadratic Forms

Given any bilinear form $b(\mathbf{x}, \mathbf{y})$ over the real numbers $\mathbb{R}$, we can write

$$b(\mathbf{x},\mathbf{y}) = \frac{1}{2}\big(b(\mathbf{x},\mathbf{y}) + b(\mathbf{y},\mathbf{x})\big) + \frac{1}{2}\big(b(\mathbf{x},\mathbf{y}) - b(\mathbf{y},\mathbf{x})\big) = s(\mathbf{x},\mathbf{y}) + t(\mathbf{x},\mathbf{y})$$

where $s(\mathbf{x},\mathbf{y}) \equiv \frac{1}{2}\big(b(\mathbf{x},\mathbf{y}) + b(\mathbf{y},\mathbf{x})\big)$ is the symmetric part and $t(\mathbf{x},\mathbf{y}) \equiv \frac{1}{2}\big(b(\mathbf{x},\mathbf{y}) - b(\mathbf{y},\mathbf{x})\big)$ is the skew-symmetric part.

**Definition 9.3.1.** By the quadratic form of a bilinear form $b(\mathbf{x},\mathbf{y})$, we mean $q(\mathbf{x}) \equiv b(\mathbf{x},\mathbf{x})$.

We see that the quadratic form of a bilinear form $b = s + t$ satisfies $q(\mathbf{x}) = s(\mathbf{x},\mathbf{x})$ because $t(\mathbf{x},\mathbf{x}) = 0$. Thus, we need only study quadratic forms of symmetric bilinear forms. Conversely, given a quadratic form $q(\mathbf{x})$, we can reconstruct the symmetric bilinear form $s(\mathbf{x},\mathbf{y})$ by writing

$$\begin{aligned} s(\mathbf{x},\mathbf{y}) &= \frac{1}{2}\big(q(\mathbf{x}+\mathbf{y}) - q(\mathbf{x}) - q(\mathbf{y})\big) \\ &= \frac{1}{2}\big(b(\mathbf{x}+\mathbf{y},\mathbf{x}+\mathbf{y}) - b(\mathbf{x},\mathbf{x}) - b(\mathbf{y},\mathbf{y})\big) \end{aligned}$$

$$= \frac{1}{2}\big(b(\mathbf{x},\mathbf{x}) + b(\mathbf{x},\mathbf{y}) + b(\mathbf{y},\mathbf{x}) + b(\mathbf{y},\mathbf{y}) - b(\mathbf{x},\mathbf{x}) - b(\mathbf{y},\mathbf{y})\big)$$
$$= \frac{1}{2}\big(b(\mathbf{x},\mathbf{y}) + b(\mathbf{y},\mathbf{x})\big).$$

## Exercises

1. Find the quadratic form $q(\mathbf{x}) = (x)_{(n)}B(x)^{\mathrm{T}}_{(n)}$ determined by the matrix
$$B = \begin{pmatrix} 1 & 2 & 3 \\ -1 & 2 & 5 \\ 2 & -1 & -2 \end{pmatrix}.$$
Show that $q(\mathbf{x}) = (x)_{(3)}S(x)^{\mathrm{T}}_{(3)}$ for the symmetric matrix $S = \frac{1}{2}(B^{\mathrm{T}} + B)$.
2. Find the symmetric matrix $S$ representing each of the following quadratic forms $q(\mathbf{x}) = (x)_{(n)}S(x)^{\mathrm{T}}_{(n)}$:
   (a) $q(\mathbf{x}) = x^2 + 6xy + 8y^2$ for $\mathbf{x} = \begin{pmatrix} x \\ y \end{pmatrix} \in \mathbb{R}^2$.
   (b) $q(\mathbf{x}) = 6xy + 4y^2$.
   (c) $q(\mathbf{x}) = x^2 + 2xy + 4xz + 5y^2 + yz + 9z^2$.
   (d) $q(\mathbf{x}) = 4xy$.
   (e) $q(\mathbf{x}) = x^2 + 6xy - 4y^2 + 4xz + z^2 - 4yz$.
   (f) $q(\mathbf{x}) = x^2 + 4xy + 9y^2$.
3. Write down the unique symmetric bilinear form determined by each of the quadratic forms given in Problem 2.
4. Show that the unique symmetric bilinear form of a quadratic form can be found by the formula
$$s(\mathbf{x},\mathbf{y}) = 1/4\big(q(\mathbf{x}+\mathbf{y}) - q(\mathbf{x}-\mathbf{y})\big).$$

## 9.4 The Normal Form

In this section, we show that the matrix of any symmetric bilinear form of a vector space $\mathbb{V}$ over a field $\mathbb{K}$ can be diagonalized if the field $\mathbb{K} \neq Z_2$.

**Theorem 9.4.1.** *Given a symmetric bilinear form* $s(\mathbf{x},\mathbf{y}) = (x)_{(n)}S(y)^{\mathrm{T}}_{(n)}$ *on* $\mathbb{V}$*, there exists a basis* $(\mathbf{a})_{(n)} = (\mathbf{e})_{(n)}A$ *of* $\mathbb{V}$ *such that* $S_a = A^{\mathrm{T}}SA$ *is diagonal.*

*Proof.* Let $q(\mathbf{x}) = s(\mathbf{x},\mathbf{x})$ be the associated quadratic form of the symmetric bilinear form $s(\mathbf{x},\mathbf{y})$. The proof is by induction on the dimension $n$ of the vector space $\mathbb{V}$.

If $n = 1$, then $S = (s(\mathbf{e}_1, \mathbf{e}_1))$ is a diagonal $1 \times 1$ matrix. Now assume true for $n-1$. If $q(\mathbf{x}) = 0$ for all $\mathbf{x} \in \mathbb{V}$, there is nothing more to prove. So suppose that $q(\mathbf{a}_1) = \alpha_1 \neq 0$ for some vector $\mathbf{a}_1 \in \mathbb{V}$ and define the linear form $f_1(\mathbf{x}) = s(\mathbf{x}, \mathbf{a}_1)$. Recalling (9.2),

$$\mathscr{H} = \{\mathbf{x} \in \mathbb{V} \mid f_1(\mathbf{x}) = 0\} \subset \mathbb{V}$$

is the $(n-1)$-hyperplane of the linear form $f_1$. Then $s(\mathbf{x}, \mathbf{y})$ is a symmetric bilinear form when restricted to the $(n-1)$-dimensional space $\mathscr{H}$. Using the induction hypothesis, we find a set of $n-1$ vectors $\{\mathbf{a}_2, \ldots, \mathbf{a}_n\}$ for which $s(\mathbf{a}_i, \mathbf{a}_j) = \delta_{ij}$ for $i, j = 2, 3, \ldots, n$. But also

$$s(\mathbf{a}_1, \mathbf{a}_j) = \delta_{1j}\alpha_1$$

for $j = 1, \ldots, n$, so in the new basis $A = (\mathbf{e})_{(n)}A$, the matrix

$$S_a = A^{\mathrm{T}}SA = \begin{pmatrix} \alpha_1 & 0 & \ldots & 0 \\ 0 & \alpha_2 & \ldots & 0 \\ \cdot & \cdot & \ldots & \cdot \\ \cdot & \cdot & \ldots & \cdot \\ 0 & 0 & \ldots & \alpha_n \end{pmatrix}$$

representing the symmetric bilinear form $s(\mathbf{x}, \mathbf{y})$ in the basis $(\mathbf{a})_{(n)}$ is diagonal. □

Note that the diagonal values $\alpha_i's$ that we found in the above theorem are not unique. We have the following:

**Corollary 9.4.2.** *If the vector space $\mathbb{V}$ is over the real numbers $\mathbb{R}$, then there exists a basis in which the matrix of a symmetric bilinear form consists only of $+1's$, $-1's$, or $0's$ down the diagonal.*

*Proof.* If $\alpha_i \neq 0$, define $\mathbf{b}_i = \frac{\mathbf{a}_i}{\sqrt{|\alpha_i|}}$ and define $\mathbf{b}_i = \mathbf{a}_i$ when $\alpha_i = 0$. Then in the basis $(\mathbf{b})_{(n)}$,

$$S_b = \begin{pmatrix} \frac{1}{\sqrt{|\alpha_1|}} & 0 & \ldots & 0 \\ 0 & \frac{1}{\sqrt{|\alpha_2|}} & \ldots & 0 \\ \cdot & \cdot & \ldots & \cdot \\ \cdot & \cdot & \ldots & \cdot \\ 0 & 0 & \ldots & \frac{1}{\sqrt{|\alpha_n|}} \end{pmatrix} S_a \begin{pmatrix} \frac{1}{\sqrt{|\alpha_1|}} & 0 & \ldots & 0 \\ 0 & \frac{1}{\sqrt{|\alpha_2|}} & \ldots & 0 \\ \cdot & \cdot & \ldots & \cdot \\ \cdot & \cdot & \ldots & \cdot \\ 0 & 0 & \ldots & \frac{1}{\sqrt{|\alpha_n|}} \end{pmatrix}$$

$$= \begin{pmatrix} \pm 1 & 0 & \ldots & 0 \\ 0 & \pm 1 & \ldots & 0 \\ \cdot & \cdot & \ldots & \cdot \\ \cdot & \cdot & \ldots & \cdot \\ 0 & 0 & \ldots & \pm 1 \; or \; 0 \end{pmatrix}.$$

□

The following theorem tells in what way the diagonal representation of a symmetric bilinear form or quadratic form is unique. It is known as *Sylvester's law of inertia* in honor of the mathematician who first discovered it. A diagonal basis can always be arranged in such a way that the first $p$ terms on the diagonal are positive, followed by $q$ diagonal terms which are negative, and the rest of the diagonal terms are zero.

**Theorem 9.4.3.** *Let $q(\mathbf{x})$ be a quadratic form of a real vector space $\mathbb{V} = \mathbb{R}^n$. Let $p$ be the number of positive terms and $q$ be the number of negative terms in the diagonal representation matrix $Q$. Then in any other diagonal representation, $p$ and $q$ are the same.*

*Proof.* Let

$$\{\mathbf{a}_1,\ldots,\mathbf{a}_p,\mathbf{a}_{p+1},\ldots,\mathbf{a}_{p+q},\ldots,\mathbf{a}_n\} \text{ and } \{\mathbf{b}_1,\ldots,\mathbf{b}_s,\mathbf{b}_{s+1},\ldots,\mathbf{a}_{s+t},\ldots,\mathbf{b}_n\}$$

be two diagonal basis of $q$. For any $\mathbf{x} \neq 0$, $\mathbf{x} = \sum_{i=1}^{p} \alpha_i \mathbf{a}_i$, then $q(\mathbf{x}) > 0$, and for any $\mathbf{y}$, $\mathbf{y} = \sum_{i=s+1}^{n} \beta_i \mathbf{b}_i$, $q(\mathbf{y}) \leq 0$. This implies that

$$(\mathbb{U} = \text{span}\{\mathbf{a}_1,\ldots,\mathbf{a}_p\}) \cap (\mathbb{W} = \text{span}\{\mathbf{b}_{s+1},\ldots,\mathbf{b}_n\}) = \{0\}.$$

Now $\dim(\mathbb{U}) = p$, $\dim(\mathbb{W}) = n - s$, and $\dim(\mathbb{U} + \mathbb{W}) \leq n$. It follows that

$$p + (n-s) = \dim(\mathbb{U}) + \dim(\mathbb{W}) = \dim(\mathbb{U}+\mathbb{W}) + \dim(\mathbb{U}\cap\mathbb{W})$$

$$= \dim(\mathbb{U}+\mathbb{W}) \leq n.$$

But this implies that $p - s \leq 0$ or $p \leq s$. Interchanging the order of the diagonal basis $(\mathbf{a})_{(n)}$ and $(\mathbf{b})_{(n)}$ gives $s \leq p$ so that $s = p$. By a similar argument, it follows that $q = t$. □

Given a symmetric bilinear form $s(\mathbf{x},\mathbf{y}) = (x)_{(n)} S (y)_{(n)}^{\mathrm{T}}$, we now give a purely algebraic method for *completing the square*, i.e., finding a basis $(\mathbf{a})_{(n)} = (\mathbf{e})_{(n)} A$ for which $S_a = A^{\mathrm{T}} S A$ is diagonal. There are two cases that need to be considered:

1. If $s_{ii} = s(\mathbf{e}_i,\mathbf{e}_i) \neq 0$, we can eliminate the variable $x_i$ from the quadratic form

   $$q(\mathbf{x}) = (x)_{(n)} S (x)_{(n)}^{\mathrm{T}}$$

   by completing the square, defining

   $$q_i = \frac{1}{4s_{ii}} \left(\frac{\partial q}{\partial x_i}\right)^2$$

   and taking $r_i = q - q_i$.
2. If $s_{ij} = s(\mathbf{e}_i,\mathbf{e}_j) \neq 0$ and $s_{ii} = s_{jj} = 0$ for $i \neq j$, we can eliminate both of the variables $x_i$ and $x_j$ from the quadratic form

   $$q(\mathbf{x}) = (x)_{(n)} S (x)_{(n)}^{\mathrm{T}}$$

by completing the square, defining

$$q_{ij} = \frac{1}{8s_{ii}}\left[\left(\frac{\partial q}{\partial x_i}+\frac{\partial q}{\partial x_j}\right)^2-\left(\frac{\partial q}{\partial x_i}-\frac{\partial q}{\partial x_j}\right)^2\right]$$

and taking $r_{ij} = q - q_{ij}$.

The process is repeated on either the remainder $r_i$ or $r_{ij}$ until the remainder is itself diagonalized. The diagonalized result is in the form

$$q = q_1 + q_2 + q_{34} + \cdots + q_{kk+1}.$$

The details of the above process are best exhibited by working through a couple of examples.

*Example 1.* Consider

$$q = (x)_{(n)}S(x)^{\mathrm{T}}_{(n)} = x_1^2 + 4x_1x_2 + x_2^2 + 4x_1x_3 - 4x_2x_3 + x_3^2$$

where

$$\mathbf{x} = \begin{pmatrix} x_1 \\ x_2 \\ x_3 \end{pmatrix}, \quad S = \begin{pmatrix} 1 & 2 & 2 \\ 2 & 1 & -2 \\ 2 & -2 & 1 \end{pmatrix}.$$

**Solution.** Since $s_{11} = 1 \neq 0$, define

$$q_1 = \frac{1}{4}\left(\frac{\partial q}{\partial x_1}\right)^2 = \frac{1}{4}(2x_1 + 4x_2 + 4x_3)^2$$

Now define

$$r_1 = q - q_1 = -3(x_2^2 + 4x_2x_3 + x_3^2).$$

Again, we have case 1, so we define

$$q_2 = \frac{-1}{12}\left(\frac{\partial r_1}{\partial x_2}\right)^2 = \frac{-3}{4}(2x_2 + 4x_3)^2.$$

Continuing, we next define

$$r_2 = r_1 - q_2 = 9x_3^2$$

Since $r_2$ is a perfect square, we set $q_3 = r_2$, and we are done.

It then follows that

$$q = q_1 + q_2 + q_3 = \frac{1}{4}(2x_1 + 4x_2 + 4x_3)^2 + \frac{-3}{4}(2x_2 + 4x_3)^2 + 9x_3^2$$

To find $A^{-1}$, we take the coefficients of the $x_i$'s in the completed squares, getting

$$A^{-1} = \begin{pmatrix} 2 & 4 & 4 \\ 0 & 2 & 4 \\ 0 & 0 & 1 \end{pmatrix},$$

and find

$$A = \frac{1}{2}\begin{pmatrix} 1 & -2 & 4 \\ 0 & 1 & -4 \\ 0 & 0 & 2 \end{pmatrix}.$$

Finally, we calculate

$$A^{\mathrm{T}}SA = \begin{pmatrix} 1/4 & 0 & 0 \\ 0 & -3/4 & 0 \\ 0 & 0 & 9 \end{pmatrix}.$$

Note that the diagonal elements of this matrix are the coefficients of the completed squares in the final expression for $q$.

*Example 2.* Consider

$$q = (x)_{(n)}S(x)^{\mathrm{T}}_{(n)} = 2x_1x_2 + 4x_1x_3 + 2x_2x_3 + 6x_1x_4 + 4x_2x_4 + 2x_3x_4$$

where

$$\mathbf{x} = \begin{pmatrix} x_1 \\ x_2 \\ x_3 \\ x_4 \end{pmatrix}, \quad S = \begin{pmatrix} 0 & 1 & 2 & 3 \\ 1 & 0 & 1 & 2 \\ 2 & 1 & 0 & 1 \\ 3 & 2 & 1 & 0 \end{pmatrix}.$$

**Solution.** Since $s_{11} = 0 = s_{22}$ and $s_{12} = 1$, we have case 2. Thus, define

$$q_{12} = \frac{1}{8s_{12}}\left\{\left(\frac{\partial q}{\partial x_1} + \frac{\partial q}{\partial x_2}\right)^2 - \left(\frac{\partial q}{\partial x_1} - \frac{\partial q}{\partial x_2}\right)^2\right\}$$

$$= \frac{1}{8}\{-(-2x_1 + 2x_2 + 2x_3 + 2x_4)^2 + (2x_1 + 2x_2 + 6x_3 + 10x_4)^2\},$$

and let

$$r_{12} = q - q_{12} = -4(x_3^2 + 3x_3x_4 + 3x_4^2),$$

which is a case 1. Thus, we now define

$$q_3 = \frac{-1}{16}\left(\frac{\partial r_{12}}{\partial x_3}\right)^2 = -(2x_3 + 3x_4)^2,$$

and

$$r_3 = r_{12} - q_3 = -3x_4$$

Since $r_3$ is a perfect square, we let $q_4 = r_3$ and we are done. Thus, we have found

$$q = q_{12} + q_3 + q_4 = \frac{1}{8}\{-(-2x_1 + 2x_2 + 2x_3 + 2x_4)^2 + (2x_1 + 2x_2 + 6x_3 + 10x_4)^2\}$$
$$- (2x_3 + 3x_4)^2 - 3x_4$$

To find $A^{-1}$, we take the coefficients of $x_i$'s in the completed squares, getting

$$A^{-1} = \begin{pmatrix} -2 & 2 & 2 & 2 \\ 2 & 2 & 6 & 10 \\ 0 & 0 & 2 & 3 \\ 0 & 0 & 0 & 1 \end{pmatrix},$$

and find

$$A = \frac{1}{4}\begin{pmatrix} -1 & 1 & -2 & -2 \\ 1 & 1 & -4 & 0 \\ 0 & 0 & 2 & -6 \\ 0 & 0 & 0 & 4 \end{pmatrix}.$$

Finally, we calculate

$$A^{\mathrm{T}}SA = \begin{pmatrix} -1/8 & 0 & 0 & 0 \\ 0 & 1/8 & 0 & 0 \\ 0 & 0 & -1 & 0 \\ 0 & 0 & 0 & -3 \end{pmatrix}.$$

Note that the diagonal elements of this matrix are the coefficients of the completed squares in the final expression for the quadratic form $q(\mathbf{x})$.

**Definition 9.4.4.** The numbers $(p, q, n-p-q)$, where $p$ is the number of positive terms, $q$ is the number of negative terms, and $n-p-q$ is the number of zeros along the diagonal representation $A^{\mathrm{T}}SA$, are called the signature of the quadratic form $q(\mathbf{x}) = s(\mathbf{x}, \mathbf{x})$ defined by the matrix $S$. The quadratic form $q(\mathbf{x})$ is said to be positive definite if $p = n$, negative definite if $q = n$, and indefinite if $p + q = n$ for nonzero $p$ and $q$. If $p + q < n$, the quadratic form is said to be degenerate.

## Exercises

1. Reduce each of the following symmetric matrices $S$ to diagonal form $S_a$ by finding a basis $(\mathbf{a})_{(n)} = (\mathbf{e})_{(n)}A$ such that $S_a = A^{\mathrm{T}}SA$ is diagonal.

(a) $\begin{pmatrix} 1 & 3 & 3 \\ 3 & 1 & -1 \\ 3 & -1 & 1 \end{pmatrix}$.

(b) $\begin{pmatrix} 1 & 2 & 3 \\ 2 & 0 & -1 \\ 3 & -1 & 1 \end{pmatrix}$.

(c) $\begin{pmatrix} 0 & 1 & -1 & 2 \\ 1 & 1 & 0 & -1 \\ -1 & 0 & -1 & 1 \\ 2 & -1 & 1 & 0 \end{pmatrix}$.

(d) $\begin{pmatrix} 0 & 1 & 3 & 4 \\ 1 & 0 & 1 & 3 \\ 3 & 1 & 0 & 1 \\ 4 & 3 & 1 & 0 \end{pmatrix}$.

2. Determine the rank and signature of each of the quadratic forms in Problem 1.
3. Show that the quadratic form $q(\mathbf{x}) = ax^2 + bxy + cy^2$ is positive definite if and only if $a > 0$ and $b^2 - 4ac < 0$.
4. Show that if $S$ is a real symmetric positive definite matrix, then there exists a real nonsingular matrix $A$ such that $S = A^{\mathrm{T}}A$. (A matrix $S$ is said to be positive definite if the quadratic form defined by $S$ is positive definite.)
5. Show that if $A$ is a real matrix and $\det A \neq 0$, then $A^{\mathrm{T}}A$ is positive definite.
6. Show that if $A$ is a real matrix, then $A^{\mathrm{T}}A$ is nonnegative semidefinite.
7. Show that if $A$ is a real matrix and $A^{\mathrm{T}}A = 0$, then $A = 0$.

# Chapter 10
# Hermitian Inner Product Spaces

*One cannot escape the feeling that these mathematical formulae have an independent existence and an intelligence of their own, that they are wiser than we are, wiser even than their discoverers, that we get more out of them than we originally put into them.*

—Heinrich Hertz

In the last chapter, we discovered that the positive definite quadratic form (3.28) used to define the inner product of the Euclidean space $\mathbb{R}^n$, and the corresponding geometric algebra $\mathbb{G}_n = \mathbb{G}(\mathbb{R}^n)$, is only one of the many possible quadratic forms. We now introduce the *pseudo-Euclidean space* $\mathbb{R}^{p,q}$ of a general *indefinite quadratic form* $q(\mathbf{x})$ and the corresponding geometric algebra $\mathbb{G}_{p,q} = \mathbb{G}(\mathbb{R}^{p,q})$. More generally, we introduce the *unitary geometric algebra* $\mathbb{U}_{p,q} = \mathbb{U}(\mathbb{H}^{p,q})$ of the Hermitian space $\mathbb{H}^{p,q}$. The complexification is obtained by assuming the existence of a new square root $\mathrm{i} = \sqrt{-1}$ which anticommutes with all the vectors of the given geometric algebra. Hermitian inner and outer products reduce to the standard inner and outer products for real multivectors. Hermitian spaces, and their corresponding unitary geometric algebras, resolve the problem of complex eigenvalues and eigenvectors. The spectral decompositions of Hermitian, normal, and unitary operators are derived, and we give a treatment of polar and singular value decompositions.[1]

[1]This chapter is based upon articles by the author that appeared in the *College Mathematics Journal* [77] and in *Advances in Applied Clifford Algebras* [90].

G. Sobczyk, *New Foundations in Mathematics: The Geometric Concept of Number*, DOI 10.1007/978-0-8176-8385-6_10,

## 10.1 Fundamental Concepts

The $n$-dimensional pseudo-Euclidean space $\mathbb{R}^{p,q}$ of an indefinite quadratic form $q(\mathbf{x})$ with signature $(p,q)$, where $n = p+q$, is characterized by the property that for each $\mathbf{x} \in \mathbb{R}^{p,q}$,

$$\mathbf{x}^2 = \mathbf{x} \cdot \mathbf{x} = q(\mathbf{x}). \tag{10.1}$$

A *standard orthonormal basis* $(\mathbf{e})_{(p+q)} = (\mathbf{e}_1, \mathbf{e}_2, \ldots, \mathbf{e}_{p+q})$ of $\mathbb{R}^{p,q}$ has the property that

$$\mathbf{e}_1^2 = \cdots = \mathbf{e}_p^2 = 1, \text{ and } \mathbf{e}_{p+1}^2 = \cdots = \mathbf{e}_{p+q}^2 = -1, \tag{10.2}$$

and the basis vectors are mutually anticommuting. In terms of this basis, any $\mathbf{x} \in \mathbb{R}^{p,q}$ has the form

$$\mathbf{x} = (x)_{(n)}(\mathbf{e})^{\mathrm{T}}_{(n)} = \begin{pmatrix} x_1 & x_2 & \cdots & x_n \end{pmatrix} \begin{pmatrix} \mathbf{e}_1 \\ \mathbf{e}_2 \\ \cdot \\ \cdot \\ \mathbf{e}_n \end{pmatrix} = \sum_{j=1}^{n} x_j \mathbf{e}_j,$$

and if $\mathbf{y} = (y)_{(n)}(\mathbf{e})^{\mathrm{T}}_{(n)} \in \mathbb{R}^{p,q}$ is a second such vector, then the *inner product* $\mathbf{x} \cdot \mathbf{y}$ for $\mathbf{x}, \mathbf{y} \in \mathbb{R}^{p,q}$ is given by

$$\mathbf{x} \cdot \mathbf{y} = \frac{1}{2}[q(\mathbf{x}+\mathbf{y}) - q(\mathbf{x}) - q(\mathbf{y})] = x_1 y_1 + \cdots + x_p y_p - x_{p+1} y_{p+1} - \cdots - x_n y_n. \tag{10.3}$$

In Chap. 7, the standard basis $(\mathbf{e})_{(n)}$ and the reciprocal basis $(\mathbf{e})^{(n)}$ of $\mathbb{R}^n$ are related by the formula (7.9), and we are able to identify $\mathbf{e}_i = \mathbf{e}^i$ for $1 \le i \le n$. The relationship between the standard basis $(\mathbf{e})_{(n)}$ and its reciprocal basis $(\mathbf{e})^{(n)}$ of $\mathbb{R}^{p,q}$ is more complicated. In this case, we have

$$\mathbf{e}^i = \frac{\mathbf{e}_i}{\mathbf{e}_i \cdot \mathbf{e}_i} = \begin{pmatrix} \mathbf{e}_i \text{ if } & 0 \le i \le p \\ -\mathbf{e}_i \text{ if } & p < i \le p+q \end{pmatrix}. \tag{10.4}$$

As we see below, the relationship (10.4) changes the way we defined the vector derivative (3.46) in Chap. 3.

In $\mathbb{R}^{p,q}$, we must carefully distinguish between the column reciprocal basis $(\mathbf{e})^{(n)}$ defined by (10.4) and the column $(\mathbf{e})^{\mathrm{T}}_{(n)}$ of the standard basis $(\mathbf{e})_{(n)}$. They are related by

$$(\mathbf{e})^{(n)} = \begin{pmatrix} \mathbf{e}^1 \\ \mathbf{e}^2 \\ \cdot \\ \cdot \\ \mathbf{e}^n \end{pmatrix} = [g] \begin{pmatrix} \mathbf{e}_1 \\ \mathbf{e}_2 \\ \cdot \\ \cdot \\ \mathbf{e}_n \end{pmatrix} = [g](\mathbf{e})^{\mathrm{T}}_{(n)}, \tag{10.5}$$

where

$$[g] = (\mathbf{e})_{(n)}^{\mathrm{T}} \cdot (\mathbf{e})_{(n)} = [\mathbf{e}_i \cdot \mathbf{e}_j] \text{ and } (\mathbf{e})^{(n)} \cdot (\mathbf{e})_{(n)} = [\mathbf{e}^i \cdot \mathbf{e}_j] = [1]_n. \tag{10.6}$$

The *metric tensor* $[g]$ of $\mathbb{R}^{p,q}$ is the diagonal matrix with the first $p$ entries $+1$ and the remaining $q$ entries $-1$, and $[1]_n$ is the identity $n \times n$ matrix.

In terms of the orthonormal and reciprocal bases $(\mathbf{e})_{(n)}$ and $(\mathbf{e})^{(n)}$, related by (10.5), any vector $\mathbf{x} \in \mathbb{R}^{p,q}$ can be expressed by

$$\mathbf{x} = (x)_{(n)}(\mathbf{e})^{(n)} = \sum_{i=1}^{n} x_i \mathbf{e}^i = (\mathbf{e})_{(n)}(x)^{(n)}, \tag{10.7}$$

where $x_i = \mathbf{x} \cdot \mathbf{e}_i$ and $x^i = \mathbf{e}^i \cdot \mathbf{x}$ for $i = 1, \dots, n$. If $\mathbf{y} = (\mathbf{e})_{(n)}(y)^{(n)}$ is a second such vector, then in agreement with (10.3), the inner product between them is

$$\begin{aligned}\mathbf{x} \cdot \mathbf{y} &= \sum_{i=1}^{n} x_i y^i \\ &= (x)_{(n)}(y)^{(n)} = (x)_{(n)}[g](y)_{(n)}^{\mathrm{T}}\end{aligned} \tag{10.8}$$

where $n = p + q$. Because of the indefinite signature of the quadratic form, a vector $\mathbf{x} \in \mathbb{R}^{p,q}$ can have negative square, $\mathbf{x}^2 = \mathbf{x} \cdot \mathbf{x} < 0$. For this reason, we define the *magnitude* $|\mathbf{x}| = \sqrt{|\mathbf{x}^2|} \geq 0$. Contrast this with the definition (3.28) in Chap. 3. In (3.44), we defined the magnitude of a $k$-vector to be always nonnegative.

Let $(\mathbf{a})_{(n)}$ be a second basis of $\mathbb{R}^{p,q}$. Then

$$(\mathbf{a})_{(n)} = (\mathbf{e})_{(n)} A, \iff (\mathbf{a})_{(n)}^{\mathrm{T}} = A^{\mathrm{T}} (\mathbf{e})_{(n)}^{\mathrm{T}}, \tag{10.9}$$

where $A$ is the matrix of transition from the standard basis $(\mathbf{e})_{(n)}$ of $\mathbb{R}^{p,q}$ to the basis $(\mathbf{a})_{(n)}$ of $\mathbb{R}^{p,q}$.

The real associative *geometric algebra* $\mathbb{G}_{p,q} = \mathbb{G}(\mathbb{R}^{p,q})$ is generated by the *geometric multiplication* of the vectors in $\mathbb{R}^{p,q}$, subjected to the rule that $\mathbf{x}^2 = q(\mathbf{x}) = \mathbf{x} \cdot \mathbf{x}$ for all vectors $\mathbf{x} \in \mathbb{R}^{p,q}$. Just as for the geometric algebra $\mathbb{G}_n$ of the Euclidean space $\mathbb{R}^n$, see (3.30), the dimension of $\mathbb{G}_{p,q}$ as a real linear space is $2^n$, where $n = p + q$, with the *standard orthonormal basis* of geometric numbers

$$\mathbb{G}_{p,q} = \operatorname{span}\{\mathbf{e}_{\lambda_{(k)}}\}_{k=0}^{n}, \tag{10.10}$$

where the $\binom{n}{k}$ $k$-vector basis elements of the form $\mathbf{e}_{\lambda_{(k)}}$ are defined by

$$\mathbf{e}_{\lambda_{(k)}} = \mathbf{e}_{\lambda_1 \dots \lambda_k} = \mathbf{e}_{\lambda_1} \cdots \mathbf{e}_{\lambda_k}$$

for each $\lambda_{(k)} = \lambda_1, \dots, \lambda_k$ where $1 \leq \lambda_1 < \cdots < \lambda_k \leq n$. When $k = 0$, $\lambda_{(0)} = 0$ and $\mathbf{e}_0 = 1$.

The vector derivative with respect to the vector $\mathbf{x} \in \mathbb{R}^{p,q}$, given in (10.7), is defined by

$$\partial_{\mathbf{x}} = \sum_{i=1}^{n} \mathbf{e}^i \mathbf{e}_i \cdot \partial_{\mathbf{x}} = \sum_{i=1}^{n} \mathbf{e}^i \frac{\partial}{\partial x^i}. \tag{10.11}$$

This is the same definition as given for the vector derivative (3.46) in $\mathbb{R}^n$, except that we must strickly use the raised indices because of the indefinite metric.

All of the algebraic and differential identities derived in Chap. 3 for the geometric algebra $\mathbb{G}_n$ remain valid in the geometric algebra $\mathbb{G}_{p,q}$, with the single exception of differential identities involving the derivative of the $|\mathbf{x}|$, given in (3.53). In $\mathbb{R}^{p,q}$, the identity (3.53) must be modified to read

$$\partial_{\mathbf{x}}|\mathbf{x}| = \begin{pmatrix} \frac{\mathbf{x}}{|\mathbf{x}|} = \hat{\mathbf{x}} \ \text{ if } \mathbf{x}^2 \geq 0 \\ -\frac{\mathbf{x}}{|\mathbf{x}|} = -\hat{\mathbf{x}} \text{ if } \mathbf{x}^2 < 0 \end{pmatrix}, \tag{10.12}$$

as easily follows by differentiating $|\mathbf{x}|^2$ and noting that

$$|\mathbf{x}|^2 = \begin{pmatrix} \mathbf{x}^2 \ \text{ if } \mathbf{x}^2 \geq 0 \\ -\mathbf{x}^2 \text{ if } \mathbf{x}^2 < 0 \end{pmatrix}.$$

## Exercises

1. Let $(\mathbf{e})_{(4)} = (\mathbf{e}_1, \mathbf{e}_2, \mathbf{e}_3, \mathbf{e}_4)$ be an orthonormal basis and let $\mathbf{a} = (1,2,-1,0)(\mathbf{e})^{\mathrm{T}}_{(4)}$ and $\mathbf{b} = (2,5,-1,1)(\mathbf{e})^{\mathrm{T}}_{(4)}$.

   (a) Compute $\mathbf{a} \cdot \mathbf{b}$ in $\mathbb{R}^4$.
   (b) Compute $\mathbf{a} \cdot \mathbf{b}$ in $\mathbb{R}^{3,1}$.
   (c) Compute $\mathbf{a} \cdot \mathbf{b}$ in $\mathbb{R}^{2,2}$.
   (d) Compute $\mathbf{a} \cdot \mathbf{b}$ in $\mathbb{R}^{1,3}$.
   (e) Compute $\mathbf{a} \cdot \mathbf{b}$ in $\mathbb{R}^{0,4}$.

2. Write down the standard orthonormal basis (10.10) for each of the geometric algebras in Problem 1.
3. (a) Show that $|\mathbf{a}+\mathbf{b}|^2 + |\mathbf{a}-\mathbf{b}|^2 = 2|\mathbf{a}|^2 + 2|\mathbf{b}|^2$ in $\mathbb{R}^n$.
   (b) Show by an example that the identity given in part (a) is not true in $\mathbb{R}^{1,1}$.
4. Let $(\mathbf{e})_{(3)} = (\mathbf{e}_1, \mathbf{e}_2, \mathbf{e}_3)$ be an orthonormal basis and let $\mathbf{a} = (1,2,-1)(\mathbf{e})^{\mathrm{T}}_{(3)}$, $\mathbf{b} = (2,5,-1)(\mathbf{e})^{\mathrm{T}}_{(3)}$, and $\mathbf{c} = (2,1,2)(\mathbf{e})^{\mathrm{T}}_{(3)}$.

   (a) Compute $\mathbf{a} \cdot (\mathbf{b} \wedge \mathbf{c})$ in $\mathbb{R}^3$.
   (b) Compute $\mathbf{a} \cdot (\mathbf{b} \wedge \mathbf{c})$ in $\mathbb{R}^{2,1}$.
   (c) Compute $\mathbf{a} \cdot (\mathbf{b} \wedge \mathbf{c})$ in $\mathbb{R}^{1,2}$.
   (d) Compute $\mathbf{a} \cdot (\mathbf{b} \wedge \mathbf{c})$ in $\mathbb{R}^{0,3}$.

5. Prove the differential identity (10.12) and find the corresponding differential identiy for $\mathbf{a} \cdot \partial_{\mathbf{x}}|\mathbf{x}|$.

## 10.2 Orthogonality Relationships in Pseudo-Euclidean Space

We now generalize the basic orthogonality relationships that we found for the Euclidean space $\mathbb{R}^n$ in Chap. 7 to the pseudo-Euclidean space $\mathbb{R}^{p,q}$, utilizing the tools of its geometric algebra $\mathbb{G}_{p,q} = \mathbb{G}(\mathbb{R}^{p,q})$. Later, we shall generalize them even further to apply to unitary spaces.

Two vectors $\mathbf{a}, \mathbf{b} \in \mathbb{R}^{p,q}$ are said to be *orthogonal* in $\mathbb{R}^{p,q}$ if $\mathbf{a} \cdot \mathbf{b} = 0$. One of the basic tasks is given a basis $(\mathbf{a})_{(n)}$ of *non-null vectors* in $\mathbb{R}^{p,q}$ to find a closely related *orthogonal basis* $(\mathbf{b})_{(n)}$ of $\mathbb{R}^{p,q}$ which satisfies the following two conditions:

$$\mathbf{b}_i \cdot \mathbf{b}_j = 0 \text{ and } \mathbf{b}_1 \wedge \cdots \wedge \mathbf{b}_k = \mathbf{a}_1 \wedge \cdots \wedge \mathbf{a}_k, \tag{10.13}$$

for each $1 \le i < j \le n$ and $1 \le k \le n$. By a *non-null* vector, we mean any vector $\mathbf{v} \in \mathbb{R}^{p,q}$ such that $\mathbf{v}^2 \neq 0$.

This task is immediately completed in *almost* the same way as for Euclidean spaces, given in (7.14), by the following recursive construction:

$$\mathbf{b}_1 = \mathbf{a}_1, \text{ and } \mathbf{b}_k = \frac{\mathbf{b}^{\dagger}_{(k-1)} \cdot (\mathbf{b}_{(k-1)} \wedge \mathbf{a}_k)}{\mathbf{b}^{\dagger}_{(k-1)} \mathbf{b}_{(k-1)}}, \tag{10.14}$$

for all $k$ such that $1 < k \le n$. Note, since the $\mathbf{b}_j$'s are orthogonal,

$$\mathbf{b}^{\dagger}_{(k-1)} \mathbf{b}_{(k-1)} = \mathbf{b}_{k-1} \cdots \mathbf{b}_1 \mathbf{b}_1 \cdots \mathbf{b}_{k-1} = \mathbf{b}_1^2 \cdots \mathbf{b}_{k-1}^2.$$

The above construction is often called the *Gram-Schmidt orthogonalization process* [27, p.369].

Let $f$ be the linear operator which takes the basis $(\mathbf{a})_{(n)}$ into the basis $(\mathbf{b})_{(n)}$, i.e., $f(\mathbf{a})_{(n)} = (\mathbf{a})_{(n)}[f] = (\mathbf{b})_{(n)}$ where $[f]$ is the matrix of $f$ with respect to the basis $(\mathbf{a})_{(n)}$. We can solve this relationship directly for the matrix $[f]$. Using the fact that $(\mathbf{b}^{-1})^{\mathrm{T}}_{(n)} \cdot (\mathbf{b})_{(n)}$ is the identity $n \times n$ matrix, we get

$$[f] = \left[(\mathbf{b}^{-1})^{\mathrm{T}}_{(n)} \cdot (\mathbf{a})_{(n)}\right]^{-1} = \begin{pmatrix} \mathbf{b}_1^{-1} \cdot \mathbf{a}_1 & \dots & \mathbf{b}_1^{-1} \cdot \mathbf{a}_n \\ \dots & \dots & \dots \\ \dots & \dots & \dots \\ \mathbf{b}_n^{-1} \cdot \mathbf{a}_1 & \dots & \mathbf{b}_n^{-1} \cdot \mathbf{a}_n \end{pmatrix}^{-1},$$

where $\mathbf{b}_j^{-1} = \frac{\mathbf{b}_j}{\mathbf{b}_j^2}$ for $j = 1, \dots, n$.

By the *Gram matrix* of the basis $(\mathbf{a})_{(n)}$, we mean the matrix

$$A = (\mathbf{a})^{\mathrm{T}}_{(n)} \cdot (\mathbf{a})_{(n)} = \begin{pmatrix} \mathbf{a}_1 \cdot \mathbf{a}_1 & \dots & \mathbf{a}_1 \cdot \mathbf{a}_n \\ \dots & \dots & \dots \\ \dots & \dots & \dots \\ \mathbf{a}_n \cdot \mathbf{a}_1 & \dots & \mathbf{a}_n \cdot \mathbf{a}_n \end{pmatrix}. \tag{10.15}$$

http://en.wikipedia.org/wiki/Gramian_matrix

But the relationship

$$(\mathbf{a})_{(n)}[f] = (\mathbf{b})_{(n)} \iff [f]^{\mathrm{T}}(\mathbf{a})^{\mathrm{T}}_{(n)} = (\mathbf{b})^{\mathrm{T}}_{(n)}$$

implies that

$$[f]^{\mathrm{T}}A[f] = [f]^{\mathrm{T}}(\mathbf{a})^{\mathrm{T}}_{(n)} \cdot (\mathbf{a})_{(n)}[f] = (\mathbf{b})^{\mathrm{T}}_{(n)} \cdot (\mathbf{b})_{(n)} = B$$

where $B$ is the diagonal Gram matrix of the orthogonal basis $(\mathbf{b})_{(n)}$. Thus, we have diagonalized the quadratic form defined by the matrix $A$. Since the relationship (10.13) implies that $\det[f] = 1$, it also follows that $\det A = \det B$. Of course, the signature of the quadratic form of the matrix $A$ is the same as signature of the quadratic form defined by the matrix $B$. This is just Sylvester's law of inertia Theorem 9.4.3 and is the same as the signature of $\mathbb{R}^{p,q}$.

Let $(\mathbf{e})_{(n)}$ be the standard basis of $\mathbb{R}^{p,q}$ and let $f(\mathbf{e})_{(n)} = (\mathbf{e})_{(n)}[f]$ define a nonsingular linear transformation on $\mathbb{R}^{p,q}$ in terms of its matrix $[f]$. We have the following important

**Definition 10.2.1.** A nonsingular linear transformation $f$ is said to be an orthogonal transformation on $\mathbb{R}^{p,q}$ if $f(\mathbf{x}) \cdot f(\mathbf{y}) = \mathbf{x} \cdot \mathbf{y}$ for all $\mathbf{x}, \mathbf{y} \in \mathbb{R}^{p,q}$.

In terms of the matrix $[f]$ of $f$ in the standard basis $(\mathbf{e})_{(n)}$, we find that for an orthogonal transformation on $\mathbb{R}^{p,q}$,

$$f(\mathbf{e})^{\mathrm{T}}_{(n)} \cdot f(\mathbf{e})_{(n)} = (\mathbf{e})^{\mathrm{T}}_{(n)} \cdot (\mathbf{e})_{(n)} = [g],$$

or

$$[f]^{\mathrm{T}}[g][f] = [g] \iff [f]^{\mathrm{T}} = [g][f]^{-1}[g]. \tag{10.16}$$

An orthogonal transformation on $\mathbb{R}^{p,q}$ is often called an *isometry* on $\mathbb{R}^{p,q}$.

The Gram determinant of the basis $(\mathbf{a})_{(n)}$ is defined to be the determinant of the Gram matrix $A$ of $(\mathbf{a})_{(n)}$. We have

$$\det A = \det[(\mathbf{a})^{\mathrm{T}}_{(n)} \cdot (\mathbf{a})_{(n)}] = \mathbf{a}^{\dagger}_{(n)} \cdot \mathbf{a}_{(n)}, \tag{10.17}$$

where $\mathbf{a}^{\dagger}_{(n)} = \mathbf{a}_n \wedge \mathbf{a}^{n-1} \wedge \cdots \wedge \mathbf{a}_1$ is the reverse of the pseudoscalar $\mathbf{a}_{(n)}$. Again, because of the properties (10.13) of the related basis $(\mathbf{b})_{(n)}$, it follows that in the geometric algebra $\mathbb{G}(\mathbb{R}^{p,q})$,

$$\mathbf{a}^{\dagger}_{(k)} \cdot \mathbf{a}_{(k)} = \mathbf{b}^{\dagger}_{(k)} \cdot \mathbf{b}_{(k)} = \mathbf{b}_1^2 \cdots \mathbf{b}_k^2,$$

for $1 \leq k \leq n$. In the case when $n = 2$, we have

$$\mathbf{a}^{\dagger}_{(2)} \cdot \mathbf{a}_{(2)} = (\mathbf{a}_2 \wedge \mathbf{a}_1) \cdot (\mathbf{a}_1 \wedge \mathbf{a}_2) = \mathbf{a}_1^2\mathbf{a}_2^2 - (\mathbf{a}_1 \cdot \mathbf{a}_2)^2 = \mathbf{b}_1^2\mathbf{b}_2^2,$$

but we no longer have the Schwarz inequality (7.15) that we had in $\mathbb{R}^n$.

When the Gram determinant of the Gram matrix $A = (\mathbf{a})_{(n)}^{\mathrm{T}} \cdot (\mathbf{a})_{(n)}$ is nonzero, the vectors $(\mathbf{a})_{(n)}$ form a basis of $\mathbb{R}^{p,q}$. In this case, the corresponding *reciprocal basis* $(\mathbf{a})^{(n)}$ of $\mathbb{R}^{p,q}$ is defined by

$$(\mathbf{a})^{(n)} = A^{-1}(\mathbf{a})_{(n)}^{\mathrm{T}}, \tag{10.18}$$

as easily follows from the relationship $[1]_n = A^{-1}A = (A^{-1}(\mathbf{a})_{(n)}^{\mathrm{T}}) \cdot (\mathbf{a})_{(n)}$.

We shall generalize these fundamental results to the unitary geometric algebra $\mathbb{U}_{p,q}$ of the unitary space $\mathbb{H}^{p,q}$ in the next section.

## Exercises

1. Prove the relationship (10.18) by showing that

$$\left(A^{-1}(\mathbf{a})_{(n)}^{\mathrm{T}}\right) \cdot (\mathbf{a})_{(n)} = A^{-1}\left((\mathbf{a})_{(n)}^{\mathrm{T}} \cdot (\mathbf{a})_{(n)}\right) = A^{-1}A = [1]_n.$$

2. Let $(\mathbf{e})_{(3)} = (\mathbf{e}_1, \mathbf{e}_2, \mathbf{e}_3)$ be the standard orthonormal basis and let $(\mathbf{a})_{(3)} = (\mathbf{e})_{(3)} \begin{pmatrix} 1 & 2 & -1 \\ 2 & 5 & -2 \\ 1 & -1 & 1 \end{pmatrix}$ be the second basis. Using (10.14), find the corresponding orthogonal basis $(\mathbf{b})_{(3)}$ satisfying (10.13) in each of the following pseudo-Euclidean spaces:

   (a) In $\mathbb{R}^3$
   (b) In $\mathbb{R}^{2,1}$
   (c) In $\mathbb{R}^{1,2}$
   (d) In $\mathbb{R}^{0,3}$

3. Let $(\mathbf{e})_{(3)} = (\mathbf{e}_1, \mathbf{e}_2, \mathbf{e}_3)$ be the standard orthonormal basis and let $(\mathbf{a})_{(3)} = (\mathbf{e})_{(3)} \begin{pmatrix} 1 & 2 & -1 \\ 2 & 5 & -2 \\ 1 & -1 & 1 \end{pmatrix}$ be the second basis. Using (10.18), find the corresponding reciprocal basis $(\mathbf{a})^{(3)}$ to $(\mathbf{a})_{(3)}$ in each of the following pseudo-Euclidean spaces:

   (a) In $\mathbb{R}^3$
   (b) In $\mathbb{R}^{2,1}$
   (c) In $\mathbb{R}^{1,2}$
   (d) In $\mathbb{R}^{0,3}$

4. Let $\mathbb{W}$ be the subspace spanned by

$$(\mathbf{a})_{(4)} = (\mathbf{e})_{(4)} \begin{pmatrix} 0 & 0 & -3 \\ 1 & 5 & -3 \\ 1 & -3 & 5 \\ 0 & -2 & 7 \end{pmatrix},$$

where $(\mathbf{e})_{(4)}$ is the standard orthonormal basis.

(a) Find an orthogonal basis for the subspace $\mathbb{W}$ in the pseudo-Euclidean space $\mathbb{R}^4$.
(b) Find an orthogonal basis for the subspace $\mathbb{W}$ in the pseudo-Euclidean space $\mathbb{R}^{0,4}$.
(c) Find an orthogonal basis for the subspace $\mathbb{W}$ in the pseudo-Euclidean space $\mathbb{R}^{3,1}$.
(d) Find an orthogonal basis for the subspace $\mathbb{W}$ in the pseudo-Euclidean space $\mathbb{R}^{2,2}$.
e) Find an orthogonal basis for the subspace $\mathbb{W}$ in the pseudo-Euclidean space $\mathbb{R}^{1,3}$.

5. Show that if $|\mathbf{x}| = |\mathbf{y}|$ for vectors $\mathbf{x},\mathbf{y} \in \mathbb{R}^{p,q}$, then in general it is no longer true that $\mathbf{x}+\mathbf{y}$ and $\mathbf{x}-\mathbf{y}$ are orthogonal. Give an example.
6. Let $\mathbf{x} = (\mathbf{e})_{(n)}(x)^{(n)}$ where $(\mathbf{e})_{(n)}$ is an orthonormal basis of $\mathbb{R}^{p,q}$. Show that

$$(x)^{(n)} = (\mathbf{e})^{(n)} \cdot \mathbf{x}$$

or $x^i = \frac{\mathbf{e}_i \cdot \mathbf{x}}{\mathbf{e}_i \cdot \mathbf{e}_i} = \mathbf{e}^i \cdot \mathbf{x}$ for $i = 1,2,\ldots,n$.
7. Let $(\mathbf{e})_{(n)}$ be an orthonormal basis of the $n$-dimensional space $\mathbb{R}^{p,q}$ and let $f : \mathbb{R}^{p,q} \to \mathbb{R}^{p,q}$ be a linear operator. Show that the matrix $[f] = (f^i_j)$ of $f$, with respect to the basis $(\mathbf{e})_{(n)}$, is defined by $f(\mathbf{e})_{(n)} = (\mathbf{e})_{(n)}[f]$ or

$$[f] = (\mathbf{e})^{(n)} \cdot f(\mathbf{e})_{(n)},$$

where

$$f^i_j = \frac{\mathbf{e}_i \cdot f(\mathbf{e}_j)}{\mathbf{e}_i \cdot \mathbf{e}_i} = \begin{pmatrix} \mathbf{e}_i \cdot f(\mathbf{e}_j) \text{ for } 1 \le i \le p \; 1 \le j \le n \\ -\mathbf{e}_i \cdot f(\mathbf{e}_j) \text{ for } p < i \le n \; 1 \le j \le n \end{pmatrix} = \mathbf{e}^i \cdot f(\mathbf{e}_j),$$

and $n = p+q$.
8. Given $(\mathbf{b})_{(n)} = (\mathbf{a})_{(n)}\triangle$ in $\mathbb{R}^{p,q}$, where $(\mathbf{b})_{(n)}$ is defined in (10.14). Show that

$$\mathbf{b}_1 = \mathbf{a}_1, \mathbf{b}_1\mathbf{b}_2 = \mathbf{a}_1 \wedge \mathbf{a}_2, \; \ldots, \; \mathbf{b}_1\mathbf{b}_2 \ldots \mathbf{b}_n = \mathbf{a}_1 \wedge \mathbf{a}_2 \wedge \ldots \wedge \mathbf{a}_n.$$

9. Show that any bilinear form $b : \mathbb{R}^n \times \mathbb{R}^n \to \mathbb{R}$ can be expressed in the form $b(\mathbf{x},\mathbf{y}) = \mathbf{x} \cdot f(\mathbf{y})$, where $f(\mathbf{x})$ is a linear operator on $\mathbb{R}^n$.
10. Verify the last equality in the relationship (10.17).

## 10.3 Unitary Geometric Algebra of Pseudo-Euclidean Space

We now show how all of the previous identities in $\mathbb{G}_{p,q}$ can be generalized to hold in a larger *complex geometric algebra* which I call the *unitary geometric algebra* $\mathbb{U}_{p,q}$.[2] We extend the geometric algebra $\mathbb{G}_{p,q}$ by a new unit vector, denoted by i, which has square minus one and anticommutes with all the vectors in $\mathbb{R}^{p,q}$. In this approach, the geometric interpretation of a complex scalar $z = x + \mathrm{i}y$ for $x, y \in \mathbb{R}$ as the sum of a real number $x$ and the distinguished vector part $y\mathrm{i}$ is unambiguous, and the Hermitian inner product arises in a natural way. This larger complexified geometric algebra is necessary in order to have all of the tools of geometric algebra available for the study of Hermitian spaces and their generalizations [63].

By a *(complex) vector* $\mathbf{x}$ in the Hermitian space $\mathbb{H}^{p,q}$, we mean the quantity

$$\mathbf{x} = \mathbf{x}_1 + \mathrm{i}\mathbf{x}_2, \quad \text{where} \quad \mathbf{x}_1, \mathbf{x}_2 \in \mathbb{R}^{p,q} \subset \mathbb{R}^{p,q+1}$$

and $\mathrm{i} := \mathbf{e}_{p+q+1} \in \mathbb{R}^{p,q+1} \subset \mathbb{G}_{p,q+1}$ with $\mathrm{i}^2 = -1$. Note that although i is just an ordinary vector in $\mathbb{R}^{p,q+1}$, we denote it by the unbold symbol i and give it the special interpretation of the *unit imaginary scalar* in the Hermitian space $\mathbb{H}^{p,q}$.

The complex vector $\mathbf{a} + \mathrm{i}\mathbf{b}$, shown in Fig. 10.1, is thus the sum of the *real vector* part $\mathbf{a} \in \mathbb{R}^{p,q}$, and the *imaginary vector* part $\mathrm{i}\mathbf{b}$ which is a bivector in $\mathbb{G}^2_{p,q+1}$. Since the distinguished unit vector $\mathrm{i} \equiv \mathbf{e}_{p+q+1}$ in $\mathbb{R}^{p,q+1}$ with square $-1$ is orthogonal to all of the basis vectors in $(\mathbf{e})_{(p+q)} \subset (\mathbf{e})_{p+q+1}$, it follows that $\mathbf{a} + \mathrm{i}\mathbf{b} = \mathbf{a} - \mathbf{b}\mathrm{i}$.

We say that $\overline{\mathbf{x}} = \mathbf{x}_1 - \mathrm{i}\mathbf{x}_2 = \mathbf{x}^\dagger$ is the *complex conjugate* of $\mathbf{x}$, where $\overline{A}$ is the operation of *complex conjugation* for $A \in \mathbb{G}_{p,q+1}$. The conjugate $\overline{A}$ of $A$ agrees with the previously defined operation of reversal when the argument $A$ is a complex vector, but it does not reverse the order of the terms in a geometric product. The conjugation $\overline{A}$ of any geometric number $A \in \mathbb{U}_{p,q}$ can most simply be obtained by replacing i by $-\mathrm{i}$ wherever i appears in $A$ and is the natural generalization of the conjugation of a complex number.[3] By a *complex scalar* $\alpha \in \mathbb{U}_{p,q}$, we mean $\alpha = a_1 + \mathrm{i}a_2$ where $a_1, a_2 \in \mathbb{R}$, and $\overline{\alpha} = a_1 - \mathrm{i}a_2$. For $A, B \in \mathbb{G}_{p,q+1}$, we have

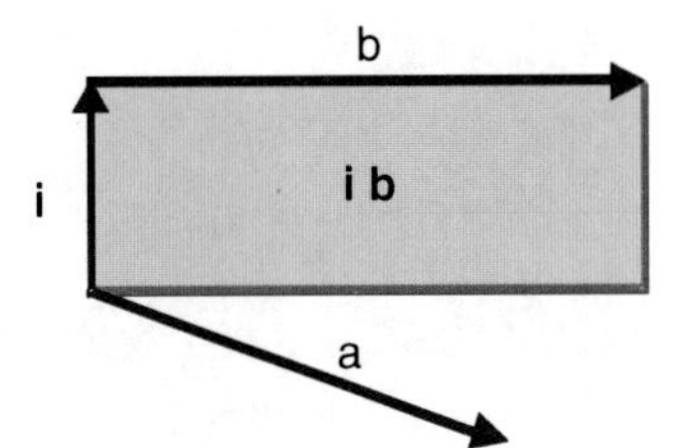


**Fig. 10.1** The complex vector $\mathbf{a} + \mathrm{i}\mathbf{b}$

[2]The material in this Section 10.3 and Section 10.4 is based upon the author's paper [88], which appeared in "Advances in Applied Clifford Algebra".

[3]In terms of the inner product of the geometric algebra $\mathbb{G}_{p,q+1}$, $\overline{A} = A + 2\mathrm{i}(\mathrm{i} \cdot A)$.

$$\overline{A+B} = \overline{A} + \overline{B}, \text{ and } \overline{AB} = \overline{A}\,\overline{B}.$$

Note that this complex conjugation is not the *Clifford conjugation* defined in (3.33).

The unitary geometric algebra

$$\mathbb{U}_{p,q} = \mathbb{U}(\mathbb{H}^{p,q}) = \mathbb{G}_{p,q+1} \tag{10.19}$$

of the Hermitian space $\mathbb{H}^{p,q} = \mathbb{R}^{p,q} + \mathrm{i}\mathbb{R}^{p,q}$ has exactly the same elements as the geometric algebra $\mathbb{G}_{p,q+1} = \mathbb{G}(\mathbb{R}^{p,q+1})$. The geometric product of the elements of $\mathbb{U}_{p,q}$ is also exactly the same as for the same elements in $\mathbb{G}(\mathbb{R}^{p,q+1})$. What is different is what we mean by *complex vectors*, how we *grade* the algebra into *complex k-vectors*, and how we define the *Hermitian inner product* and the *Hermitian outer product* of the *complex multivectors* in $\mathbb{U}_{p,q}$. By a *complex k-vector* $\mathbf{B}_k \in \mathbb{U}_{p,q}$, we mean $\mathbf{B}_k = \mathbf{B}_1 + \mathrm{i}\mathbf{B}_2$ where $\mathbf{B}_1, \mathbf{B}_2$ are $k$-vectors in $\mathbb{G}^k_{p,q}$. Alternatively, $\mathbb{U}_{p,q} = \mathbb{G}_{p,q} + \mathrm{i}\mathbb{G}_{p,q}$ can be thought of as being generated by taking all geometric sums of geometric products of the complex vectors in $\mathbb{H}^{p,q}$.

We are now ready to define the Hermitian inner and outer products of complex vectors $\mathbf{x}, \mathbf{y} \in \mathbb{H}^{p,q}$. The *Hermitian inner product* is defined by

$$\mathbf{x} \cdot \mathbf{y} = \frac{1}{2}(\mathbf{x}\mathbf{y} + \overline{\mathbf{y}}\,\overline{\mathbf{x}}) = \frac{1}{2}(\mathbf{x}\mathbf{y} + (\mathbf{x}\mathbf{y})^{\dagger}),$$

and the *Hermitian outer product* is defined by

$$\mathbf{x} \wedge \mathbf{y} = \frac{1}{2}(\mathbf{x}\mathbf{y} - \overline{\mathbf{y}}\,\overline{\mathbf{x}}) = \frac{1}{2}(\mathbf{x}\mathbf{y} - (\mathbf{x}\mathbf{y})^{\dagger}),$$

from which it follows that $\mathbf{x}\mathbf{y} = \mathbf{x} \cdot \mathbf{y} + \mathbf{x} \wedge \mathbf{y}$.

Letting $\mathbf{x} = \mathbf{x}_1 + \mathrm{i}\mathbf{x}_2$ and $\mathbf{y} = \mathbf{y}_1 + \mathrm{i}\mathbf{y}_2$ for the complex vectors $\mathbf{x}, \mathbf{y} \in \mathbb{H}^{p,q}$, we calculate

$$\begin{aligned}
\mathbf{x} \cdot \mathbf{y} &= \mathbf{x}_1 \cdot \mathbf{y}_1 + \mathbf{x}_2 \cdot \mathbf{y}_2 + \mathrm{i}(\mathbf{x}_2 \cdot \mathbf{y}_1 - \mathbf{x}_1 \cdot \mathbf{y}_2) = <\mathbf{x}\mathbf{y}>_{\mathscr{C}}, \\
\mathbf{x} \cdot \overline{\mathbf{y}} &= \mathbf{x}_1 \cdot \mathbf{y}_1 - \mathbf{x}_2 \cdot \mathbf{y}_2 + \mathrm{i}(\mathbf{x}_2 \cdot \mathbf{y}_1 + \mathbf{x}_1 \cdot \mathbf{y}_2) = <\mathbf{x}\overline{\mathbf{y}}>_{\mathscr{C}}, \\
\mathbf{x} \wedge \mathbf{y} &= \mathbf{x}_1 \wedge \mathbf{y}_1 + \mathbf{x}_2 \wedge \mathbf{y}_2 + \mathrm{i}(\mathbf{x}_2 \wedge \mathbf{y}_1 - \mathbf{x}_1 \wedge \mathbf{y}_2) = <\mathbf{x}\mathbf{y}>_{\mathscr{B}}, \\
\mathbf{x} \wedge \overline{\mathbf{y}} &= \mathbf{x}_1 \wedge \mathbf{y}_1 - \mathbf{x}_2 \wedge \mathbf{y}_2 + \mathrm{i}(\mathbf{x}_2 \wedge \mathbf{y}_1 + \mathbf{x}_1 \wedge \mathbf{y}_2) = <\mathbf{x}\overline{\mathbf{y}}>_{\mathscr{B}},
\end{aligned} \tag{10.20}$$

where $<\mathbf{x}\mathbf{y}>_{\mathscr{C}}$ and $<\mathbf{x}\mathbf{y}>_{\mathscr{B}}$ denote the complex scalar and complex bivector parts of the geometric product $\mathbf{x}\mathbf{y}$, respectively. Note also that the Hermitian inner and outer products are *Hermitian symmetric* and *Hermitian antisymmetric*, respectively, i.e.,

$$\mathbf{x} \cdot \mathbf{y} = \overline{\mathbf{y} \cdot \mathbf{x}} = \overline{\mathbf{y}} \cdot \overline{\mathbf{x}} \text{ and } \mathbf{x} \wedge \mathbf{y} = -\overline{\mathbf{y} \wedge \mathbf{x}} = -\overline{\mathbf{y}} \wedge \overline{\mathbf{x}}.$$

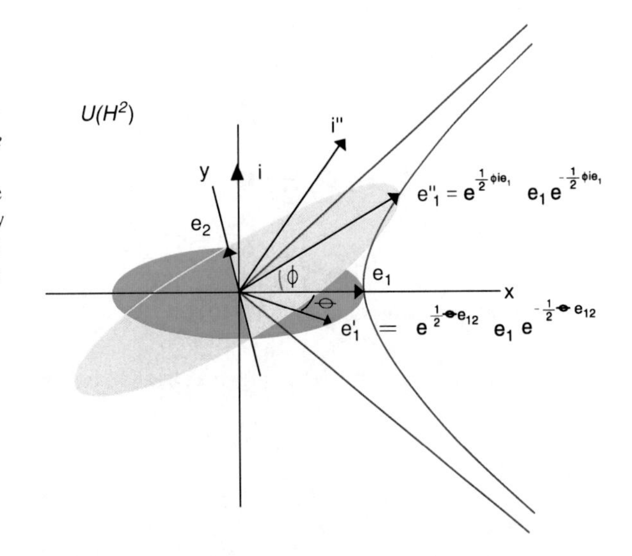


**Fig. 10.2** The vector $\mathbf{e}_1$ is rotated in the $xy$-plane of $\mathbf{e}_{12}$ into the vector $\mathbf{e}'_1$. The vector $\mathbf{e}_1$ is *boosted* into the *relative vector* $\mathbf{e}''_1$ of the *relative plane* of the *relative bivector* $\mathbf{e}''_{12} = \mathbf{e}''_1\mathbf{e}_2$. The relative plane of $\mathbf{e}''_{12}$ has the relative velocity of $\frac{\mathbf{v}}{c} = \mathbf{i}\mathbf{e}_1 \tanh\phi$ with respect to the plane of $\mathbf{e}_{12}$, where $c$ is the speed of light

The Hermitian inner and outer products reduce to the ordinary inner and outer products (10.8) in $\mathbb{G}_{p,q}$ when the complex vectors $\mathbf{x}, \mathbf{y} \in \mathbb{U}^{p,q}$ are real, i.e., $\mathbf{x} = \overline{\mathbf{x}}$ and $\mathbf{y} = \overline{\mathbf{y}}$, and the closely related identities satisfied by the Hermitian inner and outer products of complex $k$-vectors reduce to their real counterparts in $\mathbb{G}_{p,q}$. The complexification of $\mathbb{R}^2$ to $\mathbb{H}^2$ and $\mathbb{G}_2$ to $\mathbb{U}_2$ gives new geometric meaning to the notion of a complex scalar and a complex vector and the resulting complex geometric algebra. In Fig. 10.2, a rotation in the plane of the imaginary vector $\mathbf{i}\mathbf{e}_1$ takes on the interpretation of a *Lorentz boost* of the Euclidean plane having the direction of the bivector $\mathbf{e}_{12}$. We will discuss this further in the next chapter.

Let $\mathbf{x} \in \mathbb{H}^{p,q}$ be a complex vector. Using the same conventions as in (10.6) and (10.7), in the standard row basis $(\mathbf{e})_{(n)}$ of $\mathbb{H}^{p,q}$, we write

$$\mathbf{x} = (\mathbf{e})_{(n)}(x)^{(n)} = \sum_{i=1}^{n} \mathbf{e}_i x^i = \sum_{i=1}^{n} \overline{x}_i \mathbf{e}^i = (\overline{x})_{(n)}(\mathbf{e})^{(n)}, \tag{10.21}$$

where $(x)^{(n)} = [g](x)^{\mathrm{T}}_{(n)}$ is the column of complex reciprocal scalar components of $\mathbf{x}$ and $(\overline{x})_{(n)}$ is the corresponding row of the conjugated complex scalar components of $\mathbf{x}$. The relationship (10.21) reduces to (10.7) when $\overline{\mathbf{x}} = \mathbf{x}$.

Let us now explore the corresponding Hermitian versions of the identities in the unitary geometric alebra $\mathbb{U}_{p,q}$ that were given for the real geometric algebra $\mathbb{G}_n$ in Chap. 3. Just as the inner and outer products between $\mathbf{a}$ and $\mathbf{B}_k$ were defined in (3.37), we similarly define the corresponding Hermitian inner and outer products. For *odd* $k \geq 1$,

$$\mathbf{a} \cdot \mathbf{B}_k := \frac{1}{2}(\mathbf{a}\mathbf{B}_k + \overline{\mathbf{B}}_k\, \overline{\mathbf{a}}) = <\mathbf{a}\mathbf{B}_k>_{k-1},$$

$$\mathbf{a} \cdot \mathbf{B}_{k+1} := \frac{1}{2}(\mathbf{a}\mathbf{B}_{k+1} - \overline{\mathbf{B}}_{k+1}\, \mathbf{a}) = <\mathbf{a}\mathbf{B}_{k+1}>_{k}, \tag{10.22}$$

and

$$
\begin{aligned}
\mathbf{a} \wedge \mathbf{B}_k &:= \frac{1}{2}(\mathbf{a}\mathbf{B}_k - \overline{\mathbf{B}}_k\, \overline{\mathbf{a}}) = < \mathbf{a}\mathbf{B}_k >_{k+1}, \\
\mathbf{a} \wedge \mathbf{B}_{k+1} &:= \frac{1}{2}(\mathbf{a}\mathbf{B}_{k+1} + \overline{\mathbf{B}}_{k+1}\, \mathbf{a}) = < \mathbf{a}\mathbf{B}_{k+1} >_{k+2}
\end{aligned} \tag{10.23}
$$

so that $\mathbf{a}\mathbf{B}_j = \mathbf{a} \cdot \mathbf{B}_j + \mathbf{a} \wedge \mathbf{B}_j$ for all $j \geq 1$. In the case that $\mathbf{a}$ and $\mathbf{B}_k$ are real, the identities (10.22) and (10.23) reduce to (3.37).

The corresponding identity to (3.40) is

$$\mathbf{a} \cdot (\mathbf{b} \wedge \mathbf{C}_k) = (\mathbf{a} \cdot \mathbf{b})\mathbf{C}_k - \overline{\mathbf{b}} \wedge (\overline{\mathbf{a}} \cdot \mathbf{C}_k)$$

and to (3.42), when $r$ and $s$ are both odd, is

$$\mathbf{a} \cdot (\mathbf{A}_r \wedge \mathbf{B}_s) = (\mathbf{a} \cdot \mathbf{A}_r) \wedge \mathbf{B}_s - \overline{\mathbf{A}}_r \wedge (\overline{\mathbf{a}} \cdot \mathbf{B}_s) = -(\overline{\mathbf{A}}_r \wedge \overline{\mathbf{B}}_s) \cdot \mathbf{a}.$$

In the last identity, when $r$ and $s$ are both even, the identity must be modified to read

$$\mathbf{a} \cdot (\mathbf{A}_r \wedge \mathbf{B}_s) = (\mathbf{a} \cdot \mathbf{A}_r) \wedge \mathbf{B}_s + \overline{\mathbf{A}}_r \wedge (\mathbf{a} \cdot \mathbf{B}_s) = -(\overline{\mathbf{A}}_r \wedge \overline{\mathbf{B}}_s) \cdot \mathbf{a}.$$

The rules for the other two case of this identity when $r + s$ is odd are left to the reader.

From these identities, many other useful identities for the unitary geometric algebra $\mathbb{U}_{p,q}$ can be derived. For example, for complex vectors $\mathbf{a}, \mathbf{b}, \mathbf{c} \in \mathbb{H}^{p,q}$, we have

$$\mathbf{a} \cdot (\mathbf{b} \wedge \mathbf{c}) = (\mathbf{a} \cdot \mathbf{b})\mathbf{c} - \overline{\mathbf{b}}(\overline{\mathbf{a}} \cdot \mathbf{c}) = \mathbf{a} \cdot \mathbf{b}\, \mathbf{c} - \mathbf{a} \cdot \overline{\mathbf{c}}\, \overline{\mathbf{b}} = -(\overline{\mathbf{b}} \wedge \overline{\mathbf{c}}) \cdot \mathbf{a}.$$

With care, anyone familiar with the rules of real geometric algebra can quickly become adept to the closely related rules in the unitary geometric algebra $\mathbb{U}_{p,q}$.

However, there are peculiarities that must be given careful consideration. For example, in general $\mathbf{a} \wedge \mathbf{a} \neq 0$, but always $\mathbf{a} \wedge \overline{\mathbf{a}} = 0$. Also, a complex vector $\mathbf{a} \in \mathbb{H}^{p,q}$ is not always invertible, as follows from

$$\mathbf{a}\, \overline{\mathbf{a}} = \mathbf{a} \cdot \overline{\mathbf{a}} + \mathbf{a} \wedge \overline{\mathbf{a}} = \mathbf{a} \cdot \overline{\mathbf{a}},$$

which implies that the inverse

$$\mathbf{a}^{-1} = \overline{\mathbf{a}} \left( \frac{1}{\mathbf{a} \cdot \overline{\mathbf{a}}} \right) = \left( \frac{1}{\overline{\mathbf{a}} \cdot \mathbf{a}} \right) \overline{\mathbf{a}}$$

of the complex vector $\mathbf{a}$, with respect to the unitary geometric product, exists only when $\mathbf{a} \cdot \overline{\mathbf{a}} \neq 0$. In the special case that $\mathbf{a} \in \mathbb{H}^n$, the identities (10.20) show that $\mathbf{a} \cdot \mathbf{a} = 0$ only when $\mathbf{a} = 0$. However, $\mathbf{a} \wedge \mathbf{a} = 0$ for $\mathbf{a} = \mathbf{a}_1 + \mathrm{i}\mathbf{a}_2 \in \mathbb{H}^{p,q}$ only when

$\mathbf{a}_1 = \alpha \mathbf{a}_2$ for some $\alpha \in \mathbb{R}$. On the other hand, $\mathbf{a} \cdot \overline{\mathbf{a}} = 0$ for $\mathbf{a} \in \mathbb{H}^{p,q}$ if and only if $\mathbf{a}_1^2 = \mathbf{a}_2^2$ and $\mathbf{a}_1 \cdot \mathbf{a}_2 = 0$.

By the *magnitude* of a complex vector $\mathbf{a} \in \mathbb{H}^{p,q}$, we mean

$$|\mathbf{a}| = \sqrt{|\mathbf{a} \cdot \mathbf{a}|}. \tag{10.24}$$

In the case of the unitary geometric algebra $\mathbb{U}_n$ of the Euclidean space $\mathbb{R}^n$, the magnitude is positive definite, but there are still non-invertible complex vectors with positive magnitude. For example, for $\mathbf{a} = \mathbf{e}_1 + \mathrm{i}\mathbf{e}_2 \in \mathbb{H}^2$, $\mathbf{a} \cdot \mathbf{a} = 2$, but $\mathbf{a}^{-1}$ does not exist since

$$\mathbf{a} \cdot \overline{\mathbf{a}} = (\mathbf{e}_1 + \mathrm{i}\mathbf{e}_2) \cdot (\mathbf{e}_1 - \mathrm{i}\mathbf{e}_2) = \mathbf{e}_1^2 - \mathbf{e}_2^2 + 2\mathrm{i}\mathbf{e}_1 \cdot \mathbf{e}_2 = 0.$$

Sometimes, however, we write $\mathbf{a}^{-1} = \frac{\mathbf{a}}{|\mathbf{a}|^2}$ for which $\mathbf{a} \cdot \mathbf{a}^{-1} = 1$ with respect to the Hermitian inner product.

Suppose now that a column $(\mathbf{a})^{\mathrm{T}}_{(n)}$ of $n$ complex vectors in $\mathbb{H}^{p,q}$ is given. In terms of the standard basis $(\mathbf{e})_{(n)}$ of $\mathbb{H}^{p,q}$, we write

$$(\mathbf{a})^{\mathrm{T}}_{(n)} = A(\mathbf{e})^{\mathrm{T}}_{(n)} = \left[(\mathbf{e})_{(n)}A^*\right]^{\mathrm{T}}, \tag{10.25}$$

where the $k^{th}$ row of the $n \times n$ complex matrix $A$ consists of the components of the corresponding complex vectors $\mathbf{a}^k = \mathbf{a}_k$ and $A^* = \overline{A}^{\mathrm{T}}$ is the conjugate transpose of the matrix $A$. For example, for $n = 2$, we have

$$\begin{aligned}(\mathbf{a})^{\mathrm{T}}_{(2)} &= A(\mathbf{e})^{\mathrm{T}}_{(2)} = \begin{pmatrix} a_{11} & a_{12} \\ a_{21} & a_{22} \end{pmatrix}\begin{pmatrix} \mathbf{e}_1 \\ \mathbf{e}_2 \end{pmatrix} = \begin{pmatrix} a_{11}\mathbf{e}_1 + a_{12}\mathbf{e}_2 \\ a_{21}\mathbf{e}_1 + a_{22}\mathbf{e}_2 \end{pmatrix} \\ &= \begin{pmatrix} \mathbf{e}_1\overline{a}_{11} + \mathbf{e}_2\overline{a}_{12} \\ \mathbf{e}_1\overline{a}_{21} + \mathbf{e}_2\overline{a}_{22} \end{pmatrix} = \left[(\mathbf{e}_1\ \mathbf{e}_2)\begin{pmatrix} \overline{a}_{11} & \overline{a}_{21} \\ \overline{a}_{12} & \overline{a}_{22} \end{pmatrix}\right]^{\mathrm{T}} = \left[(\mathbf{e})_{(2)}A^*\right]^{\mathrm{T}}.\end{aligned} \tag{10.26}$$

We can now relate the determinant function of the matrix $A$ to the outer product of the complex vectors $(\mathbf{a})_{(n)}$. We have the rather strange looking relationship

$$\mathbf{a}_{(\overline{n})} \equiv \mathbf{a}_1 \wedge \overline{\mathbf{a}}_2 \wedge \cdots \wedge \mathbf{a}_{2k+1} \wedge \overline{\mathbf{a}}_{2k} \wedge \cdots = (\det A)\mathbf{e}_{(n)}. \tag{10.27}$$

The Hermitian character of the outer product requires that we take the conjugation of every even numbered complex vector in the product. For the 2-dimensional example of (10.26), we have

$$\begin{aligned}\mathbf{a}_1 \wedge \overline{\mathbf{a}}_2 &= (a_{11}\mathbf{e}_1 + a_{12}\mathbf{e}_2) \wedge (\overline{a}_{21}\mathbf{e}_1 + \overline{a}_{22}\mathbf{e}_2) \\ &= (a_{11}a_{22} - a_{12}a_{21})\mathbf{e}_{(2)} = \det\begin{pmatrix} a_{11} & a_{12} \\ a_{21} & a_{22} \end{pmatrix}\mathbf{e}_{12}.\end{aligned}$$

It follows that a set of complex vectors $(\mathbf{a})_{(n)}$, defined in the equation (10.25), is (complex) linearly independent iff $\det A \neq 0$.

## Exercises

1. Let $(\mathbf{e})_{(3)} = (\mathbf{e}_1, \mathbf{e}_2, \mathbf{e}_3)$ be standard orthonormal basis and let $\mathbf{a} = (1, \mathrm{i}, 1+\mathrm{i})(\mathbf{e})^{\mathrm{T}}_{(3)}$ and $\mathbf{b} = (1-\mathrm{i}, -\mathrm{i}, 2)(\mathbf{e})^{\mathrm{T}}_{(3)}$.

   (a) Compute $\mathbf{a}\cdot\mathbf{b}$ and $\mathbf{a}\cdot(\mathbf{a}\wedge\mathbf{b})$ in $\mathbb{H}^3$.
   (b) Compute $\mathbf{a}\cdot\mathbf{b}$ and $\mathbf{a}\cdot(\mathbf{a}\wedge\mathbf{b})$ in $\mathbb{H}^{2,1}$.
   (c) Compute $\mathbf{a}\cdot\mathbf{b}$ and $\mathbf{a}\cdot(\mathbf{a}\wedge\mathbf{b})$ in $\mathbb{H}^{1,2}$.
   (d) Compute $\mathbf{a}\cdot\mathbf{b}$ and $\mathbf{a}\cdot(\mathbf{a}\wedge\mathbf{b})$ in $\mathbb{H}^{0,3}$.

2. Define the complex vectors $\mathbf{a} = (1, \mathrm{i}, 2)(\mathbf{e})^{\mathrm{T}}_{(3)}, \mathbf{b} = (1, \mathrm{i}, -\mathrm{i})(\mathbf{e})^{\mathrm{T}}_{(3)}$, and $\mathbf{c} = (1, -\mathrm{i}, 0)(\mathbf{e})^{\mathrm{T}}_{(3)}$.

   (a) Compute $\mathbf{a}\wedge\overline{\mathbf{b}}\wedge\mathbf{c}$ in $\mathbb{H}^3$.
   (b) Compute $\mathbf{a}\cdot(\mathbf{b}\wedge\mathbf{c})$ and $\mathbf{a}\cdot(\overline{\mathbf{b}}\wedge\mathbf{c})$ in $\mathbb{H}^{2,1}$.
   (c) Compute $(\mathbf{a}\wedge\overline{\mathbf{b}})\cdot(\overline{\mathbf{b}}\wedge\mathbf{a})$ in $\mathbb{H}^{1,2}$.
   (d) Compute $\mathbf{a}\cdot(\overline{\mathbf{b}}\wedge\mathbf{c})$ and $(\mathbf{a}\wedge\overline{\mathbf{b}})\cdot\mathbf{c}$ in $\mathbb{H}^{0,3}$.

3. Given the complex vectors $(\mathbf{a})_{(3)} = (\mathbf{e})_{(3)} \begin{pmatrix} 2 & \mathrm{i} & 2 \\ 1 & \mathrm{i} & -\mathrm{i} \\ 1 & -\mathrm{i} & 1 \end{pmatrix}$, show the following:

   (a) Show that the vectors are linearly independent by calculating $\mathbf{a}_{(\overline{3})}$.
   (b) In $\mathbb{H}^{2,1}$, calculate $(\mathbf{a})^{\mathrm{T}}_{(3)}\cdot(\mathbf{a})_{(\overline{3})}$.
   (c) In $\mathbb{H}^{1,2}$, calculate $(\mathbf{a})^{\mathrm{T}}_{(3)}\cdot(\mathbf{a})_{(\overline{3})}$.
   (d) In $\mathbb{H}^{0,3}$, calculate $(\mathbf{a})^{\mathrm{T}}_{(3)}\cdot(\mathbf{a})_{(3)}$.

4. Show that $|\mathbf{a}+\mathbf{b}|^2 + |\mathbf{a}-\mathbf{b}|^2 = 2|\mathbf{a}|^2 + 2|\mathbf{b}|^2$ in $\mathbb{H}^n$.

## 10.4 Hermitian Orthogonality

We now generalize the results obtained in Sect. 2 for the pseudo-Euclidean space $\mathbb{R}^{p,q}$ to the unitary space $\mathbb{H}^{p,q}$. In the special case of the unitary space $\mathbb{U}_n$ of the Hermitian space $\mathbb{H}^n$, even stronger results are possible as we shall see.

Two complex vectors $\mathbf{a}, \mathbf{b} \in \mathbb{H}^{p,q}$ are said to be *Hermitian orthogonal* in $\mathbb{H}^{p,q}$ if $\mathbf{a}\cdot\mathbf{b} = 0$; they are said to be *conjugate orthogonal* if $\mathbf{a}\cdot\overline{\mathbf{b}} = 0$. The complex vectors shown in Fig. 10.3 are Hermitian orthogonal. The magnitude of a complex vector $\mathbf{a} \in \mathbb{H}^{p,q}$ has already been defined in (10.24). The complex vector $\mathbf{a}$ is said to be *non-null* if $|\mathbf{a}| \neq 0$. If $\mathbf{a} = \mathbf{a}_1 + \mathrm{i}\mathbf{a}_2$, where $\mathbf{a}_1$ and $\mathbf{a}_2$ are real vectors, then

$$|\mathbf{a}|^2 = |(\mathbf{a}_1 + \mathrm{i}\mathbf{a}_2)\cdot(\mathbf{a}_1 + \mathrm{i}\mathbf{a}_2)| = |\mathbf{a}_1\cdot\mathbf{a}_1 + \mathbf{a}_2\cdot\mathbf{a}_2| \geq 0.$$

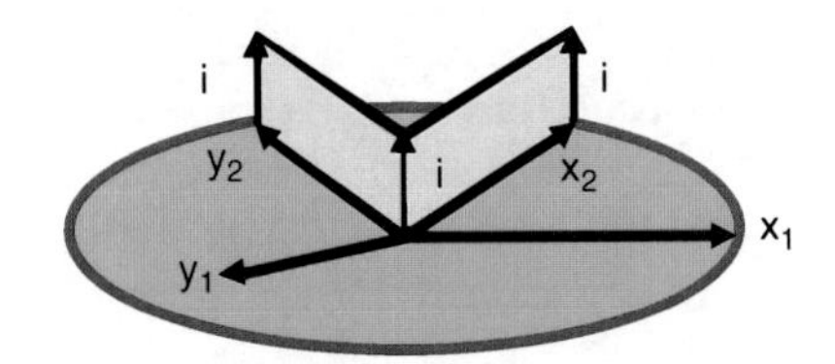


**Fig. 10.3** The complex vectors $\mathbf{x} = \mathbf{x}_1 + \mathrm{i}\mathbf{x}_2$ and $\mathbf{y} = \mathbf{y}_1 + \mathrm{i}\mathbf{y}_2$ are Hermitian orthogonal

In the special case when $\mathbf{a} \in \mathbb{H}^n$, $|\mathbf{a}|^2 = \mathbf{a} \cdot \mathbf{a} \geq 0$ so that the magnitude $|\mathbf{a}| = 0$ iff $\mathbf{a} = 0$.

A row $(\mathbf{a})_{(k)} = (\mathbf{a}_1, \ldots, \mathbf{a}_k)$ of $k$ complex vectors $\mathbf{a}_i \in \mathbb{H}^{p,q}$ is said to be *(complex) linearly independent* if for every column $(\alpha)^{\mathrm{T}}_{(k)}$ of not all zero complex scalars $\alpha_i \in \mathbb{C}$,

$$(\mathbf{a})_{(k)}(\alpha)^{\mathrm{T}}_{(k)} = \begin{pmatrix} \mathbf{a}_1 & \ldots & \mathbf{a}_k \end{pmatrix} \begin{pmatrix} \alpha_1 \\ \cdot \\ \cdot \\ \alpha_k \end{pmatrix} = \sum_{j=1}^{k} \mathbf{a}_j \alpha_j$$

$$= \sum_{j=1}^{k} \overline{\alpha}_j \mathbf{a}_j = \begin{pmatrix} \overline{\alpha}_1 & \ldots & \overline{\alpha}_k \end{pmatrix} \begin{pmatrix} \mathbf{a}_1 \\ \cdot \\ \cdot \\ \mathbf{a}_k \end{pmatrix} = (\overline{\alpha})_{(k)}(\mathbf{a})^{\mathrm{T}}_{(k)} \neq 0.$$

We can easily characterize the linear dependence or independence of a row of complex vectors by using the Hermitian outer product. But first we introduce some necessary notation. If $(\mathbf{a})_{(k)}$ is a row of complex vectors, where $k$ is even, we define the row $(\mathbf{a})_{(\overline{k})}$ by

$$(\mathbf{a})_{(\overline{k})} = (\mathbf{a}_1, \overline{\mathbf{a}}_2, \mathbf{a}_3, \overline{\mathbf{a}}_4, \ldots, \overline{\mathbf{a}}_k). \tag{10.28}$$

On the other hand, if $k$ is odd, we have

$$(\mathbf{a})_{(\overline{k})} = (\mathbf{a}_1, \overline{\mathbf{a}}_2, \mathbf{a}_3, \overline{\mathbf{a}}_4, \ldots, \mathbf{a}_k). \tag{10.29}$$

In both rows, only the even terms of the rows are conjugated. We define the corresponding alternatingly conjugated $k$-complex vectors, consistent with (10.27), by

$$\mathbf{a}_{(\overline{k})} = \wedge(\mathbf{a})_{(\overline{k})}. \tag{10.30}$$

With this notation, a row of complex vectors $(\mathbf{a})_{(k)}$ is linearly independent iff

$$\mathbf{a}_{(\overline{k})} = \wedge(\mathbf{a})_{(\overline{k})} \neq 0.$$

The *magnitude* $|\mathbf{a}_{(k)}|$ of the complex $k$-vector $\mathbf{a}_{(k)}$ is defined by

$$|\mathbf{a}_{(k)}| = \sqrt{|\overline{\mathbf{a}}^{\dagger}_{(k)} \cdot \mathbf{a}_{(k)}|} \geq 0.$$

We will shortly see that the magnitude of a complex $k$-vector is also positive definite in the special case of a $k$-vector in $\mathbb{U}_n$.

Given a basis $(\mathbf{a})_{(n)}$ of non-null complex vectors in $\mathbb{H}^{p,q}$, we want to find the closely related *Hermitian orthogonal basis* $(\mathbf{b})_{(n)}$ of $\mathbb{H}^{p,q}$ which satisfies the following two conditions:

$$\mathbf{b}_i \cdot \mathbf{b}_j = 0 \text{ and } \mathbf{b}_{(\overline{k})} = \mathbf{a}_{(\overline{k})} \tag{10.31}$$

for all $1 \le i < j \le n$ and $1 \le k \le n$.

This task is completed in much the same way as (10.14) by the following recursive construction. For $k = 1$, we set $\mathbf{b}_1 = \mathbf{a}_1$. For all $k$, $2 \le k \le n$,

$$\mathbf{b}_k = \frac{\overline{\mathbf{b}}_{(\overline{k-1})} \cdot (\mathbf{b}^{\dagger}_{(\overline{k-1})} \wedge \mathbf{a}_k)}{\overline{\mathbf{b}}_{(\overline{k-1})} \cdot \mathbf{b}^{\dagger}_{(\overline{k-1})}}. \tag{10.32}$$

Note, since the $\mathbf{b}_j$'s are orthogonal,

$$\overline{\mathbf{b}}_{(\overline{k-1})} \cdot \mathbf{b}^{\dagger}_{(\overline{k-1})} = (\mathbf{b}_1 \cdot \mathbf{b}_1) \cdots (\mathbf{b}_{k-1} \cdot \mathbf{b}_{k-1}).$$

The above construction is often called the *Gram-Schmidt orthogonalization process*, but is more general since we are dealing with an indefinite metric. In the special case of the positive definite metric of $\mathbb{H}^n$, it shows that (10.31) is positive definite. Our construction (10.32) is closely related to [29, p.258].

The method by which we constructed the corresponding orthogonal basis $(\mathbf{b})_{(n)}$ for the basis $(\mathbf{a})_{(n)}$, satisfying the conditions (10.31), makes possible the following

**Corollary 10.4.1.** *The relationship of the constructed orthogonal basis* $(\mathbf{b})_{(n)} = (\mathbf{b}_1, \mathbf{b}_2, \ldots, \mathbf{b}_n)$ *to the given basis* $(\mathbf{a})_{(n)} = (\mathbf{a}_1, \mathbf{a}_2, \ldots, \mathbf{a}_n)$ *can be expressed in the form* $(\mathbf{b})_{(n)} = (\mathbf{a})_{(n)} \triangle$ *where* $\triangle$ *is an upper triangular matrix.*

*Proof.* Notice that in the construction of the $k$th basis vector $\mathbf{b}_k$ of $(\mathbf{b})_{(n)}$, only $\mathbf{a}_1, \mathbf{a}_2, \ldots, \mathbf{a}_k$ are involved for $k = 1, 2, \ldots, n$. It is for this reason that $\triangle$ is an upper triangular matrix.

More specifically, let $f$ be the linear operator which takes the basis $(\mathbf{a})_{(n)}$ into the basis $(\mathbf{b})_{(n)}$, i.e., $f(\mathbf{a})_{(n)} = (\mathbf{a})_{(n)}[f] = (\mathbf{b})_{(n)}$ where $\triangle = [f]$ is the matrix of $f$. We can solve this relationship directly for the matrix $\triangle$, getting

$$\triangle = \left[(\mathbf{b}^{-1})^{\mathrm{T}}_{(n)} \cdot (\mathbf{a})_{(n)}\right]^{-1} = \begin{pmatrix} \mathbf{b}_1^{-1} \cdot \mathbf{a}_1 & \ldots & \mathbf{b}_1^{-1} \cdot \mathbf{a}_n \\ \ldots & \ldots & \ldots \\ \ldots & \ldots & \ldots \\ \mathbf{b}_n^{-1} \cdot \mathbf{a}_1 & \ldots & \mathbf{b}_n^{-1} \cdot \mathbf{a}_n \end{pmatrix}^{-1},$$

where each $\mathbf{b}_i^{-1} = \frac{\mathbf{b}_i}{\mathbf{b}_i \cdot \mathbf{b}_i}$. □

The Gram matrix (10.15) of the basis $(\mathbf{a})_{(n)}$ is

$$A = (\mathbf{a})^{\mathrm{T}}_{(n)} \cdot (\mathbf{a})_{(n)} = \begin{pmatrix} \mathbf{a}_1 \cdot \mathbf{a}_1 & \dots & \mathbf{a}_1 \cdot \mathbf{a}_n \\ \dots & \dots & \dots \\ \dots & \dots & \dots \\ \mathbf{a}_n \cdot \mathbf{a}_1 & \dots & \mathbf{a}_n \cdot \mathbf{a}_n \end{pmatrix}. \tag{10.33}$$

But the relationship

$$(\mathbf{a})_{(n)}[f] = (\mathbf{b})_{(n)} \quad \Longleftrightarrow \quad [f]^*(\mathbf{a})^{\mathrm{T}}_{(n)} = (\mathbf{b})^{\mathrm{T}}_{(n)},$$

from which it follows that

$$[f]^*A[f] = [f]^*(\mathbf{a})^{\mathrm{T}}_{(n)} \cdot (\mathbf{a})_{(n)}[f] = (\mathbf{b})^{\mathrm{T}}_{(n)} \cdot (\mathbf{b})_{(n)} = B$$

where $B$ is the diagonal Gram matrix of the orthogonal basis $(\mathbf{b})_{(n)}$. Thus, we have diagonalized the indefinite Hermitian quadratic form defined by the matrix $A$. Since the relationship (10.31) implies that $\det[f] = 1$, it also follows that $\det A = \det B$. Of course, the signature of the Hermitian quadratic form of the matrix $A$ is the same as the signature of the Hermitian quadratic form defined by the matrix $B$, which is just the equally valid Sylvester's law of inertia Theorem 9.4.3 applied to the complex case.

The Gram determinant of the basis $(\mathbf{a})_{(n)}$ is defined to be the determinant of Gram matrix $A$ of $(\mathbf{a})_{(n)}$. We have

$$\det A = \det[(\mathbf{a})^{\mathrm{T}}_{(n)} \cdot (\mathbf{a})_{(n)}] = \overline{\mathbf{a}}^{\dagger}_{(\overline{n})} \cdot \mathbf{a}_{(\overline{n})}.$$

In the special case of the unitary space $\mathbb{U}_n = \mathbb{U}(\mathbb{H}^n)$, because of the properties (10.31) of the related basis $(\mathbf{b})_{(n)}$, it follows that

$$\overline{\mathbf{a}}^{\dagger}_{(\overline{k})} \cdot \mathbf{a}_{(\overline{k})} = \overline{\mathbf{b}}^{\dagger}_{(\overline{k})} \cdot \mathbf{b}_{(\overline{k})} \geq 0,$$

for $1 \leq k \leq n$. When $n = 2$, this relationship reduces to the *Schwarz inequality* [66, p.218],

$$\overline{\mathbf{a}}^{\dagger}_{(\overline{2})} \cdot \mathbf{a}_{(\overline{2})} = (\overline{\mathbf{a}}_2 \wedge \mathbf{a}_1) \cdot (\mathbf{a}_1 \wedge \overline{\mathbf{a}}_2) = |\mathbf{a}_1|^2 |\mathbf{a}_2|^2 - |\mathbf{a}_1 \cdot \mathbf{a}_2|^2 \geq 0. \tag{10.34}$$

The positive definite Hermitian space $\mathbb{H}^n$, and its unitary geometric algebra $\mathbb{U}_n$, has much in common with the Euclidean space $\mathbb{R}^n$ and its geometric algebra $\mathbb{G}_n$ which it generalizes. The next theorem gives the basic properties of the distance $|\mathbf{b} - \mathbf{a}|$ between the two complex vectors $\mathbf{a}, \mathbf{b} \in \mathbb{H}^n$.

**Theorem 10.4.2.** *For vectors* $\mathbf{a},\mathbf{b}\in\mathbb{H}^n$,

1. $|\mathbf{a}-\mathbf{b}|=|\mathbf{b}-\mathbf{a}|$
2. $|\mathbf{b}-\mathbf{a}|=0$ *if and only if* $\mathbf{a}=\mathbf{b}$
3. $|\mathbf{a}+\mathbf{b}|\le|\mathbf{a}|+|\mathbf{b}|$ *(triangle inequality)*

*Proof.* The proofs of (1) and (2) follow directly from from the definitions. The proof of 3) follows from the steps

$$\begin{aligned}|\mathbf{a}+\mathbf{b}|^2 &= (\mathbf{a}+\mathbf{b})\cdot(\mathbf{a}+\mathbf{b})=\mathbf{a}\cdot\mathbf{a}+\mathbf{a}\cdot\mathbf{b}+\mathbf{b}\cdot\mathbf{a}+\mathbf{b}\cdot\mathbf{b}\\ &= |\mathbf{a}|^2+\mathbf{a}\cdot\mathbf{b}+\overline{\mathbf{a}\cdot\mathbf{b}}+|\mathbf{b}|^2\le|\mathbf{a}|^2+2|\mathbf{a}\cdot\mathbf{b}|+|\mathbf{b}|^2\\ &\le |\mathbf{a}|^2+2|\mathbf{a}||\mathbf{b}|+|\mathbf{b}|^2=(|\mathbf{a}|+|\mathbf{b}|)^2,\end{aligned}$$

with the help of Schwarz inequality (10.34). □

In the more general unitary space $\mathbb{H}^{p,q}$ where $q>0$, only the first part of the above theorem remains true. For example, let $\mathbf{a}=\mathbf{e}_1+\mathbf{e}_2$ and $\mathbf{b}=\mathbf{e}_1-\mathbf{e}_2$ in $\mathbb{H}^{1,1}$. Then $|\mathbf{a}|=|\mathbf{b}|=0$ and

$$2=|\mathbf{a}+\mathbf{b}|>|\mathbf{a}|+|\mathbf{b}|=0,$$

violating the triangle inequality, and

$$0=|\mathbf{a}|=|2\mathbf{a}-\mathbf{a}|$$

violating property (2) of the theorem.

## Exercises

1. Let $(\mathbf{e})_{(3)}=(\mathbf{e}_1,\mathbf{e}_2,\mathbf{e}_3)$ be an orthonormal basis and let
$$(\mathbf{a})_{(3)}=(\mathbf{e})_{(3)}\begin{pmatrix}1 & \mathrm{i} & 1+\mathrm{i}\\ 1-\mathrm{i} & -\mathrm{i} & 2\\ 2+\mathrm{i} & 3 & \mathrm{i}\end{pmatrix}.$$
   (a) Find the corresponding orthogonal basis $(\mathbf{b})_{(3)}$ in $\mathbb{H}^3$ given by (10.32).
   (b) Find the corresponding orthogonal basis $(\mathbf{b})_{(3)}$ in $\mathbb{H}^{2,1}$ given by (10.32).
   (c) Find the corresponding orthogonal basis $(\mathbf{b})_{(3)}$ in $\mathbb{H}^{1,2}$ given by (10.32).
   (d) Find the corresponding orthogonal basis $(\mathbf{b})_{(3)}$ in $\mathbb{H}^{0,3}$ given by (10.32).
2. Show that the vectors $\{(1,\mathrm{i},2)(\mathbf{e})^{\mathrm{T}}_{(3)},(1,\mathrm{i},-\mathrm{i})(\mathbf{e})^{\mathrm{T}}_{(3)},$ and $(1,-\mathrm{i},0)(\mathbf{e})^{\mathrm{T}}_{(3)}\}$ are orthogonal in $\mathbb{H}^3$.
3. Given the basis $(\mathbf{a})_{(3)}=\{(2,\mathrm{i},2)(\mathbf{e})^{\mathrm{T}}_{(3)},(1,\mathrm{i},-\mathrm{i})(\mathbf{e})^{\mathrm{T}}_{(3)},(1,-\mathrm{i},1)(\mathbf{e})^{\mathrm{T}}_{(3)}\}$, use the Gram-Schmidt orthogonalization process (10.32) to find a corresponding orthonormal basis $(\mathbf{b}_1,\mathbf{b}_2,\mathbf{b}_3)$ in the following Hermitian spaces:

(a) In $\mathbb{H}^3$. Also find the triangular matrix $\triangle$ such that $(\mathbf{a})_{(3)}\triangle = (\mathbf{b})_{(3)}$.
(b) In $\mathbb{H}^{2,1}$. Also find the triangular matrix $\triangle$ such that $(\mathbf{a})_{(3)}\triangle = (\mathbf{b})_{(3)}$.
(c) In $\mathbb{H}^{1,2}$. Also find the triangular matrix $\triangle$ such that $(\mathbf{a})_{(3)}\triangle = (\mathbf{b})_{(3)}$.
(d) In $\mathbb{H}^{0,3}$. Also find the triangular matrix $\triangle$ such that $(\mathbf{a})_{(3)}\triangle = (\mathbf{b})_{(3)}$.

4. Show that $|\mathbf{a}+\mathbf{b}|^2 + |\mathbf{a}-\mathbf{b}|^2 = 2|\mathbf{a}|^2 + 2|\mathbf{b}|^2$ in $\mathbb{H}^n$.
5. Let $(\mathbf{e})_{(n)}$ be an orthonormal basis of the $n$-dimensional space $\mathbb{H}^{p,q}$ and let $f : \mathbb{H}^{p,q} \to \mathbb{H}^{p,q}$ be a linear operator. Recall that the matrix $[f] = (f^i_j)$ of $f$, with respect to the basis $(\mathbf{e})_{(n)}$, is defined by $f(\mathbf{e})_{(n)} = (\mathbf{e})_{(n)}[f]$ or

$$[f] = (\mathbf{e})^{(n)} \cdot f(\mathbf{e})_{(n)}.$$

Show that

$$f^i_j = \frac{\mathbf{e}_i \cdot f(\mathbf{e}_j)}{\mathbf{e}_i \cdot \mathbf{e}_i} = \begin{pmatrix} \mathbf{e}_i \cdot f(\mathbf{e}_j) & \text{for } 1 \le i \le p \;\; 1 \le j \le n \\ -\mathbf{e}_i \cdot f(\mathbf{e}_j) & \text{for } p < i \le n \;\; 1 \le j \le n \end{pmatrix} = \mathbf{e}^i \cdot f(\mathbf{e}_j),$$

where $n = p+q$.
6. Given $(\mathbf{b})_{(n)} = (\mathbf{a})_{(n)}\triangle$ , where $(\mathbf{b})_{(n)}$ is defined in (10.32), show that

$$\mathbf{b}_1 = \mathbf{a}_1, \mathbf{b}_1\mathbf{b}_2 = \mathbf{a}_1 \wedge \mathbf{a}_2, \; \ldots, \; \mathbf{b}_1\mathbf{b}_2 \ldots \mathbf{b}_n = \mathbf{a}_1 \wedge \mathbf{a}_2 \wedge \ldots \wedge \mathbf{a}_n.$$

7. Show that any Hermitian form $h : \mathbb{H}^n \times \mathbb{H}^n \to \mathbb{C}$ can be expressed in the form $h(\mathbf{x},\mathbf{y}) = f(\mathbf{x}) \cdot \mathbf{y}$, where $f(\mathbf{x})$ is a complex linear operator on $\mathbb{H}^n$.
8. Letting $\mathbf{x} = (\overline{x})_{(n)}(A^*)^{-1}A^*(\mathbf{e})^{\mathrm{T}}_{(n)}$ and $\mathbf{y} = (\mathbf{e})_{(n)}(y)^{\mathrm{T}}_{(n)} = (\mathbf{e})_{(n)}AA^{-1}(y)^{\mathrm{T}}_{(n)}$ in $\mathbb{H}^{p,q}$, verify that the transformation rule for the Hermitian form $\mathbf{x} \cdot \mathbf{y}$ is

$$\begin{aligned} \mathbf{x} \cdot \mathbf{y} &= (\overline{x})_{(n)}(A^*)^{-1}A^*(\mathbf{e})^{\mathrm{T}}_{(n)} \cdot (\mathbf{e})_{(n)}AA^{-1}(y)^{\mathrm{T}}_{(n)} \\ &= (\overline{x})_{(n)}(A^*)^{-1}A^*[g]AA^{-1}(y)^{\mathrm{T}}_{(n)} \end{aligned}$$

in the basis $(\mathbf{a})_{(n)} = (\mathbf{e})_{(n)}A$, where $[g] = (\mathbf{e})^{\mathrm{T}}_{(n)} \cdot (\mathbf{e})_{(n)}$ is the metric tensor of $\mathbb{H}^{p,q}$. This gives the transformation rule

$$(\overline{x})_{(n)}[g](y)^{\mathrm{T}}_{(n)} = (\overline{x})_{(n)}(A^*)^{-1}A^*[g]AA^{-1}(y)^{\mathrm{T}}_{(n)}.$$

9. Let $\mathbf{x} \in \mathbb{H}^{p,q}$ and let $(\mathbf{b})_{(n)}$ be an orthogonal basis.

(a) Show that

$$\mathbf{x} = \mathbf{x} \cdot (\mathbf{b})_{(n)}(\mathbf{b}^{-1})^{\mathrm{T}}_{(n)} = \sum_{i=1}^{p+q} \frac{(\mathbf{x} \cdot \mathbf{b}_i)\mathbf{b}_i}{\mathbf{b}_i \cdot \mathbf{b}_i}.$$

(b) Show that

$$\mathbf{x} = \frac{(\mathbf{x} \cdot \mathbf{b}_{(\overline{n})}) \cdot \overline{\mathbf{b}}^{\dagger}_{(n)}}{(\mathbf{b}_1 \cdot \mathbf{b}_1) \cdots (\mathbf{b}_n \cdot \mathbf{b}_n)}.$$

## 10.5 Hermitian, Normal, and Unitary Operators

For the remaining sections of this chapter, we restrict our considerations to the positive definite Hermitian space $\mathbb{H}^n$ and its unitary geometric algebra $\mathbb{U}_n$. Whereas many of the results can be generalized to the indefinite Hermitian spaces $\mathbb{H}^{p,q}$, we will not attempt this here. The proofs given in this section were first developed in [78].

Let $\mathbf{x} \cdot \mathbf{y}$ be a Hermitian inner product on the Hermitian space $\mathbb{H}^n$. The *Hermitian adjoint* $f^*$ of an operator $f$ is uniquely defined by

$$f(\mathbf{x}) \cdot \mathbf{y} = \mathbf{x} \cdot f^*(\mathbf{y}) \tag{10.35}$$

for all vectors $\mathbf{x}, \mathbf{y} \in \mathbb{H}^n$. Now let $(\mathbf{e})_{(n)}$ be a real orthonormal basis of $\mathbb{H}^n$. We saw in Exercise 5 of the previous section that the matrix $[f] = (f_{ij})$ of $f$, satisfying $f(\mathbf{e})_{(n)} = (\mathbf{e})_{(n)}[f]$, with respect to an orthonormal basis $(\mathbf{e})_{(n)}$, is specified by

$$f_{ij} = \mathbf{e}_i \cdot f(\mathbf{e}_j).$$

From

$$f_{ij} = \mathbf{e}_i \cdot f(\mathbf{e}_j) = \overline{f(\mathbf{e}_j) \cdot \mathbf{e}_i} = \overline{\mathbf{e}_j \cdot f^*(\mathbf{e}_i)} = \overline{f_{ji}},$$

it immediately follows that the matrix of $f^*$, defined by $f^*(\mathbf{e})_{(n)} = (\mathbf{e})_{(n)}[f^*]$, is the conjugated transpose $[f^*] = \overline{[f]}^{\mathrm{T}}$ of the matrix $[f]$ of $f$. Note also that

$$\mathbf{x} \cdot f(\mathbf{y}) = \overline{f(\mathbf{y}) \cdot (\mathbf{x})} = \overline{\mathbf{y} \cdot f^*(\mathbf{x})} = f^*(\mathbf{x}) \cdot \mathbf{y}.$$

An operator $f$ is called *normal* if $ff^* = f^*f$, *Hermitian* or *self-adjoint* if $f^* = f$, and *unitary* if $f^* = f^{-1}$. It follows that Hermitian and unitary operators are normal. We will need three simple lemmas for the proof of the spectral theorem for normal operators.

**Lemma 10.5.1.** *If $f$ is a normal operator then* $\ker(f) = \ker(f^*)$.

*Proof.* For any $\mathbf{x} \in \mathbb{H}^n$ we have,

$$f(\mathbf{x}) \cdot f(\mathbf{x}) = \mathbf{x} \cdot f^*f(\mathbf{x}) = \mathbf{x} \cdot ff^*(\mathbf{x}) = f^*(\mathbf{x}) \cdot f^*(\mathbf{x}),$$

from which it follows that $f(\mathbf{x}) = 0$ if and only if $f^*(\mathbf{x}) = 0$. □

**Lemma 10.5.2.** *Every normal nilpotent operator $q$ is identically* 0.

*Proof.* Let $q$ be a normal nilpotent operator with index of nilpotency $k > 1$; that is, $q^{k-1} \neq 0$ but $q^k = 0$. Then for all $\mathbf{x}$, $qq^{k-1}\mathbf{x} = 0$, so by Lemma 1 $q^*q^{k-1}\mathbf{x} = 0$. But then

$$q^{k-1}(\mathbf{x}) \cdot q^{k-1}(\mathbf{x}) = q^*q^{k-1}(\mathbf{x}) \cdot q^{k-2}(\mathbf{x}) = 0$$

for all $\mathbf{x}$, which means that $q^{k-1} = 0$, a contradiction. Thus, a normal nilpotent operator $q$ must have index of nilpotency 1, i.e., $q = 0$. □

**Lemma 10.5.3.** *A normal projection $p$ is self-adjoint.*

*Proof.* Note that the adjoint of a projection $p$ is also a projection, for $(p^*)^2 = (p^2)^* = p^*$. The range of $1-p$ is $\ker(p)$. Thus, if $p$ is normal, then by Lemma 10.5.1

$$\ker(p) = \ker(p^*) \text{ so } 1-p = 1-p^*,$$

from which it follows that $p = p^*$. □

**Theorem 10.5.4 (The spectral decomposition theorem for a normal operator).** *If $f$ is a normal operator, then $f = \sum_{i=1}^r \lambda_i p_i$, where the $p_i$'s are self-adjoint orthogonal idempotents with $\sum p_i = 1$.*

*Proof.* By the structure Theorem 8.4.1,

$$f = \sum_{i=1}^r (\lambda_i + q_i)p_i,$$

where the operators $p_i$ and $q_i$ are all *polynomials* in $f$. But a polynomial in a normal operator is normal, so the projections $p_i$ and nilpotents $q_i$ are themselves normal operators. Thus, by Lemmas 10.5.2 and 10.5.3, each $q_i$ is zero and each $p_i$ is self-adjoint. □

The following corollary now follows trivially

**Corollary 10.5.5.** *The eigenvalues of a self-adjoint operator are real.*

*Proof.* Just compare the two expressions of $f^* = f$,

$$f = \sum_{i=1}^r \lambda_i p_i$$

where $p_i^* = p_i$ and

$$f^* = \sum_{i=1}^r \overline{\lambda_i} p_i$$

to conclude that $\overline{\lambda_i} = \lambda_i$ for $i = 1, \ldots, r$. □

## Exercises

1. Given a complex linear operator $f$, show that the Hermitian operators $f^*f$ and $ff^*$ have only positive real eigenvalues. Use this result to show that if $f^*f = 0$, then $f = 0$ also.
2. Let $f$ and $g$ be two complex linear operators. Show that $(fg)^* = g^*f^*$.
3. Given a complex linear operator $f$, show that $b(\mathbf{x},\mathbf{y}) = f(\mathbf{x}) \cdot \mathbf{y}$ is a bilinear form.
4. Show that if $\mathbb{W}$ is an invariant subspace of $f$, then $\mathbb{W}^{\perp}$ is an invariant subspace of $f^*$.
5. Show that if $f = f^*$ ($f$ is Hermitian), and $\mathbb{W}$ is invariant under $f$, then $\mathbb{W}^{\perp}$ is also invariant under $f$.
6. Let $\mathbb{W} = f(\mathbb{V})$. Show that $\mathbb{W}^{\perp}$ is the kernel of $f^*$.
7. Show that $f$ and $f^*$ have the same rank.
8. Let $\mathbb{W} = f(\mathbb{V})$. Show that $f^*(\mathbb{V}) = f^*(\mathbb{W})$.
9. Show that $f^*(\mathbb{V}) = f^*f(\mathbb{V})$ and that $f(\mathbb{V}) = ff^*(\mathbb{V})$.
10. Show that $f + f^*$ is Hermitian.
11. Show that $f - f^*$ is skew-Hermitian.
12. Show that every operator $f = s + h$ where $s$ is skew-Hermitian and $h$ is Hermitian.
13. A matrix $U$ is said to be *orthogonal* if $U^{\mathrm{T}}U = I$. Which of the following matrices are orthogonal?

$$\text{(a)} \begin{pmatrix} 1/2 & \sqrt{3}/2 \\ -\sqrt{3}/2 & 1/2 \end{pmatrix}, \quad \text{(b)} \begin{pmatrix} 1/2 & \sqrt{3}/2 \\ \sqrt{3}/2 & 1/2 \end{pmatrix}, \quad \text{(c)} \begin{pmatrix} 0.6 & 0.8 \\ 0.8 & -.6 \end{pmatrix}.$$

14. A matrix $U$ is said to be *unitary* if $U^*U = I$. Which of the following matrices are unitary?

$$\text{(a)} \begin{pmatrix} \frac{1+\mathrm{i}}{2} & \frac{1-\mathrm{i}}{2} \\ \frac{1-\mathrm{i}}{2} & \frac{1+\mathrm{i}}{2} \end{pmatrix}, \quad \text{(b)} \begin{pmatrix} 1 & \mathrm{i} \\ \mathrm{i} & 1 \end{pmatrix}, \quad \text{(c)} \begin{pmatrix} 1 & -\mathrm{i} \\ \mathrm{i} & 1 \end{pmatrix}.$$

15. Show that the matrix

$$\begin{pmatrix} \cos(\theta) & -\sin(\theta) & 0 \\ \sin(\theta) & \cos(\theta) & 0 \\ 0 & 0 & 1 \end{pmatrix}$$

is orthogonal.

16. A matrix $N$ is said to be *normal* if $N^*N = NN^*$; a matrix $H$ is *Hermitian* if $H^* = H$ and *unitary* if $U^* = U^{-1}$. Determine what kind of matrices are given below.

$$\text{(a)} \begin{pmatrix} \frac{1+\mathrm{i}}{2} & \frac{1-\mathrm{i}}{2} \\ \frac{1-\mathrm{i}}{2} & \frac{1+\mathrm{i}}{2} \end{pmatrix}, \quad \text{(b)} \begin{pmatrix} 1 & \mathrm{i} \\ \mathrm{i} & 1 \end{pmatrix}, \quad \text{(c)} \begin{pmatrix} 1 & -\mathrm{i} \\ \mathrm{i} & 1 \end{pmatrix}.$$

17. Show that if $A$ and $B$ are real symmetric matrices, and $B$ is positive definite, then the roots of $\det|B - xA| = 0$ are all real.
18. Show that every real skew-symmetric matrix $A$ has the form $A = P^{\mathrm{T}}BP$ where $P$ is orthogonal and $B^2$ is diagonal.

## *10.6 Principal Correlation

Let $\mathbb{H}^n$ be a Hermitian space with the Hermitian inner product $\mathbf{x}\cdot\mathbf{y}$ and let $(\mathbf{e})_{(n)}$ be any orthonormal basis of $\mathbb{H}^n$. Then for any $\mathbf{x} = (\mathbf{e})_{(n)}(x)^{(n)}$ and $\mathbf{y} = (\mathbf{e})_{(n)}(y)^{(n)} \in \mathbb{H}^n$, we calculate

$$\mathbf{x}\cdot\mathbf{y} = \left((\mathbf{e})_{(n)}(x)^{(n)}\right)\cdot\left((\mathbf{e})_{(n)}(y)^{(n)}\right) = \left((\overline{x})_{(n)}(\mathbf{e})^{(n)}\right)\cdot\left((\mathbf{e})_{(n)}(y)^{(n)}\right)$$
$$= (\overline{x})_{(n)}(y)^{(n)},$$

where $(\overline{x})_{(n)}$ is the conjugated row vector of $\mathbf{x}$. By noting that

$$\mathbf{x}\cdot\mathbf{y} = (\overline{x})_{(n)}(y)^{(n)} = (\overline{x})_{(n)}A^{*-1}A^{*}AA^{-1}(y)^{(n)} = (\overline{x}_a)_{(n)}A^{*}A(y_a)^{(n)},$$

we have easily found the rule for expressing $\mathbf{x}\cdot\mathbf{y}$ in terms of the components of $\mathbf{x}$ and $\mathbf{y}$ in the arbitrary basis $(\mathbf{a})_{(n)} = (\mathbf{e})_{(n)}A$.

Suppose now that $f : \mathbb{H}^n \to \mathbb{W}^m$ is a complex linear operator from the $n$-dimensional Hermitian space $\mathbb{H}^n$ with the inner product $\mathbf{x}\cdot\mathbf{y}$ to the $m$-dimensional Hermitian space $\mathbb{W}^m$ with the inner product $\mathbf{u}\cdot\mathbf{v}$. We shall denote the set of all such operators (homomorphisms) as $\mathrm{Hom}(\mathbb{H}^n, \mathbb{W}^m)$. Let $(\mathbf{e})_{(n)}$ and $(\mathbf{e})_{(m)}$ denote orthonormal bases of $\mathbb{H}^n$ and $\mathbb{W}^m$, respectively. Then for the vector $\mathbf{x} = (\mathbf{e})_{(n)}(x)^{(n)} \in \mathbb{H}^n$, we have

$$f(\mathbf{x}) = f(\mathbf{e})_{(n)}(x)^{(n)} = (\mathbf{e})_{(m)}[f]_{(m,n)}(x)^{(n)},$$

where $[f]_{(m,n)}$ is the complex $(m,n)$-matrix of $f$ with respect to the orthonormal bases $(\mathbf{e})_{(n)}$ of $\mathbb{H}^n$ and $(\mathbf{e})_{(m)}$ of $\mathbb{W}$. If $(\mathbf{a})_{(n)} = (\mathbf{e})_{(n)}A$ and $(\mathbf{b})_{(m)} = (\mathbf{e})_{(m)}B$ are new bases, then clearly

$$f(\mathbf{x}) = (\mathbf{e})_{(m)}[f](x)^{(n)} = (\mathbf{e})_{(m)}BB^{-1}[f]AA^{-1}(x)^{(n)} = (\mathbf{b})_{(m)}[f]_{ab}(x)_a^{(n)}$$

where $[f]_{ab} = B^{-1}[f]A$ is the matrix of $f$ with respect to the bases $(\mathbf{b})_{(m)}$ of $\mathbb{W}^m$ and $(\mathbf{a})_{(n)}$ of $\mathbb{H}^n$.

The adjoint operator $f^* : \mathbb{W}^m \to \mathbb{H}^n$ is defined to satisfy

$$f(\mathbf{x})\cdot\mathbf{y} = \mathbf{x}\cdot f^*(\mathbf{y})$$

for all $\mathbf{x} \in \mathbb{H}^n$ and $\mathbf{y} \in \mathbb{W}^m$. The calculations

$$f(\mathbf{e})_{(n)} = (\mathbf{e})_{(m)}[f]_{(m,n)} \quad \Longleftrightarrow \quad (\mathbf{e})^{(m)}\cdot f(\mathbf{e})_{(n)} = [f]_{(m,n)}$$

and

$$(\mathbf{e})^{(m)} \cdot f(\mathbf{e})_{(n)} = f^*(\mathbf{e})^{(m)} \cdot (\mathbf{e})_{(n)} = (\mathbf{e})^{(m)}[f^*] \cdot (\mathbf{e})_{(n)},$$

together with the fact $[f^*] = \overline{[f]}^{\mathrm{T}}$, established right before Lemma 10.5.1, show that the conjugated transpose of the matrix $[f]$ corresponds to the adjoint operator $f^*$.

We have already seen in Theorem 8.1.2, Corollary 8.1.4, that the matrix $[f]$ of $f$ becomes particularly simple when expressed in the appropriate bases $(\mathbf{a})_{(n)}$ of $\mathbb{H}^n$ and $(\mathbf{b})_{(m)}$ of $\mathbb{W}^m$. We shall state this result here as the following theorem:

**Theorem 10.6.1 (First canonical form).** *There exist a basis $(\mathbf{a})_{(n)}$ of $\mathbb{H}^n$ and a basis $(\mathbf{b})_{(m)}$ of $\mathbb{W}^m$ so that $f(\mathbf{x}) = (\mathbf{b})_{(m)}[f]_{ab}(x_a)^{(n)}$ where $[f]_{ab} = B^{-1}[f]A$ is a $m \times n$-diagonal matrix with the first $k = \mathrm{rank}(f)$ elements down the diagonal 1's and all other elements 0's.*

We will now refine this theorem by the requirement that the basis $(\mathbf{a})_{(n)}$ and $(\mathbf{b})_{(m)}$ be orthonormal in $\mathbb{H}^n$ and $\mathbb{W}^m$, respectively.

Let $(\mathbf{a})_{(n)}$ be an orthonormal basis of $\mathbb{H}^n$ and $(\mathbf{b})_{(m)}$ an orthonormal basis of $\mathbb{W}^m$. If $f(\mathbf{a})_{(n)} = (\mathbf{b})_{(m)}G$ where $G$ is a diagonal matrix with the first $k = \mathrm{rank}(f)$ elements down the diagonal positive, we say that $f$ is a *correlation* between the orthonormal basis $(\mathbf{a})_{(n)}$ of $\mathbb{H}^n$ and the orthonormal basis $(\mathbf{b})_{(m)}$ of $\mathbb{W}^m$. If $f$ is a correlation and $\mathrm{rank}(f) = \min\{n,m\}$, we say that $f$ is a *full correlation*. Finally, a full correlation $d$ between the orthonormal bases $(\mathbf{a})_{(n)}$ and $(\mathbf{b})_{(m)}$, having the property that $d(\mathbf{a})_{(n)} = (\mathbf{b})_{(m)}D$ where $D$ a full rank diagonal matrix with 1's down the diagonal, is called a *principal correlation*. A principal correlation between $\mathbb{H}^n$ and $\mathbb{W}^m$ is the generalization of the idea of a unitary operator on a vector space.

Recall that an operator $f \in \mathrm{End}(\mathbb{H}^n)$ is *normal* if $f^*f = ff^*$ and that a normal operator has the spectral form given in Theorem 10.5.4,

$$f = \sum_{i=1}^{r} \tau_i p_i$$

where the eigenvalues $\tau_i$ are distinct and where $p_i = p_i^*$ are mutually annihilating Hermitian projections which partition the identity $1 \in \mathrm{End}(\mathbb{H}^n)$. We say that an orthonormal basis $(\mathbf{a})_{(n)}$ is *compatible* with a set $\{p\}$ of mutually annihilating Hermitian idempotents which partition $1 \in \mathrm{End}(\mathbb{H}^n)$ if $p_i(\mathbf{a})_{(n)} = (\mathbf{a})_{(n)}P_i$ where $P_i$ is a diagonal matrix with a block of $n_i = \mathrm{rank}(p_i)$ 1's down the diagonal; we also require that the blocks of 1's in each of $P_1, P_2, \ldots, P_r$ run consecutively starting from the top.

Now let $f \in \mathrm{Hom}(\mathbb{H}^n, \mathbb{W}^m)$ as before. Whereas, in general, we are unable to multiply (compose) operators in $\mathrm{Hom}(\mathbb{H}^n, \mathbb{W}^m)$, the Hermitian operators $f^*f \in \mathrm{End}(\mathbb{H}^n)$ and $ff^* \in \mathrm{End}(\mathbb{W}^m)$ are always well defined and have the nonnegative eigenvalues $\lambda_i = \overline{\tau}_i \tau_i$; see Exercise 1 of 10.5.1. We can also assume that $n \leq m$; if this is not the case, interchange the names of the operators $f$ and $f^*$. If $d \in \mathrm{Hom}(\mathbb{H}^n, \mathbb{W}^m)$ is a principal correlation between $\mathbb{V}$ and $\mathbb{W}$, then $d^* \in \mathrm{Hom}(\mathbb{W}^m, \mathbb{H}^n)$ is a principal correlation between $\mathbb{W}$ and $\mathbb{V}$. When $n \leq m$,

$$d^*d = 1 \in \text{End}(\mathbb{V}) \quad \text{and} \quad p' = dd^* \in \text{End}(\mathbb{W}^m)$$

where $p'$ is a Hermitian projection with $\text{rank}(p') = m$.

We can now prove the main theorem of this section

**Theorem 10.6.2.** *(Principal correlation of a linear operator.) Given a linear operator $f \in \text{Hom}(\mathbb{H}^n, W^m)$ where $n = \dim(\mathbb{H}^n) \le m = \dim(W^m)$, there exists a principal correlation $d \in \text{Hom}(\mathbb{H}^n, W^m)$ with the property that $df^* = fd^*$ and $d^*f = f^*d$.*

*Proof.* The Hermitian operators $f^*f$ and $ff^*$ have the structure

$$f^*f = \sum_{i=1}^{r} \lambda_i p_i, \quad ff^* = \sum_{j=1}^{r'} \lambda_j p'_j$$

for $\lambda_1 > \cdots > \lambda_r \ge 0$ and where $p_i = p_i^*$, $p_i'^* = p_i'$ are mutually annihilating Hermitian idempotents. If $\lambda_r = 0$, then $r = r'$. If $\lambda_r \ne 0$ and $m = n$, then $r' = r$, but if $\lambda_r \ne 0$ and $m > n$, then $r' = r+1$ and $\lambda_{r+1} = 0$. We also have $n_i = \text{rank}(p_i) = \text{rank}(p_i')$ for all $i = 1, \ldots, r$ for which $\lambda_i > 0$. If $\lambda_k = 0$ (for $k = r$ or $k = r+1$), then $\text{rank}(p_k)$ may not be equal to $\text{rank}(p_k')$ (in fact $p_{r+1}$ is not even defined!).

The above relationships follow from the operator identity

$$ff^*f = \sum_i \lambda_i f p_i = \sum_j \lambda_j p'_j f.$$

Multiplying on the left and right sides of this identity by $p_k'$ and $p_l$, respectively, gives $\lambda_k p_k' f p_l = \lambda_l p_k' f p_l$ for all $k, l = 1, \ldots, r$. It follows that

$$p_i' f p_j = \delta_{i,j} p_i' f p_i$$

where $\delta_{i,j}$ is the Kronecker delta.

To complete the proof, we must define the principal correlation $d$. This is accomplished by first selecting any orthonormal basis $(\mathbf{a})_{(n)}$ of $\mathbb{C}^n$ compatible with the orthogonal projections $\{p\}$. The principal correlation $d : \mathbb{C}^n \to \mathbb{C}^m$ is then defined by $\mathbf{b} = d(\mathbf{a}) = \frac{1}{\lambda} f(\mathbf{a})$ where $\mathbf{a} \in \{\mathbf{a}\}$ is any representative basis element with the nonzero eigenvalue $\lambda$. In the case that $f(\mathbf{a}) = 0$, we define $d(\mathbf{a}) = \mathbf{b}$ where $p_r'\mathbf{b} = \mathbf{b}$. We then complete the orthonormal basis $(\mathbf{b})_{(m)}$ to be compatible with the projections $\{p'\}$. The constructed bases $(\mathbf{a})_{(n)}$ of $\mathbb{C}^n$ and $(\mathbf{b})_{(m)}$ of $\mathbb{C}^m$ are principally correlated by the operator $d$ satisfying $d(\mathbf{a})_{(n)} = (\mathbf{b})_{(m)} D$. It is easily verified by the construction that $fd^* = df^*$ and $d^*f = f^*d$. $\square$

From the construction of the principal correlation $d$ of the operator $f$, it is clear that $d$ is uniquely determined in all cases when $\lambda_r > 0$ and $n = m$, and not uniquely determined in all other cases. In the compatible orthonormal basis $(\mathbf{a})_{(n)}$ of $\mathbb{H}^n$ and $(\mathbf{b})_{(m)}$ of $\mathbb{W}$, $f(\mathbf{a})_{(n)} = (\mathbf{b})_{(m)} G$ where $G$ is a nonnegative diagonal $(m,n)$-matrix with decreasing distinct eigenvalues down the diagonal.

## *10.7 Polar and Singular Value Decomposition

Let $f \in \mathrm{Hom}(\mathbb{H}^n, \mathbb{W}^m)$. The *polar decomposition* of the operator $f$ follows once a principal correlation $d$ of $f$ is known. Given the principal correlation $d$ of $f$ satisfying $fd^* = df^*$ and $d^*d = 1$, the *left* and *right* polar decompositions are immediately determined by writing

$$f = (df^*)d = d(f^*d),$$

because $h = f^*d$ satisfies $h^* = d^*f = f^*d$ and so is Hermitian in $\mathbb{V}$. Similarly, $h' = df^*$ is Hermitian in $\mathbb{W}$. In the case when $\mathbb{H}^n = \mathbb{W}^m$, the principal correlation $d$ becomes a unitary operator on $\mathbb{H}^n$.

The singular value decomposition $f = ugv^*$, where $u$ and $v$ are unitary operators and $g \equiv u^*fv$ is a correlation between the standard bases $(\mathbf{e})_{(n)}$ of $\mathbb{H}^n$ and $(\mathbf{e})_{(m)}$ of $\mathbb{W}^m$, is also easily constructed from properties of the principal correlation $d$ of $f$. Let $(\mathbf{a})_{(n)}$ and $(\mathbf{b})_{(m)}$ be the orthonormal bases compatible with the principal correlation $d$ of $f$ defined in Theorem 10.6.2. The unitary operator $u \in \mathrm{End}(\mathbb{W}^m)$ is defined by $(\mathbf{b})_{(m)} = u(\mathbf{e})_{(m)}$, and the unitary operator $v \in \mathrm{End}(\mathbb{H}^n)$ is defined by $(\mathbf{a})_{(n)} = v(\mathbf{e})_{(n)}$. We then have

$$f(\mathbf{a})_{(n)} = ugv^*(\mathbf{a})_{(n)} = ug(\mathbf{e})_{(n)} = u(\mathbf{e})_{(m)}G = (\mathbf{b})_{(m)}G,$$

where $G$ is diagonal and nonnegative, exactly as required.

Finally, note that the *Moore-Penrose generalized inverse* $f^{-1\mathrm{mp}}$ of $f$ [46, page 421] can also be easily defined in terms of the principal correlation $d$ of $f$. We have

$$f^{-1\mathrm{mp}} \equiv d^*(df^*)^{-1\mathrm{mp}} = (f^*d)^{-1\mathrm{mp}}d^*$$

where the Moore-Penrose inverses $(df^*)^{-1\mathrm{mp}}$ and $(f^*d)^{-1\mathrm{mp}}$ of the Hermitian operators $df^* = \sum_i \tau_i p_i'$ and $f^*d = \sum_i \tau_i p_i$ are defined by

$$(df^*)^{-1\mathrm{mp}} \equiv \sum_{\tau_i \neq 0} \frac{1}{\tau_i} p_i'$$

and

$$(f^*d)^{-1\mathrm{mp}} \equiv \sum_{\tau_i \neq 0} \frac{1}{\tau_i} p_i,$$

where only the nonzero eigenvalues $\tau_i$ appear in the sums.

As an example, we will find a principal correlation and corresponding polar and singular value decomposition of the matrix $S$ given by

$$S = \begin{pmatrix} -1 & 2 \\ 0 & -1 \\ -2 & 0 \end{pmatrix}$$

taking $\mathbb{R}^2$ into $\mathbb{R}^3$. We calculate

$$SS^* = \begin{pmatrix} 5 & -2 & 2 \\ -2 & 1 & 0 \\ 2 & 0 & 4 \end{pmatrix} \text{ and } S^*S = \begin{pmatrix} 5 & -2 \\ -2 & 5 \end{pmatrix}$$

We find that the eigenvalues of $SS^*$ are $\tau_1^2 = 7, \tau_2^2 = 3, and \tau_3^2 = 0$, the first two, 7 and 3, also being eigenvalues of $T^*T$. The algebra $\mathbb{R}[\alpha]_{(\alpha-7)(\alpha-3)}$ of $S^*S$ is easily found, giving the spectral form

$$S^*S = \tau_1^2 P_1 + \tau_2^2 P_2$$

where

$$P_1 = \frac{S^*S - \tau_2^2}{\tau_1^2 - \tau_2^2} = 1/2 \begin{pmatrix} 1 & -1 \\ -1 & 1 \end{pmatrix}, \quad P_2 = \begin{pmatrix} 1 & 0 \\ 0 & 1 \end{pmatrix} - P_1 = 1/2 \begin{pmatrix} 1 & 1 \\ 1 & 1 \end{pmatrix}.$$

We choose the first column vectors of $P_1$ and $P_2$, normalizing them, to get the orthonormal basis $(\mathbf{a})_{(2)}$ of $\mathbb{R}^2$. The compatible orthonormal basis $(\mathbf{b})_{(3)}$ of $\mathbb{R}^3$ is defined by $\mathbf{b}_i = \frac{1}{\tau_i} f(\mathbf{a}_i)$ for the nonzero eigenvalues $\tau_1 = \sqrt{7}$ and $\tau_2 = \sqrt{3}$, and completed to form a basis of $\mathbb{R}^3$; the principal correlation $d : \mathbb{R}^2 \to \mathbb{R}^3$ is then defined by the relation

$$d(\mathbf{a})_{(2)} = (\mathbf{b})_{(3)} D \text{ where } D = \begin{pmatrix} 1 & 0 \\ 0 & 1 \\ 0 & 0 \end{pmatrix}$$

or

$$d \begin{pmatrix} \frac{1}{\sqrt{2}} & \frac{1}{\sqrt{2}} \\ \frac{-1}{\sqrt{2}} & \frac{1}{\sqrt{2}} \end{pmatrix} = \begin{pmatrix} \frac{-3}{\sqrt{14}} & \frac{1}{\sqrt{6}} & \frac{-2}{\sqrt{21}} \\ \frac{1}{\sqrt{14}} & \frac{-1}{\sqrt{6}} & \frac{-4}{\sqrt{21}} \\ \frac{-2}{\sqrt{14}} & \frac{-2}{\sqrt{6}} & \frac{1}{\sqrt{21}} \end{pmatrix} \begin{pmatrix} 1 & 0 \\ 0 & 1 \\ 0 & 0 \end{pmatrix}.$$

Multiplying both sides of the above relation by $(\mathbf{a})_{(2)}^{-1}$, we find the principal correlation

$$d = \frac{1}{2\sqrt{21}} \begin{pmatrix} \sqrt{7} - 3\sqrt{3} & \sqrt{7} + 3\sqrt{3} \\ \sqrt{3} - \sqrt{7} & -\sqrt{3} - \sqrt{7} \\ -2\sqrt{3} - 2\sqrt{7} & 2\sqrt{3} - 2\sqrt{7} \end{pmatrix}.$$

It can be directly verified that $S^*d = d^*S$ and $dS^* = Sd^*$ as required.

The polar and singular value decompositions of $S$ are immediate consequences of the existence of the principal correlation $d$ and the compatible bases $(\mathbf{a})_{(n)}$ and $(\mathbf{b})_{(m)}$. For the singular value, we find that

$$S = (\mathbf{b})_{(m)} \begin{pmatrix} \sqrt{7} & 0 \\ 0 & \sqrt{3} \\ 0 & 0 \end{pmatrix} (\mathbf{a})_{(n)}^*.$$

# Chapter 11
# Geometry of Moving Planes

*If people do not believe that mathematics is simple, it is only because they do not realize how complicated life is.*
—John von Neumann

In special relativity, the fabric of space and time becomes irrevocably knit together in the concept of space–time. The first section explains how the concept of 2-dimensional space–time in the geometric algebra $\mathbb{G}_1$, introduced in Chap. 2, generalizes to the full 4-dimensional space–time of special relativity in the geometric algebra $\mathbb{G}_3$. With only a slight loss of generality, we then restrict our attention to the study of 3-dimensional space–time in the geometric algebra $\mathbb{G}_2$. Various concepts in special relativity are considered, such as the concept of a Lorentz boost, the addition of velocities, and the novel concept of a moving plane. In the final section of this chapter, the geometric algebra $\mathbb{G}_2$ of $\mathbb{R}_2$ is split into the geometric algebra $\mathbb{G}_{1,2}$ of Minkowski space–time $\mathbb{R}^{1,2}$.

## 11.1 Geometry of Space–Time

We have already seen how well-known elementary properties of space–time can be derived in the two-dimensional hyperbolic plane in Chap. 2. In special relativity, space and time become inextricably bound together in the concept of space–time. It is not surprising that even to this day, there is a great deal of confusion surrounding the concept of space–time, considering the many different and sometimes conflicting formalisms that are employed. Geometric algebra offers powerful new tools to express profound ideas about the special relativity of space–time, with a clear geometric interpretation always close at hand.

G. Sobczyk, *New Foundations in Mathematics: The Geometric Concept of Number*, DOI 10.1007/978-0-8176-8385-6_11,

**Fig. 11.1** Rest frames are distinguished from each other by their relative velocities and relative times

Recall from Chap. 2, Sect. 2.6, that special relativity was constructed around the two postulates:

1. All coordinate systems (for measuring time and distance) moving with constant velocity relative to each other are equally valid for the formulation of the laws of physics.
2. Light propagates in every direction in a straight line and with the same speed c in every valid coordinate system.

The first postulate tells us that all *rest frames* moving with constant velocity relative to each other are equally valid. In Fig. 11.1, three men are shown moving off in different directions. We can imagine that each of these clockheads accurately measures the passage of time in each of their respective rest frames.

In the geometric algebra $\mathbb{G}_3 = \mathbb{G}(\mathbb{R}^3)$, it becomes geometrically possible to distinguish rest frames which have a constant relative velocity with respect to one another. Let $(\mathbf{e})_{(3)}$ be the standard orthonormal basis of unit vectors along the respective $(x,y,z)$ coordinate axis of $\mathbb{R}^3$. As such, $(\mathbf{e})_{(3)}$ defines a rest frame of $\mathbb{R}^3$ in $\mathbb{G}_3$. Notice that

$$(\mathbf{e}')_{(3)} = e^{\frac{\theta}{2}\mathbf{e}_{21}}(\mathbf{e})_{(3)}e^{-\frac{\theta}{2}\mathbf{e}_{21}},$$

representing a rotation of the basis vectors $(\mathbf{e})_{(3)}$ in the $(x,y)$-plane of $\mathbb{R}^3$, is an equally valid orthonormal frame of basis vectors in $\mathbb{R}^3$ and as such defines the same rest frame. That is

$$\mathbb{R}_3 = \text{span}(\mathbf{e})_{(3)} = \text{span}(\mathbf{e}')_{(3)}.$$

By an *event* $X$ in space–time, we mean $X = ct + \mathbf{x} \in \mathbb{G}_3$ where $\mathbf{x} = (\mathbf{e})_{(3)}(x)^{(3)} \in \mathbb{R}^3$. The event $X$ occurs at *time* $t$ and at the *position* $\mathbf{x} \in \mathbb{R}^3$ as measured in the rest frame $(\mathbf{e})_{(3)}$ of $\mathbb{R}^3$. In Chap. 2, Sect. 2.6, events in different rest frames were related by a transformation of the kind

$$X' = Xe^{-\phi \mathbf{e}_2} \in \mathbb{G}_3, \tag{11.1}$$

where $\tanh\phi = \frac{v}{c}$. In this case, for simplicity, we have chosen our velocity $\mathbf{v} = v\mathbf{e}_2$ to be in the direction of the $y$-axis in $\mathbb{R}^3$. The question is does the transformation (11.1) still make sense as applied in the geometric algebra $\mathbb{G}_3 = \mathbb{G}(\mathbb{R}^3)$ of the Euclidean space $\mathbb{R}^3$?

To explore this possibility, we carry out the same calculation that we did to find the famous relationships (2.24). Starting with (11.1) and $\mathbf{v} = v\hat{\mathbf{v}}$ for an arbitrary speed $v = |\mathbf{v}|$ in the direction of the unit vector $\hat{\mathbf{v}} \in \mathbb{R}^3$, we find

$$\begin{aligned} ct' + \mathbf{x}' &= (ct + \mathbf{x})(\cosh\phi - \hat{\mathbf{v}}\sinh\phi) = \cosh\phi(ct + \mathbf{x})\left(1 - \frac{\mathbf{v}}{c}\right) \\ &= \cosh\phi\left(ct - \mathbf{x}\cdot\frac{\mathbf{v}}{c}\right) + \cosh\phi\left(\mathbf{x} - t\mathbf{v} - \mathbf{x}\wedge\frac{\mathbf{v}}{c}\right) \end{aligned}$$

so that

$$t' = \frac{1}{\sqrt{1 - \frac{v^2}{c^2}}}\left(t - \frac{\mathbf{x}\cdot\mathbf{v}}{c^2}\right), \quad \text{and} \quad \mathbf{x}' = \frac{1}{\sqrt{1 - \frac{v^2}{c^2}}}\left(\mathbf{x} - t\mathbf{v} - \mathbf{x}\wedge\frac{\mathbf{v}}{c}\right), \tag{11.2}$$

where $\cosh\phi = \left(\sqrt{1 - \frac{v^2}{c^2}}\right)^{-1}$ as after (2.23) in Chap. 2.

In order to compare our result with what is in the literature, we further break down $\mathbf{x}'$ into components that arise from the parallel and perpendicular components of $\mathbf{x}$ with respect to the velocity vector $\mathbf{v}$. We write $\mathbf{x} = \mathbf{x}_{||} + \mathbf{x}_{\perp}$ where $\mathbf{x}_{||} = (\mathbf{x}\cdot\hat{\mathbf{v}})\hat{\mathbf{v}}$ and $\mathbf{x}_{\perp} = \mathbf{x} - \mathbf{x}_{||}$, and find that

$$\mathbf{x}'_{||} = \frac{1}{\sqrt{1 - \frac{v^2}{c^2}}}\left(\mathbf{x}_{||} - t\mathbf{v}\right), \quad \text{and} \quad \mathbf{x}'_{\perp} = \frac{1}{\sqrt{1 - \frac{v^2}{c^2}}}\mathbf{x}_{\perp}\left(1 - \frac{\mathbf{v}}{c}.\right) \tag{11.3}$$

The decomposition formula (11.3) is extremely interesting insofar as that it differs from what is found in all books. References [4, 9, 36, 47] are all in agreement with our formula for $\mathbf{x}'_{||}$, but they all find that $\mathbf{x}'_{\perp} = \mathbf{x}_{\perp} \in \mathbb{R}^3$. So how is our formula (11.3) possible? What is going on?

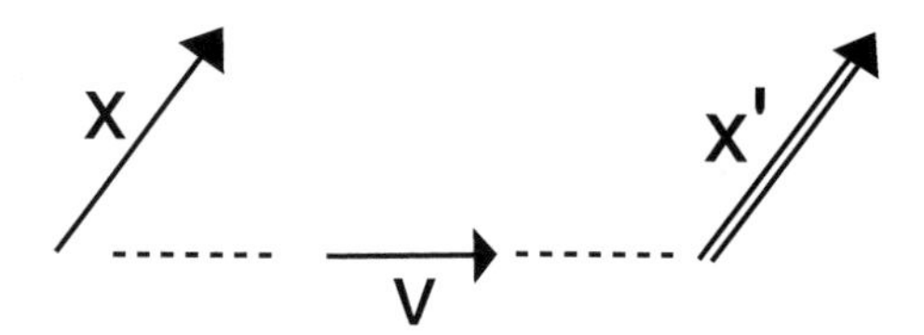


**Fig. 11.2** The relative vector $\mathbf{x}'$ is the vector $\mathbf{x}$ moving at the constant velocity $\mathbf{v}$. It is drawn as a double vector to indicate that it has picked up a bivector component

First note that ${\mathbf{x}'_\perp}^2 = \mathbf{x}_\perp^2$, which is encouraging. We also note that the *pullback*

$$e^{-\frac{\phi}{2}\hat{\mathbf{v}}}\mathbf{x}'e^{+\frac{\phi}{2}\hat{\mathbf{v}}} = e^{-\frac{\phi}{2}\hat{\mathbf{v}}}(\mathbf{x}'_{||}+\mathbf{x}'_\perp)e^{\frac{\phi}{2}\hat{\mathbf{v}}} = \mathbf{x}'_{||}+\mathbf{x}_\perp,$$

and

$$e^{-\frac{\phi}{2}\hat{\mathbf{v}}}Xe^{-\frac{\phi}{2}\hat{\mathbf{v}}} = (ct+\mathbf{x}_{||})e^{-\phi\hat{\mathbf{v}}}+\mathbf{x}_\perp. \tag{11.4}$$

We identify (11.4) the *passive Lorentz boost* found in [4, p. 60] and [36, p. 581]. This passive transformation is equivalent to the Lorentz boost coordinate transformations found in the early references [9, p. 237], [47, p. 356].

We are now ready to define what we mean by the *active Lorentz boost* of a vector $\mathbf{x} \in \mathbb{R}^3$ for the velocity vector $\mathbf{v}$.

**Definition 11.1.1.** The active Lorentz boost of the vector $\mathbf{x} \in \mathbb{R}^3 \subset \mathbb{G}_3$ is the relative vector

$$\mathbf{x}' = e^{+\frac{\phi}{2}\hat{\mathbf{v}}}\mathbf{x}e^{-\frac{\phi}{2}\hat{\mathbf{v}}} = \mathbf{x}_{||} + \frac{1}{\sqrt{1-\frac{v^2}{c^2}}}\mathbf{x}_\perp\left(1-\frac{\mathbf{v}}{c}\right) \in \mathbb{G}_3,$$

and has the interpretation of the vector $\mathbf{x}$ moving in $\mathbb{G}_3$ with velocity $\mathbf{v}$.

The relative vector $\mathbf{x}'$ is the vector $\mathbf{x}$ moving in the direction $\hat{\mathbf{v}}$ at the constant speed $v = |\mathbf{v}|$. Because of the nature of space–time, the relative vector $\mathbf{x}'$ has picked up the bivector component $-\frac{1}{\sqrt{1-\frac{v^2}{c^2}}}\frac{\mathbf{x}_\perp\mathbf{v}}{c}$, see Fig. 11.2. Notice that the parallel component $\mathbf{x}_{||}$ of $\mathbf{x}$ is unaffected by the boost.

The active Lorentz boost of the rest frame $(\mathbf{e})_{(3)}$ of $\mathbb{R}^3$, for the given velocity $\mathbf{v}$, is

$$(\mathbf{e}')_{(3)} = e^{+\frac{\phi}{2}\hat{\mathbf{v}}}(\mathbf{e})_{(3)}e^{-\frac{\phi}{2}\hat{\mathbf{v}}} \in \mathbb{R}'^3. \tag{11.5}$$

Any two reference frames $(\mathbf{e})_{(3)}$ and $(\mathbf{e}')_{(3)}$, related by (11.5), are *valid rest frames* of $\mathbb{R}^3$ and $\mathbb{R}'^3$, respectively, in the sense of postulate 1) of special relativity. For the particular case when $\mathbf{v} = v\mathbf{e}_2$, we find that

$$(\mathbf{e}')_{(3)} = \left(\mathbf{e}_1 e^{-\phi\hat{\mathbf{v}}}\ \mathbf{e}_2\ \mathbf{e}_3 e^{-\phi\hat{\mathbf{v}}}\right) \in \mathbb{R}'^3 \subset \mathbb{G}_3 \tag{11.6}$$

and is the *relative orthonormal basis* for the *relative Euclidean space* $\mathbb{R}'^3 = \text{span}(\mathbf{e}')_{(3)} \subset \mathbb{G}_3$ moving with velocity $\mathbf{v}$ with respect to the Euclidean space $\mathbb{R}^3 = \text{span}(\mathbf{e})_{(3)}$.

It is easily verified that the boosted rest frame $(\mathbf{e}')_{(3)}$ of $\mathbb{R}'^3$ obeys all of the same algebraic properties as does the orthonormal frame $(\mathbf{e})_{(3)}$ of $\mathbb{R}^3$. Just as we found for the standard orthonormal basis (3.11) of $\mathbb{G}_3$, we find that

$$\mathbf{e}_1'^2 = \mathbf{e}_2'^2 = \mathbf{e}_3'^2 = 1, \quad \text{and} \quad \mathbf{e}'_{ij} := \mathbf{e}'_i\mathbf{e}'_j = -\mathbf{e}'_j\mathbf{e}'_i$$

for $i \neq j$ and $1 \leq i, j \leq 3$. In addition,

$$i = \mathbf{e}'_{123} = \mathbf{e}'_1\mathbf{e}'_2\mathbf{e}'_3 = \mathbf{e}_{123}.$$

Except for a *regrading* of the elements of the algebra $\mathbb{G}_3$ into relative vectors and relative bivectors, which are rest frame dependent, nothing has changed. Indeed, the relative vectors can serve as equally valid generators of the algebra $\mathbb{G}_3$. Thus, every rest frame $(\mathbf{e}')_{(3)}$ of $\mathbb{R}'^3$ can serve equally well as the generators of the space–time algebra $\mathbb{G}_3$, that is, $\mathbb{G}_3 = \mathbb{G}(\mathbb{R}^3) = \mathbb{G}(\mathbb{R}'^3)$, and induce a corresponding *relative inner product* and a *relative outer product* on the elements of $\mathbb{G}_3$. What other authors have apparently failed to notice is that in special relativity, the concept of a vector and bivector are themselves observer-dependent concepts, just as are the concepts of space and time.

To finish our setup of special relativity in $\mathbb{G}_3$, we need to define what is meant by the *space–time interval* of an event $X = ct + \mathbf{x}$ in a given rest frame of $\mathbb{R}^3$. The space–time interval of an event $X$ is defined to be the quantity

$$X\widetilde{X} = (ct + \mathbf{x})(ct - \mathbf{x}) = c^2t^2 - \mathbf{x}^2 \tag{11.7}$$

in the rest frame $(\mathbf{e})_{(3)}$ of $\mathbb{R}^3$. Similarly, the space–time interval of the event $X' = ct' + \mathbf{x}'$ in the rest frame $(\mathbf{e}')_{(3)}$ of $\mathbb{R}'^3$ is given by

$$X'\widetilde{X}' = (ct' + \mathbf{x}')(ct' - \mathbf{x}') = c^2t'^2 - \mathbf{x}'^2 \tag{11.8}$$

(Recall that any geometric number $G \in \mathbb{G}_3$ has the form $G = (\alpha + \beta \mathrm{i}) + \mathbf{a} + \mathrm{i}\mathbf{b}$ where $\alpha, \beta \in \mathbb{R}$ so that $\widetilde{G} = G^{\dagger -} = (\alpha + \beta \mathrm{i}) - (\mathbf{a} + \mathrm{i}\mathbf{b})$.) The crucial fact is that the space-time interval is independent of the rest frame in which it is calculated. For example, if $X' = X\mathrm{e}^{-\phi\hat{\mathbf{v}}}$, then

$$X'\widetilde{X'} = \left(X\mathrm{e}^{-\phi\hat{\mathbf{v}}}\right)\left(\mathrm{e}^{+\phi\hat{\mathbf{v}}}\widetilde{X}\right) = X\widetilde{X}.$$

For valid rest frames, the calculation

$$X'\widetilde{X'} = c^2t'^2 - \mathbf{x}'^2 = c^2t^2 - \mathbf{x}^2 = X\widetilde{X} = 0,$$

shows that postulate (2) of special relativity is satisfied.

Early work of the author in developing special relativity in the geometric algebra $\mathbb{G}_3$ can be found in [68–70] and later with William Baylis in [5]. In the remaining sections of this chapter, we restrict our study of special relativity to the geometric algebra $\mathbb{G}_2$ of the Euclidean plane $\mathbb{R}^2$. This simplifies the discussion considerably with little loss of generality.

## Exercises

1. In (11.7) and (11.8), we used the Clifford conjugation (3.33) of $\mathbb{G}(\mathbb{R}^3)$. Show that we could equally well use the Clifford conjugation operation of $\mathbb{G}(\mathbb{R}'^3)$ or that of any other valid rest frame.
2. For the valid rest frame $(\mathbf{e}')_{(3)}$ of $\mathbb{R}'^3$ given in the example (11.6), show that

$$(\mathbf{e}')^{\mathrm{T}}_{(3)} = \begin{pmatrix} \cosh\phi & 0 & -\mathrm{i}\sinh\phi \\ 0 & 1 & 0 \\ \mathrm{i}\sinh\phi & 0 & \cosh\phi \end{pmatrix} (\mathbf{e})^{\mathrm{T}}_{(3)}.$$

   Also verify that

$$\mathbf{e}_{(3)} \equiv \mathbf{e}_1 \wedge \mathbf{e}_2 \wedge \mathbf{e}_3 = \mathbf{e}'_1\mathbf{e}'_2\mathbf{e}'_3 = \mathrm{i},$$

   so that the concept of the unit volume element i is independent of the rest frame in which it is calculated.
3. (a) Verify that $\mathbb{G}_3 = \mathbb{G}(\mathbb{R}^3) = \mathbb{G}(\mathbb{R}'^3)$ is just a regrading of the elements of $\mathbb{G}_3$ where $(\mathbf{e})_{(3)}$ and $(\mathbf{e}')_{(3)}$ are given in (11.6).
   (b) Given $\mathbf{x}' = (\mathbf{e}')_{(3)}(x')^{(3)} \in \mathbb{R}'^3$ and $\mathbf{y}' = (\mathbf{e}')_{(3)}(y')^{(3)}\mathbb{R}'^3$, calculate

$$\mathbf{x}' \cdot \mathbf{y}' \equiv \frac{1}{2}(\mathbf{x}'\mathbf{y}' + \mathbf{y}'\mathbf{x}')$$

   and

$$\mathbf{x}' \wedge \mathbf{y}' \equiv \frac{1}{2}(\mathbf{x}'\mathbf{y}' - \mathbf{y}'\mathbf{x}').$$

4. For the unit vector $\hat{\mathbf{a}} = \mathbf{e}_1\cos\alpha_x + \mathbf{e}_2\cos\alpha_y + \mathbf{e}_3\cos\alpha_z \in \mathbb{R}^3$, define the valid rest frame $(\mathbf{e}')_{(3)}$ of $\mathbb{R}'^3 = \mathrm{span}(\mathbf{e}')_{(3)}$ by

$$(\mathbf{e}')_{(3)} = \mathrm{e}^{\frac{\phi\hat{\mathbf{a}}}{2}}(\mathbf{e})_{(3)}\mathrm{e}^{-\frac{\phi\hat{\mathbf{a}}}{2}}.$$

   Find the matrix of transition $A$ such that

$$(\mathbf{e}')_{(3)} = (\mathbf{e})_{(3)}A$$

   and show that $\wedge(\mathbf{e}')_{(3)} = \wedge(\mathbf{e})_{(3)}\det A = \mathrm{i}$.

## 11.2 Relative Orthonormal Basis

We have seen in Chap. 3 that the geometric algebra $\mathbb{G}_2$ algebraically unites the complex number plane $\mathbb{C}$ and the vector plane $\mathbb{R}^2$. This unification opens up many new possibilities. Recall (3.22) that an *active rotation* $R(\mathbf{x})$ in $\mathbb{R}^2$ has the form

$$R(\mathbf{x}) = \mathrm{e}^{-\frac{1}{2}\mathrm{i}\theta}\mathbf{x}\mathrm{e}^{\frac{1}{2}\mathrm{i}\theta}, \tag{11.9}$$

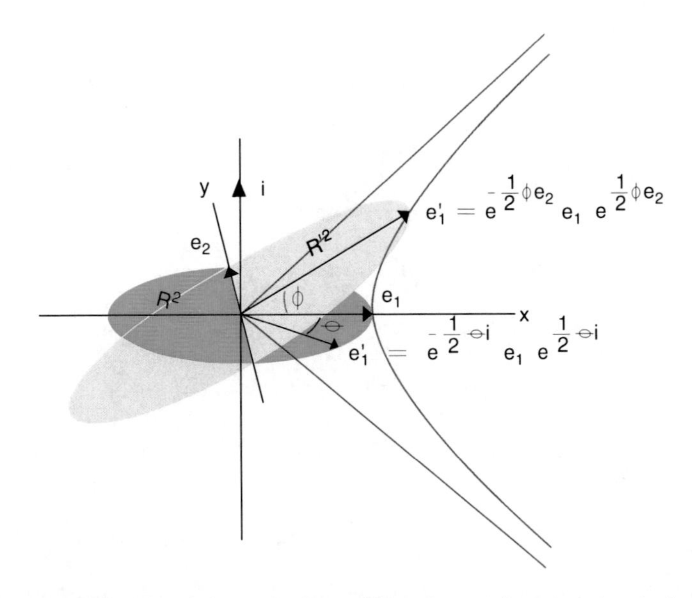


**Fig. 11.3** An active rotation and an active boost. The active rotation takes place in the Euclidean plane $\mathbb{R}^2$, and the active boost takes the Euclidean plane $\mathbb{R}^2$ into the (relative) Euclidean plane $\mathbb{R}'^2$

where the bivector $\mathrm{i} = \mathbf{e}_{12}$. Consider the transformation $L(\mathbf{x})$ defined by

$$\mathbf{x}' = L(\mathbf{x}) = \mathrm{e}^{-\frac{\phi \mathbf{a}}{2}} \mathbf{x} \mathrm{e}^{\frac{\phi \mathbf{a}}{2}} \tag{11.10}$$

for a unit vector $\mathbf{a} \in \mathbb{R}^2$. The transformation (11.10) has the same half-angle form as the rotation (11.9). We say that (11.10) defines an *active Lorentz boost* of the vector $\mathbf{x}$ into the *relative vector* $\mathbf{x}'$ moving with velocity[1]

$$\frac{\mathbf{v}}{c} = \tanh(-\phi \mathbf{a}) = -\mathbf{a} \tanh \phi \tag{11.11}$$

where $c$ is the *velocity of light*. For simplicity, we shall always take $c = 1$. An active rotation and an active boost are pictured in Fig. 11.3.

Both $R(\mathbf{x})$ and $L(\mathbf{x})$ are *inner automorphisms* on $\mathbb{G}_2$, satisfying $R(g_1 g_2) = R(g_1)R(g_2)$ and $L(g_1 g_2) = L(g_1)L(g_2)$ for all $g_1, g_2 \in \mathbb{R}^2$. Whereas $R(\mathbf{x})$ is an outermorphism, preserving the grading of the algebra $\mathbb{G}_2$ into *scalars*, *vectors*, and *bivectors*, the mapping $L(\mathbf{x})$ does not have this property as we shall shortly see.

Note that under both a Euclidean rotation (11.9) and under an active boost (11.10),

$$|\mathbf{x}'|^2 = (\mathbf{x}')^2 = \mathbf{x}^2 = |\mathbf{x}|^2,$$

[1] We have introduced the minus sign in (11.11) to keep (11.10) consistent with Definition 11.1.1.

so that the Euclidean lengths $|\mathbf{x}| = |\mathbf{x}'|$ of both the rotated vector and the boosted relative vector are preserved. Whereas the meaning of this statement is well-known for rotations, the corresponding statement for a boost needs further explanation.

The active boost (11.10) leaves invariant the direction of the boost, that is

$$L(\mathbf{a}) = \mathrm{e}^{-\frac{\phi \mathbf{a}}{2}} \mathbf{a} \mathrm{e}^{\frac{\phi \mathbf{a}}{2}} = \mathbf{a}. \tag{11.12}$$

On the other hand, for the vector $\mathbf{a}\mathrm{i}$ orthogonal to $\mathbf{a}$, we have

$$L(\mathbf{a}\mathrm{i}) = \mathrm{e}^{-\frac{\phi \mathbf{a}}{2}} \mathbf{a}\mathrm{i} \mathrm{e}^{\frac{\phi \mathbf{a}}{2}} = \mathbf{a}\mathrm{i}\mathrm{e}^{\phi \mathbf{a}} = \mathbf{a}\mathrm{i}\cosh\phi - \mathrm{i}\sinh\phi, \tag{11.13}$$

showing that the boosted relative vector $L(\mathbf{a}\mathrm{i})$ has picked up the bivector component $-\mathrm{i}\sinh\phi$.

We say that two relative vectors are *orthogonal* if they are anticommutative. From the calculation

$$L(\mathbf{a})L(\mathbf{a}\mathrm{i}) = \mathbf{a}\mathbf{a}\mathrm{i}\mathrm{e}^{\phi \mathbf{a}} = -\mathbf{a}\mathrm{i}\mathrm{e}^{\phi \mathbf{a}}\mathbf{a} = -L(\mathbf{a}\mathrm{i})L(\mathbf{a}), \tag{11.14}$$

we see that the active boost of a pair orthonormal vectors gives a pair of orthonormal relative vectors. When the active Lorentz boost is applied to the bivector $\mathrm{i} = \mathbf{e}_1\mathbf{e}_2$, we find that $j = L(\mathrm{i}) = \mathrm{i}\mathrm{e}^{\mathbf{a}\phi}$ so that a boost of the bivector i in the direction of the vector $\mathbf{a}$ gives the *relative bivector* $j = \mathrm{i}\mathrm{e}^{\mathbf{a}\phi}$. Note that

$$j^2 = \mathrm{i}\mathrm{e}^{\mathbf{a}\phi}\mathrm{i}\mathrm{e}^{\mathbf{a}\phi} = \mathrm{i}^2\mathrm{e}^{-\mathbf{a}\phi}\mathrm{e}^{\mathbf{a}\phi} = -1$$

as expected.

Using (11.12)–(11.14), we say that

$$\mathscr{B}_j := \{1, \mathbf{e}'_1, \mathbf{e}'_2, \mathbf{e}'_1\mathbf{e}'_2\}, \tag{11.15}$$

where $\mathbf{e}'_1 = \mathbf{a}$, $\mathbf{e}'_2 = \mathbf{a}\mathrm{i}\mathrm{e}^{\phi\mathbf{a}}$, and $j = \mathbf{a}\mathbf{a}\mathrm{i}\mathrm{e}^{\phi\mathbf{a}} = \mathrm{i}\mathrm{e}^{\phi\mathbf{a}}$, makes up a *relative orthonormal basis* of $\mathbb{G}_2$. Note that the defining rules for the standard basis (3.9) of $\mathbb{G}_2$ remain the same for the relative basis $\mathscr{B}_j$:

$$(\mathbf{e}'_1)^2 = (\mathbf{e}'_2)^2 = 1, \text{ and } \mathbf{e}'_1\mathbf{e}'_2 = -\mathbf{e}'_2\mathbf{e}'_1.$$

Essentially, the relative basis $\mathscr{B}_j$ of $\mathbb{G}_2$ *regrades* the algebra into relative vectors and relative bivectors moving at the velocity of $\mathbf{v} = \mathbf{a}\tanh\phi$ with respect to the standard basis $\mathscr{B}_i$. We say that $j$ defines the *direction* and *orientation* of the relative plane

$$\mathbb{R}^2_j := \{\mathbf{v}' |\ \mathbf{v}' = x'\mathbf{e}'_1 + y'\mathbf{e}'_2, \text{ for } x', y' \in \mathbb{R}\}. \tag{11.16}$$

Active rotations (3.22) and active boosts (11.10) define two different kinds of automorphisms on the geometric algebra $\mathbb{G}_2$. Whereas active rotations are well understood in Euclidean geometry, an active boost brings in concepts from non-Euclidean geometry. Since an active boost essentially regrades the geometric algebra $\mathbb{G}_2$ into relative vectors and relative bivectors, it is natural to refer to the *relative geometric algebra* $\mathbb{G}_2$ of the relative plane (11.16) when using this basis.

## Exercises

1. Find the relative orthonormal basis (11.15) in $\mathbb{G}_2$ for the active Lorentz boost defined by the velocity vector $\mathbf{v} = \mathbf{e}_1 \tanh\phi$ where $|\mathbf{v}| = \frac{1}{2}$ of the speed of light.
2. Find the relative orthonormal basis (11.15) in $\mathbb{G}_2$ for the active Lorentz boost defined by the velocity vector $\mathbf{v} = \mathbf{e}_1 \tanh\phi$ where $|\mathbf{v}| = \frac{3}{4}$ of the speed of light.

## 11.3 Relative Geometric Algebras

We have seen that both the unit bivector i and the relative unit bivector $\mathrm{j} = \mathrm{i}e^{\mathbf{a}\phi}$ have square $-1$. Let us see what can be said about the most general element $\mathrm{h} \in \mathbb{G}_2$ which has the property that $\mathrm{h}^2 = -1$. In the standard basis (3.9), $h$ will have the form

$$h = h_1\mathbf{e}_1 + h_2\mathbf{e}_2 + h_3\mathrm{i}$$

for $h_1, h_2, h_3 \in \mathbb{R}$ as is easily verified. Clearly the condition that $\mathrm{h}^2 = h_1^2 + h_2^2 - h_3^2 = -1$ will be satisfied if and only if $1 + h_1^2 + h_2^2 = h_3^2$ or $h_3 = \pm\sqrt{1 + h_1^2 + h_2^2}$. We have two cases:

1. If $h_3 \geq 0$, define $\cosh\phi = \sqrt{1 + h_1^2 + h_2^2}$, $\sinh\phi = \sqrt{h_1^2 + h_2^2}$, and the unit vector $\mathbf{a}$ such that $\mathrm{i}\mathbf{a}\sinh\phi = h_1\mathbf{e}_1 + h_2\mathbf{e}_2$ or $\mathbf{a} = \frac{h_1\mathbf{e}_2 - h_2\mathbf{e}_1}{\sqrt{1+h_1^2+h_2^2}}$. Defined in this way, $\mathrm{h} = \mathrm{i}e^{\mathbf{a}\phi}$ is a relative bivector to i.
2. If $h_3 < 0$, define $\cosh\phi = \sqrt{1 + h_1^2 + h_2^2}$, $\sinh\phi = -\sqrt{h_1^2 + h_2^2}$, and the unit vector $\mathbf{a}$ such that $\mathrm{i}\mathbf{a}\sinh\phi = -(h_1\mathbf{e}_1 + h_2\mathbf{e}_2)$ or $\mathbf{a} = \frac{h_1\mathbf{e}_2 - h_2\mathbf{e}_1}{\sqrt{1+h_1^2+h_2^2}}$. In this case, $\mathrm{h} = -\mathrm{i}e^{\mathbf{a}\phi}$ is a relative bivector to $-\mathrm{i}$.

From the above remarks, we see that any geometric number $\mathrm{h} \in \mathbb{G}_2$ with the property that $\mathrm{h}^2 = -1$ is a relative bivector to $\pm\mathrm{i}$. The set of relative bivectors to $+\mathrm{i}$,

$$\mathscr{H}^+ := \{\mathrm{i}e^{\phi\mathbf{a}} |\ \mathbf{a} = \mathbf{e}_1\cos\theta + \mathbf{e}_2\sin\theta,\ 0 \leq \theta < 2\pi,\ \phi \in \mathbb{R}\} \tag{11.17}$$

are said to be *positively oriented* . Moving relative bivectors i, j, and k are pictured in Fig. 11.4. Similarly, the set $\mathscr{H}^-$ of negatively oriented relative bivectors to $-\mathrm{i}$ can be defined.

For each positively oriented relative bivector $\mathrm{h} = \mathrm{i}e^{\mathbf{a}\phi} \in \mathscr{H}^+$, we define a positively oriented relative plane $\mathbb{R}_h^2$ by

$$\mathbb{R}_h^2 = \{\mathbf{x} |\ \mathbf{x} = x\mathbf{a} + y\mathbf{a}\mathrm{i},\ x, y \in \mathbb{R}\},$$

and the corresponding *relative basis* $\mathscr{B}_h$ of the geometric algebra $\mathbb{G}_2$:

$$\mathscr{B}_h = \{1, \mathbf{a}, \mathbf{a}\mathrm{i}e^{\mathbf{a}\phi}, \mathrm{i}e^{\mathbf{a}\phi}\}. \tag{11.18}$$

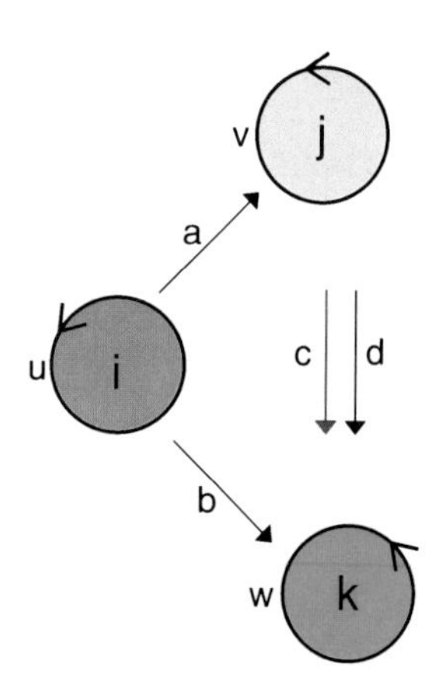


**Fig. 11.4** The relative bivectors j is moving in the direction **a**, and k is moving in the direction **b** with respect to i. Similarly, k is moving in the direction **c** with respect to j. The extra labels u, v, and w for the rest frames of these respective bivectors will be explained later, as will the significance of the direction **d**

In Fig. 11.4, we have also introduced the symbols u, v, and $w$ to label the *rest frames*, or *inertial systems*, defined by the relative bivectors i, j, and k, respectively. These symbols will later take on an algebraic interpretation as well.

For each relative plane $\mathbb{R}^2_h$, there exist a *relative inner product* and a *relative outer product*. Rather than use the relative inner and outer products on each different relative plane, we prefer to decompose the geometric product of two elements $g_1, g_2 \in \mathbb{G}_2$ into *symmetric* and *antisymmetric* parts. Thus,

$$g_1 g_2 = \frac{1}{2}(g_1 g_2 + g_2 g_1) + \frac{1}{2}(g_1 g_2 - g_2 g_1) = g_1 \odot g_2 + g_1 \boxtimes g_2 \tag{11.19}$$

where $g_1 \odot g_2 = \frac{1}{2}(g_1 g_2 + g_2 g_1)$ is called the *symmetric product* and $g_1 \boxtimes g_2 = \frac{1}{2}(g_1 g_2 - g_2 g_1)$ is called the *antisymmetric product*. We use the *boxed cross product* for the antisymmetric product to distinguish it from the closely related vector cross product.

We give here formulas for evaluating the symmetric and antisymmetric products of geometric numbers with vanishing scalar parts. Letting $A = a_1\mathbf{e}_1 + a_2\mathbf{e}_2 + a_3\mathrm{i}$, $B = b_1\mathbf{e}_1 + b_2\mathbf{e}_2 + b_3\mathrm{i}$, and $C = c_1\mathbf{e}_1 + c_2\mathbf{e}_2 + c_3\mathrm{i}$, we have

$$A \odot B = a_1 b_1 + a_2 b_2 - a_3 b_3,$$

$$A \boxtimes B = -\det \begin{pmatrix} \mathbf{e}_1 & \mathbf{e}_2 & -\mathrm{i} \\ a_1 & a_2 & a_3 \\ b_1 & b_2 & b_3 \end{pmatrix},$$

$$A \odot (B \boxtimes C) = -\det \begin{pmatrix} a_1 & a_2 & a_3 \\ b_1 & b_2 & b_3 \\ c_1 & c_2 & c_3 \end{pmatrix},$$

which bear striking resemblance to the dot and cross products of vector analysis. In general, a nonzero geometric number $A \in \mathbb{G}_2$ with vanishing scalar part is said to be a *relative vector* if $A^2 > 0$, a *nilpotent* if $A^2 = 0$, and a *relative bivector* if $A^2 < 0$.

## Exercises

1. Let $A = \mathbf{e}_1 + 2\mathbf{e}_2 + \mathrm{i}$, $B = 2\mathbf{e}_1 + \mathbf{e}_2 - \mathrm{i}$ and $C = \mathbf{e}_1 + \mathbf{e}_2 + 2\mathrm{i}$ in $\mathbb{G}_2$.
   (a) Calculate $A \odot B$.
   (b) Calculate $A \boxtimes B$.
   (c) $A \odot (B \boxtimes C)$.
2. (a) Calculate the relative basis (11.18) for the relative geometric algebra generated by the vector $\mathbf{a} = \mathbf{e}_1$ and the hyperbolic angle $\phi = 1$.
   (b) What is the velocity of this rest frame relative to the bivector $\mathrm{i} = \mathbf{e}_{12}$ of $\mathbb{G}_2$?
3. Find a nilpotent element $N \in \mathbb{G}_2$.

## 11.4 Moving Planes

Consider the set $\mathscr{H}^+$ of positively oriented relative bivectors to i. For $\mathrm{j} \in \mathscr{H}^+$, this means that $\mathrm{j} = \mathrm{i}e^{\phi\mathbf{a}}$ as given in (11.17). We say that the rest frame $v$, and its relative plane $\mathbb{R}^2_j$, defined by the bivector j, is moving with velocity $\mathbf{u}_v := \mathbf{a}\tanh\phi$ with respect to the rest frame $u$ and its relative plane $\mathbb{R}^2_\mathrm{i}$ defined by the bivector i.

Note that $\mathrm{j} = \mathrm{i}e^{\phi\mathbf{a}}$ implies that $\mathrm{i} = \mathrm{j}e^{-\phi\mathbf{a}}$ so that if j is moving with velocity $\mathbf{u}_v = \mathbf{a}\tanh\phi$ with respect to i, then i is moving with velocity $\mathbf{v}_u = -\mathbf{a}\tanh\phi$ with respect to j. Suppose now for the inertial system w that $\mathrm{k} = \mathrm{i}e^{\rho\mathbf{b}} \in \mathscr{H}^+$, where the unit vector $\mathbf{b} \in \mathbb{R}^2_\mathrm{i}$ and the hyperbolic angle $\rho \in \mathbb{R}$. Then

$$\mathrm{k} = \mathrm{i}(e^{\phi\mathbf{a}}e^{-\phi\mathbf{a}})e^{\rho\mathbf{b}} = \mathrm{j}(e^{-\phi\mathbf{a}}e^{\rho\mathbf{b}}) = \mathrm{j}e^{\omega\mathbf{c}},$$

where $e^{\omega\mathbf{c}} = e^{-\phi\mathbf{a}}e^{\rho\mathbf{b}}$ for some hyperbolic angle $\omega$ and relative unit vector $\mathbf{c} \in \mathbb{R}^2_j \cap \mathbb{R}^2_k$.

Expanding $e^{\omega\mathbf{c}} = e^{-\phi\mathbf{a}}e^{\rho\mathbf{b}}$, we get

$$\begin{aligned}\cosh\omega(1+\mathbf{v}_w) &= e^{-\phi\mathbf{a}}e^{\rho\mathbf{b}} = \cosh\phi\cosh\rho(1-\mathbf{u}_v)(1+\mathbf{u}_w)\\ &= \cosh\phi\cosh\rho\big[(1-\mathbf{u}_v\cdot\mathbf{u}_w)+(\mathbf{u}_w-\mathbf{u}_v-\mathbf{u}_v\wedge\mathbf{u}_w)\big].\end{aligned}$$

It follows that

$$\cosh\omega = (\cosh\phi\cosh\rho)(1-\mathbf{u}_v\cdot\mathbf{u}_w), \tag{11.20}$$

and

$$\mathbf{v}_w = \frac{\cosh\phi\cosh\rho}{\cosh\omega}(\mathbf{u}_w - \mathbf{u}_v - \mathbf{u}_v \wedge \mathbf{u}_w) = \frac{\mathbf{u}_w - \mathbf{u}_v - \mathbf{u}_v \wedge \mathbf{u}_w}{1 - \mathbf{u}_v \cdot \mathbf{u}_w}. \tag{11.21}$$

We have found a relative unit vector $\mathbf{c} \in \mathbb{R}^2_j \cap \mathbb{R}^2_k$, $\mathbf{c} := \frac{\mathbf{v}_w}{\tanh\omega}$ and a hyperbolic angle $\omega \geq 0$ with the property that

$$\mathrm{k} = \mathrm{j}e^{\omega\mathbf{c}} = e^{-\frac{1}{2}\omega\mathbf{c}}\mathrm{j}e^{\frac{1}{2}\omega\mathbf{c}}.$$

The relative bivector k has velocity $\mathbf{v}_w = \mathbf{c}\tanh\omega$ with respect to the j. However, the relative unit vector $\mathbf{c} \not\in \mathbb{R}^2_i$. This means that the relative vector $\mathbf{c}$ defining the direction of the velocity of the relative bivector k with respect to j is not *commensurable* with the vectors in $\mathbb{R}^2_i$.

The question arises whether or not there exists a unit vector $\mathbf{d} \in \mathbb{R}^2_i$ with the property that

$$\mathrm{k} = e^{-\frac{1}{2}\Omega\mathbf{d}}\mathrm{j}e^{\frac{1}{2}\Omega\mathbf{d}} \quad \text{for some} \quad \Omega \in \mathbb{R}. \tag{11.22}$$

Substituting $\mathrm{j} = \mathrm{i}e^{\phi\mathbf{a}}$ and $\mathrm{k} = \mathrm{i}e^{\rho\mathbf{b}}$ into this last equation gives

$$\mathrm{i}e^{\rho\mathbf{b}} = e^{-\frac{1}{2}\Omega\mathbf{d}}\mathrm{i}e^{\phi\mathbf{a}}e^{\frac{1}{2}\Omega\mathbf{d}},$$

which is equivalent to the equation

$$e^{\rho\mathbf{b}} = e^{\frac{1}{2}\Omega\mathbf{d}}e^{\phi\mathbf{a}}e^{\frac{1}{2}\Omega\mathbf{d}}. \tag{11.23}$$

The transformation $L_p : \mathbb{G}_2 - > \mathbb{G}_2$ defined by

$$L_p(\mathbf{x}) = e^{\frac{1}{2}\Omega\mathbf{d}}\mathbf{x}e^{\frac{1}{2}\Omega\mathbf{d}} \tag{11.24}$$

is called the *passive Lorentz boost* relating $\mathbb{R}^2_j$ to $\mathbb{R}^2_k$ with respect to $\mathbb{R}^2_i$.

Equation (11.23) can either be solved for $e^{\rho\mathbf{b}}$ given $e^{\Omega\mathbf{d}}$ and $e^{\phi\mathbf{a}}$, or for $e^{\Omega\mathbf{d}}$ given $e^{\phi\mathbf{a}}$ and $e^{\rho\mathbf{b}}$. Defining the velocities $\mathbf{u}_v = \mathbf{a}\tanh\phi, \mathbf{u}_w = \mathbf{b}\tanh\rho$, and $\mathbf{u}_{vw} = \mathbf{d}\tanh\Omega$, we first solve for $e^{\rho\mathbf{b}}$ given $e^{\Omega\mathbf{d}}$ and $e^{\phi\mathbf{a}}$. In terms of these velocities, equation (11.23) takes the form

$$\begin{aligned}\cosh\rho\ (1+\mathbf{u}_w) &= \cosh\phi\ e^{\frac{1}{2}\Omega\mathbf{d}}(1+\mathbf{u}_v)e^{\frac{1}{2}\Omega\mathbf{d}} = \cosh\phi\ \left(e^{\Omega\mathbf{d}} + e^{\frac{1}{2}\Omega\mathbf{d}}\mathbf{u}_v e^{\frac{1}{2}\Omega\mathbf{d}}\right)\\ &= \cosh\phi\cosh\Omega\ \left[(1+\mathbf{u}_{vw})(1+\mathbf{u}_v^{\|})\right] + \cosh\phi\ \mathbf{u}_v^{\perp}\end{aligned}$$

where $\mathbf{u}_v^{\|} = (\mathbf{u}_v \cdot \mathbf{d})\mathbf{d}$ and $\mathbf{u}_v^{\perp} = (\mathbf{u}_v \wedge \mathbf{d})\mathbf{d}$ . Equating scalar and vector parts gives

$$\cosh\rho = \cosh\phi\cosh\Omega\ (1+\mathbf{u}_v\cdot\mathbf{u}_{vw}), \tag{11.25}$$

and

$$\mathbf{u}_w = \frac{\mathbf{u}_v + \mathbf{u}_{vw} + (\frac{1}{\cosh\Omega} - 1)(\mathbf{u}_v \wedge \mathbf{d})\mathbf{d}}{1+\mathbf{u}_v\cdot\mathbf{u}_{vw}}. \tag{11.26}$$

Equation (11.26) is the (passive) composition formula for the addition of velocities of special relativity in the rest frame $u$, [36, p. 588] and [55, p. 133].

To solve (11.23) for $e^{\Omega \mathbf{d}}$ given $e^{\phi \mathbf{a}}$ and $e^{\rho \mathbf{b}}$, we first solve for the unit vector $\mathbf{d} \in \mathbb{R}^2$ by taking the antisymmetric product of both sides of (11.23) with $\mathbf{d}$ to get the relationship

$$\mathbf{d} \boxtimes \mathbf{b} \sinh\rho = e^{\frac{1}{2}\Omega \mathbf{d}} \mathbf{d} \boxtimes \mathbf{a} \sinh\phi \, e^{\frac{1}{2}\Omega \mathbf{d}} = \mathbf{a} \sinh\phi,$$

or equivalently,

$$\mathbf{d} \wedge (\mathbf{b} \sinh\rho - \mathbf{a} \sinh\phi) = 0.$$

In terms of the velocity vectors $\mathbf{u}_v$ and $\mathbf{u}_w$, we can define the unit vector $\mathbf{d}$ by

$$\mathbf{d} = \frac{\mathbf{u}_w \cosh\rho - \mathbf{u}_v \cosh\phi}{\sqrt{\mathbf{u}_v^2 \cosh^2\phi - 2\mathbf{u}_v \cdot \mathbf{u}_w \cosh\phi \cosh\rho + \mathbf{u}_w^2 \cosh^2\rho}} \tag{11.27}$$

Taking the symmetric product of both sides of (11.23) with $\mathbf{d}$ gives

$$[\mathbf{d} \odot e^{\phi \mathbf{a}}] e^{\Omega \mathbf{d}} = \mathbf{d} \odot e^{\rho \mathbf{b}},$$

or

$$(\mathbf{d} \cosh\phi + \mathbf{a} \cdot \mathbf{d} \sinh\phi) e^{\Omega \mathbf{d}} = \mathbf{d} \cosh\rho + \mathbf{b} \cdot \mathbf{d} \sinh\rho.$$

Solving this last equation for $e^{\Omega \mathbf{d}}$ gives

$$e^{\Omega \mathbf{d}} = \frac{(\mathbf{d} \cosh\rho + \mathbf{b} \cdot \mathbf{d} \sinh\rho)(\mathbf{d} \cosh\phi - \mathbf{a} \cdot \mathbf{d} \sinh\phi)}{\cosh^2\phi - (\mathbf{a} \cdot \mathbf{d})^2 \sinh^2\phi}, \tag{11.28}$$

or in terms of the velocity vectors,

$$\begin{aligned} \cosh\Omega (1 + \mathbf{u}_{vw}) &= \frac{\cosh\rho}{\cosh\phi} \left( \frac{(1 + \mathbf{u}_w \cdot \mathbf{d}\, \mathbf{d})(1 - \mathbf{u}_v \cdot \mathbf{d}\, \mathbf{d})}{1 - (\mathbf{u}_v \cdot \mathbf{d})^2} \right) \\ &= \frac{\cosh\rho}{\cosh\phi} \left( \frac{1 - (\mathbf{u}_v \cdot \mathbf{d})(\mathbf{u}_w \cdot \mathbf{d}) + (\mathbf{u}_w - \mathbf{u}_v) \cdot \mathbf{d}\, \mathbf{d}}{1 - (\mathbf{u}_v \cdot \mathbf{d})^2} \right). \end{aligned}$$

Taking scalar and vector parts of this last equation gives

$$\cosh\Omega = \frac{\cosh\rho}{\cosh\phi} \left( \frac{1 - (\mathbf{u}_v \cdot \mathbf{d})(\mathbf{u}_w \cdot \mathbf{d})}{1 - (\mathbf{u}_v \cdot \mathbf{d})^2} \right) \tag{11.29}$$

and

$$\mathbf{u}_{vw} = \frac{(\mathbf{u}_w - \mathbf{u}_v) \cdot \mathbf{d}\, \mathbf{d}}{1 - (\mathbf{u}_v \cdot \mathbf{d})(\mathbf{u}_w \cdot \mathbf{d})}. \tag{11.30}$$

We say that $\mathbf{u}_{vw}$ is the relative velocity of the passive boost (11.24) of j into k relative to i. The passive boost is at the foundation of the *algebra of physical space* formulation of special relativity [5], and a coordinate form of this passive approach was used by Einstein in his famous 1905 paper [21]. Whereas Hestenes in [38] employs the active Lorentz boost, in [36] he uses the passive form of the Lorentz boost.

The distinction between active and passive boosts continues to be the source of much confusion in the literature [62]. Whereas an active boost (11.10) mixes vectors and bivectors of $\mathbb{G}_2$, the passive boost defined by (11.24) mixes the vectors and scalars of $\mathbb{G}_2$ in the geometric algebra $\mathbb{G}_2$ of i. In the next section, we shall find an interesting geometric interpretation of this result in a closely related higher dimensional space.

## Exercises

1. Let $\mathrm{j}\,\mathrm{i}e^{\mathbf{e}_1}$ and $\mathrm{k} = \mathrm{i}e^{2\mathbf{e}_2}$.
   (a) Express the relative bivector in the form $\mathrm{k} = \mathrm{j}e^{\omega\mathbf{c}}$ by using the formulas of this section. Find the hyperbolic angle $\omega$ and the relative vector $\mathbf{c}$.
   (b) Find the hyperbolic angle $\Omega$ (or $e^{\Omega\mathbf{d}}$) and a unit vector $\mathbf{d} \in \mathbb{R}_i^2$ such that $\mathrm{k} = \mathrm{j}e^{-\frac{1}{2}\Omega\mathbf{d}}\mathrm{j}e^{\frac{1}{2}\Omega\mathbf{d}}$ as in formula (11.28).
   (c) What is the relative velocity $\mathbf{u}_v$ that the bivector j is moving with respect to i?
   (d) What is the relative velocity $\mathbf{v}_w$ that the bivector k is moving with respect to j?
   (e) What is the relative velocity $\mathbf{u}_w$ that the bivector k is moving with respect to i?

## *11.5 Splitting the Plane

Geometric insight into the previous calculations can be obtained by *splitting* or *factoring* the geometric algebra $\mathbb{G}_2$ into a larger geometric algebra $\mathbb{G}_{1,2}$. The most mundane way of accomplishing this is to factor the standard orthonormal basis vectors of (3.9) into anticommuting bivectors of a larger geometric algebra $\mathbb{G}_{1,2} = \mathbb{G}(\mathbb{R}^{1,2})$ of the pseudo-Euclidean space $\mathbb{R}^{1,2}$, called *restricted Minkowski space–time*. By Minkowski space–time, we mean the pseudo-Euclidean space $\mathbb{R}^{1,3}$, [33, p. 24]. We write

$$\mathbf{e}_1 = \gamma_0\gamma_1, \text{ and } \mathbf{e}_2 = \gamma_0\gamma_2,$$

and assume the rules $\gamma_0^2 = 1 = -\gamma_1^2 = -\gamma_2^2$, and $\gamma_\mu\gamma_\eta = -\gamma_\eta\gamma_\mu$ for all $\mu, \eta = 0, 1, 2$ and $\mu \neq \eta$.

(http://en.wikipedia.org/wiki/Minkowski_space)

The standard orthonormal basis of $\mathbb{G}_{1,2}$ consists of the eight elements

$$\{1, \gamma_0, \gamma_1, \gamma_2, \gamma_{01}, \gamma_{02}, \gamma_{21}, \gamma_{012}\}. \tag{11.31}$$

With this splitting, the standard basis elements (3.9) of $\mathbb{G}_2$ are identified with elements of the *even subalgebra*

$$\mathbb{G}_{1,2}^+ := \text{span}\{1, \mathbf{e}_1 = \gamma_{01}, \mathbf{e}_2 = \gamma_{02}, \mathbf{e}_{12} = \gamma_{21}\} \tag{11.32}$$

of $\mathbb{G}_{1,2}$. We denote the oriented unit *pseudoscalar* element by $\mathrm{s} = \gamma_{012}$. Note that $\mathrm{s} \in Z(\mathbb{G}_{1,2})$, the *center* of the algebra $\mathbb{G}_{1,2}$.

The geometric algebra $\mathbb{G}_{1,2}$ is algebraically closed in the sense that in dealing with the characteristic and minimal polynomials of the matrices which represent the elements of $\mathbb{G}_{1,2}$, we can interpret complex eigenvalues of these polynomials to be in the space–time algebra $\mathbb{G}_{1,2}$. Alternatively, we can interpret complex eigenvalues to be in $\mathbb{G}_{1,3} = \mathbb{U}(\mathbb{G}_{1,2})$ as we did in Chap. 10.

Consider now the mapping

$$\psi : \mathscr{H}^+ \longrightarrow \{\mathrm{r} \in \mathbb{G}_{1,2}^1 |\ r^2 = 1\} \tag{11.33}$$

defined by $\mathrm{r} = \psi(\mathrm{h}) = \mathrm{sh}$ for all $\mathrm{h} \in \mathscr{H}^+$. The mapping $\psi$ sets up a $1-1$ correspondence between the positively oriented unit bivectors $\mathrm{h} \in \mathscr{H}^+$ and unit timelike vectors $\mathrm{r} \in \mathbb{G}_{1,2}^1$, which are *dual* under multiplication by the pseudoscalar $s$. Suppose now that $\psi(\mathrm{i}) = \mathrm{u}$, $\psi(\mathrm{j}) = \mathrm{v}$ and $\psi(\mathrm{k}) = \mathrm{k}$. Then it immediately follows by duality that if $\mathrm{j} = \mathrm{i}e^{\phi \mathbf{a}}$, $\mathrm{k} = \mathrm{i}e^{\rho \mathbf{b}}$, and $\mathrm{k} = \mathrm{j}e^{\omega \mathbf{c}}$, then $\mathrm{v} = \mathrm{u}e^{\phi \mathbf{a}}$, $\mathrm{w} = \mathrm{u}e^{\phi \mathbf{a}}$, and $w = ve^{\omega \mathbf{c}}$, respectively. It is because of this $1-1$ correspondence that we have included the labels u, $v$, and $w$ as another way of identifying the oriented planes of the bivectors i, j, and $k$ in Fig. 11.4.

Just as vectors in $\mathbf{x}, \mathbf{y} \in \mathbb{G}_2^1$ are identified with points $\mathbf{x}, \mathbf{y} \in \mathbb{R}^2$, *Minkowski vectors* $\mathrm{x}, \mathrm{y} \in \mathbb{G}_{1,2}^1$ are identified with points $\mathrm{x}, \mathrm{y} \in \mathbb{R}^{1,2}$ the 3-dimensional pseudo-Euclidean space of restricted Minkowski space–time. A Minkowski vector $\mathrm{x} \in \mathbb{R}^{1,2}$ is said to be *timelike* if $x^2 > 0$, *spacelike* if $\mathrm{x}^2 < 0$, and *lightlike* if $\mathrm{x} \neq 0$ but $\mathrm{x}^2 = 0$. For two Minkowski vectors $\mathrm{x}, \mathrm{y} \in G_{1,2}^1$, we decompose the geometric product $xy$ into *symmetric* and *antisymmetric* parts

$$\mathrm{xy} = \frac{1}{2}(\mathrm{xy} + \mathrm{yx}) + \frac{1}{2}(\mathrm{xy} - \mathrm{yx}) = \mathrm{x} \cdot \mathrm{y} + \mathrm{x} \wedge \mathrm{y},$$

where $\mathrm{x} \cdot \mathrm{y} := \frac{1}{2}(\mathrm{xy} + \mathrm{yx})$ is called the *Minkowski inner product* and $\mathrm{x} \wedge \mathrm{y} := \frac{1}{2}(\mathrm{xy} - \mathrm{yx})$ is called the *Minkowski outer product* to distinguish these products from the corresponding inner and outer products defined in $\mathbb{G}_2$.

In [34] and [38], David Hestenes gives an active reformulation of Einstein's special relativity in the space–time algebra $\mathbb{G}_{1,3}$. In [68, 70], I show that an equivalent active reformulation is possible in the geometric algebra $\mathbb{G}_3$ of the Euclidean space $\mathbb{R}^3$. In [5], the relationship between active and passive formulations is considered.

For the two unit timelike vectors $\mathrm{u},\mathrm{v} \in \mathbb{G}_{1,2}$ where $\mathrm{u} = \psi(\mathrm{i})$ and $\mathrm{v} = \psi(\mathrm{j})$, we have

$$\mathrm{uv} = \mathrm{u}\cdot\mathrm{v} + \mathrm{u}\wedge\mathrm{v} = e^{\phi\mathbf{a}} = \cosh\phi + \mathbf{a}\sinh\phi. \tag{11.34}$$

It follows that $\mathrm{u}\cdot\mathrm{v} = \cosh\phi$ and $\mathrm{u}\wedge\mathrm{v} = \mathbf{a}\sinh\phi$, which are the hyperbolic counterparts to the geometric product (3.1) and (3.7) of unit vectors $\mathbf{a},\mathbf{b} \in \mathbb{R}^2$.

The Minkowski bivector

$$\mathbf{u}_v = \frac{\mathrm{u}\wedge\mathrm{v}}{\mathrm{u}\cdot\mathrm{v}} = \mathbf{a}\tanh\phi = -\mathbf{v}_u$$

is the *relative velocity* of the timelike vector unit vector $v$ in the rest frame of u.

Suppose now that for $\mathrm{i},\mathrm{j},\mathrm{k} \in \mathscr{H}^+$, $\psi(\mathrm{i}) = \mathrm{u}$, $\psi(\mathrm{j}) = v$, and $\psi(\mathrm{k}) = \mathrm{w}$ so that $\mathrm{uv} = e^{\phi\mathbf{a}}, \mathrm{uw} = e^{\rho\mathbf{b}}$, and $\mathrm{vw} = e^{\omega\mathbf{c}}$, respectively. Let us recalculate $vw = e^{\omega\mathbf{c}}$ in the *space–time algebra* $\mathbb{G}_{1,2}$:

$$\begin{aligned}\mathrm{vw} &= \mathrm{vuuw} = (\mathrm{vu})(\mathrm{uw}) = (v\cdot u - u\wedge v)(u\cdot\mathrm{w} + \mathrm{u}\wedge\mathrm{w})\\ &= (\mathrm{v}\cdot\mathrm{u})(\mathrm{w}\cdot\mathrm{u})(1-\mathbf{u}_v)(1+\mathbf{u}_w).\end{aligned}$$

Separating into scalar and vector parts in $\mathbb{G}_2$, we get

$$v\cdot w = (v\cdot u)(\mathrm{w}\cdot\mathrm{u})(1-\mathbf{u}_v\cdot\mathbf{u}_w) \tag{11.35}$$

and

$$(v\cdot w)\mathbf{v}_w = (\mathrm{v}\cdot\mathrm{u})(\mathrm{w}\cdot\mathrm{u})[\mathbf{u}_w - \mathbf{u}_v - \mathbf{u}_v\wedge\mathbf{u}_w], \tag{11.36}$$

identical to what we calculated in (11.20) and (11.21), respectively.

More eloquently, using (11.35), we can express (11.36) in terms of quantities totally in the algebra $\mathbb{G}_2$,

$$\mathbf{v}_w = \frac{\mathbf{u}_w - \mathbf{u}_v - \mathbf{u}_v\wedge\mathbf{u}_w}{1-\mathbf{u}_v\cdot\mathbf{u}_w} = \mathbf{c}\tanh\omega = -\mathbf{w}_v. \tag{11.37}$$

We see that the relative velocity $\mathbf{v}_w$, up to a scale factor, is the *difference* of the velocities $\mathbf{u}_w$ and $\mathbf{u}_v$ and the bivector $\mathbf{u}_v\wedge\mathbf{u}_w$ in the inertial system u. Setting $\mathrm{w} = \mathrm{v}$ in (11.35) and solving for $v\cdot u$ in terms of $\mathbf{u}_v^2$ gives

$$\cosh\phi := \mathrm{u}\cdot\mathrm{v} = \frac{1}{\sqrt{1-\mathbf{u}_v^2}}, \tag{11.38}$$

a famous expression in Einstein's theory of special relativity, [38].

Let us now carry out the calculation for (11.29) and the relative velocity (11.30) of the inertial system w with respect to v as measured in the rest frame of u. We begin by defining the bivector $D = (\mathrm{w}-\mathrm{v})\wedge \mathrm{u}$ and noting that $\mathrm{w}\wedge D = \mathrm{v}\wedge\mathrm{w}\wedge\mathrm{u} = \mathrm{v}\wedge D$. Now note that

$$\mathrm{w} = \mathrm{w}DD^{-1} = (\mathrm{w}\cdot D)D^{-1} + (\mathrm{w}\wedge D)D^{-1} = \mathrm{w}_{\|} + w_{\perp}$$

where $\mathrm{w}_{\|} = (\mathrm{w}\cdot D)D^{-1}$ is the component of *w parallel* to $D$ and $\mathrm{w}_{\perp} = (\mathrm{w}\wedge D)D^{-1}$ is the component of w *perpendicular* to $D$. Next, we calculate

$$\hat{\mathrm{w}}_{\|}\hat{\mathrm{v}}_{\|} = -\frac{(\mathrm{w}\cdot D)(\mathrm{v}\cdot D)}{|\mathrm{w}\cdot D||\mathrm{v}\cdot D|} = \frac{(\mathrm{w}\cdot D)(\mathrm{v}\cdot D)}{(\mathrm{v}\cdot D)^2} = (\mathrm{w}\cdot D)(\mathrm{v}\cdot D)^{-1}, \tag{11.39}$$

since

$$\begin{aligned}(\mathrm{w}\cdot D)^2 = (\mathrm{v}\cdot D)^2 &= [(\mathrm{w}\cdot\mathrm{v}-1)\mathrm{u} - (\mathrm{v}\cdot u)(w-v)]^2\\ &= (\mathrm{w}\cdot\mathrm{v}-1)[(\mathrm{w}\cdot\mathrm{v}-1) - 2(v\cdot u)(w\cdot u)] < 0.\end{aligned}$$

We can directly relate (11.39) to (11.29) and (11.30),

$$\begin{aligned}\hat{\mathrm{w}}_{\|}\hat{\mathrm{v}}_{\|} &= \frac{(\mathrm{w}\cdot D)\mathrm{uu}(\mathrm{v}\cdot D)}{(v\cdot D)^2} = \frac{[(\mathrm{w}\cdot D)\cdot u + (w\cdot D)\wedge u][(\mathrm{v}\cdot D)\cdot\mathrm{u} + (v\cdot D)\wedge u]}{(v\cdot D)^2}\\ &= \frac{[-(\mathrm{w}\wedge\mathrm{u})\cdot D - (w\cdot u)D][-(v\wedge u)\cdot D + (\mathrm{v}\cdot\mathrm{u})D]}{(\mathrm{v}\cdot D)^2}\\ &= -(\mathrm{w}\cdot\mathrm{u})(v\cdot u)\frac{[\mathbf{u}_w\cdot\mathbf{d}+\mathbf{d}][-\mathbf{u}_v\cdot\mathbf{d}+\mathbf{d}]}{(v\cdot\mathbf{d})^2}\\ &= -(w\cdot u)(v\cdot u)\frac{1-(\mathbf{u}_v\cdot\mathbf{d})(\mathbf{u}_w\cdot\mathbf{d}) + (\mathbf{u}_w-\mathbf{u}_v)\cdot\mathbf{d}\,\mathbf{d}}{(\mathrm{v}\cdot\mathbf{d})^2}\end{aligned}$$

where we have used the fact that $\mathbf{d} = D/|D|$, see (11.27). In the special case when $\mathrm{v} = \mathrm{w}$, the above equation reduces to $(\mathrm{v}\cdot\mathbf{d})^2 = -(\mathrm{u}\cdot\mathrm{v})^2[1-(\mathbf{u}_v\cdot\mathbf{d})^2]$. Using this result in the previous calculation, we get the desired result that

$$\hat{\mathrm{w}}_{\|}\hat{\mathrm{v}}_{\|} = \frac{(\mathrm{w}\cdot\mathrm{u})[1-(\mathbf{u}_v\cdot\mathbf{d})(\mathbf{u}_w\cdot\mathbf{d}) + (\mathbf{u}_w-\mathbf{u}_v)\cdot\mathbf{d}\,\mathbf{d}]}{(\mathrm{u}\cdot\mathrm{v})[1-(\mathbf{u}_v\cdot\mathbf{d})^2]},$$

the same expression we derived after (11.28).

Defining the active boost $L_u(\mathrm{x}) = (\hat{\mathrm{w}}_{\|}\hat{\mathrm{v}}_{\|})^{\frac{1}{2}}\mathrm{x}(\hat{\mathrm{v}}_{\|}\hat{\mathrm{w}}_{\|})^{\frac{1}{2}}$, we can easily check that it has the desired property that

$$L_u(\mathrm{v}) = L_u(\mathrm{v}_{\|}+\mathrm{v}_{\perp}) = \hat{\mathrm{w}}_{\|}\hat{\mathrm{v}}_{\|}\mathrm{v}_{\|} + \mathrm{v}_{\perp} = \mathrm{w}_{\|} + \mathrm{w}_{\perp} = \mathrm{w}.$$

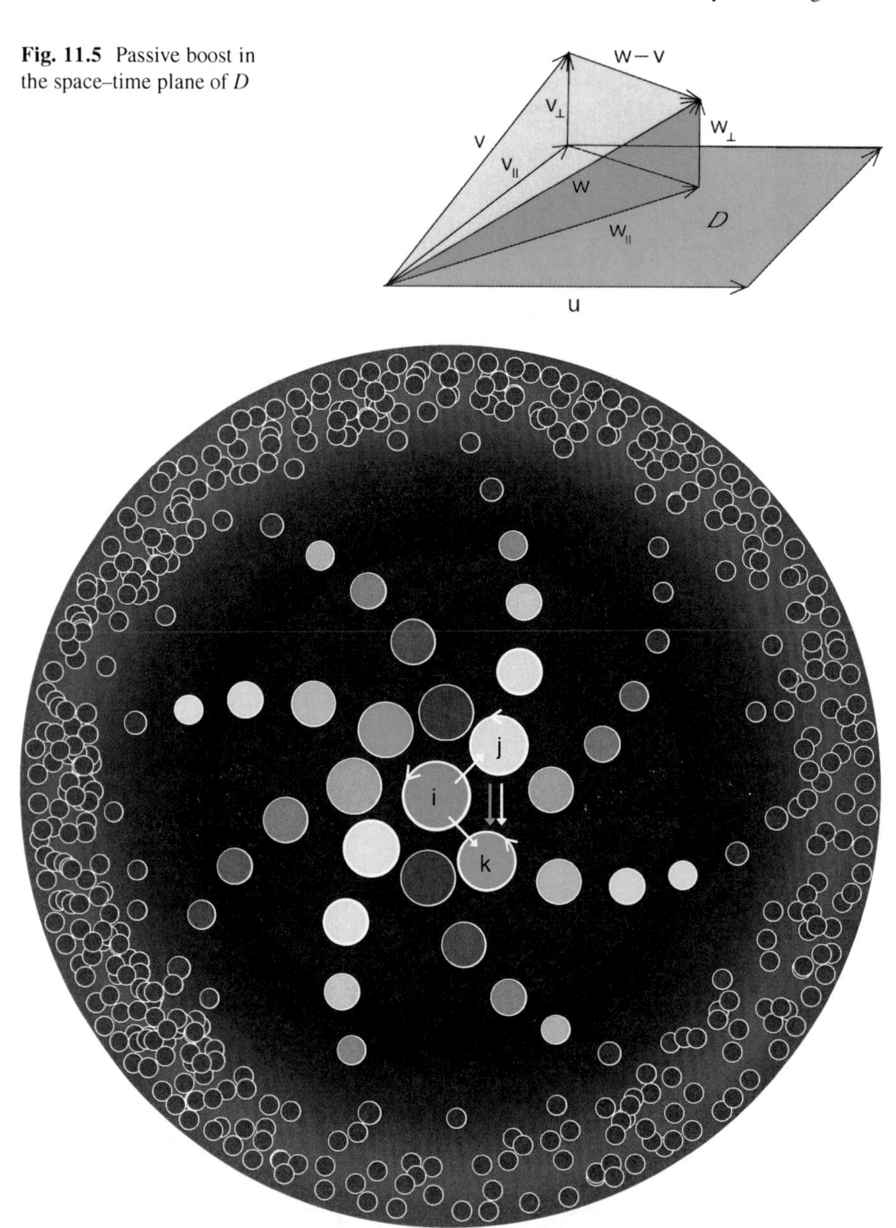


**Fig. 11.5** Passive boost in the space–time plane of $D$

**Fig. 11.6** Geometry of moving planes: Each disk represents a moving plane. The plane in the center is at rest. All the other planes are moving off in all different directions, each color representing a different direction. As the planes get further away from the center, they are moving faster and faster and turn redder and redder as they approach the speed of light on the boundary of the seen universe. Nothing can move faster than the speed of light

Thus, the active boost taking the unit timelike vector $\hat{\mathrm{v}}_{\|}$ into the unit timelike vector $\hat{\mathrm{w}}_{\|}$ is equivalent to the passive boost (11.24) in the plane of the space–time bivector $D$, see Fig. 11.4.

The above calculations show that each different rest frame u measures *passive relative velocities* between the systems v and w differently by a boost in the plane of the Minkowski bivector $D = (\mathrm{w} - \mathrm{v}) \wedge \mathrm{u}$, whereas there is a *unique* active boost (11.10) that takes the inertial system v into w in the plane of $\mathrm{v} \wedge \mathrm{w}$. The concept of a passive and active boost become equivalent when $\mathrm{u} \wedge \mathrm{v} \wedge \mathrm{w} = 0$, the case when $\mathbf{b} = \pm\mathbf{a}$ (Fig. 11.5).

Figure 11.6 is an attempt to artistically depict the idea of moving planes, as explained in the caption. It would be much harder to draw such a figure if, instead, we tried to draw the moving planes in the three-dimensional space $\mathbb{R}^3$ of $\mathbb{G}_3$.

## Exercises

1. Redo and compare the calculations done in Problems $1_{abcde}$ of the last section in the splitting geometric algebra $\mathbb{G}_{1,2}$.
2. For $\mathrm{u} = \gamma_0$ and $\mathrm{v} = \gamma_0 e^{\phi \gamma_0 \mathrm{w}}$ where $\mathrm{w} = \gamma_1 \cos\theta + \gamma_2 \sin\theta$, carry out the calculation given in (11.34), find $\mathbf{a}$, and show that $\mathbf{a}^2 = 1$.

# Chapter 12
# Representation of the Symmetric Group

*There are things which seem incredible to most men who have not studied Mathematics.*

—Archimedes of Syracuse

Over the last 112 years since Frobenius and Burnside initiated the study of representation theory, matrix algebra has played a central role with the indispensable idea of a group character [16]. Recently, point groups and crystallographic groups have been studied in geometric algebra [37, 40]. In this chapter, representations of the symmetric group are studied in the geometric algebra $\mathbb{G}_{n,n}$ of *neutral* signature. The representation depends heavily upon the definition of a new *twisted symmetric product*, which does not depend upon a matrix representation, although it is equivalent to it.

## 12.1 The Twisted Product

Let $\mathscr{A}$ be any associative algebra over the real or complex numbers and with the unity 1. For elements $a,b \in \mathscr{A}$, let $ab \in \mathscr{A}$ denote the product of $a$ and $b$ in $\mathscr{A}$. For elements $a,b \in \mathscr{A}$, we define a new product

$$a \circ b = \frac{1}{2}(1+a+b-ab) = 1 - \frac{1}{2}(1-a)(1-b), \tag{12.1}$$

which we call the *twisted product* of $a,b \in \mathscr{A}$. In the definition of the twisted product, the order of the elements in the product is always respected. This makes very general identities hold, even if its arguments are not commutative.

For any element $a \in \mathscr{A}$, we define the special symbols

$$a^+ = a \circ 0 = \frac{1}{2}(1+a), \quad a^- = (-a) \circ 0 = \frac{1}{2}(1-a), \tag{12.2}$$



G. Sobczyk, *New Foundations in Mathematics: The Geometric Concept of Number*, DOI 10.1007/978-0-8176-8385-6_12,

and note that $a^+ + a^- = 1$ and $a^+ - a^- = a$. Using these symbols, we can reexpress (12.1) in the useful alternative forms

$$a \circ b = a^+ + a^- b = b^+ + ab^-. \tag{12.3}$$

Note also the curious special cases

$$1 \circ a = 1,\ a \circ (-1) = a, \text{ and } 0 \circ 0 = \frac{1}{2}, \tag{12.4}$$

and if $a^2 = \pm 1$,

$$a \circ a = a, \quad a \circ a = 1 + a, \tag{12.5}$$

respectively.

It can be readily verified that for any $a_1, a_2, a_3 \in \mathscr{A}$ that

$$(a_1 \circ a_2) \circ a_3 = a_1 \circ (a_2 \circ a_3) = 1 - 2a_1^- a_2^- a_3^-, \tag{12.6}$$

so that the twisted product inherits the important associative property from the associative product in $\mathscr{A}$. More generally, for $n$ elements $a_1, \ldots, a_n \in \mathscr{A}$, we have

$$a_1 \circ \cdots \circ a_n = 1 - 2a_1^- \cdots a_n^- \iff 2a_1^- \cdots a_n^- = 1 - a_1 \circ \cdots \circ a_n. \tag{12.7}$$

It should also be noticed that whenever any argument is replaced by its negative on the left side of the identity, then the corresponding sign of the term must be changed on the right side. For example,

$$(-a_1) \circ a_2 \circ \cdots \circ a_n = 1 - 2a_1^+ a_2^- \cdots a_n^-.$$

For three elements $a, b, c \in \mathscr{A}$, using the definition (12.1), we derive the distributive-like property

$$a \circ (b + c) = 1 - \frac{1}{2}(1 - a)[(1 - b) + (1 - c) - 1] = a \circ b + a \circ c - a^+. \tag{12.8}$$

Letting $c = -b$ in the above identity, and using (12.2), we find the important identity that

$$2a^+ = a \circ b + a \circ (-b). \tag{12.9}$$

The *commutator* of the twisted product is also very useful. We find, using (12.7), that

$$a \circ b - b \circ a = -2(a^- b^- - b^- a^-) = -\frac{1}{2}(ab - ba), \tag{12.10}$$

as can be easily verified. Thus, the commutator of $a \circ b$ can be expressed entirely in terms of the product of $b^-$ and $a^-$. There is one other important general property that needs to be mentioned. If $a,b,g,g^{-1} \in \mathscr{A}$, where $gg^{-1} = 1$, then it immediately follows from the definition (12.1) that

$$g(a \circ b)g^{-1} = (gag^{-1}) \circ (gbg^{-1}). \tag{12.11}$$

### 12.1.1 Special Properties

Under certain conditions, the twisted product is also distributive over ordinary multiplication. Suppose for $k > 1$ that $a, b_1, \ldots, b_k \in \mathscr{A}$, $a^2 = 1$ and that $ab_i = b_i a$ for $i = 1, \ldots, k-1$. Then

$$a \circ (b_1 \cdots b_k) = \prod_{i=1}^{k} (a \circ b_i) = a^+ + a^- b_1 \cdots b_k. \tag{12.12}$$

The identity (12.12) is easily established. Using (12.3), we have

$$a \circ (b_1 \cdots b_k) = a^+ + a^- b_1 \cdots b_k$$

for the the left side of (12.12). For the right side, we use the fact that $a^+$ and $a^-$ are mutually annihilating idempotents so that

$$\prod_{i=1}^{k} (a \circ b_i) = \prod_{i=1}^{k} (a^+ + a^- b_i) = a^+ + a^- b_1 \cdots b_k.$$

In the special case that $\alpha \in \mathbb{R}$ and $a^2 = 1$, we have

$$a \circ (\alpha b) = (a \circ \alpha)(a \circ b) = (a^+ + \alpha a^-) a \circ b. \tag{12.13}$$

For $\alpha = -1$, (12.13) reduces to $a \circ (-b) = a(a \circ b)$, which further reduces to

$$a \circ (-a) = a^2 = 1 \tag{12.14}$$

when $b = a$. Curiously when $a^2 = -1$, $a \circ (-a) = 0$, a divisor of zero.

Let us suppose now that for $k \in \mathbb{N}$, we have $k$ mutually commuting elements $a_1, \ldots, a_k$ and that $a_1^2 = \cdots = a_k^2 = 1$. Then it is easy to show that

$$(a_1 \circ \cdots \circ a_k)^2 = 1, \tag{12.15}$$

as follows by induction on $k$. For $k = 1$, there is nothing to show. Suppose now that the hypothesis is true for some $k \in \mathbb{N}$ so that $(a_1 \circ \cdots \circ a_k)^2 = 1$. Then for $k+1$, using (12.3), we have

$$\begin{aligned}[(a_1 \circ \cdots \circ a_k) \circ a_{k+1}]^2 &= [a_{k+1}^+ + (a_1 \circ \cdots \circ a_k) a_{k+1}^-]^2 \\ &= (a_{k+1}^+)^2 + (a_1 \circ \cdots \circ a_k)^2 (a_{k-1}^-)^2 = a_{k-1}^+ + a_{k-1}^- = 1.\end{aligned}$$

One other striking relationship that the twisted product satisfies when $a^2 = 1$ is that

$$a^+((ab)\circ c) = a^+(b\circ c), \quad \text{and} \quad a^-((ab)\circ c) = a^-((-b)\circ c), \tag{12.16}$$

as follows directly from the steps

$$a^+((ab)\circ c) = a^+\frac{1}{2}(1+ab+c-abc) = \frac{1}{2}(a^+ + a^+b + a^+c - a^+bc)$$
$$= a^+\frac{1}{2}(1+b+c-bc) = a^+(b\circ c).$$

The other half of (12.16) follows similarly. Thus, the idempotents $a^+$ and $a^-$ *absorb* the $a$ in $(ab)\circ c$, leaving behind either 1 or $-1$ to give the stated results. In the special case that $b = 1$, (12.16) reduces to the even simpler relationship

$$a^+(a\circ c) = a^+, \quad \text{and} \quad a^-(a\circ c) = a^-c. \tag{12.17}$$

### 12.1.2 Basic Relationships

We list here basic relationships, which follow from (12.13) and (12.14), that hold between two commutative elements $a_1, a_2$ with the property that $a_1^2 = a_2^2 = 1$.

1. $(a_1\circ a_2)^2 = 1 \iff [((-a_1)\circ a_2]^2 = [a_1\circ(-a_2)]^2 = [((-a_1)\circ(-a_2)]^2 = 1$,
2. $a_1(a_1\circ a_2) = a_1\circ(-a_2) \iff (a_1\circ a_2)[a_1\circ(-a_2)] = a_1$,
3. $a_2(a_1\circ a_2) = (-a_1)\circ a_2 \iff (a_1\circ a_2)[(-a_1)\circ a_2] = a_2$,
4. $a_1a_2(a_1\circ a_2) = -(-a_1)\circ(-a_2) \iff (a_1\circ a_2)[(-a_1)\circ(-a_2)] = -a_1a_2$.

Because of the associativity of the twisted product, these relationships can be easily extended to the case of more commutative arguments which have square one. For example, the relationship 2 above for three arguments becomes

$$a_1(a_1\circ a_2\circ a_3) = a_1\circ[-(a_2\circ a_3)] \iff a_1 = (a_1\circ a_2\circ a_3)[a_1\circ(-(a_2\circ a_3))].$$

Using the absorption property (12.17), we also have

$$a_1(a_1\circ a_2\circ a_3) = (a_1^+ - a_1^-)(a_1\circ a_2\circ a_3) = a_1^+ - a_1^-(a_2\circ a_3).$$

## Exercises

1. Prove the four basic relationships given above.

## 12.2 Geometric Numbers in $\mathbb{G}_{n,n}$

Let $\mathbb{G}_{n,n}$ be a geometric algebra of the neutral signature $(n,n)$. Since these geometric algebras are algebraically isomorphic to to real matrix algebras $\mathscr{M}_{\mathbb{R}}(2^n)$ of square matrices of order $2^n \times 2^n$, they give a comprehensive geometric interpretation to these matrices. Since we carry out extensive calculations using the invaluable software package CLICAL [54], we use the conventions in that program for the definition of basis elements. We briefly review the main features below.

Let $n \geq 1$. The associative geometric algebra $\mathbb{G}_{n,n}$ is a *graded algebra*

$$\mathbb{G}_{n,n} = \mathbb{G}^0_{n,n} \oplus \mathbb{G}^1_{n,n} \oplus \cdots \oplus \mathbb{G}^{2n}_{n,n},$$

of *scalars*, *vectors*, *bivectors*, etc., as we have seen earlier in (3.30) and (10.10) in Chaps. 3 and 10. As before, the $k$-vector part of a multivector $g \in \mathbb{G}_{n,n}$ is denoted by $< g >_k \in \mathbb{G}^k_{n,n}$ for $0 \leq k \leq 2n$.

The *standard orthonormal basis* (10.10) of $\mathbb{G}_{n,n}$ is generated by taking the geometric products of the $2n$ basis vectors to get

$$\mathbb{G}_{n,n} = \{e_{\lambda_{(k)}}\}_{k=0}^{2n}, \tag{12.18}$$

where $e_i^2 = 1$ for $1 \leq i \leq n$ and $e_i^2 = -1$ for $n < i \leq 2n$. Recall that $e_{\lambda_{(k)}} = e_{\lambda_1,\ldots,\lambda_k} = e_{\lambda_1} \cdots e_{\lambda_k}$, where $1 \leq \lambda_1 < \cdots < \lambda_k \leq 2n$. By the *index set* of the geometric algebra $\mathbb{G}_{n,n}$, we mean the set of all indices $\mathscr{I}_{n,n} = \{\lambda_{(k)}\}_{k=0}^{2n}$ which define the basis elements $e_{\lambda_{(k)}}$ of $\mathbb{G}_{n,n}$. By $(\mathbb{G}^\dagger_{n,n})^{\mathrm{T}}$ we mean the *reverse* of the basis elements of $\mathbb{G}_{n,n}$ written as a column matrix. Thus,

$$(\mathbb{G}^\dagger_{n,n})^{\mathrm{T}} = (\{e^\dagger_{\lambda_{(k)}}\}_{k=0}^{2n})^{\mathrm{T}}.$$

We now construct a $2^n \times 2^n$ *real matrix representation* of the geometric algebra $\mathbb{G}_{n,n}$. First, we define the $2n$ idempotents $u_i^\pm = \frac{1}{2}(1 \pm e_{i,n+i})$ for $1 \leq i \leq n$, which we use to construct the $2^n$ *primitive idempotents*

$$u^{\pm\pm\ldots\pm}_{1\,2\ldots n} = u_1^\pm u_2^\pm \cdots u_n^\pm. \tag{12.19}$$

Letting $\mathscr{U}_n^+ = u_1^+ \cdots u_n^+$, all of the other primitive idempotents can then be defined succinctly by

$$\mathscr{U}^{sn(\alpha)} = e^\dagger_\lambda \mathscr{U}_n^+ e_\lambda$$

for $\lambda \in \mathscr{I}_n$ the index set for the geometric algebra $\mathbb{G}_n$.

By the *spectral basis* of the geometric algebra $\mathbb{G}_{n,n}$, we mean

$$(\mathbb{G}^\dagger_{n,n})^{\mathrm{T}} \mathscr{U}_n^+ \mathbb{G}_{n,n}, \tag{12.20}$$

where the convention of matrix row–column multiplication is utilized. Given a geometric number $g \in \mathbb{G}_{n,n}$, the relationship between $g$ and its matrix representation $[g]$ with respect to the spectral basis (12.20) is given by

$$g = \mathbb{G}_n \mathscr{U}_n^+ [g]^{\mathrm{T}} (\mathbb{G}_n^\dagger)^{\mathrm{T}} \iff g = \mathbb{G}_n^\dagger \mathscr{U}_n^+ [g] \mathbb{G}_n^{\mathrm{T}}, \tag{12.21}$$

where $[g]^{\mathrm{T}}$ denotes the transpose of the matrix $[g]$. If we denote the entries in the matrix $[g]$ by $g_{\alpha\beta}$ where $\alpha, \beta \in \mathscr{I}_n$, then we also have the relationship

$$g_{\beta\alpha} \mathscr{U}_n^+ = \mathscr{U}_n^+ e_\alpha^{-1} g e_\beta \mathscr{U}_n^+. \tag{12.22}$$

More details of this construction can be found in [71, 82, 86].

We have already encountered the construction of the spectral basis for $\mathbb{G}_{1,1}$, given in (4.7). For $n = 2$, we have

$$\begin{pmatrix} 1 \\ e_1 \\ e_2 \\ e_{21} \end{pmatrix} \mathscr{U}_2^+ \begin{pmatrix} 1 & e_1 & e_2 & e_{12} \end{pmatrix} = \begin{pmatrix} u_{++} & e_1 u_{-+} & e_2 u_{+-} & e_{12} u_{--} \\ e_1 u_{++} & u_{-+} & e_{12} u_{+-} & e_2 u_{--} \\ e_2 u_{++} & e_{21} u_{-+} & u_{+-} & -e_1 u_{--} \\ e_{21} u_{++} & e_2 u_{-+} & -e_1 u_{+-} & u_{--} \end{pmatrix}. \tag{12.23}$$

for the geometric algebra $\mathbb{G}_{2,2}$. The relationship (12.21) for an element $g \in \mathbb{G}_{2,2}$ and its matrix $[g]$ is

$$g = \mathbb{G}_2 \mathscr{U}_2^+ [g]^{\mathrm{T}} (\mathbb{G}_2^\dagger)^{\mathrm{T}} = \begin{pmatrix} 1 & e_1 & e_2 & e_{12} \end{pmatrix} u_1^+ u_2^+ [g]^{\mathrm{T}} \begin{pmatrix} 1 \\ e_1 \\ e_2 \\ e_{21} \end{pmatrix}. \tag{12.24}$$

## Exercises

All of the following problems are in the geometric algebra $\mathbb{G}_{2,2}$ for which (12.23) and (12.24) apply.

1. (a) Find the matrix $[e_k]$ of the unit vectors $e_k$ for $k = 1, 2, 3, 4$.
   (b) Show that the matrix $[1]_4$ that represents the identity element $g = 1$ is the $4 \times 4$ identity matrix.
   (c) Find the matrix representing the pseudoscalar element $\mathrm{i} = e_{1234}$.
   (d) More generally, show that $[g_1 + g_2] = [g_1] + [g_2]$ and $[g_1 g_2] = [g_1][g_2]$ for any elements $g_1, g_2 \in \mathbb{G}_{2,2}$.
2. Find the matrix $[e_1 + ie_3]$ which represents the element $e_1 + ie_3$.
3. Find the inverse $(5 + 3e_1)^{-1}$ of the element $5 + 3e_1$, and show that

$$[(5 + 3e_1)^{-1}] = [5 + 3e_1]^{-1}.$$

4. Find the matrix $[5+3e_1]$ of the element $5+3e_1 \in \mathbb{G}_{2,2}$.

   (a) What is the characteristic equation and eigenvalues for this matrix?
   (b) What is trace$[5+3e_1]$ and det$[5+3e_1]$?

5. Show that the $e_1$ conjugation, defined in (4.10), satisfies the properties

$$(g_1+g_2)^{e_1} = g_1^{e_1} + g_2^{e_1}, \text{ and } (g_1 g_2)^{e_1} = g_1^{e_1} g_2^{e_1}$$

   where $g_1, g_2 \in \mathbb{G}_{2,2}$. What is the most general element $g \in \mathbb{G}_{2,2}$ for which $g^{e_1} = g$?

## 12.3 The Twisted Product of Geometric Numbers

Let $\mathbb{G} = \mathbb{G}_{n,n}$ and $a, b \in \mathbb{G}^1$ be vectors. From the definition (12.1),

$$a \circ b = \frac{1}{2}(1+a+b-ab), \text{ and } b \circ a = \frac{1}{2}(1+a+b-ba).$$

Using (12.10) and the definition of the exterior product of vectors, we find that

$$a \circ b - b \circ a = \frac{1}{2}(ba-ab) = 2(b^- a^- - a^- b^-) = b \wedge a. \tag{12.25}$$

A similar calculation shows that for vectors $a, b, c \in \mathbb{G}^1$,

$$a \circ (b \wedge c) - (b \wedge c) \circ a = \frac{1}{2}[(b \wedge c)a - a(b \wedge c)] = (b \wedge c) \cdot a. \tag{12.26}$$

Note that if the vector $a$ is orthogonal to the plane of $b \wedge c$, then $a \cdot (b \wedge c) = 0$ so that $a \circ (b \wedge c) = (b \wedge c) \circ a$ in this case.

It is interesting to consider the meaning of the *symmetric part* of the twisted product. For vectors $a, b \in \mathbb{G}^1$, we find that

$$a \circ b + b \circ a = 1 + a + b - a \cdot b, \tag{12.27}$$

and for the vector $a$ and bivector $b \wedge c$, we find that

$$a \circ (b \wedge c) + (b \wedge c) \circ a = 1 + a + b \wedge c - a \wedge (b \wedge c). \tag{12.28}$$

If $a, b, c \in \mathbb{G}$ are geometric numbers with the property that $ac = -ca$ and $cb = bc$, then

$$c[a \circ b] = [(-a) \circ b]c. \tag{12.29}$$

If in addition $cb = -bc$, then we have

$$c[a \circ b] = [(-a) \circ (-b)]c. \tag{12.30}$$

These relationships are easily established directly from the definition (12.3) and can easily be extended to the twisted product of more arguments.

Let $a$ and $b$ be any geometric numbers with the property that $ab=-ba$. Then from the alternative definition of the twisted product (12.3), we find that $ab^+=b^-a$ and

$$a^+b^+=\frac{1}{2}(1+a)b^+=\frac{1}{2}(b^++b^-a)=\frac{1}{2}b\circ a. \tag{12.31}$$

Compare this with the general relationship (12.7) for arbitrary elements. Under the same condition that $ab=-ab$, we also have the similar relationships

$$a^-b^+=\frac{1}{2}b\circ(-a),\ \ a^+b^-=\frac{1}{2}(-b)\circ a,\ \text{ and }\ a^-b^-=\frac{1}{2}(-b)\circ(-a). \tag{12.32}$$

We also have

$$a\circ b=1-(-b)\circ(-a)\quad\Longleftrightarrow\quad a^+b^+=\frac{1}{2}-b^-a^- \tag{12.33}$$

which follows from (12.31), (12.32), and (12.7), and the simple steps

$$a\circ b=\frac{1}{2}b^+a^+=1-(1-2b^+a^+)=1-(-b)\circ(-a).$$

Identities (12.32) and (12.33) are striking because they allow us to reverse the order of the twisted symmetric product by changing the signs of its respective arguments.

For three geometric numbers $a,b,c$ with the properties that $a^2=1$ and $ab=-ba$, we find that

$$(a\circ b)(a\circ c)=(a^++a^-b)(a^++a^-c)=a^++a^-b=a\circ b. \tag{12.34}$$

If, instead, we have the condition that $a^2=1$ and $ac=ca$, then

$$(a\circ c)(a\circ b)=(a^++a^-c)(a^++a^-b)=a^++a^-cb=a\circ(cb). \tag{12.35}$$

We see in the above relationships that the properties of the twisted product $a\circ b$ heavily depend upon the commutativity of $a$ and $b$. Suppose now that we have two pairs of commutative arguments, $a_1,a_2$ and $b_1,b_2$. Whereas by assumption $a_1a_2=a_2a_1$ and $b_1b_2=b_2b_1$, we need some way of conveniently expressing the commutativity or anticommutativity relationships between them. To this end, we define the *entanglement table* $tangle(a_1,a_2;b_1,b_2)$.

**Definition 12.3.1.** If $a_1b_1=-b_1a_1, a_1b_2=b_2a_1, a_2b_1=-b_1a_2$ and $a_2b_2=-b_2a_2$, we say that the *entanglement table* $tangle(a_1,a_2;b_1,b_2)$ is

| | $b_1$ | $b_2$ |
|---|---|---|
| $a_1$ | $-$ | $+$ |
| $a_2$ | $-$ | $-$ |

and similarly for other possible commutativity relationships between $a_1,a_2$ and $b_1,b_2$.

We are interested in doing calculations for commutative pairs of arguments $a_1, a_2$ and $b_1, b_2$ which satisfy the following four entanglement tables:

*Table 1.*

| | $b_1$ | $b_2$ |
|---|---|---|
| $a_1$ | $-$ | $+$ |
| $a_2$ | $-$ | $-$ |

*Table 2.*

| | $b_1$ | $b_2$ |
|---|---|---|
| $a_1$ | $+$ | $-$ |
| $a_2$ | $-$ | $+$ |

*Table 3.*

| | $b_1$ | $b_2$ |
|---|---|---|
| $a_1$ | $+$ | $-$ |
| $a_2$ | $+$ | $+$ |

*Table 4.*

| | $b_1$ | $b_2$ |
|---|---|---|
| $a_1$ | $-$ | $-$ |
| $a_2$ | $-$ | $-$ |

Thus, suppose that we are given two pairs of commutative elements $a_1, a_2 \in \mathbb{G}$ and $b_1, b_2 \in \mathbb{G}$ that satisfy the entanglement Table 1. In addition, we assume that $a_i^2 = b_i^2 = 1$ for $i = 1,2$. Then, using (12.11) and the basic relationships found in Sect 12.2, we calculate

$$(a_1 \circ a_2)(b_1 \circ b_2)(a_1 \circ a_2) = [(a_1 \circ a_2)b_1(a_1 \circ a_2)] \circ [(a_1 \circ a_2)b_2(a_1 \circ a_2)]$$
$$= \big[(a_1 \circ a_2)[(-a_1) \circ (-a_2)]b_1\big] \circ \big[(a_1 \circ a_2)[a_1 \circ (-a_2)]b_2\big]$$
$$= (-a_1 a_2 b_1) \circ (a_1 b_2) = (a_1 b_2) \circ (-a_2 b_1 b_2).$$

A similar calculation shows that

$$(b_1 \circ b_2)(a_1 \circ a_2)(b_1 \circ b_2) = (a_1 b_2) \circ (-a_2 b_1 b_2).$$

It follows from the above calculations that for Table 1 entanglement and square one arguments,

$$(a_1 \circ a_2)(b_1 \circ b_2)(a_1 \circ a_2) = (b_1 \circ b_2)(a_1 \circ a_2)(b_1 \circ b_2). \tag{12.36}$$

This is the basic relationship between adjacent 2-cycles in the symmetric group. Indeed, the relationship (12.36) shows that the element $(a_1 \circ a_2)(b_1 \circ b_2)$ has order 3.

For Table 2 entanglement and square one arguments, we find that

$$(a_1 \circ a_2)(b_1 \circ b_2)(a_1 \circ a_2) = (a_1 b_1) \circ (a_2 b_2) = (b_1 \circ b_2)(a_1 \circ a_2)(b_1 \circ b_2), \tag{12.37}$$

so the element $(a_1 \circ a_2)(b_1 \circ b_2)$ with entanglement Table 2 will also have order 3. Calculations for Tables 3 and 4 entanglement with square one arguments proceed in exactly the same way and give for Table 3

$$(a_1 \circ a_2)(b_1 \circ b_2)(a_1 \circ a_2) = b_1 \circ (a_2 b_2) = (a_2 \circ b_1)(b_1 \circ b_2), \tag{12.38}$$

and for Table 4

$$(a_1 \circ a_2)(b_1 \circ b_2)(a_1 \circ a_2) = [(-a_1) \circ (-a_2)](b_1 \circ b_2). \tag{12.39}$$

From (12.38), we see directly that

$$[(a_1 \circ a_2)(b_1 \circ b_2)]^2 = a_2 \circ b_1. \tag{12.40}$$

For this case if $b_1 = -a_2$, then $(a_1 \circ a_2)(b_1 \circ b_2)$ will have order 2; otherwise it will have order 4. In the case of Table 4 entanglement, we have

$$[(a_1 \circ a_2)(b_1 \circ b_2)]^2 = (-a_1) \circ (-a_2), \tag{12.41}$$

so $(a_1 \circ a_2)(b_1 \circ b_2)$ will have order 4.

There is one very important property about higher-degree twisted products that we will need. Suppose that $a_1 \circ a_2$ and $b_1 \circ b_2$ are pairs of commutative square one arguments and let $c$ be an element with square one and such that $ca_i = a_i c$ and $cb_i = b_i c$ for $i = 1, 2$. Then the order of the element $(a_1 \circ a_2 \circ c)(b_1 \circ b_2 \circ c)$ will be the same as the element $(a_1 \circ a_2)(b_1 \circ b_2)$; in symbols,

$$|(a_1 \circ a_2 \circ c)(b_1 \circ b_2 \circ c)| = |(a_1 \circ a_2)(b_1 \circ b_2)|, \tag{12.42}$$

as easily follows from the steps

$$\begin{aligned}(a_1 \circ a_2 \circ c)(b_1 \circ b_2 \circ c) &= [c^+ + (a_1 \circ a_2)c^-][c^+ + c^-(b_1 \circ b_2)] \\ &= c^+ + c^-(a_1 \circ a_2)(b_1 \circ b_2),\end{aligned}$$

and the fact that $c^+$ and $c^-$ are mutually annihilating idempotents.

Although we are far from exhausting the many interesting identities that can be worked out, we now have a sufficient arsenal to continue our study of the symmetric groups in the geometric algebra $\mathbb{G}_{n,n}$.

## 12.4 Symmetric Groups in Geometric Algebras

Our objective is to construct geometric numbers which represent the elements of the various subgroups $\mathscr{S}$ of the symmetric group $S_n$ as a *group algebra* $\mathscr{R}$ in the geometric algebra $\mathbb{G} = \mathbb{G}_{n,n}$. The construction naturally leads to the interpretation of $\mathscr{R}$ as a *real regular* $\mathscr{R}$*-module*, [48, p. 55–56]. To accomplish this, we construct elements which represent the successive entangled 2-cycles

$$(12), (13), \dots, (1n), \tag{12.43}$$

which are the generators of the symmetric group $S_n$. Although there are many such representations, we consider here only elements in $\mathbb{G}_{n,n}$ of the form

$$g = a_1 \circ a_2 \circ \cdots \circ a_n, \tag{12.44}$$

where $a_1,\dots,a_n \in \mathbb{G}_{n,n}$ are *mutually commuting blades* such that $a_i^2 = 1$ for $1 \le i \le n$. Since by (12.15), $g^2 = 1$, the element $g$ has order 2 and is a good candidate for representing a 2-cycle, where $1 \in \mathbb{G}$ represents the group identity element. Once we have a way of representing all 2-cycles in $S_n$ as elements of $\mathbb{G}_{n,n}$, we can also construct the elements of any of its subgroups.

Let $\mathscr{R} = \{g_1,\dots,g_k\}$ be a representation of a finite group all of whose elements are of the form (12.44) or a product of such elements in $\mathbb{G}_{n,n}$. Let $g \in \mathscr{S}$. By the *group character* $\chi(g)$ (with respect to its embedding in $\mathbb{G}_{n,n}$), we mean

$$\chi(g) = 2^n < g >_0 \tag{12.45}$$

where $< g >_0$ is the scalar part of the geometric number $g \in \mathbb{G}_{n,n}$. We have defined the group character function $\chi$ on $\mathbb{G}_{n,n}$ to agree with its matrix equivalent, since $\mathbb{G}_{n,n}$ is real isomorphic to the real matrix algebra $\mathscr{M}_{\mathbb{R}}(2^n \times 2^n)$. In so doing, we can directly incorporate all of the results from the rich theory of group representations that has evolved over the years since April, 1896, when F. G. Frobenius wrote to R. Dedekind about his new ideas regarding finite groups [50]. Indeed, we fully expect that a new richer representation theory can be constructed based upon geometric algebra.

### 12.4.1 The Symmetric Group $S_4$ in $\mathbb{G}_{4,4}$

We represent the symmetric group $S_1$ by $\mathscr{R}_1 = \{1\} = \mathbb{G}_{0,0} \subset \mathbb{G}_{4,4}$, where $\mathbb{G}_{0,0} = \mathbb{R}$ is the degenerate geometric algebra of real numbers. The symmetric group $S_2$ is represented by

$$\mathscr{R}_2 = \text{gen}\{e_1\}, \tag{12.46}$$

where $e_1 \in \mathbb{G}_{1,1} \subset \mathbb{G}_{4,4}$. Using (12.20), we find that

$$1 \simeq \begin{pmatrix} 1 & 0 \\ 0 & 1 \end{pmatrix}, \quad e_1 \simeq \begin{pmatrix} 0 & 1 \\ 1 & 0 \end{pmatrix}. \tag{12.47}$$

Let us now carefully construct the representation $\mathscr{R}_4$ of $S_4$ in $\mathbb{G}_{2,2}$ from (12.46) given above, since it is the $k = 1$ step in our recursive construction of the representation $\mathscr{R}_{2^{k+1}}$ of $S_{2^{k+1}}$ in $\mathbb{G}_{k+1,k+1}$, given the representation $\mathscr{R}_{2^k}$ of $S_{2^k}$ in $\mathbb{G}_{k,k}$. We use the conventions for the basis elements of $\mathbb{G}_{4,4}$ that we previously established in (12.18). All calculations can be checked using Lounesto's Clical [54].

The construction of the representation of $S_{2^2}$ from the representation of $S_2$ breaks down into three steps.

**Step 1.** We map the generator(s) $e_1$ of $S_2$ into the larger geometric algebra $\mathbb{G}_{2,2}$ by taking the twisted product of $e_1$ with the element $e_{62}$ to produce $e_1 \circ e_{62}$, which will be the new element representing the 2-cycle $(12)$ in $S_4$.

**Step 2.** The element representing the 2-cycle $(13)$ in $S_4$ is created from $e_1 \circ e_{62}$ by introducing the orthogonal transformation represented by $\pi_2$ which *permutes* the basis vectors $e_1, e_2, e_5, e_6$ to give

$$e_1 \to e_2 \to e_1, \quad e_5 \to e_6 \to e_5. \tag{12.48}$$

Thus,

$$e_2 \circ e_{51} = \pi_2(e_1 \circ e_{62}) = [\pi_2(e_1)] \circ [\pi_2(e_{61})].$$

We can construct one more entangled 2-cycle in the geometric algebra $\mathbb{G}_{2,2}$ to get a regular representation $\mathscr{R}_4$ of $S_4$, which constitutes the last step 3 of the construction.

**Step 3.** We multiply the vector $e_1$ in (12.46) on the left by $e_6$ to get the bivector $e_{61}$ which anticommutes with both $e_1$ and $e_{62}$ used in step 1 of the construction. Using this bivector, we then construct

$$e_{25} = -\pi_2(e_{61}) = -\pi_2(e_6)\pi_2(e_1) = -e_5 e_2.$$

The element that represents $(14)$ in $\mathbb{G}_{2,2}$ is given by

$$e_{61} \circ [-\pi_2(e_{61})] = e_{61} \circ e_{25}. \tag{12.49}$$

We now have a complete set of generators for our representation $\mathscr{R}_4$ of $S_4$ in $\mathbb{G}_{2,2}$.

$$\begin{pmatrix} e_1 \circ e_{62} \simeq (12) \\ e_2 \circ e_{51} \simeq (13) \\ e_{61} \circ e_{25} \simeq (14) \end{pmatrix} \tag{12.50}$$

Indeed, the matrix representations $[e_1 \circ e_{62}], [e_2 \circ e_{51}], [e_{61} \circ e_{25}]$ of these elements given by (12.23) are exactly the matrices obtained by interchanging the first and second rows, the first and third rows, and the first and fourth rows of the $4 \times 4$ identity matrix, respectively. By appealing to the matrix-geometric algebra isomorphism, we have found the corresponding regular representation of $S_4$ in the geometric algebra $\mathbb{G}_{2,2}$. The reason for the minus sign that appears in step 3 will become evident later.

Let us consider in detail the structure of the representation $\mathscr{R}_4$ of $S_4$ that we have found, since, in turn, it will determine the structure of the faithful representations of all higher order symmetric groups that we consider. In order to facilitate and clarify our discussion, we write

$$r_1 = e_1 \circ e_{62} = a_1 \circ a_2, \; r_2 = e_2 \circ e_{51} = b_1 \circ b_2, \; r_3 = e_{61} \circ e_{25} = c_1 \circ c_2.$$

The elements $a_1, a_2, b_1, b_2, c_1, c_2 \in \mathbb{G}_{4,4}$ are not all independent. It can be easily checked that $c_1 = -a_1 a_2 b_1$ and $c_2 = a_1 b_1 b_2$. For convenience, we produce here the entanglement table $tangle(a_1, a_2, b_1, b_2, c_1, c_2)$ for the elements defining $r_1, r_2, r_3$:

| | $a_1$ | $a_2$ | $b_1$ | $b_2$ | $c_1$ | $c_2$ |
|---|---|---|---|---|---|---|
| $a_1$ | + | + | − | − | − | + |
| $a_2$ | + | + | − | + | − | − |
| $b_1$ | − | − | + | + | + | − |
| $b_2$ | − | + | + | + | − | − |
| $c_1$ | − | − | + | − | + | + |
| $c_2$ | + | − | − | − | + | + |

We give here a complete table of the geometric numbers representing the 24 elements of $S_4$. As generators, we choose the elements which represent $(12), (13)$, and $(14)$, respectively.

$$\begin{pmatrix} S_4 & \simeq & g \in \mathbb{G}_{2,2} \subset \mathbb{G}_{4,4} & \chi(g) \\ (1)(2)(3) & \simeq & 1 = 1 \circ 1 & 4 \\ (12) & \simeq & r_1 = a_1 \circ a_2 & 2 \\ (13) & \simeq & r_2 = b_1 \circ b_2 & 2 \\ (14) & \simeq & r_3 = c_1 \circ c_2 = (-a_2 b_2) \circ (-a_1 b_1 b_2) & 2 \\ (23) & \simeq & r_1 r_2 r_1 = c_1 \circ (-c_2) = (a_2 b_2) \circ (-a_1 a_2 b_1) & 2 \\ (24) & \simeq & r_1 r_3 r_1 = b_1 \circ (-b_2) & 2 \\ (34) & \simeq & r_2 r_3 r_2 = (-a_1) \circ (-a_2) & 2 \\ (12)(34) & \simeq & r_1 r_2 r_3 r_2 = -a_1 a_2 & 0 \\ (13)(24) & \simeq & r_2 r_1 r_3 r_1 = b_1 & 0 \\ (14)(23) & \simeq & r_3 r_1 r_2 r_1 = c_1 = -a_1 a_2 b_1 & 0 \\ (123) & \simeq & r_2 r_1 = (b_1 \circ b_2)(a_1 \circ a_2) & 1 \\ (132) & \simeq & r_1 r_2 = (a_1 \circ a_2)(b_1 \circ b_2) & 1 \\ (124) & \simeq & r_3 r_1 = (c_1 \circ c_2)(a_1 \circ a_2) & 1 \\ (142) & \simeq & r_1 r_3 = (a_1 \circ a_2)(c_1 \circ c_2) & 1 \\ (134) & \simeq & r_3 r_2 = (c_1 \circ c_2)(b_1 \circ b_2) & 1 \\ (143) & \simeq & r_2 r_3 = (b_1 \circ b_2)(c_1 \circ c_2) & 1 \\ (234) & \simeq & r_1 r_3 r_2 r_1 = -a_1 a_2 r_3 r_1 & 1 \\ (243) & \simeq & r_1 r_2 r_3 r_1 = -a_1 a_2 r_2 r_1 & 1 \\ (1234) & \simeq & r_3 r_2 r_1 = c_1 r_2 = -a_1 a_2 b_1 r_2 & 0 \\ (1243) & \simeq & r_2 r_3 r_1 = b_1 r_3 & 0 \\ (1324) & \simeq & r_3 r_1 r_2 = c_1 r_1 = -a_1 a_2 b_1 r_1 & 0 \\ (1342) & \simeq & r_1 r_3 r_2 = -a_1 a_2 r_3 & 0 \\ (1423) & \simeq & r_2 r_1 r_3 = b_1 r_1 & 0 \\ (1432) & \simeq & r_1 r_2 r_3 = -a_1 a_2 r_2 & 0 \end{pmatrix}. \tag{12.51}$$

Let us summarize in what we have accomplished so far.

$$(1) \quad \text{and} \quad \begin{pmatrix} 1 & 62 \\ 2 & 51 \\ 61 & 25 \end{pmatrix} \tag{12.52}$$

In (12.52), we have abbreviated the elements $e_i$ and $e_{ij}$ by using only their indices which serve to completely define the elements in the twisted symmetric product. Thus, $S_2$ is generated by the element $e_1 \in \mathbb{G}_{1,1}$, and $S_4$ is generated by the elements

$$\{e_1 \circ e_{62}, e_2 \circ e_{51}, e_{61} \circ e_{25}\} \subset \mathbb{G}_{2,2},$$

under the geometric product. The matrices of our elements correspond exactly to the regular $4 \times 4$ matrix representation.

## Exercises

1. The elements in table (12.52) generate the symmetric group $S_4$. Find the matrix representation of each of these generators using (12.24).
2. Write down a complete multiplication table for the elements of the symmetric group $S_3$ as a subgroup of $S_4$. Find the $4 \times 4$ matrix representation for each of six elements in $S_3$.

### *12.4.2 The Geometric Algebra* $\mathbb{G}_{4,4}$

In order to see more clearly what is involved in this recursive construction, let us continue the construction for $k = 2$ and $k = 3$ and find the representations $\mathscr{R}_8$ of $S_8$ and $\mathscr{R}_{16}$ of $S_{16}$ in $\mathbb{G}_{4,4}$. As a subalgebra of $\mathbb{G}_{4,4}$, the geometric algebra $\mathbb{G}_{3,3}$ is generated by the basis vectors $e_1, e_2, e_3, e_5, e_6, e_7$.

We begin by applying **step 1** to the table (12.52) with the element $e_{73}$ to produce the new table

$$\begin{pmatrix} 1 & 62 & 73 \\ 2 & 51 & 73 \\ 61 & 25 & 73 \end{pmatrix}. \tag{12.53}$$

For **step 2**, we define the orthogonal transformation $\pi_3 = \pi_{k+1}$ on the basis elements of $\mathbb{G}_{3,3}$ $\{e_1, e_2, e_3, e_5, e_6, e_7\}$ by

$$e_1 \to e_2 \to e_3 \to e_1, \quad e_5 \to e_6 \to e_7 \to e_5. \tag{12.54}$$

Applying this transformation successively twice to (12.53) gives us the additional two copies

$$\begin{pmatrix} 2 & 73 & 51 \\ 3 & 62 & 51 \\ 72 & 36 & 51 \end{pmatrix} \simeq \begin{pmatrix} 3 & 51 & 62 \\ 1 & 73 & 62 \\ 53 & 17 & 62 \end{pmatrix}, \tag{12.55}$$

which represent different but equivalent representations of $S_4$ in $\mathbb{G}_{3,3}$ considered as a subalgebra of $\mathbb{G}_{4,4}$. Next, we paste together the distinct elements found in these 3 tables to get

$$\begin{pmatrix} 1 & 62 & 73 \\ 2 & 51 & 73 \\ 61 & 25 & 73 \\ 3 & 62 & 51 \\ 72 & 36 & 51 \\ 35 & 71 & 62 \end{pmatrix}. \tag{12.56}$$

Notice that we have changed the sign of the first two elements of the row $\begin{pmatrix}53 & 17 & 62\end{pmatrix}$. The reason for this will be explained later. This completes step 2.

In (12.56), we have constructed 6 of the 7 elements that represent the 2-cycles of $S_8$ in $\mathbb{G}_{3,3}$. We use **step 3** to construct the last element by taking the first two elements $\begin{pmatrix}61 & 25\end{pmatrix}$ from the last row of (12.52) and place a 3 in front of them to get $\begin{pmatrix}361 & 325\end{pmatrix}$. Step 3 is completed by applying $\pi_3$ to $e_{361}$ to get $\pi_3(e_{361}) = e_{172}$. Notice in this case we do not change the sign of this element as we did in (12.49). In general, the sign for $\pi_{k+1}$ in this step is $(-1)^k$. The element representing the 7th 2-cycle $(18)$ is thus found to be

$$e_{361} \circ e_{325} \circ e_{172} \simeq (18). \tag{12.57}$$

Thus, the table

$$\begin{pmatrix} 1 & 62 & 73 \\ 2 & 51 & 73 \\ 61 & 25 & 73 \\ 3 & 62 & 51 \\ 72 & 36 & 51 \\ 35 & 71 & 62 \\ 361 & 325 & 172 \end{pmatrix} \tag{12.58}$$

gives the regular representation $\mathscr{R}_8$ of $S_8$ in $\mathbb{G}_{3,3} \subset \mathbb{G}_{4,4}$.

We continue the construction for $k = 3$ of the representation $\mathscr{R}_{16}$ of $S_{16}$ in $\mathbb{G}_{4,4}$.

**Step 1.** We extend the table (12.58) by adding a 4th column consisting of the element $(84)$, which represents $e_{84}$ which has square 1 and commutes with all of the other elements in the table, getting

$$\begin{pmatrix} 1 & 62 & 73 & 84 \\ 2 & 51 & 73 & 84 \\ 61 & 25 & 73 & 84 \\ 3 & 62 & 51 & 84 \\ 72 & 36 & 51 & 84 \\ 35 & 71 & 62 & 84 \\ 361 & 325 & 172 & 84 \end{pmatrix} \tag{12.59}$$

which is a faithful but no longer regular representation of $S_8$ because it is no longer maximal in $\mathbb{G}_{4,4}$.

**Step 2.** Next, by applying the orthogonal transformation $\pi_{k+1}=\pi_4=(1234)(5678)$, or

$$e_1 \to e_2 \to e_3 \to e_4 \to e_1, \quad \text{and} \quad e_5 \to e_6 \to e_7 \to e_8 \to e_5 \tag{12.60}$$

repeatedly to the table (12.59), we find the additional representations of the following *distinct* 2-cycles of $S_8$ in $\mathbb{G}_{4,4}$:

$$\begin{pmatrix} 4 & 73 & 62 & 51 \\ 46 & 82 & 73 & 51 \\ 83 & 47 & 62 & 51 \\ 472 & 436 & 283 & 51 \\ 45 & 81 & 73 & 62 \\ 183 & 147 & 354 & 62 \\ 542 & 182 & 614 & 73 \end{pmatrix}. \tag{12.61}$$

We now paste together the above representations of the 14 distinct 2-cycles of $S_8$ in (12.59) and (12.61), getting the table

$$\begin{pmatrix} 1 & 62 & 73 & 84 \\ 2 & 51 & 73 & 84 \\ 61 & 25 & 73 & 84 \\ 3 & 62 & 51 & 84 \\ 72 & 36 & 51 & 84 \\ 35 & 71 & 62 & 84 \\ 361 & 325 & 172 & 84 \\ 4 & 73 & 62 & 51 \\ 46 & 82 & 73 & 51 \\ 83 & 47 & 62 & 51 \\ 472 & 436 & 283 & 51 \\ 45 & 81 & 73 & 62 \\ 183 & 147 & 354 & 62 \\ 542 & 182 & 614 & 73 \end{pmatrix}. \tag{12.62}$$

This completes step 2 of the construction.

**Step 3.** The last additional row to the table (12.62) is constructed by taking the first *three* numbers $\big(361\ 325\ 172\big)$ of the last row in (12.59) and placing the number 8 in front of them to get $\big(8361\ 8325\ 8172\big)$. The fourth entry is obtained by applying $-\pi_4$ to the first number 8361 to get $-\pi_4(8361) = (4572)$. We thus get the $15^{th}$ and last row $\big(8361\ 8325\ 8172\ 4572\big)$. This completes the last step of the construction of the elements in $\mathscr{R}_{16}$ representing $S_{16}$ in $\mathbb{G}_{4,4}$.

We give a summary of our results in the combined tables

$$(1) \quad \begin{pmatrix} 1 & 62 \\ 2 & 51 \\ 61 & 25 \end{pmatrix} \quad \begin{pmatrix} 1 & 62 & 73 \\ 2 & 51 & 73 \\ 61 & 25 & 73 \\ 3 & 62 & 51 \\ 72 & 36 & 51 \\ 35 & 71 & 62 \\ 361 & 325 & 172 \end{pmatrix} \quad \begin{pmatrix} 1 & 62 & 73 & 84 \\ 2 & 51 & 73 & 84 \\ 61 & 25 & 73 & 84 \\ 3 & 62 & 51 & 84 \\ 72 & 36 & 51 & 84 \\ 35 & 71 & 62 & 84 \\ 361 & 325 & 172 & 84 \\ 4 & 73 & 62 & 51 \\ 46 & 82 & 73 & 51 \\ 83 & 47 & 62 & 51 \\ 472 & 436 & 283 & 51 \\ 45 & 81 & 73 & 62 \\ 183 & 147 & 354 & 62 \\ 542 & 182 & 614 & 73 \\ 8361 & 8325 & 8172 & 4572 \end{pmatrix} \qquad (12.63)$$

## *12.4.3 The General Construction in $\mathbb{G}_{n,n}$*

For $k = 0$ and $\mathscr{R}_1 = \{1\}$, we are given $\mathscr{R}_2 = \{1, e_1\}$, the representation of $S_2$ in $\mathbb{G}_{1,1}$ considered as a subalgebra of $\mathbb{G}_{n,n}$. We give here the general rule for the recursive construction of the representation $\mathscr{R}_{2^{k+1}}$ of $S_{2^{k+1}}$ in $\mathbb{G}_{k+1,k+1}$, given that the representation of $\mathscr{R}_{2^k}$ of $S_{2^k}$ in $\mathbb{G}_{k,k}$ for $1 \le k < n$ has been constructed, where both $\mathbb{G}_{k,k}$ and $\mathbb{G}_{k+1,k+1}$ are considered as subalgebras of $\mathbb{G}_{n,n}$.

**Step 1.** Each $r_\alpha \in \mathscr{R}_{2^k}$ is mapped into $r_\alpha \circ e_{n+k+1,k+1} \in \mathscr{R}_{2^{k+1}}$ .

**Step 2.** The orthogonal transformation represented by $\pi_{k+1} = (1, \ldots, k+1)(n+1, \ldots, n+k+1)$ is applied repeatedly to each of the elements of $\mathscr{R}_{2^k} \circ e_{n+k+1,k+1}$ until no more distinct new elements are formed. Taken together, this will form a table of $2^{k+1} - 1$ distinct elements of $\mathscr{R}_{2^{k+1}}$.

**Step 3.** If $k$ is *odd*, the last element of $\mathscr{R}_{2^{k+1}}$ is formed by placing the vector $e_{n+k+1}$ in front of the first $k$ elements of the last entry in $\mathscr{R}_{2^k}$. Thus, if this last entry in $\mathscr{R}_{2^k}$ is $e_{\gamma_1} \circ \cdots \circ e_{\gamma_k}$, this operation gives

$$e_{n+k+1,\gamma_1} \circ \cdots \circ e_{n+k+1,\gamma_k}.$$

The last element in $\mathscr{R}_{2^{k+1}}$ is then formed by applying $-\pi_{k+1}$ to $e_{n+k+1,\gamma_1}$ to get $-\pi_{k+1}(e_{n+k+1,\gamma_1})$. The last element is then given by

$$e_{n+k+1,\gamma_1} \circ \cdots \circ e_{n+k+1,\gamma_k} \circ [-\pi_{k+1}(e_{n+k+1,\gamma_1})].$$

If $k$ is *even*, this step is modified by placing the vector $e_{k+1}$ in front of the first $k$ elements, instead of $e_{n+k+1}$, and applying the orthogonal transformation $\pi_{k+1}$ *without* the minus sign.

## *12.5 The Heart of the Matter

Let us formalize the regular representations of the symmetric groups that we have found in the previous section. We start by reorganizing the rows of (12.63) and identifying the primitive idempotent $\mathscr{U}_4^+ = e_{15}^+ e_{26}^+ e_{37}^+ e_{48}^+$ and the identity element $r_0 = 1 \circ \cdots \circ 1 = 1$ of our representation $\mathscr{R}_{16}$ of the group $S_{16}$. Let us also explicitly define the elements $r_\alpha$ of our representation by writing

$$\begin{pmatrix} r_0 \\ r_1 \\ r_2 \\ r_3 \\ r_4 \\ r_{21} \\ r_{31} \\ r_{41} \\ r_{32} \\ r_{42} \\ r_{43} \\ r_{321} \\ r_{421} \\ r_{431} \\ r_{432} \\ r_{4321} \end{pmatrix} \mathscr{U}_4^+ = \begin{pmatrix} 1 & 1 & 1 & 1 \\ 1 & 62 & 73 & 84 \\ 2 & 51 & 73 & 84 \\ 3 & 62 & 51 & 84 \\ 4 & 73 & 62 & 51 \\ 61 & 25 & 73 & 84 \\ 35 & 71 & 62 & 84 \\ 45 & 81 & 73 & 62 \\ 72 & 36 & 51 & 84 \\ 46 & 82 & 73 & 51 \\ 83 & 47 & 62 & 51 \\ 361 & 325 & 172 & 84 \\ 542 & 182 & 614 & 73 \\ 183 & 147 & 354 & 62 \\ 472 & 436 & 283 & 51 \\ 8361 & 8325 & 8172 & 4572 \end{pmatrix} \mathscr{U}_4^+ = \begin{pmatrix} e_0 = 1 \\ e_1 \\ e_2 \\ e_3 \\ e_4 \\ e_{21} \\ e_{31} \\ e_{41} \\ e_{32} \\ e_{42} \\ e_{43} \\ e_{321} \\ e_{421} \\ e_{431} \\ e_{432} \\ e_{4321} \end{pmatrix} \mathscr{U}_4^+. \qquad (12.64)$$

The matrix identities in (12.64) reflect everything we have accomplished in this paper and also what is left to be done in order to *prove* that our construction produces a regular representation as claimed. First, we are identifying and renaming the geometric numbers which represent the entangled 2-cycles of $S_{16}$. Thus, for example, the first and second equalities of the ninth row of our matrix equation express that

$$r_{32} = e_{72} \circ e_{36} \circ e_{51} \circ e_{84}, \quad \text{and} \quad r_{32}\mathscr{U}_4^+ = e_{32}\mathscr{U}_4^+, \qquad (12.65)$$

respectively.

Multiplying both sides of the last expression in (12.65) on the left by $r_{32}$ gives

$$\mathscr{U}_4^+ = r_{32} e_{32} \mathscr{U}_4^+$$

since $r_{32}^2 = 1$, showing that $r_{32}$ flips the first (the row containing $e_0$) and the ninth rows of (12.64). This is more elegantly stated for the representation $\mathscr{R}_{2^n}$ of $S_{2^n}$ in $\mathbb{G}_{n,n}$ by saying that

$$r_\alpha \mathscr{U}_n^+ = e_\alpha \mathscr{U}_n^+ \iff \mathscr{U}_n^+ = r_\alpha e_\alpha \mathscr{U}_n^+ \tag{12.66}$$

for all indexes $\alpha \in \mathscr{I}_n$ which define the basis elements $e_\alpha$ of the geometric algebra $\mathbb{G}_n$. For a regular representation, we also need to show that such an operation leaves all the other rows of (12.64) fixed. In other words, we must show in general that for all nonzero indices $\alpha, \beta \in \mathscr{I}_n$, and $\beta \neq \alpha$, that

$$r_\alpha e_\beta \mathscr{U}_n^+ = e_\beta \mathscr{U}_n^+ \iff r_\alpha \mathscr{U}_n^{sgn(\beta)} = r_\alpha e_\beta \mathscr{U}_n^+ e_\beta^{-1} = \mathscr{U}_n^{sgn(\beta)}. \tag{12.67}$$

The usual regular matrix representation of $\mathscr{S}_n$ follows immediately from (12.66) and (12.67), giving for each $\alpha \in \mathscr{I}_n$,

$$r_\alpha = r_\alpha \sum_{\beta \in \mathscr{I}_N} \mathscr{U}_n^{sn(\beta)} = r_\alpha \mathscr{U}_n^+ + \sum_{\beta \in \mathscr{I}_n, \beta \neq 0} e_\alpha^{-\delta_{\alpha,\beta}} \mathscr{U}_n^{sn(\beta)},$$

where $\delta_{\alpha,\beta} = 0$ for $\alpha \neq \beta$ and $\delta_{\alpha,\beta} = 1$ when $\alpha = \beta$.

Examining the right side of (12.64), we see that each 2-cycle representative $r_\alpha$ can be classified into *k-vector types*, where $k \geq 1$. For example, we say that the $r_\alpha$ in the sixth through eleventh rows of (12.64) are of bivector or 2-vector type. Thus, each of the representative elements $r_\alpha$ of $S_{16}$, as given in (12.64), is of $k$-vector type for some $1 \leq k \leq 4$. Since in the steps 1, 2, and 3 of our construction, each $k$-vector is related by the orthogonal transformation $\pi_4$ to each of the other representatives of the same $k$-vector type, we need only prove our assertions for one representative element of each $k$-vector type. Note also that the orthogonal transformation $\pi_4$ leaves the primitive idempotent $\mathscr{U}_4^+$ invariant, that is, $\pi_4(\mathscr{U}_4^+) = \mathscr{U}_4^+$. This means that for all $\alpha \in \mathscr{I}_4$,

$$\pi_4(r_\alpha \mathscr{U}_4^+) = \pi_4(r_\alpha)\mathscr{U}_4^+ = \pi_4(e_\alpha)\mathscr{U}_4^+. \tag{12.68}$$

The property (12.68) justifies our change of the sign in the last row of the table (12.56). Indeed, if we adjusted the indices of standard basis (12.18) to reflect the order obtained in the construction process, no change of sign would be required.

More generally, let $\mathscr{R}_{2^n}$ be the representation obtained for $S_{2^n}$ by the three steps outlined in Sect. 4.1 and let $\pi_n$ denote the orthogonal transformation defined in the second step. Then $\pi_n(\mathscr{U}_n^+) = \mathscr{U}_n^+$, since it only *permutes* the idempotent factors of $\mathscr{U}_n^+$ among themselves. In addition, because $\pi_n$ is an orthogonal transformation, it permutes $k$-vector representatives among themselves so that

$$\pi_n(r_\alpha \mathscr{U}_n^+) = \pi_n(r_\alpha)\mathscr{U}_n^+ = \pi_n(e_\alpha)\mathscr{U}_n^+. \tag{12.69}$$

We are now ready to state and prove the only theorem of this chapter.

**Theorem 12.5.1.** *For each n, there exists a regular representation $\mathscr{R}_{2^n}$ of $S_{2^n}$ in the geometric algebra $\mathbb{G}_{n,n}$ which acts on the set of elements $\mathscr{E}_n^{\dagger}\mathscr{U}_n^+$. The representations $\mathscr{R}_{2^n}$ are constructed recursively using the steps 1, 2, and 3, as explained in Sect. 4.3 for $\mathbb{G}_{n,n}$ and satisfy the relationships (12.66), (12.67), and (12.69).*

*Proof.* Suppose that $n > 1$ and that the representation $\mathscr{R}_{2^n}$ has been constructed according to three steps outlined in Sect. 4.1. By construction,

$$\mathscr{U}_n^+ = \prod_{i=1}^{n} e_{i,n\,|\,i}^+ .$$

We will show, by induction on $k$ for $1 \le k \le n-1$, that the first representative of each $k+1$-vector type satisfies the relationships (12.66) and (12.67). Whereas the recursive construction begins with $k=1$, we must first show that the element $r_1 = e_1 = e_1^+ + e_1^- e_1$, which occurs when $k=0$ and which has vector type 1, satisfies (12.66) and (12.67).

Thus, for $k=0$, we are given that $r_1 = e_1 \circ e_{n+2,2} \circ \cdots \circ e_{2n,n}$. By the repeated application of the absorption properties (12.16) and (12.17), we find that

$$r_1 \mathscr{U}_n^+ = (e_1 \circ e_{n+2,2} \circ \cdots \circ e_{2n,n}) \prod_{i=1}^{n} e_{i,n+i}^+ = e_1 \prod_{i=1}^{n} e_{i,n+i}^+ = e_1 \mathscr{U}_n^+,$$

so that the relationship (12.66) is satisfied for all of the 1-vector types $r_1, \ldots, r_n$. To show that (12.67) is satisfied, let $\beta \in \mathscr{I}_n$, $\beta \neq 1$, and $\beta \neq 0$. It follows that for some $i$, $1 < i \le n$, we have $\mathscr{U}_n^{sn(i)} = e_{n+i,i}$ and $\mathscr{U}_n^{sn(i)} \mathscr{U}_n^{\beta} = \mathscr{U}_n^{\beta}$. It then easily follows from the absorption property (12.17) that

$$r_1 \mathscr{U}_n^{\beta} = r_1 \mathscr{U}_n^{sn(i)} \mathscr{U}_n^{\beta} = \mathscr{U}_n^{\beta}. \tag{12.70}$$

For $k=1$, the recursive construction produces the element

$$r_{21} = (e_{n+2,1} \circ e_{2,n+1}) \circ e_{n+3,3} \circ \cdots \circ e_{2n,n},$$

which is the first element with 2-vector type. To show (12.66), we calculate with the repeated help of the absorption property (12.17), as well as (12.33),

$$\begin{aligned} r_{21} \mathscr{U}_n^+ &= [(e_{n+2,1} \circ e_{2,n+1}) \circ e_{n+3,3} \circ \cdots \circ e_{2n,n}] \mathscr{U}_n^+ \\ &= (e_{n+2,1} \circ e_{2,n+1}) \mathscr{U}_n^+ = (e_{2,n+1}^+ + e_{2,n+1}^- e_{n+2,1}) \mathscr{U}_n^+ \\ &= (e_{21}^+ + e_{211(n+1)}^- e_{1,n+1}^- e_{2,2,n+2,1}) \mathscr{U}_n^+ = (e_{21}^+ + e_{21}^+ e_{21}) \mathscr{U}_n^+ \\ &= (e_{21}^+ - e_{21}^-) \mathscr{U}_n^+ = e_{21} \mathscr{U}_n^+ . \end{aligned}$$

A similar calculation to (12.70) shows that (12.67) is true, i.e., that

$$r_{21}\mathscr{U}_n^{sn(\beta)} = \mathscr{U}_n^{sn(\beta)}$$

for any index $\beta \in \mathscr{I}_n$ such that $\beta \neq 0$ and $\beta \neq 21$.

Suppose now that the $j$-vector representative $r_{j...1}$ has the properties (12.66) and (12.67) for all positive integers $j$ less than or equal to $k$ where $1 \leq k < n$. Then we must prove that the same is true for the representative element $r_{k+1,k,...,1}$. Recall that $r_{k+1,k,...,1}$ is constructed in step 3 from

$$r_{k,...,1} = e_{\gamma_1} \circ \cdots \circ e_{\gamma_k}$$

where $e_{\gamma_i}$, for $1 \leq i \leq k$, are commuting square one $k$-vectors in $\mathbb{G}_{n,n}$ obtained in the construction process. Then, by the induction assumption, we know that

$$r_{k,...,1}\mathscr{U}_n^+ = e_{k,...,1}\mathscr{U}_n^+.$$

According to step 3 given in Sect. 4.3,

$$r_{k+1,k,...,1} = e_{n+k+1,\gamma_1} \circ \cdots \circ e_{n+k+1,\gamma_k} \circ e_{\gamma_{k+1}},$$

where $e_{\gamma_{k+1}} = (-1)^k \pi_{k+1}(e_{n+k+1,\gamma_1})$.

We now verify (12.66) for $r_{k+1,...,1}$, getting

$$r_{k+1,...,1}\mathscr{U}_n^+ = (e_{n+k+1,\gamma_1} \circ \cdots \circ e_{n+k+1,\gamma_k} \circ e_{\gamma_{k+1}}) \prod_{i=1}^n e_{i,n+i}^+$$

$$= [e_{\gamma_{k+1}}^+ + e_{\gamma_{k+1}}^- (e_{n+k+1,\gamma_1} \circ \cdots \circ e_{n+k+1,\gamma_k})] \prod_{i=1}^n e_{i,n+i}^+. \tag{12.71}$$

There are basically four cases of (12.71) that need to be verified, depending upon whether $k$ is *even* or *odd* and whether $e_{1...k+1}^2 = \pm 1$.

*Case i)* $k$ is odd and $e_{1...k+1}^2 = -1$. In this case, with the help of (12.16) and (12.33), (12.71) simplifies to

$$= [e_{(k+1),...,1}^+ + e_{(k+1),...,1}^+ e_{(k+1),...,1}] \prod_{i=1}^n e_{i,n+i}^+ = e_{(k+1),...,1} \prod_{i=1}^n e_{i,n+i}^+,$$

since $e_{(k+1),...,1}^+ e_{(k+1),...,1} = -e_{(k+1),...,1}^-$.

*Case ii)* $k$ is even and $e_{1...k+1}^2 = -1$. In this case, with the help of (12.16) and (12.33), (12.71) simplifies in exactly the same way as case (i).

*Case iii)* $k$ is odd and $e_{1\ldots k+1}^2 = 1$. In this case, with the help of (12.16) and (12.33), (12.71) simplifies to

$$= [e_{(k+1),\ldots,1}^{+} + e_{(k+1),\ldots,1}^{-} e_{(k+1),\ldots,1}] \prod_{i=1}^{n} e_{i,n+i}^{+} = e_{(k+1),\ldots,1} \prod_{i=1}^{n} e_{i,n+i}^{+},$$

since $e_{(k+1),\ldots,1}^{-} e_{(k+1),\ldots,1} = -e_{(k+1),\ldots,1}^{-}$.

*Case iv)* $k$ is even and $e_{1\ldots k+1}^2 = 1$. In this case, with the help of (12.16) and (12.33), (12.71) simplifies in exactly the same way as case (iii). □

# Chapter 13
# Calculus on *m*-Surfaces

*But mathematics is the sister, as well as the servant, of the arts and is touched with the same madness and genius.*
—Harold Marston Morse

We apply all of the machinery of linear algebra developed in the preceding chapters to the study of calculus on an $m$-surface. The concept of the boundary of a surface is defined, and the classical integration theorems in any dimension follow from a single *fundamental theorem of calculus*.[1]

## 13.1 Rectangular Patches on a Surface

Let $R = \times_{i=1}^{m}[a^i, b^i]$ be a closed $m$-rectangle in $\mathbb{R}^m$ where $m \geq 1$. By the *interior* $R^*$ of $R$, we mean the *open rectangle* $R^* = \times_{i=1}^{m}(a^i, b^i)$. We denote the points $\mathbf{s} \in R$ by the *position vectors* $\mathbf{s} = \sum_i s^i \mathbf{e}_i$ where $(\mathbf{e})_{(m)}$ is the standard orthonormal basis of $\mathbb{R}^m$ and $a^i \leq s^i \leq b^i$. The *boundary* of the rectangle is the $(m-1)$*-chain*

$$\beta(R) = \oplus_{i=1}^{m}(R_-^i \oplus R_+^i) \tag{13.1}$$

where the faces $R_\pm^i$ are defined by $R_+^i = R(s^i = b^i)$ and $R_-^i = R(s^i = a^i)$, respectively. For $m = 1$, the boundary $\beta(R)$ consists of the end points of the interval $[a^1, b^1]$,

$$\beta(R) = R_-^1 \oplus R_+^1 = \{a^1\} \oplus \{b^1\}.$$

[1] This chapter is based upon an article written by the author and Omar León Sanchez [88].

G. Sobczyk, *New Foundations in Mathematics: The Geometric Concept of Number*, DOI 10.1007/978-0-8176-8385-6_13,

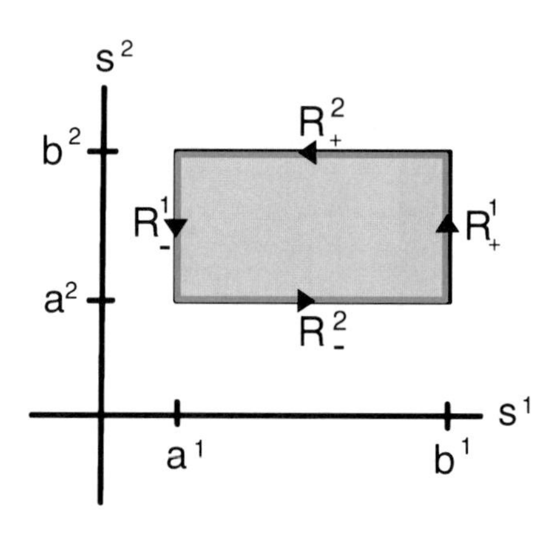


**Fig. 13.1** The 2-rectangle $R = [a^1, b^1] \times [a^2, b^2]$ is shown together with the oriented 1-chain $R^2_- \oplus R^1_+ \oplus R^2_+ \oplus R^1_-$ that is its boundary

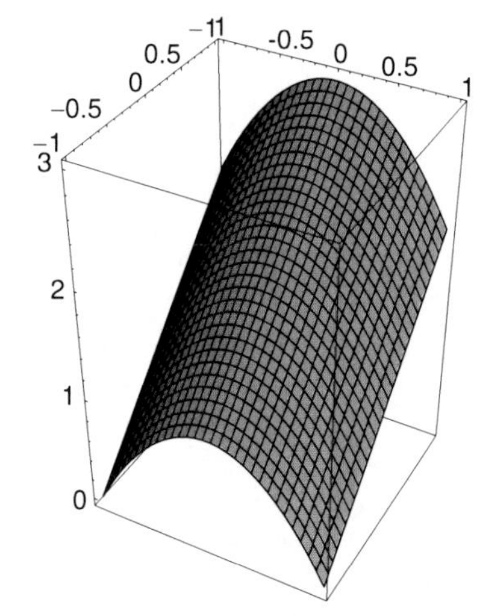


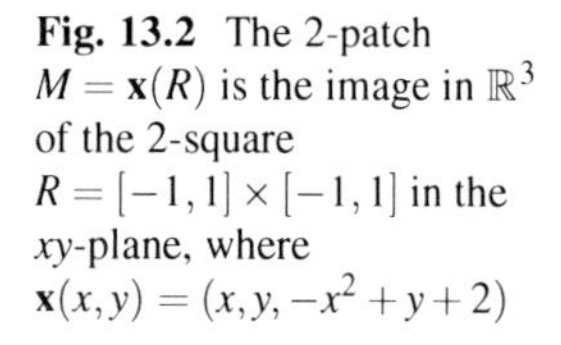

**Fig. 13.2** The 2-patch $M = \mathbf{x}(R)$ is the image in $\mathbb{R}^3$ of the 2-square $R = [-1,1] \times [-1,1]$ in the $xy$-plane, where $\mathbf{x}(x,y) = (x, y, -x^2 + y + 2)$

In Fig. 13.1, a 2-rectangle is shown together with the 1-chain that defines its boundary. As we shall see, the concept of a chain is needed to properly define the concept of integration over $m$-dimensional surfaces and their boundaries.

Let $\mathscr{M}$ be an $m$-surface in $\mathbb{R}^n$ which is of class $C^p$, where $p \geq 2$ and $m \leq n$.[2] We first study the properties of an $m$-dimensional $C^p$ *regular rectangular patch* $M$ on the surface $\mathscr{M}$. For example, a curve is the image of a 1-rectangle or interval. The graph of the image in $\mathbb{R}^3$ of the 2-rectangle $R = [-1,1] \times [-1,1]$ where

$$\mathbf{x}(x,y) = (x, y, -x^2 + y + 2),$$

is shown in Fig. 13.2.

The tangent vectors $(\mathbf{x})_{(m)}$ at an interior point $\mathbf{x} \in M$ are defined by

$$\mathbf{x}_i = \frac{\partial \mathbf{x}}{\partial s^i}, \qquad i = 1, \ldots, m, \tag{13.2}$$

[2]The concept of an $m$-surface will be more rigorously defined in Chap. 15.

and make up a local basis of the *tangent vector space* $\mathscr{T}_{\mathbf{x}}^1$ at the point $\mathbf{x} \in M$. A $C^p$ rectangular $m$-patch $M$ is said to be *regular* if

$$\mathbf{x}_{(m)} = \wedge(\mathbf{x})_{(m)} = \mathbf{x}_1 \wedge \cdots \wedge \mathbf{x}_m \neq 0 \tag{13.3}$$

at all interior points $\mathbf{x} \in M$. We now give the formal definition of a $C^p$ regular rectangular patch in $\mathscr{M}$.

**Definition 13.1.1.** A set $M = \mathbf{x}(R) \subset \mathbb{R}^n$ is a *regular rectangular m-patch* of class $C^p$, where $p \geq 1$, if $\mathbf{x}\colon R \to M$ is bicontinuous at all points in $M$ and of class $C^p$ at all interior points $\mathbf{x} \in M$.

The *boundary* of $M$ is the $(m-1)$-chain

$$\beta(M) = \mathbf{x}(\beta(R)) = \oplus_{i=1}^m (M_-^i \oplus M_+^i) \tag{13.4}$$

where the faces $M_\pm^i = \mathbf{x}(R_\pm^i)$, respectively, for $i = 1, \ldots, m$.

Piecing together regular rectangular patches on an $m$-surface is technically more difficult than piecing together the rectangular strips that are used to define the *Riemann integral* for the area under a curve in elementary calculus, but the idea is the same. The problem is matching up the points on the boundary in such a way that the local properties, such as orientation and smoothness, are preserved when moving from one patch to another.

For a regular patch, the tangent $m$-vector $I_{\mathbf{x}} = \mathbf{x}_{(m)}$, given in (13.3), is continuous and does not vanish at any interior point $\mathbf{x} \in M$. The *reciprocal basis* $(\mathbf{x})^{(m)}$ to $(\mathbf{x})_{(m)}$ at the point $\mathbf{x} \in M$ is defined by

$$\mathbf{x}^i = (-1)^{i-1} \mathbf{x}_{(m)}^i \cdot \mathbf{x}_{(m)}^{-1} \tag{13.5}$$

where

$$\mathbf{x}_{(m)}^i = \mathbf{x}_1 \wedge \cdots \wedge \mathbf{x}_{i-1} \wedge \mathbf{x}_{i+1} \wedge \cdots \wedge \mathbf{x}_m,$$

and satisfies the relations $\mathbf{x}^i \cdot \mathbf{x}_j = \delta^i{}_j$. For example, for $i = 1, 2$ we have

$$\mathbf{x}^1 = (\mathbf{x}_2 \wedge \cdots \wedge \mathbf{x}_m) \cdot \mathbf{x}_{(m)}^{-1}, \text{ and } \mathbf{x}^2 = -(\mathbf{x}_1 \wedge \mathbf{x}_3 \wedge \cdots \wedge \mathbf{x}_m) \cdot \mathbf{x}_{(m)}^{-1}.$$

Just as we did in Chaps. 7 and 10, we can turn the problem of finding a reciprocal basis into a corresponding problem of finding the inverse of a matrix. Thus, let $(\mathbf{x})_{(m)} = \begin{pmatrix}\mathbf{x}_1 & \mathbf{x}_2 & \ldots & \mathbf{x}_m\end{pmatrix}$ be the row matrix made up of the vectors $\mathbf{x}_i$. We wish to find a column matrix of vectors

$$(\mathbf{x})^{(m)} = \begin{pmatrix} \mathbf{x}^1 \\ \mathbf{x}^2 \\ \cdot \\ \cdot \\ \mathbf{x}^m \end{pmatrix}$$

with the property that $(\mathbf{x})^{(m)} \cdot (\mathbf{x})_{(m)} = [1]_m$ where $[1]_m$ is the $m \times m$ identity matrix. The $m \times m$ symmetric Gram matrix $[g] = [g_{ij}]$ of the inner products $g_{ij} = \mathbf{x}_i \cdot \mathbf{x}_j$ of the tangent vectors $\mathbf{x}_i$ is given by

$$[g] = (\mathbf{x})^{\mathrm{T}}_{(m)} \cdot (\mathbf{x})_{(m)} = \begin{pmatrix} \mathbf{x}_1 \\ \mathbf{x}_2 \\ \cdot \\ \cdot \\ \mathbf{x}_m \end{pmatrix} \cdot \begin{pmatrix} \mathbf{x}_1 & \mathbf{x}_2 & \cdots & \mathbf{x}_m \end{pmatrix} = \begin{pmatrix} \mathbf{x}_1 \cdot \mathbf{x}_1 & \dots & \mathbf{x}_1 \cdot \mathbf{x}_m \\ \cdot & & \cdot \\ \cdot & & \cdot \\ \cdot & & \cdot \\ \mathbf{x}_m \cdot \mathbf{x}_1 & \dots & \mathbf{x}_m \cdot \mathbf{x}_m \end{pmatrix}. \quad (13.6)$$

Just as we saw in (7.12) and (10.18), the reciprocal basis is found by multiplying both sides of (13.6) on the left by $[g]^{-1}$, and simplifying, to give $(\mathbf{x})^{(m)} = [g]^{-1} (\mathbf{x})^{\mathrm{T}}_{(m)}$.

For example, consider the matrix of column vectors in $\mathbb{R}^3$

$$(\mathbf{x})_{(2)} = \begin{pmatrix} 1 & 2 \\ 2 & 1 \\ 1 & 1 \end{pmatrix}.$$

The Gram matrix of inner products is

$$[g] = (\mathbf{x})^{\mathrm{T}}_{(2)} \cdot (\mathbf{x})_{(2)} = \begin{pmatrix} 6 & 5 \\ 5 & 6 \end{pmatrix},$$

and has the inverse $[g]^{-1} = \frac{1}{11}\begin{pmatrix} 6 & -5 \\ -5 & 6 \end{pmatrix}$. It follows that the reciprocal basis of row vectors is given by

$$\begin{pmatrix} \mathbf{x}^1 \\ \mathbf{x}_2 \end{pmatrix} = [g]^{-1} \begin{pmatrix} \mathbf{x}_1^{\mathrm{T}} \\ \mathbf{x}_2^{\mathrm{T}} \end{pmatrix} = \frac{1}{11} \begin{pmatrix} -4 & 7 & 1 \\ 7 & -4 & 1 \end{pmatrix}.$$

## Exercises

1. Consider the image $M = \mathbf{x}(R)$ of the 2-rectangle $R = [-1,1] \times [-1,1]$ in the $xy$-plane, where

$$\mathbf{x}(x,y) = (x, y, -x^2 + y + 2),$$

see Fig. 13.2.

(a) Find $\mathbf{x}_1 = \frac{\partial \mathbf{x}}{\partial x}$ and $\mathbf{x}_2 = \frac{\partial \mathbf{x}}{\partial y}$
(b) Find the reciprocal basis $\{\mathbf{x}^1, \mathbf{x}^2\}$ at the point $(x,y) \in R$.

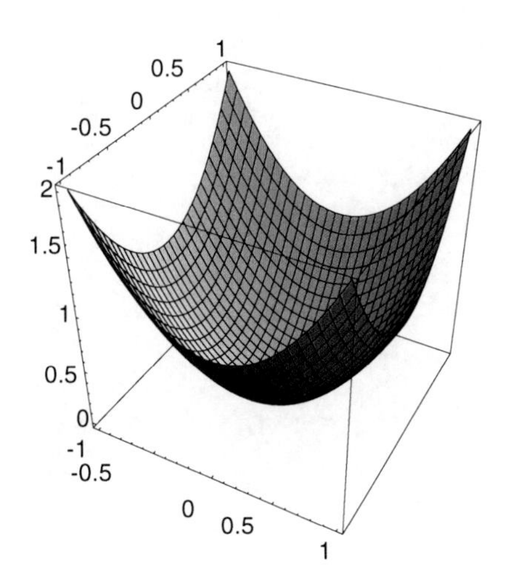


**Fig. 13.3** The 2-patch $M = \mathbf{x}(R)$ is the image of the 2-square $R = [-1,1] \times [-1,1]$ in $\mathbb{R}^3$, where $\mathbf{x}(x,y) = (x,y,x^2+y^2)$

(c) Find the 2-chain $\beta(M)$ which is the boundary to $M$.

(d) Find the tangent vector to the boundary of surface $M$ at the points

$$\mathbf{x} = \mathbf{x}\left(\frac{1}{2},-1\right), \mathbf{x}\left(1,\frac{1}{2}\right), \mathbf{x}\left(\frac{1}{2},1\right), \mathbf{x}\left(-1,\frac{1}{2}\right).$$

2. Let $R = [-1,1] \times [-1,1]$ be a 2-square in $\mathbb{R}^2$ and let $M = \{\mathbf{x}(R)\} \subset \mathbb{R}^3$ where $\mathbf{x}(x,y) = (x,y,x^2+y^2)$. See Fig. 13.3.

   (a) Find $\mathbf{x}_1 = \frac{\partial \mathbf{x}}{\partial x}$ and $\mathbf{x}_2 = \frac{\partial \mathbf{x}}{\partial y}$

   (b) Find the reciprocal basis $\{\mathbf{x}^1, \mathbf{x}^2\}$ at the point $\mathbf{x}(x,y) \in M$.

$$Ans{:}\ \mathbf{x}^1 = \left(\frac{4y^2+1}{4x^2+4y^2+1}, -\frac{4xy}{4x^2+4y^2+1}, \frac{2x}{4x^2+4y^2+1}\right),$$

$$\mathbf{x}^2 = \left(-\frac{4xy}{4x^2+4y^2+1}, \frac{4x^2+1}{4x^2+4y^2+1}, \frac{2y}{4x^2+4y^2+1}\right).$$

   (c) Find the tangent vector to the boundary of the surface $M$ at the points $\mathbf{x} = \mathbf{x}(1,0), \mathbf{x}(0,1), \mathbf{x}(-1,0), \mathbf{x}(0,-1)$.

   (d) Find the 2-chain $\beta(M)$ which is the boundary to $M$.

3. Let $R = [-1,1] \times [-1,1]$ be a 2-square in $\mathbb{R}^2$ and let $M = \{\mathbf{x}(R)\} \subset \mathbb{R}^3$ where $\mathbf{x}(x,y) = (x,y,x^2-y^2+1)$. See Fig. 13.4.

   (a) Find $\mathbf{x}_1 = \frac{\partial \mathbf{x}}{\partial x}$ and $\mathbf{x}_2 = \frac{\partial \mathbf{x}}{\partial y}$

   (b) Find the reciprocal basis $\{\mathbf{x}^1, \mathbf{x}^2\}$ at the point $\mathbf{x}(x,y) \in M$.

$$ans{:}\ \mathbf{x}^1 = \left\{\frac{4y^2+1}{4x^2+4y^2+1}, \frac{4xy}{4x^2+4y^2+1}, \frac{2x}{4x^2+4y^2+1}\right\},$$

$$\mathbf{x}^2 = \left\{\frac{4xy}{4x^2+4y^2+1}, \frac{4x^2+1}{4x^2+4y^2+1}, -\frac{2y}{4x^2+4y^2+1}\right\}.$$

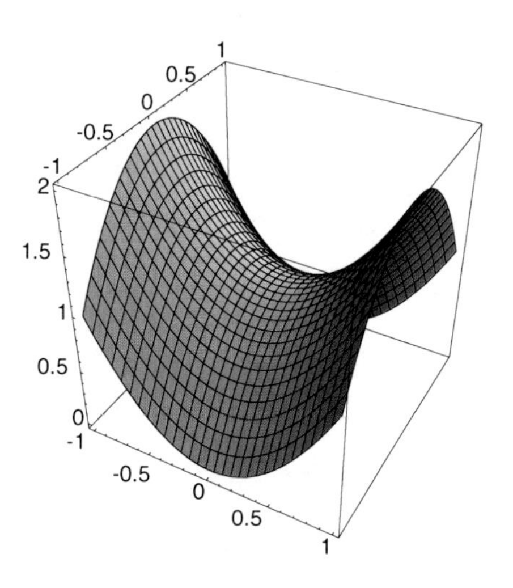


**Fig. 13.4** The 2-patch $M = \mathbf{x}(R)$ is the image of the 2-square $R = [-1,1] \times [-1,1]$ in $\mathbb{R}^3$, where $\mathbf{x}(x,y) = (x, y, x^2 - y^2 + 1)$

(c) Find the tangent vector to the boundary of the surface $M$ at the points $\mathbf{x} = \mathbf{x}(1,0), \mathbf{x}(0,1), \mathbf{x}(-1,0), \mathbf{x}(0,-1)$.

(d) Find the 2-chain $\beta(M)$ which is the boundary to $M$.

4. Let $\mathbf{A} = \begin{pmatrix} \mathbf{a}_1 \\ \mathbf{a}_2 \\ \cdot \\ \cdot \\ \mathbf{a}_m \end{pmatrix}$, $\mathbf{B} = \begin{pmatrix} \mathbf{b}_1 & \mathbf{b}_2 & \cdots & \mathbf{b}_m \end{pmatrix}$, and $C$ be a real $m \times k$ matrix. Show that $(\mathbf{A} \cdot \mathbf{B})C = \mathbf{A} \cdot (\mathbf{B}C)$ and $C(\mathbf{A} \cdot \mathbf{B}) = (C\mathbf{A}) \cdot \mathbf{B}$, where the product $\mathbf{A} \cdot \mathbf{B}$ is defined as in (13.6).

5. Let us consider cylindrical coordinates defined in $\mathbb{R}^3$ by

$$\mathbf{x} = \mathbf{x}(r, \theta, z) = \mathbf{e}_1 r \cos\theta + \mathbf{e}_2 r \sin\theta + \mathbf{e}_3 z.$$

(a) Find

$$\mathbf{x}_r = \frac{\partial \mathbf{x}}{\partial r}, \quad \mathbf{x}_\theta = \frac{\partial \mathbf{x}}{\partial \theta}, \quad \mathbf{x}_z = \frac{\partial \mathbf{x}}{\partial z}.$$

(b) Find the corresponding reciprocal vectors $\mathbf{x}^r, \mathbf{x}^\theta, \mathbf{x}^z$.

(c) The vector derivative (3.46) on $\mathbb{R}^3$ can be defined by

$$\partial_{\mathbf{x}} = \mathbf{x}^r \frac{\partial}{\partial r} + \mathbf{x}^\theta \frac{\partial}{\partial \theta} + \mathbf{x}^z \frac{\partial}{\partial z}.$$

Show that $\partial_{\mathbf{x}} \wedge \partial_{\mathbf{x}} = 0$, just as we found in (3.54).

6. Let $\mathbf{x}(r, \theta, \phi) = (r\cos\theta \sin\theta, r\sin\theta \sin\phi, r\cos\phi)$ where $r > 0$, $0 \le \theta \le 2\pi$, and $0 \le \phi \le \pi$ be *spherical coordinates* in $\mathbb{R}^3$.

(a) Find

$$\mathbf{x}_r = \frac{\partial \mathbf{x}}{\partial r}, \quad \mathbf{x}_\theta = \frac{\partial \mathbf{x}}{\partial \theta}, \quad \mathbf{x}_\phi = \frac{\partial \mathbf{x}}{\partial \phi}.$$

(b) Find the corresponding reciprocal vectors $\mathbf{x}^r, \mathbf{x}^\theta, \mathbf{x}^\phi$.

(c) The vector derivative (3.46) on $\mathbb{R}^3$ can be defined by

$$\partial_{\mathbf{x}} = \mathbf{x}^r \frac{\partial}{\partial r} + \mathbf{x}^\theta \frac{\partial}{\partial \theta} + \mathbf{x}^\phi \frac{\partial}{\partial \phi}.$$

Show that $\partial_{\mathbf{x}} \wedge \partial_{\mathbf{x}} = 0$, just as we found in (3.54).

## 13.2 The Vector Derivative and the Directed Integral

Let $\mathscr{M}$ be a regular $m$-surface in $\mathbb{R}^n$ of class $C^p$, where $p \geq 2$, containing the $m$-patch $M$. Just as we defined the vector derivative on $\mathbb{R}^n$ in (3.46), we now define the *two-sided vector derivative* at an interior point $\mathbf{x}$ in the $m$-patch $M \subset \mathscr{M}$. By "two sided" we mean that the vector derivative acts both to the left and to the right. This is convenient because the geometric product is not, in general, commutative. Let $f, g \colon M \to \mathbb{G}_n$ be $\mathbb{G}_n$-valued functions of class $C^p$ on the interior of the $m$-patch $M \subset \mathscr{M}$, where $p \geq 2$.

**Definition 13.2.1.** The *two-sided vector derivative* $\dot{g}(\mathbf{x})\dot{\partial}_{\mathbf{x}}\dot{f}(\mathbf{x})$ of $f(\mathbf{x})$ and $g(\mathbf{x})$ on $M = \mathbf{x}(R)$ at the point $\mathbf{x} \in M$ is defined indirectly in terms of the partial derivatives on the interior of the $m$-rectangle $R$ by

$$\dot{g}(\mathbf{x})\dot{\partial}_{\mathbf{x}}\dot{f}(\mathbf{x}) = \sum_{i=1}^{m} (\mathbf{x}_i \cdot \dot{\partial}_{\mathbf{x}})(\dot{g}(\mathbf{x})\mathbf{x}^i \dot{f}(\mathbf{x})) = \sum_{i=1}^{m} \frac{\dot{\partial}}{\partial s^i}(\dot{g}(\mathbf{x}(\mathbf{s}))\mathbf{x}^i \dot{f}(\mathbf{x}(\mathbf{s}))), \tag{13.7}$$

where the derivatives act only on the dotted arguments.

In the above definition, we are using the *chain rule*,

$$\partial_{\mathbf{x}} = \sum_{i=1}^{m} \mathbf{x}^i \frac{\partial \mathbf{x}}{\partial s^i} \cdot \partial_{\mathbf{x}} = \sum_{i=1}^{m} \mathbf{x}^i \frac{\partial}{\partial s^i},$$

but all the formulas for the vector derivative (3.46) in $\mathbb{R}^m$ given in Chap. 3, except those depending upon the integrability condition (3.54), remain valid for the vector derivative on an $m$-surface. The integrability condition (3.54) for an $m$-surface $\mathscr{M} \subset \mathbb{R}^n$ must be replaced by the weaker condition that

$$(\mathbf{a} \wedge \mathbf{b}) \cdot (\partial_{\mathbf{x}} \wedge \partial_{\mathbf{x}}) = 0 \tag{13.8}$$

for all *tangent vector fields* $\mathbf{a} = \mathbf{a}(\mathbf{x}), \mathbf{b} = \mathbf{b}(\mathbf{x})$ in $\mathscr{T}_{\mathbf{x}}$ at all interior points $\mathbf{x} \in \mathscr{M}$.

We can express the reciprocal vectors $\mathbf{x}^i$ in terms of the vector derivative. We have $\mathbf{x}^i = \partial_{\mathbf{x}} s^i$ for $i = 1, \ldots, m$. The *Leibniz product rule* for partial derivatives gives

$$\dot{g}(\mathbf{x})\dot{\partial}_{\mathbf{x}}\dot{f}(\mathbf{x}) = \dot{g}(\mathbf{x})\dot{\partial}_{\mathbf{x}}f(\mathbf{x}) + g(\mathbf{x})\dot{\partial}_{\mathbf{x}}\dot{f}(\mathbf{x}).$$

For $m \geq 2$, the oriented tangent $(m-1)$-vector at a given point $\mathbf{x}$ on a given face $M_{\pm}^i$ of the boundary of $M$ is defined to be

$$\mathbf{x}_{(m-1)}^i = \pm\mathbf{x}_{(m)}\mathbf{x}^i. \tag{13.9}$$

The vector $\mathbf{x}^i$ is reciprocal to $\mathbf{x}_i$, and the $\pm$ sign is chosen on a given face in such a way that $\mathbf{n} = \pm\mathbf{x}^i$ defines the direction of the *outward normal* to the face at a given point $\mathbf{x}$ on the boundary. In the case when $m = 1$, the oriented tangent 0-vector on each face of $M_{\pm}^1$ is taken to be $-1$ on $M_{-}^1 = \{\mathbf{x}(a^1)\}$ and $+1$ on $M_{+}^1 = \{\mathbf{x}(b^1)\}$, respectively.

The points of an $m$-rectangle $R$, when assigned an *orientation*, make up an *m-chain*. An $m$-chain $R$ has the *directed content*

$$\mathscr{D}(R) = \prod_{i=1}^{m}(b^i - a^i)\mathbf{e}_{(m)}, \tag{13.10}$$

which is the magnitude of the $m$-rectangle $R$ times the unit $m$-vector

$$\mathbf{e}_{(m)} = \frac{\mathrm{d}\mathbf{s}_{(m)}}{\mathrm{d}s^{(m)}} = \frac{\partial\mathbf{s}}{\partial s^1} \wedge \cdots \wedge \frac{\partial\mathbf{s}}{\partial s^m},$$

giving it its *direction*.

The $m$-rectangular patch $M$, the image of the $m$-rectangle $R$, is also an $m$-chain inheriting its orientation from the $m$-chain $R$. We now define the concept of the *directed integral* over the $m$-chain $M$. Let $f, g\colon M \to \mathbb{G}_n$ be continuous $\mathbb{G}_n$-valued functions on the $m$-rectangular patch $M \subset \mathscr{M}$.

**Definition 13.2.2.** The directed integral over $M$, for $m \geq 1$, is defined by

$$\int_M g(\mathbf{x})\mathrm{d}\mathbf{x}_{(m)}f(\mathbf{x}) = \int_R g(\mathbf{x}(\mathbf{s}))\mathbf{x}_{(m)}f(\mathbf{x}(\mathbf{s}))\mathrm{d}s^{(m)}, \tag{13.11}$$

where $\mathbf{s} = \sum_{i=1}^m s^i\mathbf{e}_i$, $\mathrm{d}s^{(m)} = \mathrm{d}s^1\cdots\mathrm{d}s^m$ and

$$\mathbf{x}_{(m)} = \frac{\mathrm{d}\mathbf{x}_{(m)}}{\mathrm{d}s^{(m)}} = \frac{\partial\mathbf{x}}{\partial s^1} \wedge \cdots \wedge \frac{\partial\mathbf{x}}{\partial s^m}.$$

We use the directed integral to define the directed content of the $m$-chain $M$, for $m \geq 1$. We have

$$\mathscr{D}(M) = \int_M \mathrm{d}\mathbf{x}_{(m)}. \tag{13.12}$$

A directed integral over a surface is shown in Fig. 13.5.

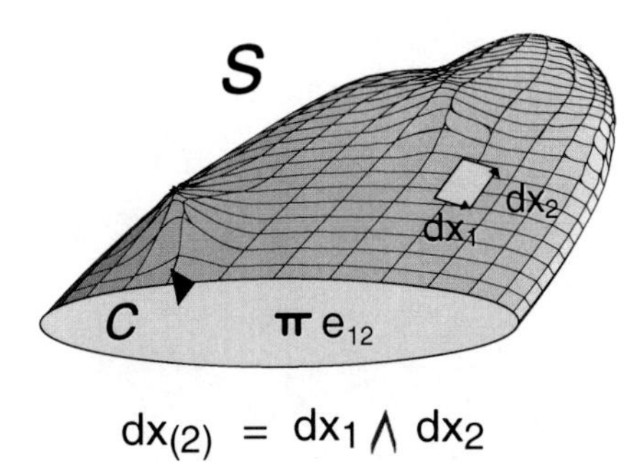


**Fig. 13.5** Choosing $f(\mathbf{x}) = g(\mathbf{x}) = 1$, when the element of directed area $\mathrm{d}\mathbf{x}_{(2)} = \mathrm{d}\mathbf{x}_1 \wedge \mathrm{d}\mathbf{x}_2$ is integrated over the smooth-oriented surface $S$ bounded by the oriented unit circle $C$ in the $xy$-plane, we get $\int_S \mathrm{d}\mathbf{x}_{(2)} = \pi \mathbf{e}_{12}$

We also need to define the *directed integral* over the boundary $\beta(M)$ of the $m$-chain $M$.

**Definition 13.2.3.** The directed integral over the boundary, for $m \geq 2$, is

$$\int_{\beta(M)} g(\mathbf{x})\mathrm{d}\mathbf{x}_{(m-1)} f(\mathbf{x}) = \sum_{i=1}^{m} \int_{M_+^i \oplus M_-^i} g(\mathbf{x})\mathrm{d}\mathbf{x}_{(m-1)} f(\mathbf{x})$$
$$= \sum_{i=1}^{m} \int_{R_+^i \oplus R_i^-} g(\mathbf{x}(\mathbf{s}))\mathbf{x}_{(m-1)} f(\mathbf{x}(\mathbf{s}))\mathrm{d}s^{(m-1)} \tag{13.13}$$

where $\mathbf{s} = \sum_{i=1}^{m} s^i \mathbf{e}_i$ and $\mathrm{d}s^{(m-1)} = \frac{\mathrm{d}s^{(m)}}{\mathrm{d}s^i} = \mathrm{d}s^1 \cdots \mathrm{d}s^{i-1}\mathrm{d}s^{i+1} \cdots \mathrm{d}s^m$ in $R_\pm^i$. For $m = 1$, the directed integral over the boundary is given by

$$\int_{\beta(M)} g(\mathbf{x})\mathrm{d}\mathbf{x}_{(0)} f(\mathbf{x}) = \int_{M_+^1 \oplus M_-^1} g(\mathbf{x})\mathrm{d}\mathbf{x}_{(0)} f(\mathbf{x})$$
$$= g(\mathbf{x}(b^1))f(\mathbf{x}(b^1)) - g(\mathbf{x}(a^1))f(\mathbf{x}(a^1)). \tag{13.14}$$

Our objective now is to prove the *fundamental theorem of calculus* relating the integral over an $m$-chain $M$ to the integral over its boundary $\beta(M)$. One direct consequence of this important theorem will be that

$$\mathscr{D}(\beta(M)) = 0, \tag{13.15}$$

where $\mathscr{D}(M)$ was defined in (13.12). In addition, all of the classical integration theorems such as Stoke's theorem, Green's theorem, and Gauss' divergence theorem will follow as special cases.

We will need the following lemma.

**Lemma 13.2.4.** *Let $M$ be a rectangular $m$-patch of class $C^2$ in $\mathbb{R}^n$. Then*

$$\sum_{i=1}^{j} \frac{\partial}{\partial s^i} \mathbf{x}_{(j)} \cdot \mathbf{x}^i = 0,$$

*for $j = 1, \ldots, m$. In the case when $j = m$, the "$\cdot$" can be removed.*

*Proof.* For $j = 1$, clearly $\frac{\partial}{\partial s^1}\mathbf{x}_1 \cdot \mathbf{x}^1 = 0$. We now inductively assume that the lemma is true for all $j < m$ and calculate

$$\sum_{i=1}^{m} \frac{\partial}{\partial s^i}\mathbf{x}_{(m)} \cdot \mathbf{x}^i = \sum_{i=1}^{m} \frac{\partial}{\partial s^i}[(\mathbf{x}_{(m-1)} \wedge \mathbf{x}_m) \cdot \mathbf{x}^i]$$

$$= \frac{\partial}{\partial s^m}\mathbf{x}_{(m-1)} - \sum_{i=1}^{m-1} \frac{\partial}{\partial s^i}[(\mathbf{x}_{(m-1)} \cdot \mathbf{x}^i) \wedge \mathbf{x}_m]$$

$$= \mathbf{x}_{(m-1)m} - \sum_{i=1}^{m-1} (\mathbf{x}_{(m-1)} \cdot \mathbf{x}^i) \wedge \mathbf{x}_{mi} = 0.$$

In the last step, we have used the fact that partial derivatives commute so that

$$\mathbf{x}_{mi} = \frac{\partial \mathbf{x}_m}{\partial s^i} = \frac{\partial \mathbf{x}_i}{\partial s^m} = \mathbf{x}_{im}.$$

□

We now have everything necessary to prove the important fundamental theorem of calculus. This theorem, as mentioned earlier, will give all of the classical integration theorems such as *Green's theorem* for the plane, *Stokes' theorem* for a surface in 3-dimensional space, *Gauss' divergence theorem*, and higher dimensional analogues.

**Fundamental Theorem of Calculus 13.2.5.** *Let $M$ be a rectangular m-patch of class $C^2$ and $f, g\colon M \to \mathbb{G}_n$ of class $C^1$. Then*

$$\int_M g(\mathbf{x})\mathrm{d}\mathbf{x}_{(m)}\partial_{\mathbf{x}} f(\mathbf{x}) = \int_{\beta(M)} g(\mathbf{x})\mathrm{d}\mathbf{x}_{(m-1)} f(\mathbf{x}) \tag{13.16}$$

*Proof.*

$$\int_{\beta(M)} g\mathrm{d}\mathbf{x}_{(m-1)} f$$

$$= \sum_{i=1}^{m} \int_{M_+^i \oplus M_-^i} g\mathrm{d}\mathbf{x}_{(m-1)} f = \sum_{i=1}^{m} \int_{R_+^i \oplus R_-^i} g\mathbf{x}_{(m-1)} f \mathrm{d}s^{(m-1)}$$

$$= \sum_{i=1}^{m} \int_{R_+^i \oplus R_-^i} \left[\int_{a_i}^{b_i} \frac{\partial}{\partial s^i}(g\mathbf{x}_{(m)}\mathbf{x}^i f)\mathrm{d}s^i\right] \frac{\mathrm{d}s^{(m)}}{\mathrm{d}s^i}$$

$$= \sum_{i=1}^{m} \int_R \left[\frac{\dot\partial}{\partial s^i}(\dot g\mathbf{x}_{(m)}\mathbf{x}^i \dot f)\right] \mathrm{d}s^{(m)}$$

$$= \sum_{i=1}^{m} \int_M (\mathbf{x}_i \cdot \dot\partial)\dot g\mathrm{d}\mathbf{x}_{(m)}\mathbf{x}^i \dot f = \int_M g\mathrm{d}\mathbf{x}_{(m)}\partial f.$$

□

Because this theorem is so important, giving all of the classical integration theorems, it is worthwhile to go through the steps of its proof in detail for the case when $m = 2$. This should also help the reader understand the notation that is being used. We begin with the right side (13.16) of the theorem. By the Definition (13.13),

$$\int_{\beta(M)} g\mathrm{d}\mathbf{x}_{(1)} f = \int_{M_+^1 \oplus M_-^1} g\mathrm{d}\mathbf{x}_{(1)} f + \int_{M_+^2 \oplus M_-^2} g\mathrm{d}\mathbf{x}_{(1)} f$$

with the help of the chain rule,

$$= \int_{R_+^1 \oplus R_-^1} g\mathbf{x}_{(1)} f \mathrm{d}s^{(1)} + \int_{R_+^2 \oplus R_-^2} g\mathbf{x}_{(1)} f \mathrm{d}s^{(1)}$$

with the help of (13.9), Fubini's theorem [95, p. 919], and the one-dimensional fundamental theorem of calculus,

$$= \int_{a^2}^{b^2} \left[ \int_{a^1}^{b^1} \frac{\partial}{\partial s^1} (g\mathbf{x}_{(1)}\mathbf{x}^1 f)\mathrm{d}s^1 \right] \frac{\mathrm{d}s^{(2)}}{\mathrm{d}s^1} + \int_{a^1}^{b^1} \left[ \int_{a^2}^{b^2} \frac{\partial}{\partial s^1} (g\mathbf{x}_{(1)}\mathbf{x}^1 f)\mathrm{d}s^1 \right] \frac{\mathrm{d}s^{(2)}}{\mathrm{d}s^2}$$

again, with the help of Fubini's theorem, Lemma 13.2.4, and the chain rule,

$$= \int_{M_+^1 \oplus M_-^1} \mathbf{x}_1 \cdot \dot{\partial} \dot{g} \mathrm{d}\mathbf{x}_{(2)} \mathbf{x}^1 \dot{f} + \int_{M_+^1 \oplus M_-^1} \mathbf{x}_2 \cdot \dot{\partial} \dot{g} \mathrm{d}\mathbf{x}_{(2)} \mathbf{x}^2 \dot{f}$$

and finally with the help of Definition (13.11)

$$= \int_M g\mathrm{d}\mathbf{x}_{(2)} \partial f,$$

which completes the proof.

Choosing $g(\mathbf{x}) = f(\mathbf{x}) = 1$, the constant functions, in the fundamental theorem immediately gives the following:

**Corollary 13.2.6.** *Let $M$ be a rectangular $m$-patch of class $C^2$, then*

$$\mathscr{D}(M) = \int_{\beta(M)} \mathrm{d}\mathbf{x}_{(m-1)} = 0.$$

Directed integration over $m$-chains and their boundaries can be extended to any *simple m-surface* $\mathscr{S}$ in $\mathbb{R}^n$ obtained by gluing together the images of $m$-rectangles, making sure that proper orientations are respected on their boundaries [45, p.101–103]. While technically more difficult than evaluating a Riemann integral in the plane, since surfaces and directed quantities are involved, the idea is basically the same.

In the application of the fundamental theorem of calculus 13.2.5, the convention of the *outward normal* can be very confusing. It is worthwhile to look at this

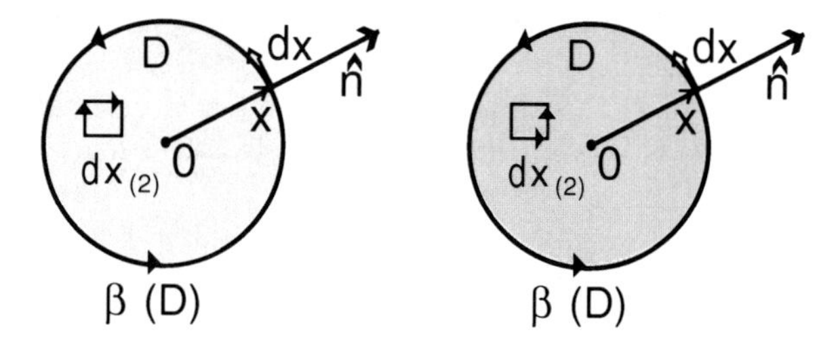


**Fig. 13.6** Showing the two possible orientations for the unit disk $D$ bounded by the positively oriented unit circle $\beta(D)$ centered at origin

convention in detail for $n = 2$. Let us consider in detail this theorem applied to the unit disk $D$ centered at the origin, where $g(\mathbf{x}) = 1$ and $f(\mathbf{x}) = \mathbf{x}$. We get

$$2\int_D \mathrm{d}\mathbf{x}_{(2)} = \int_D \mathrm{d}\mathbf{x}_{(2)}\partial_{\mathbf{x}}\mathbf{x} = \int_{\beta(D)} \mathrm{d}\mathbf{x}\mathbf{x}.$$

This equation implies that

$$\int_D \mathrm{d}\mathbf{x}_{(2)} = \frac{1}{2}\int_{\beta(D)} \mathrm{d}\mathbf{x}\mathbf{x} = -\frac{1}{2}\int_{\beta(D)} \mathbf{x}\mathrm{d}\mathbf{x}. \tag{13.17}$$

One hard fast convention is that when integrating over a simple closed curve in the plane, the *counterclockwise direction* is always considered to be positive. But we also have the *outward normal* convention (13.9) that needs to be satisfied. This convention implies that we should choose

$$\mathrm{d}\mathbf{x}_{(2)} = |\mathbf{e}_{21}|\mathbf{e}_{21} = -\mathrm{i}|\mathbf{e}_{21}| \text{ and } \mathrm{d}\mathbf{x} = \hat{\mathbf{n}}|\mathrm{d}\mathbf{x}| \tag{13.18}$$

where $\mathrm{i} = \mathbf{e}_{12}$ and $\hat{\mathbf{n}}$ is the outward normal unit vector at the point $\mathbf{x}$ on the boundary of $D$. With these convention, multiplying equation (13.17) on the right by i gives

$$\int_D |\mathrm{d}\mathbf{x}_{(2)}| = \frac{1}{2}\int_{\beta(D)} \mathrm{i}\mathrm{d}\mathbf{x}\mathbf{x} = \frac{1}{2}\int_{\beta(D)} \hat{\mathbf{n}}\mathbf{x}|\mathrm{d}\mathbf{x}|.$$

It seems more natural to choose $\mathrm{d}\mathbf{x}_{(2)} = \mathbf{e}_{12}|\mathrm{d}\mathbf{x}_{(2)}| = \mathrm{i}|\mathrm{d}\mathbf{x}_{(2)}|$. We can do this if $\mathrm{d}\mathbf{x}\mathrm{i} = \hat{\mathbf{n}}|\mathrm{d}\mathbf{x}|$. In this case, multiplying (13.17) by $-\mathrm{i}$ gives

$$\int_D |\mathrm{d}\mathbf{x}_{(2)}| = \frac{1}{2}\int_{\beta(D)} \mathbf{x}\mathrm{d}\mathbf{x}\mathrm{i} = \frac{1}{2}\int_{\beta(D)} \mathbf{x}\hat{\mathbf{n}}|\mathrm{d}\mathbf{x}|.$$

The problem is that there is no one convention that works well in all dimensions. The two different conventions are summarized in Fig. 13.6.

## Exercises

1. Consider the image $M = \mathbf{x}(R)$ of the 2-rectangle $R = [-1,1] \times [-1,1]$, where

$$\mathbf{x}(s^1, s^2) = (s^1, s^2, -(s^1)^2 + s^2 + 2).$$

See Fig. 13.2.

(a) Find the area $|M|$ of the patch $M$ by calculating the integral $\int_M |\mathrm{d}\mathbf{x}_{(2)}| = \int_R |\mathbf{x}_{(m)}| \mathrm{d}s^{(2)}$, where $\mathrm{d}s^{(2)} = \mathrm{d}s^1 \mathrm{d}s^2$.

(b) Find the length $|\partial M|$ of the boundary $\partial M$ of $M$ by evaluating the integral

$$|\partial M| = \int_{\beta(M)} |\mathrm{d}\mathbf{x}|.$$

2. Let $R = [-1,1] \times [-1,1]$ be a 2-square in $\mathbb{R}^2$ and let $M = \{\mathbf{x}(R)\} \subset \mathbb{R}^3$ where $\mathbf{x}(s^1, s^2) = (s^1, s^2, (s^1)^2 + (s^2)^2)$, see Fig. 13.3.

(a) Find the area $|M|$ of the patch $M$ by calculating the integral

$$\int_M |\mathrm{d}\mathbf{x}_{(2)}| = \int_R |\mathbf{x}_{(2)}| \mathrm{d}s^{(2)} = \int\int_R \sqrt{g} \mathrm{d}s^1 \mathrm{d}s^2,$$

where $g = -(\mathbf{x}_1 \wedge \mathbf{x}_2)^2 = \det[g]$ and $[g]$ is the Gram matrix.

(b) Find the length $|\partial M|$ of the boundary $\partial M$ of $M$ by evaluating the integral

$$|\partial M| = \int_{\beta(M)} |\mathrm{d}\mathbf{x}|.$$

3. Let $R = [-1,1] \times [-1,1]$ be a 2-square in $\mathbb{R}^2$ and let $M = \{\mathbf{x}(R)\} \subset \mathbb{R}^3$ where $\mathbf{x}(s^1, s^2) = (s^1, s^2, (s^1)^2 (s^2)^2 + 1)$, see Fig. 13.4.

(a) Find the area $|M|$ of the patch $M$ by calculating the integral $\int_M |\mathrm{d}\mathbf{x}_{(2)}| = \int_R |\mathbf{x}_{(2)}| \mathrm{d}s^{(2)}$, where $\mathrm{d}s^{(2)} = \mathrm{d}s^1 \mathrm{d}s^2$. *Ans.* $4 - \frac{1}{3}\tan^{-1}\left(\frac{4}{3}\right) + \frac{7\log(5)}{3} = 7.44626$.

(b) Find the length $|\partial M|$ of the boundary $\partial M$ of $M$ by evaluating the integral

$$|\partial M| = \int_{\beta(M)} |\mathrm{d}\mathbf{x}|.$$

*Ans.* $4\sqrt{5} + 2\sinh^{-1}(2) = 11.8315$.

4. Use the fundamental theorem of calculus to show that

(a) $\int_M \mathrm{d}\mathbf{x}_{(2)} = \frac{1}{2}\int_{\beta(M)} \mathrm{d}\mathbf{x} \wedge \mathbf{x}$.

(b) $\int_{\beta(M)} \mathrm{d}\mathbf{x} \cdot \mathbf{x} = 0$.

5. Choosing $\mathbf{a} = \mathbf{e}_i$ and $\mathbf{b} = \mathbf{e}_j$, show that (13.8) is a consequence of the property that partial derivative commute.

6. (a) Whereas the integrability condition (13.8) holds for an $m$-surface, show that the stronger integrability condition (3.54) is not valid for the vector derivative on a 2-dimensional cylinder. What is the relationship of the vector derivative on the cylinder to the vector derivative in $\mathbb{R}^3$ expressed in cylindrical coordinates discussed in Problem 5 of Sect. 13.1?
   (b) Similarly, show that the stronger integrability condition (3.54) is not valid for the vector derivative on the 2-dimensional unit sphere. What is the relationship to the vector derivative in $\mathbb{R}^3$ expressed in spherical coordinates discussed in Problem 6 of Sect. 13.1?
7. Show that the integrability condition (13.8) implies that

$$[\mathbf{b},\mathbf{a}] = \mathbf{b}\cdot\partial_{\mathbf{x}}\mathbf{a} - \mathbf{a}\cdot\partial_{\mathbf{x}}\mathbf{b} \in \mathscr{T}_{\mathbf{x}}$$

for all tangent vector fields $\mathbf{a},\mathbf{b} \in \mathscr{T}_{\mathbf{x}}$.

## 13.3 Classical Theorems of Integration

Let $\mathbf{x} = \mathbf{x}(s)$, for $s \in R = [a,b]$, define a 1-dimensional curve $M = \mathbf{x}(R) \subset \mathbb{R}^n$ and let $f(\mathbf{x}), g(\mathbf{x})$ be $C^1$ functions mapping the curve $M \to \mathbb{G}_n$. With the help of (13.14), the fundamental theorem of calculus 13.2.5 gives

$$\int_M \dot{g}(\mathbf{x})\mathrm{d}\mathbf{x}\cdot\dot{\partial}\dot{f}(\mathbf{x}) = \int_{\beta(M)} g(\mathbf{x})\mathrm{d}\mathbf{x}_{(0)}f(\mathbf{x}) = |_a^b g(\mathbf{x}(s))f(\mathbf{x}(s))$$
$$= g(\mathbf{b})f(\mathbf{b}) - g(\mathbf{a})f(\mathbf{a}),$$

where $\mathbf{a} = \mathbf{x}(a)$ and $\mathbf{b} = \mathbf{x}(b)$. Note, with the help of the chain rule, that

$$\mathrm{d}\mathbf{x}\partial = \mathrm{d}\mathbf{x}\cdot\partial_{\mathbf{x}} = \mathrm{d}s\frac{\mathrm{d}}{\mathrm{d}s}.$$

This gives the condition of when an integral along a curve $M$ connecting the points $\mathbf{a},\mathbf{b} \in \mathbb{R}^n$ is independent of the path $\mathbf{x}(s)$. Note also that the vector derivative $\partial = \partial_{\mathbf{x}}$ differentiates both to the left and to the right, as indicated by the dots over the arguments.

For a 2-dimensional surface $\mathscr{S}$ embedded in $\mathbb{R}^n$, where the boundary $\beta(\mathscr{S})$ of $\mathscr{S}$ is a simple closed curve, the fundamental theorem of calculus 13.2.5 gives

$$\int_{\mathscr{S}} g(\mathbf{x})\mathrm{d}\mathbf{x}_{(2)}\partial f(\mathbf{x}) = \int_{\mathscr{S}} g(\mathbf{x})\mathrm{d}\mathbf{x}_{(2)}\cdot\partial f(\mathbf{x}) = \int_{\beta(\mathscr{S})} g(\mathbf{x})\mathrm{d}\mathbf{x}f(\mathbf{x}). \tag{13.19}$$

This general integration theorem includes the classical Green's theorem in the plane and Stokes' theorem for a surface in $\mathbb{R}^3$ as special cases [91, p.109], [92, p.134]. If $\mathscr{S} \subset \mathbb{R}^2$ is a 2-surface in the plane, and choosing $g(\mathbf{x}) = 1$ and $\mathbf{f} = f(\mathbf{x})$ to be a $C^1$ vector-valued function in $\mathbb{R}^2$, we have

$$\int_{\mathscr{S}} d\mathbf{x}_{(2)} \partial \wedge \mathbf{f} = \int_{\beta(\mathscr{S})} d\mathbf{x} \cdot \mathbf{f}, \tag{13.20}$$

which is equivalent to the standard Green's theorem.

Now let $\mathscr{S}$ be a 2-surface in $\mathbb{R}^3$, having the simple closed curve $\beta(\mathscr{S})$ as its boundary. For $g(\mathbf{x}) = 1$ and $\mathbf{f} = f(\mathbf{x})$, a vector-valued function in $\mathbb{R}^3$, (13.20) becomes equivalent to Stokes' theorem for a simple surface in $\mathbb{R}^3$.

If $\mathscr{V}$ is a region in $\mathbb{R}^3$ bounded by a simple 2-surface $\beta(\mathscr{V})$, and $g(\mathbf{x}) = 1$ and $\mathbf{f} = f(\mathbf{x})$ is a vector-valued function in $\mathbb{R}^3$, then Theorem 13.2.5 gives

$$\int_{\mathscr{V}} d\mathbf{x}_{(3)} \partial \cdot \mathbf{f} = \int_{\beta(\mathscr{V})} d\mathbf{x}_{(2)} \wedge \mathbf{f}, \tag{13.21}$$

which is equivalent to the Gauss' divergence theorem. This becomes more evident when we multiply both sides of the above integral equation by $I^{-1} = \mathbf{e}_{321}$, giving

$$\int_{\mathscr{V}} \partial \cdot \mathbf{f}\, |d\mathbf{x}_{(3)}| = \int_{\beta(\mathscr{V})} \hat{\mathbf{n}} \cdot \mathbf{f}\, |d\mathbf{x}_{(2)}|, \tag{13.22}$$

where $\hat{\mathbf{n}}$ is the unit outward normal to the surface element $d\mathbf{x}_{(2)}$ at the point $\mathbf{x} \in \beta(\mathscr{V})$.

Over the last several decades, a powerful *geometric analysis* has been developed using geometric algebra. This geometric analysis generalizes the concept of an analytic function in the complex plane to a *monogenic function* in $\mathbb{R}^n$, [10].

**Definition 13.3.1.** A differentiable geometric-valued function $f(\mathbf{x})$ is said to be *monogenic* on an open region $M \subset \mathbb{R}^n$ if $\partial_{\mathbf{x}} f(\mathbf{x}) = 0$ for all $\mathbf{x} \in M$.

Let $\mathscr{M} \subset \mathbb{R}^n$ be an open region bounded by $\mathscr{S}$ an $(n-1)$-dimensional simple closed surface. Let $f = f(\mathbf{x})$ be a differentiable geometric-valued function in the interior of $\mathscr{M}$ and continuous on its boundary $\mathscr{S}$. Then for any interior point $\mathbf{x}' \in M$,

$$f(\mathbf{x}') = \frac{1}{\Omega} \int_M |d\mathbf{x}_{(n)}| \frac{(\mathbf{x}' - \mathbf{x})}{|\mathbf{x}' - \mathbf{x}|^n} \partial_{\mathbf{x}} f(\mathbf{x}) - \frac{1}{\Omega} \int_{\mathscr{S}} |d\mathbf{x}_{(n-1)}| \frac{(\mathbf{x}' - \mathbf{x})}{|\mathbf{x}' - \mathbf{x}|^n} \hat{\mathbf{n}} f(\mathbf{x}), \tag{13.23}$$

where $\hat{\mathbf{n}}$ is the unit outward normal to the surface $\mathscr{S}$ and $\Omega = \frac{2\pi^{n/2}}{\Gamma(n/2)}$ is the area of the $(n-1)$-dimensional unit sphere in $\mathbb{R}^n$. This important result is a consequence of Theorem 13.2.5 by choosing $g(\mathbf{x})$ to be the *Cauchy kernel function* $g(\mathbf{x}') = \frac{(\mathbf{x}'-\mathbf{x})}{|\mathbf{x}'-\mathbf{x}|^n}$, which has the property that $\partial_{\mathbf{x}'} g(\mathbf{x}') = 0$ for all $\mathbf{x}' \neq \mathbf{x} \in \mathbb{R}^n$. For a proof of this general result, see [43, p. 259–262]. For a monogenic function $f(\mathbf{x})$ in $\mathscr{M}$ with boundary $\mathscr{S}$, it follows from (13.23) that

$$f(\mathbf{x}') = -\frac{1}{\Omega} \int_{\mathscr{S}} |d\mathbf{x}_{(n-1)}| \frac{(\mathbf{x}' - \mathbf{x})}{|\mathbf{x}' - \mathbf{x}|^n} \hat{\mathbf{n}} f(\mathbf{x}). \tag{13.24}$$

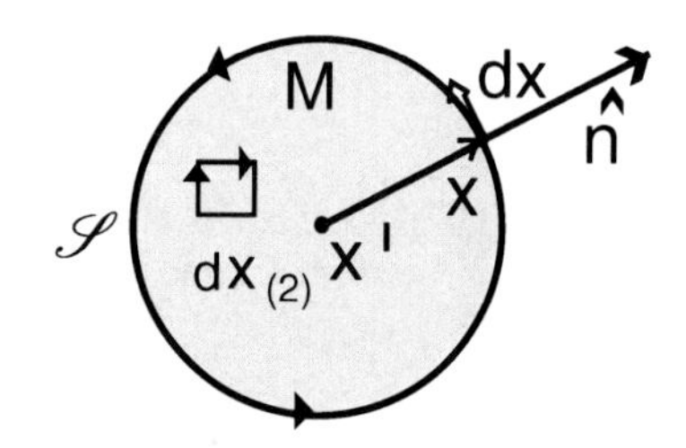


**Fig. 13.7** The oriented unit disk $M$ is bounded by the oriented unit circle $\mathscr{S}$ centered at the point $\mathbf{x}'$

This means that a monogenic function $f(\mathbf{x})$ in a region $\mathscr{M} \subset \mathbb{R}^n$ is completely determined by its values on its boundary $\mathscr{S}$. This generalizes Cauchy's famous integral theorem for analytic functions in the complex number plane [2, p.119].

Let us work through the details of the derivation of the important formulas (13.23) and (13.24) when $n = 2$ and where $M \subset \mathbb{R}^2$ is the unit disk bounded by the unit circle $\mathscr{S}$ centered at the point $\mathbf{x}'$, see Fig. 13.7. Let $f = f(\mathbf{x})$ and $g = g(\mathbf{x})$ be differentiable geometric functions on $M$ with values in the geometric algebra $\mathbb{G}_2$. Applying the fundamental theorem of calculus (13.16), we have

$$\int_M g\mathrm{d}\mathbf{x}_{(2)}\partial_{\mathbf{x}} f = \int_{\mathscr{S}} g\mathrm{d}\mathbf{x} f. \tag{13.25}$$

We now choose $g(\mathbf{x}) = \frac{\mathbf{x}-\mathbf{x}'}{|\mathbf{x}-\mathbf{x}'|^2}$ and note that $\partial_{\mathbf{x}} g = 0$ at all points $\mathbf{x} \in M$, except at the singular point $\mathbf{x} = \mathbf{x}'$. Expanding the left side of this equation gives

$$\begin{aligned}\int_M g\mathrm{d}\mathbf{x}_{(2)}\partial_{\mathbf{x}} f &= \int_M g\mathrm{d}\mathbf{x}_{(2)}(\dot{\partial}_{\mathbf{x}}\dot{f}) - \int_M (\dot{g}\dot{\partial}_{\mathbf{x}})\mathrm{d}\mathbf{x}_{(2)} f \\ &= \int_M g\mathrm{d}\mathbf{x}_{(2)}(\dot{\partial}_{\mathbf{x}}\dot{f}) - \Big(\int_M \mathrm{d}\mathbf{x}_{(2)}\partial_{\mathbf{x}}\cdot g(\mathbf{x})\Big) f(\mathbf{x}').\end{aligned}$$

Applying the fundamental theorem of calculus to the second term gives

$$\Big(\int_M \mathrm{d}\mathbf{x}_{(2)}\partial_{\mathbf{x}}\cdot g(\mathbf{x})\Big) f(\mathbf{x}') = \int_{\mathscr{S}} \mathrm{d}\mathbf{x}\frac{\mathbf{x}-\mathbf{x}'}{|\mathbf{x}-\mathbf{x}'|} f(\mathbf{x}') = \mathbf{e}_{21}\int_{\mathscr{S}} |\mathrm{d}\mathbf{x}| f(\mathbf{x}') = 2\pi\mathbf{e}_{21} f(\mathbf{x}').$$

Equating what we have found for the left side with the right side of (13.25) gives

$$f(\mathbf{x}') = \frac{1}{2\pi \mathrm{i}}\int_{\mathscr{S}} \frac{1}{\mathbf{x}-\mathbf{x}'}\mathrm{d}\mathbf{x} f(\mathbf{x}) - \frac{1}{2\pi \mathrm{i}}\int_M \frac{1}{\mathbf{x}-\mathbf{x}'}\mathrm{d}\mathbf{x}_{(2)}(\partial_{\mathbf{x}} f(\mathbf{x}))$$

or

$$f(\mathbf{x}') = \frac{1}{2\pi \mathrm{i}}\int_{\mathscr{S}} \frac{1}{\mathbf{x}-\mathbf{x}'}\mathrm{d}\mathbf{x} f(\mathbf{x}) - \frac{1}{2\pi}\int_M \frac{1}{\mathbf{x}-\mathbf{x}'}(\partial_{\mathbf{x}} f(\mathbf{x}))|\mathrm{d}\mathbf{x}_{(2)}|, \tag{13.26}$$

where $\mathrm{i} = \mathbf{e}_{12}$ and $\mathrm{d}\mathbf{x}_{(2)} = -\mathrm{i}|\mathrm{d}\mathbf{x}_{(2)}|$. This very general formula applies to any $\mathbb{G}_2$-valued function $f(\mathbf{x})$ which is differentiable at the interior points of $M$ and continuous on the boundary $\mathscr{S}$ of $M$. It expresses the value of $f$ at any interior point

$\mathbf{x}' \in M$ in terms of the sum of an integral of $f(\mathbf{x})$ on the boundary $\mathscr{S}$ and an integral of $\partial_{\mathbf{x}} f(\mathbf{x})$ over the interior of $M$. Making the natural identification $z = \mathbf{e}_1\mathbf{x} = x + \mathrm{i}y$, we can write (13.26) in the equivalent complex form,

$$f(\mathbf{e}_1 z') = \frac{1}{2\pi\mathrm{i}}\int_{\mathscr{S}} \frac{1}{z-z'}\mathrm{d}z f(\mathbf{e}_1 z) - \frac{1}{\pi}\int_M \frac{1}{z-z'}\left(\frac{\mathrm{d}}{\mathrm{d}\bar{z}} f(\mathbf{e}_1 z)\right)|\mathrm{d}\mathbf{x}_{(2)}|, \tag{13.27}$$

where we have identified $\frac{\mathrm{d}}{\mathrm{d}\bar{z}} = \frac{1}{2}\mathbf{e}_1\partial_{\mathbf{x}}$. If in $M$ $g(z) = f(\mathbf{e}_1 z)\mathbf{e}_1$ is a function of $z$ and not $\bar{z}$, or equivalently $\frac{\mathrm{d}}{\mathrm{d}\bar{z}} g(z) = 0$, then $g(z)$ is said to be an *analytic function* in $M$.

More generally, if $f(\mathbf{x})$ is a differentiable vector-valued function in $M$, then $g(z) = f(\mathbf{e}_1 z)\mathbf{e}_1$ is the corresponding differentiable complex valued function, and

$$z - z' = r\mathrm{e}^{\mathrm{i}\theta} \iff \frac{1}{z-z'} = \frac{\mathrm{e}^{-\mathrm{i}\theta}}{r}.$$

If in addition $r = 1$, then $\mathrm{d}z = \mathrm{i}\mathrm{e}^{\mathrm{i}\theta}\mathrm{d}\theta$. Using these relationships, (13.27) can be written in the interesting complex form

$$g(z') = \frac{1}{2\pi\mathrm{i}}\int_0^{2\pi} g(z' + \mathrm{e}^{\mathrm{i}\theta})\mathrm{d}\theta - \frac{1}{\pi}\int_0^{2\pi}\int_0^1 \mathrm{e}^{-\mathrm{i}\theta}\frac{\mathrm{d}}{\mathrm{d}\bar{z}} g(z)\mathrm{d}r\mathrm{d}\theta. \tag{13.28}$$

If in the disk $M$

$$\partial_{\mathbf{x}} f(\mathbf{x}) = 0 \iff \frac{\mathrm{d}g(z)}{\mathrm{d}\bar{z}} = 0,$$

then (13.26)–(13.28) all reduce to

$$f(\mathbf{x}') = \frac{1}{2\pi\mathrm{i}}\int_{\mathscr{S}} \frac{1}{\mathbf{x}-\mathbf{x}'}\mathrm{d}\mathbf{x} f(\mathbf{x}) \iff g(z') = \frac{1}{2\pi\mathrm{i}}\int_{\mathscr{S}} \frac{g(z)}{z-z'}\mathrm{d}z, \tag{13.29}$$

where $g(z) = f(\mathbf{e}_1 z)\mathbf{e}_1$ is the corresponding analytic function in the disk $M$. This is Cauchy's famous integral formula for an analytic function $g(z)$ in the unit disk $M$.

Formula (13.28) is considerable more general than (13.29). For example, if $g(z) = z\bar{z} = x^2 + y^2$ so that $\frac{\mathrm{d}g(z)}{\mathrm{d}\bar{z}} = z$, then (13.28) gives the valid relationship

$$z'\bar{z}' = \frac{1}{2\pi\mathrm{i}}\int_0^{2\pi} (z' + \mathrm{e}^{\mathrm{i}\theta})(\bar{z}' + \mathrm{e}^{-\mathrm{i}\theta})\mathrm{d}\theta - \frac{1}{\pi}\int_0^{2\pi}\int_0^1 \mathrm{e}^{-\mathrm{i}\theta}(z' + r\mathrm{e}^{\mathrm{i}\theta})\mathrm{d}r\mathrm{d}\theta.$$

In evaluating the left-hand side of Eq. (13.25), we had to make use of the fundamental theorem of calculus as applied to *generalized functions*. Whereas integrating over the singularity $\mathbf{x} = \mathbf{x}'$ presents difficulties, the integration over the boundary of the disk, the unit circle, is well defined. The theory of *generalized functions* or *distributions* makes precise what is going on here [30].

Another way of constructing geometric calculus on a $k$-surface is by defining the $k$-surface to be the limit set of a sequence of $k$-chains [73].

## Exercises

1. If the acceleration of a particle at any time $t > 0$ is

$$\mathbf{a} = \frac{\mathrm{d}\mathbf{v}}{\mathrm{d}t} = 5\cos 2t\mathbf{e}_1 - \sin t\mathbf{e}_2 + 2t\mathbf{e}_3,$$

and the velocity $\mathbf{v}$ and position vector $\mathbf{x}$ are 0 at $t = 0$, find $\mathbf{v}(t)$ and $\mathbf{x}(t)$ at any time $t > 0$.

2. Noting that $\frac{\mathrm{d}}{\mathrm{d}t}\left(\mathbf{a} \wedge \frac{\mathrm{d}\mathbf{a}}{\mathrm{d}t}\right) = \mathbf{a} \wedge \frac{\mathrm{d}^2\mathbf{a}}{\mathrm{d}t^2}$, evaluate the integral $\int_0^{\pi/2} \mathbf{a} \wedge \frac{\mathrm{d}^2\mathbf{a}}{\mathrm{d}t^2}\mathrm{d}t$ for the acceleration $\mathbf{a}$ given in Problem 1.

3. Let $\mathbf{a} = (3x+2y)\mathbf{e}_1 + 2yz\mathbf{e}_2 + 5xz^2\mathbf{e}_3$.
   (a) Calculate $\int_C \mathbf{a} \cdot \mathrm{d}\mathbf{x}$ from the point $(0,0,0)$ to the point $(1,1,1)$ where $C$ is the curve defined by $x = t$, $y = -t^2 + 2t$, and $z = 2t^2 - t$.
   (b) Calculate the integral in (a) where $C$ is the straight line joining $(0,0,0)$ and $(1,1,1,)$.

4. If $\mathbf{a} = \partial_{\mathbf{x}}\phi(\mathbf{x})$, show that $\int_{\mathbf{x}_1}^{\mathbf{x}_2} \mathbf{a} \cdot \mathrm{d}\mathbf{x}$ is independent of the curve $C$ from the point $\mathbf{x}_1$ to the point $\mathbf{x}_2$.

5. Evaluate the integral

$$\int\int_{\mathscr{S}} \mathbf{a} \cdot \mathbf{n}|\mathrm{d}\mathbf{x}_{(2)}|,$$

where $\mathbf{a} = 2z\mathbf{e}_1 + \mathbf{e}_2 + 3y\mathbf{e}_3$, $\mathscr{S}$ is that part of the plane $2x + 3y + 6z = 12$ which is bounded by the coordinate planes, and $\mathbf{n}$ is the normal unit vector to this plane.

6. Let $\mathbf{f}(\mathbf{x}) = M(\mathbf{x})\mathbf{e}_1 + N(\mathbf{x})\mathbf{e}_2$, where $M(\mathbf{x}), N(\mathbf{x})$ are $C^1$ real-valued functions for $\mathbf{x} \in R \subset \mathbb{R}^2$, where $R$ is a closed region in $\mathbb{R}^2$ bounded by the simple curve $C$. Using (13.20), show that

$$\int_C \mathbf{f} \cdot \mathrm{d}\mathbf{x} = \int\int_R (\partial_{\mathbf{x}} \wedge \mathbf{f})\mathrm{d}\mathbf{x}_{(2)} = \int\int \left(\frac{\partial N}{\partial x} - \frac{\partial M}{\partial y}\right)\mathrm{d}x\mathrm{d}y$$

which is classically known as Green's theorem in the plane. Note that $\mathrm{d}\mathbf{x} = \mathrm{d}x\mathbf{e}_1 + \mathrm{d}y\mathbf{e}_2$ and $\mathrm{d}\mathbf{x}_{(2)} = \mathrm{d}x\mathrm{d}y\mathbf{e}_{21}$ since we are integrating *counterclockwise* around the curve $C$ in the $xy$-plane.

7. Let $S$ be an open-oriented 2-surface in $\mathbb{R}^3$ bounded by a closed simple curve $C$ and let $\mathbf{f}(\mathbf{x})$ be a $C^1$ vector-valued function on $S$. Using (13.19), show that

$$\int_C \mathbf{f} \cdot \mathrm{d}\mathbf{x} = \int\int_R (\partial_{\mathbf{x}} \wedge \mathbf{f})\mathrm{d}\mathbf{x}_{(2)} = \int\int_R (\partial_{\mathbf{x}} \times \mathbf{f}) \cdot \hat{\mathbf{n}}|\mathrm{d}\mathbf{x}_{(2)}|,$$

which is classically known as Stokes' theorem for a surface $S \subset \mathbb{R}^3$.

8. Let $V$ be a volume bounded by a closed surface $S$ in $\mathbb{R}^3$ and $\mathbf{f}(\mathbf{x}) \in C^1$ be a vector-valued function. Show that (13.21) implies (13.22),

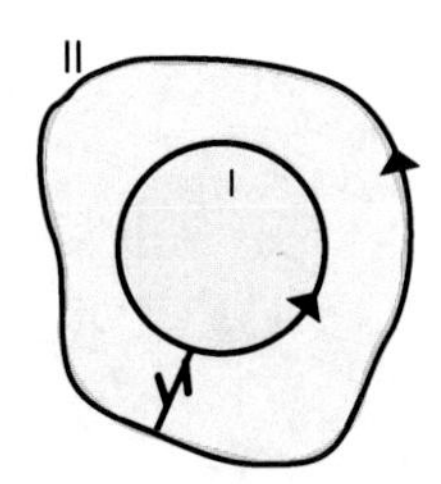


**Fig. 13.8** The path integrals $\int_I f(\mathbf{x})\mathrm{d}\mathbf{x}$ and $\int_{II} f(\mathbf{x})\mathrm{d}\mathbf{x}$ over the simple closed curves $I$ and $II$ are equal if $f(\mathbf{x})$ is monogenic within the enclosed regions. The path integrals in opposite directions over the connecting paths between them cancel out

$$\int\int_S \mathbf{f}\cdot\hat{\mathbf{n}}|\mathrm{d}\mathbf{x}_{(2)}| = \int\int\int_V (\partial_{\mathbf{x}}\cdot\mathbf{f})|\mathrm{d}\mathbf{x}_{(3)}|$$

which is classically known as Gauss' divergence theorem in $\mathbb{R}^3$.

9. Using the fundamental theorem of calculus, show that the vector derivative $\partial_{\mathbf{x}} f(\mathbf{x})$ of a geometric-valued function $f(\mathbf{x}) \in \mathbb{G}_3$ can be defined by

$$\partial_{\mathbf{x}} f(\mathbf{x}) = \lim_{\Delta V\to 0}\frac{1}{\Delta V}\int\int_{\Delta S}|\mathrm{d}\mathbf{x}_{(2)}|\hat{\mathbf{n}}f(\mathbf{x}),$$

where the small volume element $\Delta V$ around the point $\mathbf{x}$ is bounded by the closed surface $\Delta S$ and $\hat{\mathbf{n}}$ is the outward unit normal vector at each point $\mathbf{x}\in\Delta S$.

10. Show that the path integrals $\int_I f(\mathbf{x})\mathrm{d}\mathbf{x}$ and $\int_{II} f(\mathbf{x})\mathrm{d}\mathbf{x}$ over the simple closed curves $I$ and $II$ shown in Fig. 13.8 are equal if $f(\mathbf{x})$ is monogenic within the enclosed regions in $\mathbb{R}^2$. Conclude that the unit circle centered at the singular point or *pole* at $\mathbf{x}'$ can be replaced by any simple closed curve with the same orientation containing the pole $\mathbf{x}'$ without changing the value $f(\mathbf{x}')$ in the path integral given in (13.29).

11. (a) Show that $g(z) = f(\mathbf{e}_1\mathbf{z})\mathbf{e}_1$ is analytic iff $f(\mathbf{x})$ is a monogenic vector-valued function.

    (b) Show that an analytic function is $g(z)$ is monogenic but all monogenic functions are not analytic.

# Chapter 14
# Differential Geometry of Curves

*Cowboys have a way of trussing up a steer or a pugnacious bronco which fixes the brute so that it can neither move nor think. This is the hog-tie, and it is what Euclid did to geometry.*
—Eric Bell, *The Search for Truth*

The study of the differential geometry of surfaces rightly begins with the study of curves, since a curve can be considered to be a 1-dimensional surface, and a more general surface can be considered to be the set of all curves which belong to the surface. In particular, the classical formulas of Frenet-Serret are derived for the *moving frame* along a curve. The study of the calculus of a $k$-surface, begun in the previous chapter, is also a part of differential geometry.

## 14.1 Definition of a Curve

A mapping $\mathbf{x} : [a,b] \to \mathbb{R}^n$ is said to be a *regular curve of class* $C^k$ on $[a,b]$, if the $j^{th}$ derivative $\frac{\mathrm{d}^j\mathbf{x}}{\mathrm{d}t^j}$ is defined and continuous for all $t \in (a,b)$ and all $1 \le j \le k$ and also $\frac{\mathrm{d}\mathbf{x}}{\mathrm{d}t} \ne 0$ for all $t \in (a,b)$. The particular parameterization of a curve is generally unimportant. By a *regular reparameterization* of the $C^k$ curve $\mathbf{x}(t)$, we mean a $C^k$ function $t = t(s)$ for all $c \le s \le d$ such that $\frac{\mathrm{d}t}{\mathrm{d}s} \ne 0$ for all $s \in (c,d)$. This condition guarantees that the function $t(s)$ is one-to-one and onto on the interval $[c,d]$ and is therefore invertible, $s = t^{-1}(t)$ for all $t \in [a,b]$.

For example, $\mathbf{x}(t) = (t, t^2, \sqrt[3]{t^4}) \in \mathbb{R}^3$ is a regular $C^1$ curve on any interval $[a,b] \subset \mathbb{R}$. This curve is shown on the interval $[-1,1]$ in the Fig. 14.1.

G. Sobczyk, *New Foundations in Mathematics: The Geometric Concept of Number*, DOI 10.1007/978-0-8176-8385-6_14,

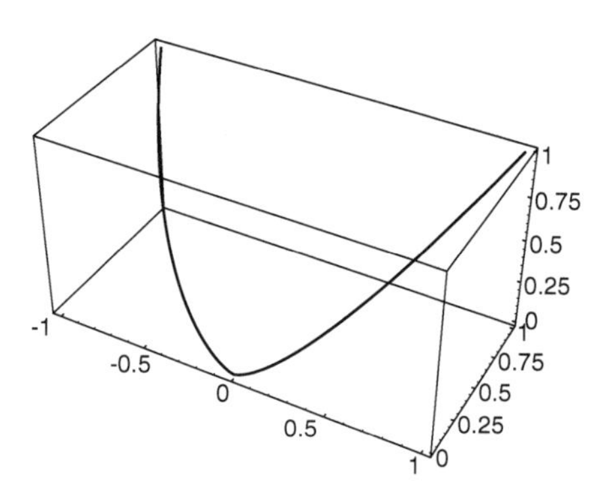


**Fig. 14.1** The curve $\mathbf{x}(t) = (t, t^2, \sqrt[3]{t^4}) \in \mathbb{R}^3$ for $t \in [-1, 1]$

Let $\mathbf{x} = \mathbf{x}(t)$ be a $C^k$ regular curve on the interval $[a,b]$ and let $t \in [a,b]$. The *arclength* of the curve $\mathbf{x}$ from $a$ to $t$ is given by the integral

$$s(t) = \int_{\mathbf{x}} |\mathrm{d}\mathbf{x}| = \int_a^t \left| \frac{\mathrm{d}\mathbf{x}}{\mathrm{d}t} \right| \mathrm{d}t. \tag{14.1}$$

By the fundamental theorem of calculus (13.2.5),

$$\frac{\mathrm{d}s}{\mathrm{d}t} = \frac{\mathrm{d}}{\mathrm{d}t} \int_a^t \left| \frac{\mathrm{d}\mathbf{x}}{\mathrm{d}t} \right| \mathrm{d}t = \left| \frac{\mathrm{d}\mathbf{x}}{\mathrm{d}t} \right| \neq 0,$$

so that $\mathbf{x}(s) = \mathbf{x}(t(s))$ is a regular reparameterization of the curve $\mathbf{x}(t)$ on the interval $[a,b]$. The regular reparameterization $\mathbf{x}(s)$ is called the *natural reparameterization* or *unit speed reparameterization* and is very useful tool for deriving the theoretical properties of curves. The only problem is that it is often impossible to find a convenient form for the inverse function (14.1) involved in its definition.

For example, for the curve $\mathbf{x}(t) = (t, t^2, \sqrt[3]{t^4}) \in \mathbb{R}^3$ on the interval $[-1, 1]$, the arclength is given by

$$s(t) = \int_{-1}^t \left| \frac{\mathrm{d}\mathbf{x}}{\mathrm{d}t} \right| \mathrm{d}t = \int_a^t \sqrt{1 + 4t^2 + \frac{16}{9} t^{\frac{2}{3}}}\, \mathrm{d}t.$$

But this integral is solvable only by using elliptic integrals. Thus, there is no good expression for the natural reparameterization $\mathbf{x}(t(s))$ of this curve. However, the arclength $s(t)$ is well defined on the interval $[-1, 1]$ as shown in Fig. 14.2.

## Exercises

1. Find the arclength of the curve $\mathbf{x}(\theta) = (\cos\theta, \sin\theta, 1 - \sin\theta - \cos\theta)$ for $0 \leq \theta \leq 2\pi$.
2. Let $h > 0$ and $r > 0$ be constants.

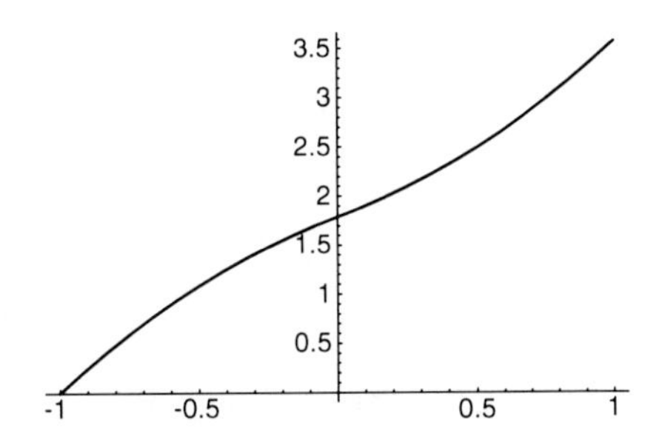


**Fig. 14.2** Graph of the arclength of the curve $\mathbf{x}(t) = (t, t^2, \sqrt[3]{t^4}) \in \mathbb{R}^3$ for $t \in [-1, 1]$

(a) Find the arclength of the circular helix $\mathbf{x}(t) = (r\cos t, r\sin t, ht)$ for $0 \le t \le 2\pi$.
(b) Reparameterize the right circular helix by arclength.

3. (a) Find the arclength of the curve $\mathbf{x}(t) = (e^t \cos t, e^t \sin t, e^t)$ for $0 \le t \le 1$.
   (b) Reparameterize the curve $\mathbf{x}(t)$ by arclength.
4. (a) Find the arclength of the curve $\mathbf{x}(t) = (\cosh t, \sinh t, t)$ for $0 \le t \le 1$.
   (b) Reparameterize the curve $\mathbf{x}(t)$ by arclength.
5. Show that $t = \frac{\theta^2}{\theta^2+1}$ is a regular reparameterization on $0 < \theta < \infty$ and takes the interval $0 < \theta < \infty$ onto $0 < t < 1$.
6. Let $\mathbf{x}(s) = \frac{1}{2}(s + \sqrt{s^2+1}, (s + \sqrt{s^2+1})^{-1}, \sqrt{2}\log(s + \sqrt{s^2+1}))$. Show that $|\frac{\mathrm{d}}{\mathrm{d}s}\mathbf{x}(s)| = 1$ and conclude that $s$ is the natural parameter of arclength.
7. (a) Show that $\mathbf{x} = (t, t^2+1, (t-1)^3)$ is a regular curve for all $t \in \mathbb{R}$.
   (b) Find the arclength of this curve from $t = 1$ to $t = 2$.
8. Show that $\mathbf{x}(s) = \frac{1}{3}\left(((1+s)^{3/2}, (1-s)^{3/2}, \frac{3s}{\sqrt{2}})\right)$ is a unit speed curve.
9. (a) Show that $\mathbf{x}(\theta) = (1+\cos\theta, \sin\theta, 2\sin(\theta/2))$, for $-2\pi \le \theta \le 2\pi$, is regular and lies on the sphere of radius 2 and the cylinder $(x_1 - 1)^2 + x_2^2 = 1$.
   (b) Find the arclength of this curve from $t = -2\pi$ to $t = 2\pi$.

## 14.2 Formulas of Frenet–Serret

Let $\mathbf{x}(t)$ be the trajectory of a particle in $\mathbb{R}^3$ and $\mathbf{v} = \mathbf{x}'(t) = \frac{\mathrm{d}\mathbf{x}}{\mathrm{d}t}$ be its velocity. Assuming that the time $t$ is a regular parameterization of the curve $\mathbf{x}(t)$, we can consider the reparameterization $\mathbf{x}(s)$ of the curve in terms of the natural parameter of arclength $s$. The *unit tangent* to the curve $\mathbf{x}(s)$ is defined by $\mathbf{T} = \frac{\mathrm{d}\mathbf{x}}{\mathrm{d}s} = \dot{\mathbf{x}}$. We also have

$$\frac{\mathrm{d}\mathbf{x}}{\mathrm{d}t} = \frac{\mathrm{d}\mathbf{x}}{\mathrm{d}s}\frac{\mathrm{d}s}{\mathrm{d}t} = \mathbf{T}\frac{\mathrm{d}s}{\mathrm{d}t},$$

where $|\mathbf{T}| = 1$ and $\frac{\mathrm{d}s}{\mathrm{d}t} = \left|\frac{\mathrm{d}\mathbf{x}}{\mathrm{d}t}\right|$ is the *speed* of the particle.

The *curvature* $\kappa$ is defined by $\kappa = |\frac{\mathrm{d}\mathbf{T}}{\mathrm{d}s}| = |\dot{\mathbf{T}}|$. We will use the dot notation to indicate differentiation with respect to the natural parameter $s$. We have the following formula for the calculation the curvature $\kappa$ directly in terms of the parameter $t$,

$$\kappa = |\dot{\mathbf{T}}(s)| = \frac{|\mathbf{x}' \times \mathbf{x}''|}{|\mathbf{x}'|^3}.$$

This can be verified by noting that $\mathbf{x}' = \mathbf{T}\frac{\mathrm{d}s}{\mathrm{d}t}$ so that

$$\mathbf{x}'' = \mathbf{T}'\frac{\mathrm{d}s}{\mathrm{d}t} + \mathbf{T}\frac{\mathrm{d}^2 s}{\mathrm{d}t^2} = \frac{\mathrm{d}\mathbf{T}}{\mathrm{d}s}\left(\frac{\mathrm{d}s}{\mathrm{d}t}\right)^2 + \mathbf{T}\frac{\mathrm{d}^2 s}{\mathrm{d}t^2},$$

from which the result easily follows.

The *principal normal vector* to the curve $\mathbf{x}(t)$ is defined by $\mathbf{N}(t) := \frac{\mathbf{T}'}{|\mathbf{T}'|}$. Clearly $\mathbf{T}\cdot\mathbf{N} = 0$, since $\mathbf{T}\cdot\mathbf{T} = 1$ implies by differentiation that

$$\frac{\mathrm{d}(\mathbf{T}\cdot\mathbf{T})}{\mathrm{d}t} = 2\mathbf{T}'\cdot\mathbf{T} = 0.$$

The unit *binormal* $\mathbf{B}$ to the curve $\mathbf{x}(t)$ is defined by $\mathbf{B} = \mathbf{T}\times\mathbf{N}$ so that $\{\mathbf{T},\mathbf{N},\mathbf{B}\}$ forms a *right-handed orthonormal frame*. This orthonormal frame is called the *comoving Frenet frame* and is defined at each point along the curve $\mathbf{x}(t)$.

Since $\mathbf{B}\cdot\mathbf{B} = 1$ implies $\dot{\mathbf{B}}\cdot\mathbf{B} = 0$, and $\mathbf{B} = \mathbf{T}\times\mathbf{N}$, we have

$$\dot{\mathbf{B}} = \dot{\mathbf{T}}\times\mathbf{N} + \mathbf{T}\times\dot{\mathbf{N}} = \mathbf{T}\times\dot{\mathbf{N}}.$$

It follows that $\frac{\mathrm{d}\mathbf{B}}{\mathrm{d}s} = -\tau\mathbf{N}$ for some $\tau = \tau(s) \in \mathbb{R}$. We say that $\tau(s)$ is the *torsion* of the curve $\mathbf{x}(s)$. To complete the formulas for the derivatives of the Frenet frame of a curve $\mathbf{x}(s)$, we must calculate $\dot{\mathbf{N}}$. From $\mathbf{N} = \mathbf{B}\times\mathbf{T}$, we get $\dot{\mathbf{N}} = \dot{\mathbf{B}}\times\mathbf{T} + \mathbf{B}\times\dot{\mathbf{T}}$ so that

$$\dot{\mathbf{N}} = -\tau\mathbf{N}\times\mathbf{T} + \kappa\mathbf{B}\times\dot{\mathbf{N}} = -\tau\mathbf{T} + \tau\mathbf{B}.$$

The formulas just derived, known as the *Frenet-Serret equations*, can be nicely summarized in the matrix form

$$\frac{\mathrm{d}}{\mathrm{d}s}\begin{pmatrix}\mathbf{T}\\ \mathbf{N}\\ \mathbf{B}\end{pmatrix} = \begin{pmatrix}0 & \kappa & 0\\ -\kappa & 0 & \tau\\ 0 & -\tau & 0\end{pmatrix}\begin{pmatrix}\mathbf{T}\\ \mathbf{N}\\ \mathbf{B}\end{pmatrix} = \begin{pmatrix}\kappa\mathbf{N}\\ -\kappa\mathbf{T}+\tau\mathbf{B}\\ -\tau\mathbf{N}\end{pmatrix}. \tag{14.2}$$

Defining what is known as the *Darboux bivector*

$$\Omega = \frac{1}{2}\begin{pmatrix}\mathbf{T} & \mathbf{N} & \mathbf{B}\end{pmatrix}\wedge\begin{pmatrix}\dot{\mathbf{T}}\\ \dot{\mathbf{N}}\\ \dot{\mathbf{B}}\end{pmatrix} = \kappa\mathbf{N}\wedge\mathbf{T} + \tau\mathbf{B}\wedge\mathbf{N},$$

of the curve $\mathbf{x}(s)$, the Frenet-Serret equations can be put in the simple form

$$\frac{\mathrm{d}}{\mathrm{d}s}\begin{pmatrix}\mathbf{T}\\ \mathbf{N}\\ \mathbf{B}\end{pmatrix} = \Omega \cdot \begin{pmatrix}\mathbf{T}\\ \mathbf{N}\\ \mathbf{B}\end{pmatrix}, \tag{14.3}$$

as can be easily verified. The Darboux bivector $\Omega$ defines the *angular velocity* of the Frenet frame along the curve $\mathbf{x}(s)$. Shortly, we will see how these equations completely determine the intrinsic structure of a curve $\mathbf{x}(s)$ in $\mathbb{R}^3$.

## Exercises

1. Find the equations of the tangent line and the normal plane (the plane whose normal vector is tangent) to the curve $\mathbf{x}(t) = (1+t, t^2, 1-t^3)$ at the point $t = 1$.
2. Let $\mathbf{x}(s) = (r\cos(s/r), r\sin(s/r), 0)$ be the circle of radius $r$ in the $xy$-plane. Find the Frenet frame $(\mathbf{T}, \mathbf{N}, \mathbf{B})$ and $\kappa$ and $\tau$ for $\mathbf{x}(s)$.
3. Let $\mathbf{x}(t)$ be a regular curve in $\mathbb{R}^3$. Show that the following formulas are valid:

   (a) $\mathbf{T} = \frac{\mathbf{x}'(t)}{|\mathbf{x}'(t)|}$
   (b) $\mathbf{B} = \frac{\mathbf{x}'(t)\times\mathbf{x}''(t)}{|\mathbf{x}'(t)\times\mathbf{x}''(t)|}$
   (c) $\mathbf{N} = \mathbf{B}\times\mathbf{N}$
   (d) $\kappa = \frac{|\mathbf{x}'(t)\times\mathbf{x}''(t)|}{|\mathbf{x}'|^3}$
   (e) $\tau = \frac{\mathbf{x}'\cdot(\mathbf{x}''\times\mathbf{x}''')}{|\mathbf{x}'(t)\times\mathbf{x}''(t)|^2}$

4. Find the Frenet frame, curvature, and torsion for the circular helix $\mathbf{x}(t) = (r\cos t, r\sin t, ht)$, where $r$ and $h$ are constant.
5. Find the Frenet frame, curvature, and torsion for the curve $\mathbf{x}(t) = (1+t^2, t, t^3)$, where $r$ and $h$ are constant.
6. Given the curve $\mathbf{x}(s) = \left(\cos hs \ \sin hs \ \sqrt{1-h^2}s\right)$ where $0 < h < 1$ is a constant,

   (a) Show that $\mathbf{x}(s)$ is a unit speed curve. *Hint: Show that* $|\dot{\mathbf{x}}(s)| = 1$.
   (b) Find the Frenet comoving frame for $\mathbf{x}(s)$.
   (c) Show that the curvature $\kappa = h^2$ and that the torsion $\tau = h\sqrt{1-h^2}$.
   (d) Show that the angular velocity $\Omega = -h\mathbf{e}_1\mathbf{e}_2$ and verify that

$$\frac{\mathrm{d}}{\mathrm{d}s}\begin{pmatrix}\mathbf{T}\\ \mathbf{N}\\ \mathbf{B}\end{pmatrix} = \Omega \cdot \begin{pmatrix}\mathbf{T}\\ \mathbf{N}\\ \mathbf{B}\end{pmatrix}.$$

7. For the unit speed curve

$$\mathbf{x}(s) = \frac{1}{2}(s + \sqrt{s^2+1}, (s+\sqrt{s^2+1})^{-1}, \sqrt{2}\log(s+\sqrt{s^2+1})),$$

find its Frenet-Serret frame, curvature, and torsion.
8. Find the unit speed parameterization for the curve $\mathbf{y}(t) = (\cosh t, \sinh t, t)$. Using this, find its Frenet-Serret frame, curvature, and torsion.
9. Find the unit speed parameterization for the curve $\mathbf{y}(t) = (e^t \cos t, e^t \sin t, e^t)$. Using this, find its Frenet-Serret frame, curvature, and torsion.
10. Given the unit speed curve, $\mathbf{x}(s) = \frac{1}{3}\left((1+s)^{3/2}, (1-s)^{3/2}, \frac{3s}{\sqrt{2}}\right)$, find its Frenet-Serret frame, curvature, and torsion.

## 14.3 Special Curves

We will now investigate curves in $\mathbb{R}^3$ that have special properties, such as being a straight line, lying in a plane, or lying on a sphere.

Clearly, if $\mathbf{T} = \frac{\mathrm{d}\mathbf{x}}{\mathrm{d}s}$ is a constant unit vector, then

$$\mathbf{x}(s) = \int_0^s \mathbf{T}\mathrm{d}s = s\mathbf{T},$$

which is the equation of a straight line through the origin and having the unit tangent vector $\mathbf{T}$.

Let us now determine the conditions for $\mathbf{x}(s)$ to be a *planar curve* lying in the plane of the bivector $\mathbf{T}\wedge\mathbf{N}$. Using the Frenet-Serret formulas, we calculate

$$\frac{\mathrm{d}^3\mathbf{x}}{\mathrm{d}s^3} = \frac{\mathrm{d}}{\mathrm{d}s}(\kappa\mathbf{N}) = \frac{\mathrm{d}\kappa}{\mathrm{d}s}\mathbf{N} + \kappa\frac{\mathrm{d}\mathbf{N}}{\mathrm{d}s} = \frac{\mathrm{d}\kappa}{\mathrm{d}s}\mathbf{N} - \kappa^2\mathbf{T} + \kappa\tau\mathbf{B}.$$

From this calculation, it is clear that the unit speed curve $\mathbf{x}(s)$, with $\kappa \neq 0$ for all $s$, will lie in the plane of $\mathbf{T}\wedge\mathbf{N}$ if and only if $\tau = 0$ for all $s$. But since $\frac{\mathrm{d}\mathbf{B}}{\mathrm{d}s} = -\tau\mathbf{N}$, it follows that $\mathbf{x}(s)$ being a plane curve is equivalent to $\mathbf{B}$ being a constant vector.

A *helix* is a regular curve $\mathbf{x}(t)$ such that for some fixed unit vector $\mathbf{a}$, called the *axis* of the helix, $\mathbf{T}\cdot\mathbf{a}$ is constant. We will now show that a unit speed curve $\mathbf{x}(s)$ with $\kappa \neq 0$ for all $s$ is a helix if and only if there is a constant $c$ such that $\tau = c\kappa$. This result is due to Lancret (1802).

Following [56, p.32], first, we assume that $\mathbf{x}(s)$ is a helix so that $\mathbf{a}\cdot\mathbf{T} = \cos\theta$ is a constant. We have $\theta \neq k\pi$, for otherwise $\mathbf{T} = \pm\mathbf{a}$ would imply that $\kappa - 0$. It then follows that

$$\frac{\mathrm{d}}{\mathrm{d}s}(\mathbf{a}\cdot\mathbf{T}) = \kappa\mathbf{a}\cdot\mathbf{N} = 0,$$

so that $\mathbf{a} = \cos\theta\mathbf{T} + \sin\theta\mathbf{B}$. Since $\mathbf{a}$ is constant,

$$\frac{d\mathbf{a}}{ds} = \kappa \cos\theta \mathbf{N} - \tau \sin\theta \mathbf{N} = 0$$

from which it follows that $\tau = \kappa \cot\theta \neq 0$. If we now assume that $\tau = \kappa \cot\theta \neq 0$ and define $\mathbf{a} = \cos\theta \mathbf{T} + \sin\theta \mathbf{B}$, we can essentially reverse the above argument to conclude that $\mathbf{a}$ is constant. Finally, since $\mathbf{a} \cdot \mathbf{T} = \cos\theta$ is a constant, $\mathbf{x}(s)$ is a helix.

A unit speed curve $\mathbf{x}(s)$ on a sphere can be characterized by the condition that for some fixed point $\mathbf{c}$, called the *center* of the sphere, $(\mathbf{x} - \mathbf{c}) \cdot \mathbf{T} = 0$ for all $s$. Noting that

$$\frac{d}{ds}(\mathbf{x} - \mathbf{c})^2 = 2(\mathbf{x} - \mathbf{c}) \cdot \mathbf{T},$$

it follows that $\mathbf{x}(s)$ is a unit speed curve on a sphere if and only if $(\mathbf{x} - \mathbf{c})^2 = r^2$ for some constant $r$ which is the radius of the sphere.

## Exercises

1. Show that $\kappa\tau = -\mathbf{T}' \cdot \mathbf{B}'$.
2. If $\mathbf{x}(s)$ is a unit speed curve, show that $\mathbf{x}' \cdot (\mathbf{x}'' \times \mathbf{x}''') = \kappa^2 \tau$.
3. Find the equation of the *tangent plane* (having the direction of the bivector $\mathbf{T} \wedge \mathbf{N}$) to the unit speed curve $\mathbf{x}(s)$ at the point $\mathbf{x} = \mathbf{x}(s)$.
4. Show that $\mathbf{TN} = i\mathbf{B}$ implies that $\mathbf{N} = i\mathbf{TB} = -\mathbf{T} \times \mathbf{B}$ and $\mathbf{T} = i\mathbf{BN} = \mathbf{N} \times \mathbf{B}$, where $i = \mathbf{e}_{123}$.
5. Let $\mathbf{x}(s)$ be a unit speed curve with $\kappa\tau \neq 0$ and $\rho = 1/\kappa$, $\sigma = 1/\tau$ and suppose that $\rho^2 + (\rho'\sigma)^2 = r^2$ where $r$ is a constant. Show that $\mathbf{x}(s)$ lies on a sphere with radius $r$ and center $\mathbf{c} = \mathbf{x} + \rho\mathbf{N} + \rho'\sigma\mathbf{B}$.
6. Prove that $\mathbf{x}(t)$ is a straight line if and only if $\mathbf{x}' \wedge \mathbf{x}'' = 0$.

## 14.4 Uniqueness Theorem for Curves

We have seen how special properties of curves are determined by properties of their curvature and torsion as long as $\kappa \neq 0$ for all values of the unit parameter $s$. The following theorem tells to what extent curvature and torsion determines a curve.

**Fundamental Theorem of Curves 14.4.1.** *Any regular curve $\mathbf{x}(t)$ in $\mathbb{R}^3$ with $\kappa > 0$ is completely determined, up to position, by its curvature $\kappa(t)$ and torsion $\tau(t)$. More precisely, given an initial position $\mathbf{x}_0$ and initial directions $\mathbf{T}_0, \mathbf{N}_0, \mathbf{B}_0$, the regular curve $\mathbf{x}(t)$ is uniquely determined by its curvature $\kappa(t)$ and torsion $\tau(t)$.*

The proof of the Fundamental Theorem of Curves depends on the existence and uniqueness of a solution to a system of *ordinary differential equations*, in this case the Frenet-Serret equations (14.2), together with the initial position $\mathbf{x}_0$ and the initial directions $\mathbf{T}_0, \mathbf{N}_0, \mathbf{B}_0$. See [56, p.42–44] for an elementary discussion of the issues involved. A more direct approach to the problem, involving the Darboux bivector (14.3), is given in the problems below.

## Exercises

1. Show that $\frac{\mathrm{d}}{\mathrm{d}s}\exp\left(\frac{1}{2}\mathrm{i}\hat{\mathbf{c}}\theta\right) = \Omega \exp\left(\frac{1}{2}\mathrm{i}\hat{\mathbf{c}}\theta\right)$ where

$$\Omega = \left[\frac{\mathrm{d}}{\mathrm{d}s}\exp\left(\frac{1}{2}\mathrm{i}\hat{\mathbf{c}}\theta\right)\right]\exp\left(-\frac{1}{2}\mathrm{i}\hat{\mathbf{c}}\theta\right) = \frac{1}{2}\left[\mathrm{i}\frac{\mathrm{d}\theta}{\mathrm{d}s} + (1-\exp(\mathrm{i}\hat{\mathbf{c}}\theta)\frac{\mathrm{d}\hat{\mathbf{c}}}{\mathrm{d}s}\right]\hat{\mathbf{c}}$$

   and $\mathrm{i} = \mathbf{e}_{123}$.
2. Show that

$$\begin{pmatrix}\mathbf{T}\\ \mathbf{N}\\ \mathbf{B}\end{pmatrix} = \exp\left(\frac{1}{2}\mathrm{i}\hat{\mathbf{c}}\theta\right)\begin{pmatrix}\mathbf{T}_0\\ \mathbf{N}_0\\ \mathbf{B}_0\end{pmatrix}\exp\left(-\frac{1}{2}\mathrm{i}\hat{\mathbf{c}}\theta\right)$$

   is the solution of the Frenet-Serret equations (14.3), if $\hat{\mathbf{c}}$ and $\theta$ of Problem 1 are chosen so that $\Omega = -\frac{\mathrm{i}}{2}(\kappa\mathbf{B} + \tau\mathbf{T})$. The unique unit speed curve $\mathbf{x}(s)$ is then specified by $\mathbf{x}(s) = \int_0^s \mathbf{T}\mathrm{d}s$ with the initial conditions that $\mathbf{T}(0) = \mathbf{T}_0, \mathbf{N}(0) = \mathbf{N}_0, \mathbf{B}(0) = \mathbf{B}_0$, and $\mathbf{x}(0) = \mathbf{x}_0$.
3. Show that the equation

$$\begin{pmatrix}\mathbf{T}\\ \mathbf{N}\\ \mathbf{B}\end{pmatrix} = \exp\left(\frac{1}{2}\mathrm{i}\hat{\mathbf{c}}\theta\right)\begin{pmatrix}\mathbf{T}_0\\ \mathbf{N}_0\\ \mathbf{B}_0\end{pmatrix}\exp\left(-\frac{1}{2}\mathrm{i}\hat{\mathbf{c}}\theta\right)$$

   can be directly solved for $\hat{\mathbf{c}}$ and $\theta$ by using

$$(\mathbf{T}_0, \mathbf{N}_0, \mathbf{B}_0)\begin{pmatrix}\mathbf{T}\\ \mathbf{N}\\ \mathbf{B}\end{pmatrix} = \partial_{\mathbf{a}}\exp\left(\frac{1}{2}\mathrm{i}\hat{\mathbf{c}}\theta\right)\mathbf{a}\exp\left(-\frac{1}{2}\mathrm{i}\hat{\mathbf{c}}\theta\right)$$

   to get

$$\mathrm{e}^{\mathrm{i}\hat{\mathbf{c}}\theta} = \frac{1}{2}(\mathbf{T}\mathbf{T}_0 + \mathbf{N}\mathbf{N}_0 + \mathbf{B}\mathbf{B}_0 - 1)$$

   or, equivalently,

$$\cos\theta = \frac{1}{2}(\mathbf{T}\cdot\mathbf{T}_0 + \mathbf{N}\cdot\mathbf{N}_0 + \mathbf{B}\cdot\mathbf{B}_0 - 1)$$

and

$$\hat{\mathbf{c}}\sin\theta = \frac{1}{2}(\mathbf{T}\times\mathbf{T}_0 + \mathbf{N}\times\mathbf{N}_0 + \mathbf{B}\times\mathbf{B}_0).$$

4. By equating the expressions found for $\Omega$ in Problems 2 and 3, show that $\theta = \theta(s)$ and $\hat{\mathbf{c}} = \hat{\mathbf{c}}(s)$ satisfy the equivalent differential equation

$$\frac{\mathrm{d}\theta}{\mathrm{d}s}\hat{\mathbf{c}} + (1 - \mathrm{e}^{\mathrm{i}\hat{\mathbf{c}}\theta})\hat{\mathbf{c}}\times\frac{\mathrm{d}\hat{\mathbf{c}}}{\mathrm{d}s} = -(\kappa\mathbf{B}_0 + \tau\mathbf{T}_0) = -2\mathrm{i}\mathrm{e}^{-\frac{1}{2}\mathrm{i}\hat{\mathbf{c}}\theta}\frac{\mathrm{d}}{\mathrm{d}s}\left[\mathrm{e}^{\frac{1}{2}\mathrm{i}\hat{\mathbf{c}}\theta}\right]$$

or, equivalently,

$$\frac{\mathrm{d}\theta}{\mathrm{d}s}\hat{\mathbf{c}} - \sin\theta\frac{\mathrm{d}\hat{\mathbf{c}}}{\mathrm{d}s} + (1-\cos\theta)\hat{\mathbf{c}}\times\frac{\mathrm{d}\hat{\mathbf{c}}}{\mathrm{d}s} = -(\kappa\mathbf{B}_0 + \tau\mathbf{T}_0).$$

This shows that the differential equation for the Frenet-Serret frame is completely determined by the initial directions $\mathbf{T}_0, \mathbf{N}_0, \mathbf{B}_0$ and the scalar functions $\kappa(s)$ and $\tau(s)$ for curvature and torsion along the unit speed curve $\mathbf{x}(s)$.

5. By squaring the equation found in Problem 4, show that

$$\left(\frac{\mathrm{d}\theta}{\mathrm{d}s}\right)^2 + 2(1-\cos\theta)\left(\frac{\mathrm{d}\hat{\mathbf{c}}}{\mathrm{d}s}\right)^2 = \kappa^2 + \tau^2.$$

Use this to show that $\frac{\mathrm{d}\theta}{\mathrm{d}s}|_{s=0} = \sqrt{\kappa_0^2 + \tau_0^2}$.

6. By differentiating the last equation given in Problem 3, show that

$$\frac{\mathrm{d}\hat{\mathbf{c}}}{\mathrm{d}s}\sin\theta + \hat{\mathbf{c}}\cos\theta\frac{\mathrm{d}\theta}{\mathrm{d}s} = \frac{1}{2}\left[\kappa\mathbf{N}\times\mathbf{T}_0 + (-\kappa\mathbf{T} + \tau\mathbf{B})\times\mathbf{N}_0 - \tau\mathbf{N}\times\mathbf{B}_0\right].$$

# Chapter 15
# Differential Geometry of $k$-Surfaces

*The essence of mathematics is not to make simple things complicated, but to make complicated things simple.*
—S. Gudder

We have discussed the calculus of a $k$-surface, or a $k$-manifold $\mathscr{M}$ embedded in $\mathbb{R}^n$ in Chap. 13, utilizing the basic building block of a $k$-rectangle. In differential geometry, a $k$-surface is rigorously defined by an *atlas* of *charts* which maps the points of open sets in $\mathbb{R}^k$ onto regions in the manifold $\mathscr{M}$ in a one-to-one continuous and differentiable manner, in much the same way that the maps of an atlas represent the surface of the Earth. The tricky part of the definition is to guarantee that in regions where the charts on $\mathscr{M}$ overlap, that they do so in a consistent way.

In 1936, Hassler Whitney and, later in 1956, John Nash proved a number of embedding theorems which show that an abstract $k$-manifold can always be isometrically (preserving distances) embedded in $\mathbb{R}^n$ if $n \geq k^2 + 5k + 3$. In light of the Whitney-Nash embedding theorems [57, 58, 96, 97],

http://en.wikipedia.org/wiki/Nash_embedding_theorem
http://www.math.poly.edu/~yang/papers/gunther.pdf

little or nothing is lost by assuming that a general manifold is embedded in $\mathbb{R}^n$ or even more generally in $\mathbb{R}^{p,q}$. We call such manifolds *vector manifolds*. By making such an assumption, all the tools of the geometric algebra $\mathbb{G}_{p,q}$ of $\mathbb{R}^{p,q}$ become available in the study of the *extrinsic* properties of $\mathscr{M}$, alongside the *intrinsic* properites of $\mathscr{M}$ which are independent of the particular embedding in $\mathbb{R}^n$.

The most basic object on a $k$-surface is its *tangent geometric algebra* $\mathscr{T}_{\mathbf{x}}$ at each point $\mathbf{x}$ on the surface. Since a $k$-surface is embedded in $\mathbb{R}^n$, the concepts of differentiation and integration carry over in a natural way from differentiation and integration on $\mathbb{R}^n$. Whereas $\mathbb{R}^n$ is flat, a general $k$-surface has many different

G. Sobczyk, *New Foundations in Mathematics: The Geometric Concept of Number*, DOI 10.1007/978-0-8176-8385-6_15,

kinds of *curvature*. We study the *principle*, *Gaussian*, *mean*, and *normal* curvatures of a $k$-surface. The *shape* and *curvature* tensors of a $k$-surface are also defined and studied.

## 15.1 The Definition of a $k$-Surface $\mathscr{M}$ in $\mathbb{R}^n$

Let $\mathbb{R}^n$ denote Euclidean $n$-space and $\mathbb{G}_n = \mathbb{G}(\mathbb{R}^n)$ the corresponding geometric algebra of $\mathbb{R}^n$. By a *regular r-differentiable* $k$-dimensional surface $\mathscr{M}$ in $\mathbb{R}^n$, we mean an *alas* of $r$-differentiable *charts* (coordinate patches) which taken together cover $\mathscr{M}$.

Let

$$\mathbf{x}(\mathbf{s}) = \mathbf{x}(s^1,\ldots,s^k) \text{ and } \mathbf{y}(\mathbf{s}) = \mathbf{y}(s^1,\ldots,s^k) \tag{15.1}$$

be two $r$-differentiable charts on $\mathscr{M}$, pictured in Fig. 15.1. This means that the charts are $C^r$ differentiable in each of the local coordinates $s^i$ of the point $\mathbf{x} \in \mathscr{M}$. In addition, the composition maps $\mathbf{x} \circ \mathbf{y}^{-1}(\mathbf{s})$ and $\mathbf{y} \circ \mathbf{x}^{-1}(\mathbf{s})$ must be $C^r$ differentiable in the overlapping regions on $\mathscr{M}$. When $r > 0$, $\mathscr{M}$ is called a *differentiable* or *smooth manifold*. A more careful discussion of these fundamental concepts, with examples, is found in [45, p. 1–5] or recently in [52, p. 11–22].

In Chap. 13, a coordinate patch $M = \mathbf{x}(R)$ was chosen to be the image of a closed $k$-rectangle $R = \times_{i=1}^{k}[a^i, b^i]$ in $\mathscr{M}$. We now need only require that the closed rectangular patches lie within the open regions on $\mathscr{M}$ defined by the charts covering $\mathscr{M}$. The *tangent space* $\mathscr{T}_{\mathbf{x}}^1$ of tangent vectors at each interior point $\mathbf{x}$ of a regular $k$-surface $\mathscr{M}$ is spanned by the linearly independent tangent vectors $(\mathbf{x})_{(k)}$,

$$\mathscr{T}_{\mathbf{x}}^1 := \operatorname{span}(\mathbf{x})_{(k)} = \operatorname{span}\left(\mathbf{x}_1\ \mathbf{x}_2\ \ldots\ \mathbf{x}_k\right) \tag{15.2}$$

where $\mathbf{x}_i = \frac{\partial \mathbf{x}}{\partial s^i}$. The vectors of the tangent space $\mathscr{T}_{\mathbf{x}}^1$, under the geometric product in $\mathbb{G}_n$, generate the tangent geometric algebra $\mathscr{T}_{\mathbf{x}}$ to the surface $\mathscr{M}$ at the point $\mathbf{x}$. Of course, the tangent geometric algebra $\mathscr{T}_{\mathbf{x}}$ is a $2^k$-dimensional geometric subalgebra of the geometric algebra $\mathbb{G}_n$.

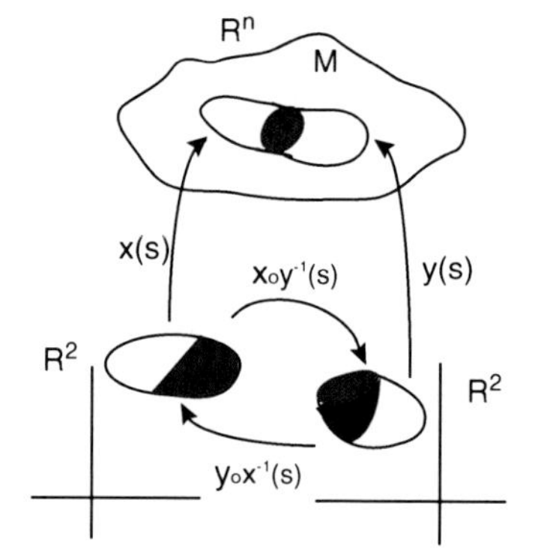


**Fig. 15.1** The images of two coordinate charts on $M$ must be compatible on their overlapping ranges

Let

$$I_{\mathbf{x}} = \wedge(\mathbf{x})_{(k)} = \mathbf{x}_1 \wedge \cdots \wedge \mathbf{x}_k = \alpha_{\mathbf{x}} \widehat{I}_{\mathbf{x}} \in \mathcal{T}_{\mathbf{x}}, \tag{15.3}$$

be the tangent $k$-dimensional *pseudoscalar element* to the $k$-surface at the point $\mathbf{x} \in \mathcal{M}$ and let

$$\alpha_{\mathbf{x}} = |I_{\mathbf{x}}| = \sqrt{I_{\mathbf{x}} I_{\mathbf{x}}^{\dagger}} = I_{\mathbf{x}} \widehat{I}_{x}^{\dagger} = \sqrt{\det g} \tag{15.4}$$

be its magnitude. The pseudoscalar element $I_{\mathbf{x}}$ determines whether or not the $k$-surface is orientable.

**Definition 15.1.1.** The $k$-surface $\mathcal{M}$ is orientable if the unit pseudoscalar element $\widehat{I}_{\mathbf{x}}$ is single valued at all points $\mathbf{x} \in \mathcal{M}$. Otherwise, the $k$-surface is non-orientable.

In the case $k = n-1$, the $k$-surface $\mathcal{M}$ is a hypersurface in $\mathbb{R}^n$. Letting $\mathrm{i} = \wedge(\mathbf{e})_{(n)}$ be the pseudoscalar element of $\mathbb{R}^n$ defined by the standard basis $(\mathbf{e})_{(n)} = (\mathbf{e}_1 \; \mathbf{e}_2 \; \ldots \; \mathbf{e}_n)$,

$$\mathbf{n}_{\mathbf{x}} = \widehat{I}_{\mathbf{x}} \mathrm{i}^{-1} = \frac{\mathbf{x}_1 \wedge \mathbf{x}_2 \wedge \cdots \wedge \mathbf{x}_k}{\alpha_{\mathbf{x}}} (\mathbf{e}_n \wedge \mathbf{e}_{n-1} \wedge \cdots \wedge \mathbf{e}_1) \tag{15.5}$$

is the unit vector normal to the surface $\mathcal{M}$ at the point $\mathbf{x}$. If $\mathcal{M}$ is an orientable closed hypersurface, then the standard basis can be chosen in such a way that $\mathbf{n}_{\mathbf{x}}$ is the outward normal to the surface at each point $\mathbf{x} \in \mathcal{M}$.

As discussed before in (13.6), the *metric tensor* $[g] = [g_{ij}]$ of the surface $\mathcal{M}$ is defined by the matrix

$$[g_{ij}] = \begin{pmatrix} \mathbf{x}_1 \\ \mathbf{x}_2 \\ \cdot \\ \cdot \\ \mathbf{x}_k \end{pmatrix} \cdot (\mathbf{x}_1 \; \mathbf{x}_2 \; \ldots \; \mathbf{x}_k) = \begin{pmatrix} \mathbf{x}_1 \cdot \mathbf{x}_1 & \ldots & \mathbf{x}_1 \cdot \mathbf{x}_k \\ \ldots & \ldots & \ldots \\ \ldots & \ldots & \ldots \\ \mathbf{x}_k \cdot \mathbf{x}_1 & \ldots & \mathbf{x}_k \cdot \mathbf{x}_k \end{pmatrix}, \tag{15.6}$$

where $g_{ij} = \mathbf{x}_i \cdot \mathbf{x}_j$. The *reciprocal metric tensor* $[g]^{-1} = [g^{ij}]$ is defined by $[g^{ij}] = [g_{ij}]^{-1}$. The *reciprocal basis vectors* $(\mathbf{x}^1 \; \mathbf{x}^2 \; \ldots \; \mathbf{x}^k)^{\mathrm{T}}$, in the notation used in (7.12), is given by

$$(\mathbf{x})^{(k)} = [g]^{-1} (\mathbf{x})_{(k)}.$$

Also, $\mathbf{x}^j = g^{ji} \mathbf{x}_i$ where the Einstein summation convention over repeated upper and lower indices is being used.

A *geometric-valued function* $F : \mathcal{M} \to \mathbb{G}_n$ assigns a value $F = F(\mathbf{x}) \in \mathbb{G}_n$ for each point $\mathbf{x} \in \mathcal{M}$. The function $F(\mathbf{x})$ is said to be *continuous* in the induced topology from $\mathbb{R}^n$ at the point $\mathbf{x}$ if $\lim_{\mathbf{y} \to 0} F(\mathbf{x} + \mathbf{y}) = F(\mathbf{x})$, where $\mathbf{y}$ is in an open set $\mathcal{N}_{\mathbf{x}} \subset \mathcal{M}$ containing the point $\mathbf{x}$. By the *vector derivative* $\partial_{\mathbf{x}} F(\mathbf{x})$ of $F(\mathbf{x})$ at the point $\mathbf{x} \in \mathcal{M}$, we mean

$$\partial_{\mathbf{x}} F(\mathbf{x}) := \sum_{i=1}^{k} \mathbf{x}^i \frac{\partial F(\mathbf{x})}{\partial s^i}, \tag{15.7}$$

if the derivative exists. Note that the reciprocal basis vectors $\mathbf{x}^i$, as before, satisfy (or are defined by) the property that $\mathbf{x}^i = \partial_{\mathbf{x}} s^i$.

By the *directional* **a**-*derivative* of $F(\mathbf{x})$ we mean

$$F_{\mathbf{a}}(\mathbf{x}) := \mathbf{a} \cdot \partial_{\mathbf{x}} F(\mathbf{x}), \tag{15.8}$$

provided the derivative exists. This is the same vector derivative and **a**-derivative that were defined in Definition 13.2.1 and in (3.45) and (3.46) in Chap. 3. More formally, we say that $F(\mathbf{x})$ is *differentiable* at $\mathbf{x} \in \mathcal{M}$ if there is a linear map $F_{\mathbf{a}} : \mathcal{T}_{\mathbf{x}} \to \mathbb{G}_n$ such that

$$\lim_{\mathbf{v} \to 0, \mathbf{v} \in \mathcal{T}_{\mathbf{x}}^1} \frac{|F(\mathbf{x}+\mathbf{v}) - F(\mathbf{x}) - F_{\mathbf{v}}(\mathbf{x})|}{|\mathbf{v}|} = 0. \tag{15.9}$$

This definition will be refined in an exercise after the projection operator onto the tangent space has been defined. The map $F_{\mathbf{a}}(\mathbf{x})$ is linear in the tangent vector $\mathbf{a} \in \mathcal{T}_{\mathbf{x}}^1$ and uniquely determined by the local properties of $F$ at the point $\mathbf{x} \in \mathcal{M}$, [92, p.16]. The vector derivative at the point $\mathbf{x}$ can be directly expressed in terms of the vector derivative in the tangent space $\mathcal{T}_{\mathbf{x}}$ by

$$\partial_{\mathbf{x}} F(\mathbf{x}) = \partial_{\mathbf{v}} \Big( \mathbf{v} \cdot \partial_{\mathbf{x}} F(\mathbf{x}) \Big), \tag{15.10}$$

where $\mathbf{v}$ ranges over the tangent space $\mathcal{T}_{\mathbf{x}}^1$ at the point $\mathbf{x}$.

It is worthwhile to review the crucial integrability property (13.8) of the vector derivative. For a $C^2$ differentiable function $F(\mathbf{x})$, we have

$$(\mathbf{a} \wedge \mathbf{b}) \cdot (\partial_{\mathbf{x}} \wedge \partial_{\mathbf{x}}) F(\mathbf{x}) = 0, \tag{15.11}$$

where $\mathbf{a}, \mathbf{b}$ are tangent vector fields in $\mathcal{T}_{\mathbf{x}}^1$. Expanding (15.11),

$$(\mathbf{a} \wedge \mathbf{b}) \cdot (\partial_{\mathbf{x}} \wedge \partial_{\mathbf{x}}) F(\mathbf{x}) = [\mathbf{b} \cdot \partial_{\mathbf{x}}, \mathbf{a} \cdot \partial_{\mathbf{x}}] F(\mathbf{x}) - [\mathbf{b}, \mathbf{a}] \cdot \partial_{\mathbf{x}} F(\mathbf{x}),$$

where $[\mathbf{b} \cdot \partial_{\mathbf{x}}, \mathbf{a} \cdot \partial_{\mathbf{x}}] := \mathbf{b} \cdot \partial_{\mathbf{x}} \mathbf{a} \cdot \partial_{\mathbf{x}} - \mathbf{a} \cdot \partial_{\mathbf{x}} \mathbf{b} \cdot \partial_{\mathbf{x}}$ and $[\mathbf{b}, \mathbf{a}] := \mathbf{b} \cdot \partial_{\mathbf{x}} \mathbf{a} - \mathbf{a} \cdot \partial_{\mathbf{x}} \mathbf{b}$. The quantity $[\mathbf{a}, \mathbf{b}]$ is called the *Lie bracket* of the vector fields $\mathbf{a}, \mathbf{b} \in \mathcal{T}_{\mathbf{x}}$. We also express this integrability condition of the differentiable function $F(\mathbf{x})$ more directly by writing

$$F_{\mathbf{a},\mathbf{b}}(\mathbf{x}) - F_{\mathbf{b},\mathbf{a}}(\mathbf{x}) = 0, \tag{15.12}$$

where $F_{\mathbf{a},\mathbf{b}}(\mathbf{x}) := \mathbf{b} \cdot \partial_{\mathbf{x}} F_{\mathbf{a}}(\mathbf{x}) - F_{\mathbf{b} \cdot \partial \mathbf{a}}(\mathbf{x})$.

An important class of functions defined on a $k$-surface $\mathcal{M}$ are *geometric-valued tensors* or *g-tensors*.

**Definition 15.1.2.** A homogeneous geometric-valued $r$-tensor

$$H_{\mathbf{x}} : \mathcal{M} \times \mathbb{G}_n^r \longrightarrow \mathbb{G}_n^r$$

is a linear $r$-vector-valued function $H_{\mathbf{x}}(A_r)$ in $r$-vector argument $A_r \in \mathbb{G}_n^r$ at each point $\mathbf{x} \in \mathcal{M}$.

Thus, a homogeneous $r$-tensor assigns a linear $r$-vector-valued function to each point in an open neighborhood of the point $\mathbf{x} \in \mathcal{M}$. We will shortly see that we can generalize this definition to include additional linear arguments.

The directional $\mathbf{a}$-derivative and vector derivative can be extended to $g$-tensors as given below.

**Definition 15.1.3.** The *directional* $\mathbf{a}$*-derivative* of the $g$-tensor $H_{\mathbf{x}}(g)$ at the point $\mathbf{x} \in \mathcal{M}$ is defined by

$$H_{\mathbf{a}}(g) = \mathbf{a} \cdot \partial_{\mathbf{x}} H_{\mathbf{x}}(g) - H_{\mathbf{x}}(\mathbf{a} \cdot \partial_{\mathbf{x}} g) \tag{15.13}$$

for all $g \in \mathbb{G}_n$. The vector derivative of $H_{\mathbf{x}}(g)$ is specified by

$$\dot{\partial}_{\mathbf{x}} \dot{H}_{\mathbf{x}}(g) = \partial_{\mathbf{v}} H_{\mathbf{v}}(g) = \sum_{i=1}^{k} \mathbf{x}^i H_{\mathbf{x}_i}(g). \tag{15.14}$$

Of course, if in (15.13) $g \in \mathbb{G}_n$ is a constant and not a function of $\mathbf{x} \in \mathcal{M}$, the definition simplifies to $H_{\mathbf{a}}(g) = \mathbf{a} \cdot \partial_{\mathbf{x}} H_{\mathbf{x}}(g)$. Indeed, the purpose of the second term in (15.13) of Definition 15.1.3 is to guarantee that the $\mathbf{a}$-directional derivative of a $g$-tensor depends only upon the value of $g$ at the point $\mathbf{x} \in \mathcal{M}$ and so $g$ can always be treated as a constant. Higher-order $\mathbf{a}$-derivatives of $H_{\mathbf{x}}(g)$ are defined in much the same way in order to preserve the linearity in its tensor arguments. For example, for $\mathbf{a}, \mathbf{b} \in \mathcal{T}_{\mathbf{x}}$,

$$H_{\mathbf{a},\mathbf{b}}(g) = \mathbf{b} \cdot \partial_{\mathbf{x}} H_{\mathbf{a}}(g) - H_{\mathbf{b} \cdot \partial \mathbf{a}}(g) - H_{\mathbf{a}}(\mathbf{b} \cdot \partial_{\mathbf{x}} g). \tag{15.15}$$

Much of the local differential geometry of a $k$-surface $\mathcal{M}$ in $\mathbb{R}^n$ depends upon how the pseudoscalar element $I_{\mathbf{x}}$ behaves at the point $\mathbf{x}$. Important in this study is the projection operator, $P_{\mathbf{x}} : \mathbb{G}_n \longrightarrow \mathcal{T}_{\mathbf{x}}$, which projects the geometric algebra $\mathbb{G}_n$ into the tangent geometric algebra $\mathcal{T}_{\mathbf{x}}$ at each point $\mathbf{x} \in \mathcal{M}$. Let

$$A = \sum_{r=0}^{n} < A >_r = \sum_{r=0}^{n} A_r \in \mathbb{G}_n$$

be a general multivector in $\mathbb{G}_n$. The projection operator $P_{\mathbf{x}}(A)$ is our first example of a geometric-valued tensor on $\mathcal{M}$.

**Definition 15.1.4.** The projection operator $P_{\mathbf{x}} : \mathbb{G}_n \longrightarrow \mathcal{T}_{\mathbf{x}}$ is specified by

$$P_{\mathbf{x}}(A) := P_{\mathbf{x}}(\sum_{r=0}^{n} A_r) = A_0 + \sum_{r=1}^{k} (A_r \cdot I_{\mathbf{x}}) I_{\mathbf{x}}^{-1},$$

for all multivectors $A \in \mathbb{G}_n$.

The projection operator is an outermorphism at each point $\mathbf{x} \in \mathcal{M}$ and so satisfies the outermorphism rule 7.1.1. Of course, if $\mathbf{a} \in \mathcal{T}_{\mathbf{x}}^{1}$, then $P_{\mathbf{x}}(\mathbf{a}) = \mathbf{a}$. The projection $P_{\mathbf{x}}$ is a symmetric operator, satisfying

$$P_{\mathbf{x}}(A_r) \cdot B_r = P_{\mathbf{x}}(A_r) \cdot P_{\mathbf{x}}(B_r) = A_r \cdot P_{\mathbf{x}}(B_r), \tag{15.16}$$

for all $r$-vectors $A_r, B_r \in \mathbb{G}_n^r$. This follows from the steps

$$\begin{aligned} P_{\mathbf{x}}(A_r) \cdot B_r &= < A_r \cdot I_{\mathbf{x}} I_{\mathbf{x}}^{-1} B_r >_0 = < A_r \cdot I_{\mathbf{x}} I_{\mathbf{x}}^{-1} \cdot B_r >_0 \\ &= < A_r \cdot I_{\mathbf{x}} I_{\mathbf{x}}^{-1} I_{\mathbf{x}}^{-1} I_{\mathbf{x}} \cdot B_r >_0 = P_{\mathbf{x}}(A_r) \cdot P_{\mathbf{x}}(B_r). \end{aligned}$$

The directional $\mathbf{a}$-derivative of the projection operator $P_{\mathbf{x}}$ at the point $\mathbf{x} \in \mathcal{M}$ is given by

$$P_{\mathbf{a}}(A) = \mathbf{a} \cdot \partial_{\mathbf{x}} P_{\mathbf{x}}(A) - P_{\mathbf{x}}(\mathbf{a} \cdot \partial_{\mathbf{x}} A) \tag{15.17}$$

for all $A \in \mathbb{G}_n$. Taking the directional $\mathbf{a}$-derivative of (15.16), gives

$$P_{\mathbf{a}}(A_r) \cdot B_r = P_{\mathbf{a}}(A_r) \cdot P_{\mathbf{x}}(B_r) + P_{\mathbf{x}}(A_r) \cdot P_{\mathbf{a}}(B_r) = A_r \cdot P_{\mathbf{a}}(B_r) \tag{15.18}$$

for all $A_r, B_r \in \mathbb{G}_n^r$, showing that the operator $P_{\mathbf{a}}$ is also symmetric on $\mathbb{G}_n^r$. However, $P_{\mathbf{a}}$ is not an outermorphism because it obeys, instead, the *derivation rule*. Taking the directional $\mathbf{a}$-derivative of the equation

$$P_{\mathbf{x}}(A_r \wedge B_s) = P_{\mathbf{x}}(A_r) \wedge P_{\mathbf{x}}(B_s),$$

for the $r$-vector $A_r$ and $s$-vector $B_s$, we find that that

$$P_{\mathbf{a}}(A_r \wedge B_s) = P_{\mathbf{a}}(A_r) \wedge P_{\mathbf{x}}(B_s) + P_{\mathbf{x}}(A_r) \wedge P_{\mathbf{a}}(B_s). \tag{15.19}$$

Since $P_{\mathbf{x}}$ is a projection, it satisfies the basic property that $P_{\mathbf{x}}^2(A_r) = P_{\mathbf{x}}(A_r)$ for all $r$-vectors $A_r \in \mathbb{G}_n^r$. Taking the directional $\mathbf{b}$-derivative of this relationship gives

$$P_{\mathbf{b}}\left(P_{\mathbf{x}}(A_r)\right) + P_{\mathbf{x}}\left(P_{\mathbf{b}}(A_r)\right) = P_{\mathbf{b}}(A_r). \tag{15.20}$$

From this it immediately follows that if $A_r \in \mathcal{T}_{\mathbf{x}}^{r}$, then $P_{\mathbf{x}}\left(P_{\mathbf{b}}(A_r)\right) = 0$, and if $P_{\mathbf{x}}(A_r) = 0$, then $P_{\mathbf{x}}\left(P_{\mathbf{b}}(A_r)\right) = P_{\mathbf{b}}(A_r)$. This is important in determining the basic curvature relationships of a $k$-surface.

Now define the *matrix of second derivatives* of the coordinate tangent vectors $\mathbf{x}_i$ to $\mathcal{M}$ at $\mathbf{x}$ by

$$(\mathbf{x}_{ij}) := \begin{pmatrix} \frac{\partial \mathbf{x}_1}{\partial s^1} & \cdots & \frac{\partial \mathbf{x}_1}{\partial s^k} \\ \cdots & \cdots & \cdots \\ \cdots & \cdots & \cdots \\ \frac{\partial \mathbf{x}_k}{\partial s^1} & \cdots & \frac{\partial \mathbf{x}_k}{\partial s^k} \end{pmatrix}, \tag{15.21}$$

where $\mathbf{x}_{ij} = \frac{\partial \mathbf{x}_i}{\partial s^j}$. Since partial derivatives commute,

$$\mathbf{x}_{ij} = \frac{\partial^2 \mathbf{x}}{\partial s^j \partial s^i} = \frac{\partial^2 \mathbf{x}}{\partial s^i \partial s^j} = \mathbf{x}_{ji},$$

the matrix of second derivatives $(\mathbf{x}_{ij})$ is symmetric.

The matrix $(\mathbf{x}_{ij})$ can be decomposed into *intrinsic* and *extrinsic* vector matrix parts:

$$(\mathbf{x}_{ij}) = (\mathbf{K}_{ij}) + (\mathbf{L}_{ij}). \tag{15.22}$$

The *Christoffel vectors* $\mathbf{K}_{ij} := P_{\mathbf{x}}(\mathbf{x}_{ij})$ are tangent to the surface, and the normal vectors $\mathbf{L}_{ij}$ are orthogonal to the surface at the point $\mathbf{x} \in \mathscr{M}$. Since $P_{\mathbf{x}}(\mathbf{x}_i) = \mathbf{x}_i$, it follows that

$$\frac{\partial P_{\mathbf{x}}(\mathbf{x}_i)}{\partial s^j} = P_j(\mathbf{x}_i) + P_{\mathbf{x}}(\mathbf{x}_{ij}) = \mathbf{x}_{ij},$$

which implies that $\mathbf{L}_{ij} = P_j(\mathbf{x}_i)$, where $P_j := \frac{\partial P_{\mathbf{x}}}{\partial s^j}$. The *Christoffel components* are defined by $\Gamma_{ij}^k = \mathbf{K}_{ij} \cdot \mathbf{x}^k$. When $\mathscr{M}$ is an oriented $(n-1)$-dimensional hypersurface in $\mathbb{R}^n$, with the outward unit normal vector $\mathbf{n}_{\mathbf{x}}$ at the point $\mathbf{x} \in \mathscr{M}$, the *normal components* $L_{ij}$ are defined by $\mathbf{L}_{ij} = L_{ij}\mathbf{n}_{\mathbf{x}}$, or

$$L_{ij} = \mathbf{L}_{ij} \cdot \mathbf{n}_{\mathbf{x}} = \mathbf{x}_{ij} \cdot \mathbf{n}_{\mathbf{x}} = L_{ji}. \tag{15.23}$$

By applying the subscript notation to $\mathbf{a}$-derivatives, as we did in (15.15), we have

$$\mathbf{L}_{\mathbf{a},\mathbf{b}} = P_{\mathbf{a}}(P_{\mathbf{x}}(\mathbf{b})) = P_{\mathbf{b}}(P_{\mathbf{x}}(\mathbf{a})) = \mathbf{L}_{\mathbf{b},\mathbf{a}}, \tag{15.24}$$

where $\mathbf{a}, \mathbf{b} \in \mathscr{T}_{\mathbf{x}}^1$. When $\mathbf{a} = \mathbf{x}_i$ and $\mathbf{b} = \mathbf{x}_j$, $\mathbf{L}_{\mathbf{a},\mathbf{b}} = \mathbf{L}_{ij}$ as expected. Noting that $\mathbf{x}_{\mathbf{a}} = \mathbf{a} \cdot \partial_{\mathbf{x}} \mathbf{x} = \mathbf{a}$, and

$$\mathbf{x}_{\mathbf{a},\mathbf{b}} = \mathbf{b} \cdot \partial_{\mathbf{x}} \mathbf{x}_{\mathbf{a}} - \mathbf{x}_{\mathbf{b} \cdot \partial \mathbf{a}} = \mathbf{b} \cdot \partial_{\mathbf{x}} \mathbf{x}_{\mathbf{a}} - P_{\mathbf{x}}(\mathbf{b} \cdot \partial \mathbf{a}),$$

it also makes sense to write $\mathbf{K}_{\mathbf{a},\mathbf{b}} = P_{\mathbf{x}}(\mathbf{x}_{\mathbf{a},\mathbf{b}})$.

It is easy to verify that

$$\mathbf{K}_{ij} = \frac{1}{2}\mathbf{x}^k \left( \frac{\partial g_{ik}}{\partial s^j} - \frac{\partial g_{ij}}{\partial s^k} + \frac{\partial g_{kj}}{\partial s^i} \right). \tag{15.25}$$

We have

$$\mathbf{K}_{ij} = \mathbf{x}^k \mathbf{x}_k \cdot \mathbf{x}_{ij} = \mathbf{x}^k (\partial_j \mathbf{x}_k \cdot \mathbf{x}_i - \mathbf{x}_{jk} \cdot \mathbf{x}_i) = \mathbf{x}^k (\partial_j g_{ik} - \mathbf{x}_{jk} \cdot \mathbf{x}_i).$$

Noting that $\mathbf{K}_{ij} = \mathbf{K}_{ji}$, we get

$$\mathbf{K}_{ij} = \frac{1}{2}(\mathbf{K}_{ij} + \mathbf{K}_{ji}) = \mathbf{x}^k \left( \partial_j g_{ik} - \mathbf{x}_{jk} \cdot \mathbf{x}_i + \partial_i g_{jk} - \mathbf{x}_{ik} \cdot \mathbf{x}_j \right).$$

Substituting $-\mathbf{x}_{jk} \cdot \mathbf{x}_i = -\partial_k g_{ij} + \mathbf{x}_j \cdot \mathbf{x}_{ik}$ into the last equation gives the desired result (15.25). This shows that $\mathbf{K}_{ij}$ are completely determined by the metric tensor $g$ at the point $\mathbf{x} \in \mathscr{M}$.

## Exercises

Let $\mathcal{M} \subset \mathbb{R}^n$ and $P_{\mathbf{x}}$ be the projection onto the tangent algebra $\mathcal{T}_{\mathbf{x}}$ at $\mathbf{x} \in \mathcal{M}$.

1. Show that $g_{ij,k} := \mathbf{x}_k \cdot \partial_{\mathbf{x}} g_{ij} = \frac{\partial g_{ij}}{\partial s^k} = \mathbf{x}_{ik} \cdot \mathbf{x}_j + \mathbf{x}_i \cdot \mathbf{x}_{jk} = g_{lj}\Gamma_{ik}^{\;l} + g_{li}\Gamma_{jk}^{\;l}$.
2. Show that $g^{ij}g_{jk} = \delta_k^i$ implies $g^{ij}_{\;\;k} = -g^{il}\Gamma^j_{lk} - g^{lj}\Gamma^i_{lk}$.
3. For $\mathbf{a} \in \mathbb{R}^n$, write $\mathbf{a} = \mathbf{a}_{\|} + \mathbf{a}_{\perp}$ where $\mathbf{a}_{\|} := P_{\mathbf{x}}(\mathbf{a})$. Show that

$$P_i(\mathbf{a}) = P_i(\mathbf{a}_{\|}) + P_i(\mathbf{a}_{\perp}) = \mathbf{a} \cdot \mathbf{L}_{ij}\mathbf{x}^j + \mathbf{a} \cdot \mathbf{x}^j \mathbf{L}_{ij}.$$

Let $\mathbf{x} = \mathbf{x}(s^1, s^2)$ be a 2-surface in $\mathbb{R}^3$. Then the *chain rule* gives the 1-differential

$$\mathrm{d}\mathbf{x} = \frac{\partial \mathbf{x}}{\partial s^1}\mathrm{d}s^1 + \frac{\partial \mathbf{x}}{\partial s^2}\mathrm{d}s^2 = \mathbf{x}_1 \mathrm{d}s^1 + \mathbf{x}_2 \mathrm{d}s^2$$

in the direction defined by $(\mathrm{d}s^1, \mathrm{d}s^2)$.

4. Show that $\mathrm{d}s^2 := \mathrm{d}\mathbf{x}^2 = g_{11}(\mathrm{d}s^1)^2 + 2g_{12}\mathrm{d}s^1\mathrm{d}s^2 + g_{22}(\mathrm{d}s^2)^2$. The expression $\mathrm{d}s^2$ is called the *first fundamental form* of the surface $\mathbf{x}(s^1, s^2)$ in the direction defined by $(\mathrm{d}s^1, \mathrm{d}s^2)$.
5. Show that $g_{11} > 0$, $g_{22} > 0$ and $g_{11}g_{22} - g_{12}^2 = (\mathbf{x}_1 \times \mathbf{x}_2)^2 > 0$.
6. Let $\mathbf{x} = \mathbf{x}(s^1, s^2)$ be a 2-surface in $\mathbb{R}^3$. Show that the $2^{nd}$ *differential*, in the direction $(\mathrm{d}s^1, \mathrm{d}s^2)$, is given by

$$\mathrm{d}^2\mathbf{x} = \mathbf{x}_{11}(\mathrm{d}s^1)^2 + 2\mathbf{x}_{12}\mathrm{d}s^1\mathrm{d}s^2 + \mathbf{x}_{22}(\mathrm{d}s^2)^2.$$

7. Let $\mathbf{x} = \mathbf{x}(s^1, s^2)$ be a 2-surface in $\mathbb{R}^3$ and $\hat{\mathbf{n}} = \frac{\mathbf{x}_1 \times \mathbf{x}_2}{|\mathbf{x}_1 \times \mathbf{x}_2|}$ be the unit orthonormal vector to $\mathbf{x}(s^1, s^2)$. It follows that $\mathrm{d}\mathbf{x} \cdot \hat{\mathbf{n}} = 0$. Using Problem 6, show that

$$-\mathrm{d}\mathbf{x} \cdot \mathrm{d}\hat{\mathbf{n}} = L_{11}(\mathrm{d}s^1)^2 + 2L_{12}\mathrm{d}s^1\mathrm{d}s^2 + L_{22}(\mathrm{d}s^2)^2,$$

which is called the *second fundamental form* of $\mathbf{x}(s^1, s^2)$.

8. Show that Taylor's theorem at the point $\mathbf{x}(s_0^1, s_0^2)$ in the direction of $(\mathrm{d}s^1, \mathrm{d}s^2)$ can be expressed as

$$\mathbf{x}(s_0^1 + \mathrm{d}s^1, s_0^2 + \mathrm{d}s^2) = \mathbf{x}(s_0^1, s_0^2) + \sum_{i=1}^{k} \frac{1}{k!} d^k \mathbf{x}(\mathrm{d}s^1, \mathrm{d}s^2) + o[((\mathrm{d}s^1)^2 + (\mathrm{d}s^2)^2)^{\frac{k}{2}}],$$

where $\mathrm{d}^k\mathbf{x}$ is the $k$th differential of $\mathbf{x}$ at the point $\mathbf{x}_0 = \mathbf{x}(s_0^1, s_0^2)$ in the direction of $(\mathrm{d}s^1, \mathrm{d}s^2)$, and the "small oh" notation $\mathbf{f}(\mathbf{x}) = o[((\mathrm{d}s^1)^2 + (\mathrm{d}s^2)^2)^{\frac{k}{2}}]$ means that

$$\lim_{\mathbf{x} \to \mathbf{x}_0} \frac{\mathbf{f}(\mathbf{x})}{((\mathrm{d}s^1)^2 + (\mathrm{d}s^2)^2)^{\frac{k}{2}}} = 0.$$

9. Recalling Problem 7 of Sect. 13.2, show that the integrability condition (13.8) implies that $P_{\mathbf{a}}(\mathbf{b}) = P_{\mathbf{b}}(\mathbf{a})$ for all tangent vector fields $\mathbf{a}, \mathbf{b} \in \mathcal{T}_{\mathbf{x}}$.

10. Let $P_x$ be the projection onto the tangent algebra $\mathcal{T}_x$ at the point $x \in \mathcal{M}$. We say that $F(x)$ is differentiable at $x$ if

$$\lim_{y \to x, y \in \mathcal{M}} \frac{|F(y) - F(x) - F_v(x)|}{|v|} = 0$$

where $v = P_x(y - x)$. Why is this refinement of (15.9) necessary?

## 15.2 The Shape Operator

Let $\mathcal{M}$ be a regular differentiable $k$-surface in $\mathbb{R}^n$. The most general *shape operator* $L(g)$ at a point $\mathbf{x} \in \mathcal{M}$ is the linear operator $L : \mathbb{G}_n \longrightarrow \mathbb{G}_n$, defined by

$$L(g) = \dot{\partial}_{\mathbf{x}} \dot{P}_{\mathbf{x}}(g) = \partial_{\mathbf{v}} P_{\mathbf{v}}(g) \tag{15.26}$$

where $g$ is any geometric number in $\mathbb{G}_n$. Of course when $g \in G_n$ is a constant, $L(g) = \partial_{\mathbf{x}} P_{\mathbf{x}}(g)$. In the case that $g = \alpha \in \mathbb{R}$ is a scalar, $L(\alpha) = 0$, since $P_{\mathbf{x}}(\alpha) = \alpha$, so we need only study $L(A_r)$ for $r$-vectors $A_r \in \mathbb{G}_n^r$ for $r \geq 1$. The shape operator plays an important role in determining both the intrinsic and the extrinsic properties of the surface $\mathcal{M}$. In particular, we are interested in studying the shape operator evaluated at tangent vectors $\mathbf{a} \in \mathcal{T}_{\mathbf{x}}^1$.

Let $A_r, B_s$ be an $r$-vector and a $s$-vector in $\mathbb{G}_n$. Using (15.19), we find that

$$L(A_r \wedge B_s) = \partial_{\mathbf{v}} P_{\mathbf{v}}(A_r \wedge B_s) = \partial_{\mathbf{v}} P_{\mathbf{v}}(A_r) \wedge P_{\mathbf{x}}(B_s) + \partial_{\mathbf{v}} P_{\mathbf{x}}(A_r) \wedge P_{\mathbf{v}}(B_s).$$

From (15.20), we know that

$$P_{\mathbf{v}}\left(P_{\mathbf{x}}(A_r)\right) + P_{\mathbf{x}}\left(P_{\mathbf{v}}(A_r)\right) = P_{\mathbf{v}}(A_r)$$

for any $\mathbf{v} \in \mathcal{T}_{\mathbf{x}}^1$. Taking the vector derivative $\partial_{\mathbf{v}}$ of this equation gives

$$\begin{aligned} L(A_r) = \partial_{\mathbf{v}} P_{\mathbf{v}}(A_r) &= \partial_{\mathbf{v}} P_{\mathbf{v}}\left(P_{\mathbf{x}}(A_r)\right) + \partial_{\mathbf{v}} P_{\mathbf{x}}\left(P_{\mathbf{v}}(A_r)\right) \\ &= \partial_{\mathbf{v}} P_{\mathbf{v}}\left(P_{\mathbf{x}}(A_r)\right) + P_{\mathbf{x}}\left(\partial_{\mathbf{v}}(P_{\mathbf{v}}(A_r))\right). \end{aligned}$$

It follows that if $P_{\mathbf{x}}(A_r) = A_r$, then $P_{\mathbf{x}}(L(A_r)) = 0$, and if $P_{\mathbf{x}}(A_r) = 0$, then $L(A_r) = P_{\mathbf{x}}\left(L(A_r)\right)$. This is important in determining the basic curvature relationships below.

In order to prove the following theorem, we need

**Lemma 15.2.1.** *If $P_{\mathbf{x}}(A_r) = A_r$ for $r \geq 1$, then for all $\mathbf{a} \in \mathcal{T}_{\mathbf{x}}^1$*

$$P_{\mathbf{a}}(A_r) = P_{\mathbf{a}}(\mathbf{v}) \wedge (\partial_{\mathbf{v}} \cdot A_r).$$

*Proof.* The proof is by induction on $r$. For $r = 1$, and $A_1 = \mathbf{b} \in \mathcal{T}_{\mathbf{x}}^1$, we have

$$P_{\mathbf{a}}(\mathbf{b}) = P_{\mathbf{a}}(\mathbf{v}) \partial_{\mathbf{v}} \cdot \mathbf{b}.$$

Assuming now for $r > 1$ that $P_{\mathbf{a}}(A_r) = P_{\mathbf{a}}(\mathbf{v}) \wedge (\partial_{\mathbf{v}} \cdot A_r)$, for $r+1$ we have

$$P_{\mathbf{a}}(\mathbf{b} \wedge A_r) = P_{\mathbf{a}}(\mathbf{b}) \wedge A_r + \mathbf{b} \wedge P_{\mathbf{a}}(A_r) = P_{\mathbf{a}}(\mathbf{b}) \wedge A_r + \mathbf{b} \wedge [P_{\mathbf{a}}(\mathbf{v}) \wedge (\partial_{\mathbf{v}} \cdot A_r)]$$
$$= P_{\mathbf{a}}(\mathbf{b}) \wedge A_r - P_{\mathbf{a}}(\mathbf{v}) \wedge \mathbf{b} \wedge (\partial_{\mathbf{v}} \cdot A_r)$$
$$= P_{\mathbf{a}}(\mathbf{v}) \wedge [\partial_{\mathbf{v}} \cdot (\mathbf{b} \wedge A_r)].$$

□

We have the following

**Theorem 15.2.2.** *Let $A_r \in \mathbb{G}_n^r$, where $1 \leq r \leq n$.*

*i) If $A_r \in \mathscr{T}_{\mathbf{x}}^r$, then*

$$L(A_r) = \partial_{\mathbf{v}} P_{\mathbf{v}}(A_r) = \partial_{\mathbf{v}} \wedge P_{\mathbf{v}}(A_r).$$

*ii) If $P_{\mathbf{x}}(A_r) = 0$, then*

$$L(A_r) = \partial_{\mathbf{v}} P_{\mathbf{v}}(A_r) = \partial_{\mathbf{v}} \cdot P_{\mathbf{v}}(A_r).$$

*Proof.* i) The proof is by induction on $r \geq 1$. For $r = 1$ and $\mathbf{a} \in \mathscr{T}_{\mathbf{x}}^1$, we have

$$L(\mathbf{a}) = \partial_{\mathbf{v}} P_{\mathbf{v}}(\mathbf{a}) = \partial_{\mathbf{v}} \cdot P_{\mathbf{v}}(\mathbf{a}) + \partial_{\mathbf{v}} \wedge P_{\mathbf{v}}(\mathbf{a}) = \partial_{\mathbf{v}} \wedge P_{\mathbf{v}}(\mathbf{a}),$$

since $P_{\mathbf{x}}(P_{\mathbf{v}}(\mathbf{a})) = 0$. Assume now that $L(A_r) = \partial_{\mathbf{v}} \wedge P_{\mathbf{v}}(A_r)$ for $A_r \in \mathscr{T}_{\mathbf{x}}^r$. This implies that $\partial_{\mathbf{v}} \cdot P_{\mathbf{v}}(A_r)$=0. Then for $r+1$ and $\mathbf{a} \in \mathscr{T}_{\mathbf{x}}^1$, we have

$$L(\mathbf{a} \wedge A_r) = \partial_{\mathbf{v}} P_{\mathbf{v}}(\mathbf{a}) \wedge A_r + \partial_{\mathbf{v}} \mathbf{a} \wedge P_{\mathbf{v}}(A_r)$$
$$= -P_{\mathbf{v}}(\mathbf{a}) \wedge (\partial_{\mathbf{v}} \cdot P_{\mathbf{v}}(A_r)) + \partial_{\mathbf{v}} \wedge P_{\mathbf{v}}(\mathbf{a}) \wedge A_r + P_{\mathbf{a}}(A_r)$$
$$-\mathbf{a} \wedge (\partial_{\mathbf{v}} \cdot P_{\mathbf{v}}(A_r)) + \partial_{\mathbf{v}} \wedge \mathbf{a} \wedge P_{\mathbf{v}}(A_r)$$
$$= \partial_{\mathbf{v}} \wedge P_{\mathbf{v}}(\mathbf{a}) \wedge A_r + \partial_{\mathbf{v}} \wedge \mathbf{a} \wedge P_{\mathbf{v}}(A_r) = \partial_{\mathbf{v}} \wedge P_{\mathbf{v}}(\mathbf{a} \wedge A_r).$$

ii) Let $A_r \in \mathbb{G}_n^r$ be any $r$-vector such that $P_{\mathbf{x}}(A_r) = 0$. We must show that $\partial_{\mathbf{v}} \wedge P_{\mathbf{v}}(A_r) = 0$. The proof is by induction on $r \geq 1$. For $r = 1$, let $A_1 = \mathbf{a}$ and suppose that $P_{\mathbf{x}}(\mathbf{a}) = 0$. We now calculate

$$(\mathbf{b} \wedge \mathbf{c}) \cdot (\partial_{\mathbf{v}} \wedge P_{\mathbf{v}}(\mathbf{a})) = P_{\mathbf{c}}(\mathbf{a}) \cdot \mathbf{b} - P_{\mathbf{b}}(\mathbf{a}) \cdot \mathbf{c}$$
$$= [P_{\mathbf{c}}(\mathbf{b}) - P_{\mathbf{b}}(\mathbf{c})] \cdot \mathbf{a} = 0$$

for all $\mathbf{b}, \mathbf{c} \in \mathscr{T}_{\mathbf{x}}^1$, since by Problem 9 of Sect. 15.1, $P_{\mathbf{c}}(\mathbf{b}) = P_{\mathbf{b}}(\mathbf{c})$.

Now assume that for $r > 1$, $\partial_{\mathbf{v}} \wedge P_{\mathbf{v}}(A_r) = 0$ for all $A_r \in \mathbb{G}_n^r$ such that $P_{\mathbf{x}}(A_r) = 0$. Now let $\mathbf{a} \wedge A_r \in \mathbb{G}_n^{r+1}$ such that $P_{\mathbf{x}}(\mathbf{a} \wedge A_r) = 0$. Then

$$\partial_{\mathbf{v}} \wedge P_{\mathbf{v}}(\mathbf{a} \wedge A_r) = \partial_{\mathbf{v}} \wedge P_{\mathbf{v}}(\mathbf{a}) \wedge P_{\mathbf{x}}(A_r) + \partial_{\mathbf{v}} \wedge P_{\mathbf{x}}(\mathbf{a}) \wedge P_{\mathbf{v}}(A_r) = 0.$$

□

For tangent vectors $\mathbf{a} \in \mathscr{T}_{\mathbf{x}}^1$, the shape operator determines the *shape bivector.* With the help of (15.20), we find that

$$L(\mathbf{a}) = \dot{\partial}_{\mathbf{x}} \dot{P}_{\mathbf{x}}(\mathbf{a}) = \dot{\partial}_{\mathbf{x}} \wedge \dot{P}_{\mathbf{x}}(\mathbf{a}) = \partial_{\mathbf{v}} \wedge P_{\mathbf{v}}(\mathbf{a}), \tag{15.27}$$

since $P_i(\mathbf{a}) = P_{\mathbf{a}}(\mathbf{x}_i)$ for $1 \le i \le k$. Equation (15.27) shows that the shape bivector is a measure of how the surface changes at the point $\mathbf{x}$ when moving off in the direction of the vector $\mathbf{a}$.

We can get another expression for the shape bivector $L(\mathbf{a})$ by evaluating $(\mathbf{a} \cdot \partial_{\mathbf{x}} I_{\mathbf{x}}) I_{\mathbf{x}}^{-1}$, for $I_{\mathbf{x}} = \mathbf{x}_1 \wedge \cdots \wedge \mathbf{x}_k$ and $I_k^{-1} = \mathbf{x}^k \wedge \cdots \wedge \mathbf{x}^1$. Recalling (15.4), $I_{\mathbf{x}} = \alpha_g \widehat{I_{\mathbf{x}}}$ where

$$\alpha_g := I_{\mathbf{x}} \widehat{I}_{\mathbf{x}}^{\dagger} = \sqrt{\det(g)} > 0.$$

It follows that

$$\mathbf{a} \cdot \partial_{\mathbf{x}} I_{\mathbf{x}} = (\mathbf{a} \cdot \partial_{\mathbf{x}} \alpha_g) \widehat{I}_{\mathbf{x}} + \alpha_g (\mathbf{a} \cdot \partial_{\mathbf{x}} \widehat{I}_{\mathbf{x}}) = \frac{\mathbf{a} \cdot \partial_{\mathbf{x}} \alpha_g}{\alpha_g} I_{\mathbf{x}} + \alpha_g P_{\mathbf{a}}(\widehat{I}_{\mathbf{x}})$$
$$= (\mathbf{a} \cdot \partial_{\mathbf{x}} \log(\alpha_g)) I_{\mathbf{x}} + P_{\mathbf{a}}(I_{\mathbf{x}}),$$

which implies that

$$(\mathbf{a} \cdot \partial_{\mathbf{x}} I_{\mathbf{x}}) I_{\mathbf{x}}^{-1} = \mathbf{a} \cdot \partial_{\mathbf{x}} \log(\alpha_g) - L(\mathbf{a}), \tag{15.28}$$

since

$$P_{\mathbf{a}}(I_{\mathbf{x}}) I_{\mathbf{x}}^{-1} = P_{\mathbf{a}}(\mathbf{x}_1 \wedge \cdots \wedge \mathbf{x}_k) \mathbf{x}^k \wedge \cdots \wedge \mathbf{x}^1 = -\partial_{\mathbf{x}} \wedge P_{\mathbf{a}}(\mathbf{x}) = -L(\mathbf{a}).$$

Thus, the shape bivector $L(\mathbf{a})$ is completely determined by the directional derivative of the pseudoscalar element $I_{\mathbf{x}}$.

Dotting (15.27) with a second tangent vector $\mathbf{b}$ gives

$$\mathbf{b} \cdot L(\mathbf{a}) = P_{\mathbf{b}}(\mathbf{a}) = P_{\mathbf{a}}(\mathbf{b}) = \mathbf{a} \cdot L(\mathbf{b}) = \mathbf{L}_{\mathbf{a},\mathbf{b}}. \tag{15.29}$$

For the tangent coordinate vectors $\mathbf{x}_i, \mathbf{x}_j$, this gives

$$\mathbf{x}_i \cdot L(\mathbf{x}_j) = \mathbf{x}_i \cdot (\dot{\partial}_{\mathbf{v}} \wedge \dot{P}_j(\mathbf{v})) = P_j(\mathbf{x}_i) = \mathbf{L}_{ij},$$

so the shape operator (15.26) is a generalization of the normal vectors $\mathbf{L}_{ij}$ defined in (15.22) and $\mathbf{L}_{\mathbf{a},\mathbf{b}}$ defined in (15.24).

Let $\mathbf{n}$ be the unit normal to the $(n-1)$-hypersurface $\mathscr{M}$ in $\mathbb{R}^n$. The *Weingarten map* $\underline{\mathbf{n}} : \mathscr{T}_{\mathbf{x}}^1 \longrightarrow \mathscr{T}_{\mathbf{x}}^1$ is defined by

$$\underline{\mathbf{n}}(\mathbf{a}) := \mathbf{a} \cdot \partial_{\mathbf{x}} \mathbf{n} = -P_{\mathbf{a}}(\mathbf{n}), \tag{15.30}$$

and can be extended to an outermorphism on the tangent algebra $\mathscr{T}_{\mathbf{x}}$. Dotting the shape bivector $L(\mathbf{a})$, given in (15.27), with $\mathbf{n}$ and using (15.18), we find that

$$\begin{aligned}\mathbf{n}\cdot L(\mathbf{a}) &= \mathbf{n}\cdot(\partial_{\mathbf{v}}\wedge P_{\mathbf{a}}(\mathbf{v})) = -\partial_{\mathbf{v}}\mathbf{n}\cdot P_{\mathbf{a}}(\mathbf{v})\\ &= -\partial_{\mathbf{v}}\mathbf{v}\cdot P_{\mathbf{a}}(\mathbf{n}) = -P_{\mathbf{a}}(\mathbf{n}) = \underline{\mathbf{n}}(\mathbf{a}),\end{aligned}$$

so in the case of a hypersurface, the shape bivector also completely determines the Weingarten map. In terms of the tangent vectors $\mathbf{x}_i$, we get the normal components defined in (15.23),

$$L_{ij} = \mathbf{n}\cdot\mathbf{L}_{ij} = (\mathbf{n}\wedge\mathbf{x}_i)\cdot L(\mathbf{x}_j) = -\mathbf{x}_i\cdot[\mathbf{n}\cdot L(\mathbf{x}_j)] = -\mathbf{x}_i\cdot\underline{\mathbf{n}}(\mathbf{x}_j).$$

We will use the second fundamental form, defined in Problem 7, and Taylor's theorem in Problem 8 of the last section to study the local properties of a surface in $\mathbb{R}^3$. Let $\mathbf{x} = \mathbf{x}(s^1,s^2)$ be a point on a surface of class $C^2$ or greater, and let $\mathbf{y} = \mathbf{x}(s^1+\mathrm{d}s^1, s^2+\mathrm{d}s^2)$ be a nearby point. Then

$$r = (\mathbf{y}-\mathbf{x})\cdot\hat{\mathbf{n}},$$

the projection of the vector from the point $\mathbf{x}$ to the point $\mathbf{y}$ onto the unit normal vector $\hat{\mathbf{n}}$, will indicate which side of the tangent plane to the surface at $\mathbf{x}$ that the point $\mathbf{y}$ is on.

Taylor's theorem gives

$$\mathbf{y}-\mathbf{x} = \mathrm{d}\mathbf{x} + \frac{1}{2}\mathrm{d}^2\mathbf{x} + o((\mathrm{d}s^1)^2+(\mathrm{d}s^2)^2).$$

Since $\mathrm{d}\mathbf{x}$ is in the tangent plane, $\mathrm{d}\mathbf{x}\cdot\hat{\mathbf{n}} = 0$, so $\mathrm{d}^2\mathbf{x}\cdot\hat{\mathbf{n}} = -\mathrm{d}\mathbf{x}\cdot\mathrm{d}\hat{\mathbf{n}}$. It follows that

$$\begin{aligned} r &= \frac{1}{2}\mathrm{d}^2\mathbf{x}\cdot\hat{\mathbf{n}} + o((\mathrm{d}s^1)^2+(\mathrm{d}s^2)^2)\\ &= \frac{1}{2}\Big(L_{11}(\mathrm{d}s^1)^2 + 2L_{12}\mathrm{d}s^1\mathrm{d}s^2 + L_{22}(\mathrm{d}s^2)^2\Big) + o((\mathrm{d}s^1)^2+(\mathrm{d}s^2)^2)\end{aligned}$$

The function $f(\mathrm{d}s^1,\mathrm{d}s^2) = \frac{1}{2}\Big(L_{11}(\mathrm{d}s^1)^2 + 2L_{12}\mathrm{d}s^1\mathrm{d}s^2 + L_{22}(\mathrm{d}s^2)^2\Big)$ is called the *osculating paraboloid* at $\mathbf{x}$. The nature of this paraboloid is determined by the *discriminant* $L_{11}L_{22}-L_{12}^2$ and approximates the surface $\mathbf{x}(R)$ at the point $\mathbf{x}(s^1,s^2)$. We have the following four cases:

1. If $L_{11}L_{22}-L_{12} > 0$, we say that the surface is *elliptic* at the point $\mathbf{x}(s^1,s^2)$.
2. If $L_{11}L_{22}-L_{12} = 0$, we say that the surface is *parabolic* at the point $\mathbf{x}(s^1,s^2)$.
3. If $L_{11}L_{22}-L_{12} < 0$, we say that the surface is *hyperbolic* at the point $\mathbf{x}(s^1,s^2)$.
4. If $L_{11} = L_{22} = L_{12} = 0$, we say that the surface is *planar* at the point $\mathbf{x}(s^1,s^2)$.

Consider the following example: Let

$$\mathbf{x}(s^1,s^2) = \big(s^1, s^2, (s^1)^2 s^2 + s^1 s^2 - (s^2)^2 + 2\big)$$

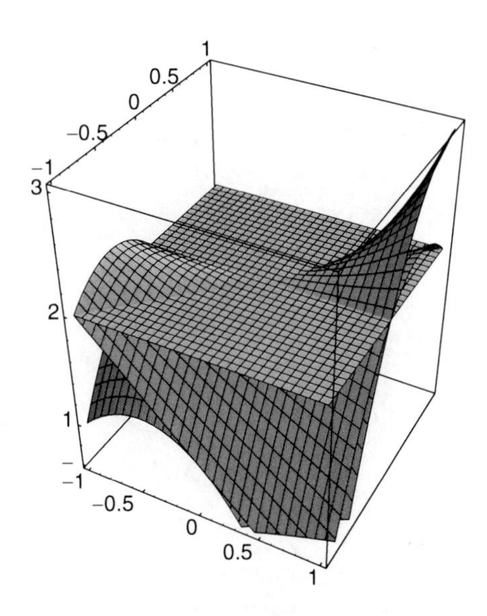


**Fig. 15.2** The surface $\mathbf{x}(s^1,s^2) = (s^1,s^2,(s^1)^2s^2 + s^1s^2 - (s^2)^2 + 2)$, its tangent plane, and osculating paraboloid at the point $(0,0,2)$

for $-1 < s^1 < 1$ and $-1 < s^2 < 1$. We calculate

$$\mathbf{x}_1 = (1,0,s^2+2s^1s^2),\ \mathbf{x}_2 = (0,1,s^1+(s^1)^2-2s^2),\ \mathbf{x}_{11} = (0,0,2s^2),$$
$$\mathbf{x}_{12} = (0,0,2),\ \text{and}\ \mathbf{x}_{22} = (0,0,-2)$$

The tangent plane to this surface at the point $\mathbf{x}(0,0) = (0,\ 0,\ 2)$ is

$$\mathbf{z} = (0,0,2) + s^1\mathbf{x}_1 + s^2\mathbf{x}_2 = (s^1,s^2,2)$$

The osculating paraboloid at $(0,0,2)$ is given by

$$\mathbf{y} = (0,0,2) + \mathrm{d}\mathbf{x}(s^1,s^2) + \frac{1}{2}\mathrm{d}^2\mathbf{x} = \left(s^1,\ s^2,\ (s^1)^2s^2 - (s^2)^2 + s^1s^2 + 2\right)$$

The graphs of the surface $\mathbf{x}(s^1,s^2)$, its tangent plane $\mathbf{y}(s^1,s^2)$, and its osculating paraboloid $\mathbf{z}(s^1,s^2)$ are given in Fig. 15.2. We further calculate $L_{11} = 0,\ L_{12} = 1$, $L_{22} = -2$, and the discriminant

$$L_{11}L_{22} - L_{12}^2 = -4 < 0$$

at the point $(0,0,2)$. It follows that the surface $\mathbf{x}(s^1,s^2)$ is hyperbolic at the point $\mathbf{x}(0,0) = (0,0,2)$.

## Exercises

1. Given the surface $\mathbf{x}(s^1,s^2) = (s^1,s^2,(s^1)^2s^2 + s^1s^2 - (s^2)^2 + 2)$ of the last example.

   (a) Calculate the tangent vectors $\mathbf{x}_1,\mathbf{x}_2$ and the mixed derivatives $\mathbf{x}_{11},\mathbf{x}_{12}$, and $\mathbf{x}_{22}$ at the point $\mathbf{x}(\frac{1}{2},0)$.
   (b) Calculate the tangent plane at the point $\mathbf{x}(\frac{1}{2},0)$.
   (c) Calculate the osculating paraboloid at the point $\mathbf{x}(\frac{1}{2},0)$.
   (d) Classify the surface at the point $\mathbf{x}(\frac{1}{2},0)$.

2. For the 2-rectangle $R = [-1,1] \times [-1,1]$, define the surface $\mathbf{x}(R)$ by

$$\mathbf{x}(s^1,s^2) = (s^1,s^2,-(s^1)^2 + s^2 + 2),$$

   see Fig. 13.2 of Sect. 13.1.

   (a) Calculate the tangent vectors $\mathbf{x}_1,\mathbf{x}_2$ and the mixed derivatives $\mathbf{x}_{11},\mathbf{x}_{12}$, and $\mathbf{x}_{22}$ at the point $\mathbf{x}(0,0)$.
   (b) Calculate the tangent plane at the point $\mathbf{x}(0,0)$.
   (c) Calculate the osculating paraboloid at the point $\mathbf{x}(0,0)$.
   (d) Classify the surface at the point $\mathbf{x}(0,0)$.

3. For the 2-rectangle $R = [-1,1] \times [-1,1]$, define the surface $\mathbf{x}(R)$ by

$$\mathbf{x}(x,y) = (x,y,x^2 + y^2),$$

   see Fig. 13.3 of Sect. 13.1.

   (a) Calculate the tangent vectors $\mathbf{x}_1,\mathbf{x}_2$ and the mixed derivatives $\mathbf{x}_{11},\mathbf{x}_{12}$, and $\mathbf{x}_{22}$ at the point $\mathbf{x}(0,0)$.
   (b) Calculate the tangent plane at the point $\mathbf{x}(0,0)$.
   (c) Calculate the osculating paraboloid at the point $\mathbf{x}(0,0)$.
   (d) Classify the surface at the point $\mathbf{x}(0,0)$.

4. For the 2-rectangle $R = [-1,1] \times [-1,1]$, define the surface $\mathbf{x}(R)$ by

$$\mathbf{x}(x,y) = (x,y,x^2 - y^2 + 1),$$

   see Fig. 13.4 of Sect. 13.1.

   (a) Calculate the tangent vectors $\mathbf{x}_1,\mathbf{x}_2$ and the mixed derivatives $\mathbf{x}_{11},\mathbf{x}_{12}$, and $\mathbf{x}_{22}$ at the point $\mathbf{x}(0,0)$.
   (b) Calculate the tangent plane at the point $\mathbf{x}(0,0)$.
   (c) Calculate the osculating paraboloid at the point $\mathbf{x}(0,0)$.
   (d) Classify the surface at the point $\mathbf{x}(0,0)$.

## 15.3 Geodesic Curvature and Normal Curvature

Let $\mathscr{M} \subset \mathbb{R}^n$ be a regular differentiable $k$-surface with the projection $P_{\mathbf{x}}$ onto the tangent space $\mathscr{T}_{\mathbf{x}}$ at the point $\mathbf{x} \in \mathscr{M}$, and let $\mathbf{x} = \mathbf{x}(s)$ be a unit speed curve on $\mathscr{M}$ with the tangent unit vector $\mathbf{T} = \frac{\mathrm{d}\mathbf{x}}{\mathrm{d}s} = P_{\mathbf{x}}(\mathbf{T})$. Then

$$\frac{\mathrm{d}^2\mathbf{x}}{\mathrm{d}s^2} = \frac{\mathrm{d}\mathbf{T}}{\mathrm{d}s} = \dot{\mathbf{T}} = P_{\mathbf{x}}\left(\frac{\mathrm{d}\mathbf{T}}{\mathrm{d}s}\right) + P_{\mathbf{T}}(\mathbf{T}) = \kappa_{\mathrm{g}}\mathbf{S} + \kappa_{\mathrm{n}}\mathbf{K} \tag{15.31}$$

where $\mathbf{S} = \frac{P_{\mathbf{x}}\left(\frac{\mathrm{d}\mathbf{T}}{\mathrm{d}s}\right)}{\left|P_{\mathbf{x}}\left(\frac{\mathrm{d}\mathbf{T}}{\mathrm{d}s}\right)\right|}$ and $\mathbf{K} = \frac{P_{\mathbf{T}}(\mathbf{T})}{|P_{\mathbf{T}}(\mathbf{T})|}$. The intrinsic *absolute geodesic curvature* is $\kappa_{\mathrm{g}} = \left|P_{\mathbf{x}}\left(\frac{\mathrm{d}\mathbf{T}}{\mathrm{d}s}\right)\right|$ and the extrinsic *absolute normal curvature* is $\kappa_{\mathrm{n}} = |P_{\mathbf{T}}(\mathbf{T})|$. Both the intrinsic absolute geodesic curvature $\kappa_{\mathrm{g}}$ and the absolute normal curvature $\kappa_{\mathrm{n}}$ are positive quantities.

However, if $\mathscr{M}$ is an orientable $(n-1)$-hypersurface surface in $\mathbb{R}^n$, with the unit normal vector $\mathbf{n}$, the *normal curvature* $\kappa_{\mathrm{n}} := \mathbf{n} \cdot P_{\mathbf{T}}(\mathbf{T})$ can take on negative values with respect to the normal vector $\mathbf{n}$. In the particular case of an orientable 2-surface $\mathscr{M}$ in $\mathbb{R}^3$, we define the *geodesic curvature* with respect to the normal vector $\mathbf{n}$ to be

$$\kappa_{\mathrm{g}} = (\mathbf{n} \times \mathbf{T}) \cdot \dot{\mathbf{T}}, \tag{15.32}$$

which can also be negative. In these cases, the signs of the vectors $\mathbf{S}$ and $\mathbf{K}$ in (15.31) must be adjusted accordingly.

In terms of the coordinate vectors, we find that

$$P_{\mathbf{x}}(\mathbf{T}) = \mathbf{T} = \frac{\mathrm{d}\mathbf{x}}{\mathrm{d}s} = \sum_{i=1}^{k} \frac{\mathrm{d}s^i}{\mathrm{d}s}\mathbf{x}_i,$$

and differentiating a second time with respect to $s$ gives

$$\begin{aligned} P_{\mathbf{x}}\left(\frac{\mathrm{d}\mathbf{T}}{\mathrm{d}s}\right) + P_{\mathbf{T}}(\mathbf{T}) &= \frac{\mathrm{d}\mathbf{T}}{\mathrm{d}s} = \frac{\mathrm{d}^2 s^i}{\mathrm{d}s^2}\mathbf{x}_i + \frac{\mathrm{d}s^i}{\mathrm{d}s}\frac{\mathrm{d}s^j}{\mathrm{d}s}\mathbf{x}_{ij} \\ &= \frac{\mathrm{d}^2 s^i}{\mathrm{d}s^2}\mathbf{x}_i + \frac{\mathrm{d}s^i}{\mathrm{d}s}\frac{\mathrm{d}s^j}{\mathrm{d}s}(\mathbf{K}_{ij} + \mathbf{L}_{ij}), \end{aligned} \tag{15.33}$$

where we are utilizing the summation convention on repeated upper and lower indices. It follows that

$$\kappa_{\mathrm{g}}\mathbf{S} = \frac{\mathrm{d}^2 s^i}{\mathrm{d}s^2}\mathbf{x}_i + \frac{\mathrm{d}s^i}{\mathrm{d}s}\frac{\mathrm{d}s^j}{\mathrm{d}s}\mathbf{K}_{ij} \tag{15.34}$$

and

$$\kappa_{\mathrm{n}}\mathbf{K} = \frac{\mathrm{d}s^i}{\mathrm{d}s}\frac{\mathrm{d}s^j}{\mathrm{d}s}\mathbf{L}_{ij}. \tag{15.35}$$

**Definition 15.3.1.** The curve $\mathbf{x}(s)$ is a *geodesic* if $\kappa_{\mathrm{g}} = 0$ along $\mathbf{x}(s)$. Equivalently, $\mathbf{x}(s)$ is a geodesic if

$$P_{\mathbf{x}}\left(\frac{\mathrm{d}^2\mathbf{x}}{\mathrm{d}s^2}\right)=P_{\mathbf{x}}(\dot{\mathbf{T}})=0. \tag{15.36}$$

A geodesic $\mathbf{x}(s)$ on a surface plays the role of a straight line in flat Euclidean space. For example, on the surface of a sphere, the great circles are geodesics and give the shortest distance between any two points on the surface.

For a non-unit speed regular curve, we have $\frac{\mathrm{d}\mathbf{x}}{\mathrm{d}t}=\frac{\mathrm{d}\mathbf{x}}{\mathrm{d}s}\frac{\mathrm{d}s}{\mathrm{d}t}$, and

$$\frac{\mathrm{d}^2\mathbf{x}}{\mathrm{d}t^2}=\frac{\mathrm{d}\mathbf{T}}{\mathrm{d}s}\left(\frac{\mathrm{d}s}{\mathrm{d}t}\right)^2+\mathbf{T}\frac{\mathrm{d}^2s}{\mathrm{d}t^2}, \tag{15.37}$$

which implies that $(\frac{\mathrm{d}\mathbf{x}}{\mathrm{d}t})^2=(\frac{\mathrm{d}s}{\mathrm{d}t})^2$ and $\frac{\mathrm{d}\mathbf{x}}{\mathrm{d}t}\cdot\frac{\mathrm{d}^2\mathbf{x}}{\mathrm{d}t^2}=\frac{\mathrm{d}s}{\mathrm{d}t}\frac{\mathrm{d}^2s}{\mathrm{d}t^2}$. Taking the projection of (15.37) gives

$$P_{\mathbf{x}}\left(\frac{\mathrm{d}^2\mathbf{x}}{\mathrm{d}t^2}\right)=P_{\mathbf{x}}\left(\frac{\mathrm{d}\mathbf{T}}{\mathrm{d}s}\right)\left(\frac{\mathrm{d}s}{\mathrm{d}t}\right)^2+\mathbf{T}\frac{\mathrm{d}^2s}{\mathrm{d}t^2}.$$

We see from this equation and (15.36), that a non-unit speed regular curve $\mathbf{x}(t)$ is a geodesic iff

$$P_{\mathbf{x}}\left(\frac{\mathrm{d}^2\mathbf{x}}{\mathrm{d}t^2}\right)=\mathbf{T}\frac{\mathrm{d}^2s}{\mathrm{d}t^2} \iff P_{\mathbf{x}}\left(\frac{\mathrm{d}\mathbf{v}}{\mathrm{d}t}\wedge\mathbf{v}\right)=0. \tag{15.38}$$

A geometric-valued function $F(\mathbf{x})$ on $\mathscr{M}$ is said to be a *multivector field* on $\mathscr{M}$ if $P_{\mathbf{x}}(F(\mathbf{x}))=F(\mathbf{x})\in\mathscr{T}_{\mathbf{x}}$ for all $\mathbf{x}\in\mathscr{M}$. Thus, the values of a multivector field $F$ on $\mathscr{M}$ at the point $\mathbf{x}$ are restricted to the tangent algebra $\mathscr{T}_{\mathbf{x}}$. We have already seen in (15.17), and elsewhere, that neither the vector derivative $\partial_{\mathbf{x}}$ nor the directional derivative $\mathbf{a}\cdot\partial_{\mathbf{x}}$ preserve a tangent field on $\mathscr{M}$. We now define a *coderivative* and *directional coderivative* that preserves the tangency of a field on $\mathscr{M}$.

**Definition 15.3.2.** By the *vector coderivative* $\nabla_{\mathbf{x}}F(\mathbf{x})$ of a multivector field $F(\mathbf{x})$, induced by the vector derivative $\partial_{\mathbf{x}}$ of the $k$-surface $\mathscr{M}$, we mean

$$\nabla_{\mathbf{x}}F(\mathbf{x})=P_{\mathbf{x}}\left(\partial_{\mathbf{x}}P_{\mathbf{x}}(F(\mathbf{x}))\right)=P_{\mathbf{x}}\left(\partial_{\mathbf{x}}F(\mathbf{x})\right), \tag{15.39}$$

and the corresponding *directional* $\mathbf{a}$*-coderivative* is defined by

$$F_{/\!\!a}(\mathbf{x})=\mathbf{a}\cdot\nabla_{\mathbf{x}}F(\mathbf{x})=P_{\mathbf{x}}\left(\mathbf{a}\cdot\partial_{\mathbf{x}}P_{\mathbf{x}}(F(\mathbf{x}))\right)=P_{\mathbf{x}}\left(\mathbf{a}\cdot\partial_{\mathbf{x}}F(\mathbf{x})\right). \tag{15.40}$$

In terms of the vector coderivative, the condition (15.36) for a unit speed curve $\mathbf{x}(s)$ to be a geodesic becomes $\mathbf{T}\cdot\nabla_{\mathbf{x}}\mathbf{T}=0$ or $\mathbf{T}_{/\!\!T}=0$.

Every vector field $\mathbf{v}=\mathbf{v}(\mathbf{x})$ on $\mathscr{M}$ satisfies the identity

$$\mathbf{v}\cdot\nabla_{\mathbf{x}}\mathbf{v}=\mathbf{v}\cdot(\nabla\wedge\mathbf{v})+\frac{1}{2}\nabla\mathbf{v}^2. \tag{15.41}$$

A tangent vector field $\mathbf{a}=\mathbf{a}(\mathbf{x})$ is said to be *parallel along a curve* $\mathbf{x}=\mathbf{x}(t)$ if

$$\mathbf{v}\cdot\nabla_{\mathbf{x}}\mathbf{a}=P_{\mathbf{x}}\left(\frac{\mathrm{d}\mathbf{a}}{\mathrm{d}t}\right)=\lambda(t)\mathbf{a},$$

where $\mathbf{v} = \frac{\mathrm{d}\mathbf{x}}{\mathrm{d}t}$. In the case of a unit speed curve $\mathbf{x}(s)$, this equation simplifies to $\mathbf{T} \cdot \nabla_{\mathbf{x}} \mathbf{a} = 0$ where $\mathbf{T} = \frac{\mathrm{d}\mathbf{x}}{\mathrm{d}s}$.

If $\mathbf{a} \wedge (\mathbf{v} \cdot \nabla_{\mathbf{x}} \mathbf{a}) = 0$ for the directional coderivative in *every direction* $\mathbf{v} \in \mathscr{T}_{\mathbf{x}}^k$, including $\mathbf{v} = \mathbf{a}$, then it follows from (15.41) that $\nabla_{\mathbf{x}} \wedge \mathbf{a} = 0 = \nabla_{\mathbf{x}} |\mathbf{a}|^2$, so that

$$\mathbf{v} \cdot \nabla_{\mathbf{x}} \mathbf{a} = (\mathbf{v} \cdot \mathbf{b})\mathbf{a}$$

for some constant vector $\mathbf{b}$. In this case, we say that the tangent vector field $\mathbf{a}(\mathbf{x})$ is everywhere parallel to a *geodesic spray*, [52, p.545], [43, p.206].

http://en.wikipedia.org/wiki/Geodesic#Geodesic_flow

## Exercises

1. Let $\mathbf{x}(s)$ be a unit speed curve on $\mathscr{M}$. Prove that $\kappa^2 = \kappa_{\mathrm{n}}^2 + \kappa_{\mathrm{g}}^2$, where $\kappa$ is the curvature, $\kappa_{\mathrm{g}}$ is the intrinsic geodesic curvature, and $\kappa_{\mathrm{n}}$ is the extrinsic normal curvature of the unit speed curve.
2. For the surface $\mathbf{x}(x,y) = (x,y,x^2+y^2)$ in $\mathbb{R}^3$, find the normal curvature $\kappa_{\mathrm{n}}$ of the curve $\mathbf{x}(t^2,t)$ at $t = 1$.
3. The surface $\mathbf{x}(x,y) = (x,y,\sqrt{1-x^2-y^2})$ for $x^2+y^2 \le 1$ is the upper hemisphere above the $xy$-plane. For the unit speed curve $\mathbf{y}(s) = (\cos s, 0, -\sin s)$ on $\mathbf{x}(x,y)$, show that $\kappa = 1$, $\kappa_{\mathrm{n}} = -1$, and $\kappa_{\mathrm{g}} = 0$. Since $\kappa_{\mathrm{g}} = 0$, $\mathbf{y}(s)$ is a great circle on $\mathbf{x}(x,y)$.
4. Let $\mathbf{x}(s^1,s^2)$ be a surface with the property that $g_{11} = 1$ and $g_{12} = 0$. Let $a$ be a constant, and let $\mathbf{y}(s) = \mathbf{x}(s,a)$ be a curve on $\mathbf{x}(s^1,s^2)$. Show that $\mathbf{y}(s)$ is a geodesic on $\mathbf{x}(s^1,s^2)$.
5. Let $\mathbf{x}(x,y)$ be a surface, $\mathbf{x}(s) = \mathbf{x}(x(s),y(s))$ be a unit speed curve on $\mathbf{x}(s)$, and let $\mathbf{x}_1 = \frac{\partial \mathbf{x}}{\partial x}$ and $\mathbf{x}_2 = \frac{\partial \mathbf{x}}{\partial y}$.

   (a) Show that $\frac{\mathrm{d}\mathbf{x}}{\mathrm{d}s} = \frac{\mathrm{d}x}{\mathrm{d}s}\mathbf{x}_1 + \frac{\mathrm{d}y}{\mathrm{d}s}\mathbf{x}_2$.
   (b) Show that

$$P_{\mathbf{x}}\left(\frac{\mathrm{d}^2\mathbf{x}}{\mathrm{d}s^2}\right) = \left(\frac{\mathrm{d}x}{\mathrm{d}s}\right)^2 \mathbf{K}_{11} + 2\frac{\mathrm{d}x}{\mathrm{d}s}\frac{\mathrm{d}y}{\mathrm{d}s}\mathbf{K}_{12} + \left(\frac{\mathrm{d}y}{\mathrm{d}s}\right)^2 \mathbf{K}_{22} + \mathbf{x}_1\frac{\mathrm{d}^2x}{\mathrm{d}s^2} + \mathbf{x}_2\frac{\mathrm{d}^2y}{\mathrm{d}s^2}.$$

   (c) Prove that $\mathbf{x}(s)$ is a geodesic if and only if $x(s)$ and $y(s)$ satisfy

$$\frac{\mathrm{d}^2x}{\mathrm{d}s^2} + \Gamma_{11}^1\left(\frac{\mathrm{d}x}{\mathrm{d}s}\right)^2 + 2\Gamma_{12}^1\frac{\mathrm{d}x}{\mathrm{d}s}\frac{\mathrm{d}y}{\mathrm{d}s} + \Gamma_{22}^1\left(\frac{\mathrm{d}y}{\mathrm{d}s}\right)^2 = 0$$

   and

$$\frac{\mathrm{d}^2y}{\mathrm{d}s^2} + \Gamma_{11}^2\left(\frac{\mathrm{d}x}{\mathrm{d}s}\right)^2 + 2\Gamma_{12}^2\frac{\mathrm{d}x}{\mathrm{d}s}\frac{\mathrm{d}y}{\mathrm{d}s} + \Gamma_{22}^2\left(\frac{\mathrm{d}y}{\mathrm{d}s}\right)^2 = 0.$$

## 15.4 Gaussian, Mean, and Principal Curvatures of $\mathscr{M}$

Let $\mathscr{M}$ be an $(n-1)$-hypersurface surface in $\mathbb{R}^n$, with the unit normal vector $\mathbf{n}$. The *Gaussian curvature* $K_G = \det(\underline{\mathbf{n}})$ and the *mean curvature* $\kappa_m$ are defined by

$$\underline{\mathbf{n}}(\mathbf{x}_1) \wedge \underline{\mathbf{n}}(\mathbf{x}_2) \wedge \cdots \wedge \underline{\mathbf{n}}(\mathbf{x}_{n-1}) = \det(\underline{\mathbf{n}}) I_{\mathbf{x}} = K_G I_{\mathbf{x}},$$

where $I_{\mathbf{x}}$ is the unit tangent $(n-1)$-vector, or *pseudoscalar* to the $k$-surface at $\mathbf{x} \in \mathscr{M}$, and

$$\kappa_m = \frac{1}{n-1} \partial_{\mathbf{v}} \cdot \underline{\mathbf{n}}(\mathbf{v}), \tag{15.42}$$

respectively. The Gaussian curvature can also be expressed in the form

$$K_G = \frac{1}{k!} (\partial_{\mathbf{v}_k} \wedge \cdots \wedge \partial_{\mathbf{v}_1}) \cdot \underline{\mathbf{n}}(\mathbf{v}_1 \wedge \cdots \wedge \mathbf{v}_k). \tag{15.43}$$

For a 2-surface $\mathbf{x}(s^1, s^2)$ in $\mathbb{R}^3$, the Gaussian curvature $K_G$ is

$$\underline{\mathbf{n}}(\mathbf{x}_1 \wedge \mathbf{x}_2) = \underline{\mathbf{n}}(\mathbf{x}_1) \wedge \underline{\mathbf{n}}(\mathbf{x}_2) = K_G\, \mathbf{x}_1 \wedge \mathbf{x}_2.$$

Since $g = \det(g_{ij}) = (\mathbf{x}_1 \wedge \mathbf{x}_2) \cdot (\mathbf{x}_2 \wedge \mathbf{x}_1) = |\mathbf{x}_1 \times \mathbf{x}_2|^2$, a unit trivector is defined by

$$i = [(\mathbf{x}_1 \wedge \mathbf{x}_2)\mathbf{n}]/\sqrt{g}.$$

Multiplying the above equation by $-i$ gives the result

$$\underline{\mathbf{n}}(\mathbf{x}_1) \times \underline{\mathbf{n}}(\mathbf{x}_2) = -iK_G\, \mathbf{x}_1 \wedge \mathbf{x}_2 = K_G\, \mathbf{x}_1 \times \mathbf{x}_2 = K_G \sqrt{g}\, \mathbf{n}.$$

The second fundamental form for a 2-surface $\mathbf{x}(s^1, s^2)$ in $\mathbb{R}^3$ can be expressed by

$$L_{\mathbf{a},\mathbf{b}} = \mathbf{L}_{\mathbf{a},\mathbf{b}} \cdot \mathbf{n} = P_{\mathbf{a}}(\mathbf{b}) \cdot \mathbf{n} = \mathbf{b} \cdot P_{\mathbf{a}}(\mathbf{n}). \tag{15.44}$$

The *principal curvatures* $\kappa_1, \kappa_2$ for the 2-surface $\mathbf{x}(s^1, s^2)$ are defined to be the *eigenvalues* of the symmetric linear operator $P_{\mathbf{v}}(\mathbf{n})$ for $\mathbf{v} \in \mathscr{T}_{\mathbf{x}}^1$. The corresponding *eigenvectors* are said to be the *principal directions* at the point $\mathbf{x}(s^1, s^2)$.

### Exercises

1. Given the surface

$$\mathbf{x}(s^1, s^2) = (s^1, s^2, (s^1)^2 s^2 + s^1 s^2 - (s^2)^2 + 2),$$

calculate the Gaussian, mean, and principal curvatures at the point $\mathbf{x}(s^1, s^2)$. See Fig. 15.2.

2. For the 2-rectangle $R=[-1,1]\times[-1,1]$, define the surface $\mathbf{x}(R)$ by

$$\mathbf{x}(s^1,s^2)=(s^1,s^2,-(s^1)^2+s^2+2),$$

see Fig. 13.2 of Sect. 13.1. Calculate the Gaussian, mean, and principal curvatures at the point $\mathbf{x}(s^1,s^2)$.

3. For the 2-rectangle $R=[-1,1]\times[-1,1]$, define the surface $\mathbf{x}(R)$ by

$$\mathbf{x}(x,y)=(x,y,x^2+y^2),$$

see Fig. 13.3 of Sect. 13.1. Calculate the Gaussian, mean, and principal curvatures at the point $\mathbf{x}(x,y)$.

4. For the 2-rectangle $R=[-1,1]\times[-1,1]$, define the surface $\mathbf{x}(R)$ by

$$\mathbf{x}(x,y)=(x,y,x^2-y^2+1),$$

see Fig. 13.4 of Sect. 13.1. Calculate the Gaussian, mean, and principal curvatures at the point $\mathbf{x}(x,y)$.

## 15.5 The Curvature Bivector of a $k$-Surface $\mathcal{M}$

We now calculate $\nabla_{\mathbf{x}}\wedge\nabla_{\mathbf{x}}\mathbf{C}_r$ where $\mathbf{C}_r$ is any tangent $r$-vector field on $\mathcal{M}$ and discover that it is directly connected to the *Riemann curvature bivector* $R(\mathbf{a}\wedge\mathbf{b})$ of $\mathcal{M}$ at the point $\mathbf{x}$. We first note, by a simple calculation, that for tangent vectors $\mathbf{a},\mathbf{b}$,

$$(\mathbf{a}\wedge\mathbf{b})\cdot(\nabla_{\mathbf{x}}\wedge\nabla_{\mathbf{x}})=[\mathbf{b}\cdot\nabla_{\mathbf{x}},\mathbf{a}\cdot\nabla_{\mathbf{x}}]-[\mathbf{b},\mathbf{a}]\cdot\nabla_{\mathbf{x}},$$

where $[\mathbf{b}\cdot\nabla_{\mathbf{x}},\mathbf{a}\cdot\nabla_{\mathbf{x}}]=\mathbf{b}\cdot\nabla_{\mathbf{x}}\mathbf{a}\cdot\nabla_{\mathbf{x}}-\mathbf{a}\cdot\nabla_{\mathbf{x}}\mathbf{b}\cdot\nabla_{\mathbf{x}}$ and $[\mathbf{b},\mathbf{a}]=\mathbf{b}\cdot\nabla_{\mathbf{x}}\mathbf{a}-\mathbf{a}\cdot\nabla_{\mathbf{x}}\mathbf{b}$. Here, the vector coderivatives $\nabla_{\mathbf{x}}$ differentiate only to the right.

We now calculate,

$$\begin{aligned}\mathbf{b}\cdot\nabla_{\mathbf{x}}\mathbf{a}\cdot\nabla_{\mathbf{x}}\mathbf{C}_r &=P_{\mathbf{x}}(\mathbf{b}\cdot\partial_{\mathbf{x}}P_{\mathbf{x}}(\mathbf{a}\cdot\partial_{\mathbf{x}}\mathbf{C}_r))=P_{\mathbf{x}}(P_{\mathbf{b}}(\mathbf{a}\cdot\partial_{\mathbf{x}}\mathbf{C}_r))+P_{\mathbf{x}}(\mathbf{b}\cdot\partial_{\mathbf{x}}\mathbf{a}\cdot\partial_{\mathbf{x}}\mathbf{C}_r)\\ &=P_{\mathbf{b}}(P_{\mathbf{a}}(\mathbf{C}_r))+P_{\mathbf{x}}(\mathbf{b}\cdot\partial_{\mathbf{x}}\mathbf{a}\cdot\partial_{\mathbf{x}}\mathbf{C}_r).\end{aligned}$$

It follows that

$$\begin{aligned}(\mathbf{a}\wedge\mathbf{b})\cdot(\nabla_{\mathbf{x}}\wedge\nabla_{\mathbf{x}})\mathbf{C}_r &=[\mathbf{b}\cdot\nabla_{\mathbf{x}},\mathbf{a}\cdot\nabla_{\mathbf{x}}]\mathbf{C}_r-[\mathbf{b},\mathbf{a}]\cdot\nabla_{\mathbf{x}}\mathbf{C}_r=P_{\mathbf{b}}(P_{\mathbf{a}}(\mathbf{C}_r))\\ &\quad-P_{\mathbf{a}}(P_{\mathbf{b}}(\mathbf{C}_r))+P_{\mathbf{x}}(\mathbf{b}\cdot\partial_{\mathbf{x}}\mathbf{a}\cdot\partial_{\mathbf{x}}\mathbf{C}_r)\\ &\quad-P_{\mathbf{x}}(\mathbf{a}\cdot\partial_{\mathbf{x}}\mathbf{b}\cdot\partial_{\mathbf{x}}\mathbf{C}_r)-P_{\mathbf{x}}([\mathbf{b},\mathbf{a}]\cdot\partial_{\mathbf{x}}\mathbf{C}_r)\\ &=[P_{\mathbf{b}},P_{\mathbf{a}}](\mathbf{C}_r)+(\mathbf{a}\wedge\mathbf{b})\cdot(\partial_{\mathbf{x}}\wedge\partial_{\mathbf{x}})\mathbf{C}_r=[P_{\mathbf{b}},P_{\mathbf{a}}](\mathbf{C}_r),\end{aligned}\tag{15.45}$$

where $[P_{\mathbf{b}},P_{\mathbf{a}}]=P_{\mathbf{b}}P_{\mathbf{a}}-P_{\mathbf{a}}P_{\mathbf{b}}$. Applying (15.45) to the $(r+s)$-tangent vector field $A_r\wedge B_s\in\mathcal{T}_{\mathbf{x}}^s$, with the help of (15.19) and (15.20), we find that

$$(\mathbf{a}\wedge\mathbf{b})\cdot(\nabla_{\mathbf{x}}\wedge\nabla_{\mathbf{x}})A_r\wedge B_s=[P_{\mathbf{b}},P_{\mathbf{a}}](A_r\wedge B_s)$$
$$=\Big([P_{\mathbf{b}},P_{\mathbf{a}}](A_r)\Big)\wedge B_s+A_r\wedge\Big([P_{\mathbf{b}},P_{\mathbf{a}}](B_s)\Big). \tag{15.46}$$

We can now define a new quantity, called the *Riemann curvature bivector.*

**Definition 15.5.1.** The Riemann curvature bivector of the $k$-surface $\mathscr{M}$ is given by

$$R(\mathbf{a}\wedge\mathbf{b})=\partial_{\mathbf{v}}\wedge P_{\mathbf{a}}P_{\mathbf{b}}(\mathbf{v})$$

for all tangent vectors $\mathbf{a},\mathbf{b}\in\mathscr{T}_{\mathbf{x}}^1$.

The Riemann curvature bivector is intimately connected to the shape bivector (15.27), as is evident from

$$P_{\mathbf{a}}(L(\mathbf{b}))=P_{\mathbf{a}}\Big(\partial_{\mathbf{v}}\wedge P_{\mathbf{v}}(\mathbf{b})\Big)=\partial_{\mathbf{v}}\wedge P_{\mathbf{a}}(P_{\mathbf{v}}(\mathbf{b}))=R(\mathbf{a}\wedge\mathbf{b}),$$

with the help of (15.18) and the derivation rule (15.19).

The simple calculation for the tangent vector field $\mathbf{c}\in\mathscr{T}_{\mathbf{x}}^1$,

$$R(\mathbf{a}\wedge\mathbf{b})\cdot\mathbf{c}=[\partial_{\mathbf{v}}\wedge P_{\mathbf{a}}P_{\mathbf{b}}(\mathbf{v})]\cdot\mathbf{c}=\partial_{\mathbf{v}}P_{\mathbf{a}}(P_{\mathbf{b}}(\mathbf{v}))\cdot\mathbf{c}-P_{\mathbf{a}}(P_{\mathbf{b}}(\mathbf{c}))$$
$$=P_{\mathbf{b}}(P_{\mathbf{a}}(\mathbf{c}))-P_{\mathbf{a}}(P_{\mathbf{b}}(\mathbf{c}))=[P_{\mathbf{b}},P_{\mathbf{a}}](\mathbf{c})=(\mathbf{a}\wedge\mathbf{b})\cdot(\nabla_{\mathbf{x}}\wedge\nabla_{\mathbf{x}})\mathbf{c}, \tag{15.47}$$

together with (15.46), shows that for the $(r+s)$-vector field $A_r\wedge B_s\in\mathscr{T}_{\mathbf{x}}^{r+s}$,

$$(\mathbf{a}\wedge\mathbf{b})\cdot(\nabla_{\mathbf{x}}\wedge\nabla_{\mathbf{x}})A_r\wedge B_s=R(\mathbf{a}\wedge\mathbf{b})\boxtimes(A_r\wedge B_s)$$
$$=\Big(R(\mathbf{a}\wedge\mathbf{b})\boxtimes A_r\Big)\wedge B_s+A_r\wedge\Big(R(\mathbf{a}\wedge\mathbf{b})\boxtimes B_s\Big), \tag{15.48}$$

where $A\boxtimes B=\frac{1}{2}(AB-BA)$ denotes the antisymmetric part of the geometric product $AB$. Dotting both sides of (15.47) on the right by $\mathbf{d}$, and using the symmetry property (15.18), gives the further useful relationships

$$R(\mathbf{a}\wedge\mathbf{b})\cdot(\mathbf{c}\wedge\mathbf{d})=P_{\mathbf{a}}(\mathbf{c})\cdot P_{\mathbf{b}}(\mathbf{d})-P_{\mathbf{a}}(\mathbf{d})\cdot P_{\mathbf{b}}(\mathbf{c}). \tag{15.49}$$

and

$$R(\mathbf{a}\wedge\mathbf{b})=\frac{1}{2}\partial_{\mathbf{v}}\wedge\partial_{\mathbf{u}}\Big(P_{\mathbf{u}}(\mathbf{a})\cdot P_{\mathbf{v}}(\mathbf{b})-P_{\mathbf{v}}(\mathbf{a})\cdot P_{\mathbf{u}}(\mathbf{b})\Big). \tag{15.50}$$

We have the following.

**Theorem 15.5.2.** *The Riemann curvature bivector satisfies the following identities:*

*i)* $\partial_{\mathbf{u}}\wedge R(\mathbf{a}\wedge\mathbf{u})=\partial_{\mathbf{u}}\wedge\partial_{\mathbf{v}}\wedge P_{\mathbf{a}}P_{\mathbf{u}}(\mathbf{v})=0$. *($1^{\text{st}}$ Bianchi identity).*
*ii)* $\dot{\nabla}_{\mathbf{x}}\wedge\dot{R}(\mathbf{a}\wedge\mathbf{b})=0$. *($2^{\text{nd}}$ Bianchi identity).*
*iii)* $R(\mathbf{a}\wedge\mathbf{b})\cdot(\mathbf{c}\wedge\mathbf{d})=(\mathbf{a}\wedge\mathbf{b})\cdot R(\mathbf{c}\wedge\mathbf{d})$. *(Symmetric bivector operator)*

*Proof.* i) The proof of this part follows from the symmetry property (15.18), which gives $P_{\mathbf{u}}(\mathbf{v})=P_{\mathbf{v}}(\mathbf{u})$ for all tangent vectors $\mathbf{u},\mathbf{v}\in\mathscr{T}_{\mathbf{x}}^1$.

ii) Taking the $\mathbf{w}$-coderivative of both sides of (15.50) gives

$$R_{\not{w}}(\mathbf{a}\wedge\mathbf{b}) = \frac{1}{2}\partial_{\mathbf{v}}\wedge\partial_{\mathbf{u}}\Big(P_{\mathbf{u},\mathbf{w}}(\mathbf{a})\cdot P_{\mathbf{v}}(\mathbf{b}) + P_{\mathbf{u}}(\mathbf{a})\cdot P_{\mathbf{v},\mathbf{w}}(\mathbf{b})$$
$$-P_{\mathbf{v},\mathbf{w}}(\mathbf{a})\cdot P_{\mathbf{u}}(\mathbf{b}) - P_{\mathbf{v}}(\mathbf{a})\cdot P_{\mathbf{u},\mathbf{w}}(\mathbf{b})\Big).$$

If we now take the curl of $R_{\not{w}}$ with respect to $\partial_{\mathbf{w}}$, the identity follows by noting that

$$\partial_{\mathbf{w}}\wedge R_{\not{w}}(\mathbf{a}\wedge\mathbf{b}) = \dot{\nabla}\wedge\dot{R}(\mathbf{a}\wedge\mathbf{b})$$

on the left side of the equation and where we are using the integrability condition that $P_{\mathbf{u},\mathbf{w}} = P_{\mathbf{w},\mathbf{u}}$ and $P_{\mathbf{v},\mathbf{w}} = P_{\mathbf{w},\mathbf{v}}$ on the right side of the equation.

iii) The fact that $R(\mathbf{a}\wedge\mathbf{b})$ is a symmetric bivector operator follows from (15.49) and is left as an exercise.

□

The classical *Riemann curvature tensor* is defined by

$$R_{ijkl} = R(\mathbf{x}_i\wedge\mathbf{y}_j)\cdot(\mathbf{x}_k\wedge\mathbf{x}_l). \tag{15.51}$$

Taking the partial $k$-derivative of the relation $P_{\mathbf{x}}(\mathbf{L}_{ij}) = 0$ gives

$$P_k(\mathbf{L}_{ij}) + P_{\mathbf{x}}(\mathbf{L}_{ij,k}) = 0$$

or

$$P_{\mathbf{x}}(\mathbf{L}_{ij,k}) = -P_k(\mathbf{L}_{ij}) = -P_kP_i(\mathbf{x}_j).$$

From this relationship, we get

$$P_{\mathbf{x}}(\mathbf{L}_{kj,i}) - P_{\mathbf{x}}(\mathbf{L}_{ki,j}) = P_jP_i(\mathbf{x}_k) - P_iP_j(\mathbf{x}_k) = R(\mathbf{x}_i\wedge\mathbf{x}_j)\cdot\mathbf{x}_k.$$

In the case of a $(n-1)$-hypersurface in $\mathbb{R}^n$, with the normal vector $\mathbf{n}_{\mathbf{x}}$ at the point $\mathbf{x}$, by taking the $k$-derivative of the second fundamental form (15.23), we get

$$L_{ij,k} = \frac{\partial L_{ij}}{\partial s^k} = P_{jk}(\mathbf{x}_i)\cdot\mathbf{n} + P_j(\mathbf{x}_{ik})\cdot\mathbf{n}.$$

Using this relationship, after cancelations, we arrive at the classical *Codazzi-Mainardi equations*,

$$L_{ij,k} - L_{ik,j} = \mathbf{x}_{ik}\cdot P_j(\mathbf{n}) - \mathbf{x}_{ij}\cdot P_k(\mathbf{n}) = P_j(\mathbf{x}_{ik})\cdot\mathbf{n} - P_k(\mathbf{x}_{ij})\cdot\mathbf{n}$$
$$= (\mathbf{K}_{ik}\cdot\mathbf{x}^l)(\mathbf{x}_l\cdot P_j(\mathbf{n})) - (\mathbf{K}_{ij}\cdot\mathbf{x}^l)(\mathbf{x}_l\cdot P_k(\mathbf{n})) = \Gamma^l_{ik}L_{lj} - \Gamma^l_{ij}L_{lk}$$

where we are employing the Einstein summation convention, [45, p.76], [56, p.142].

## Exercises

1. Using the symmetry property (15.18) of the operator $P_{\mathbf{c}}$ for $\mathbf{c} \in \mathscr{T}_{\mathbf{x}}^k$, show that

$$([P_{\mathbf{b}}, P_{\mathbf{a}}](\mathbf{c})) \cdot \mathbf{c} = 0.$$

2. Prove that the classical Riemann curvature tensor (15.51) satisfies:
   (a) $R(\mathbf{x}_i \wedge \mathbf{x}_j) \cdot (\mathbf{x}_k \wedge \mathbf{x}_l) = (\partial_{\mathbf{v}} \wedge P_i P_j(\mathbf{v})) \cdot (\mathbf{x}_k \wedge \mathbf{x}_l) = \mathbf{x}_k \cdot P_i P_j(\mathbf{x}_l) - \mathbf{x}_l \cdot P_i P_j(\mathbf{x}_k)$.
   (b) For an orientable $(n-1)$-hypersurface in $\mathbb{R}^n$, show that

$$\begin{aligned} R(\mathbf{x}_i \wedge \mathbf{x}_j) \cdot (\mathbf{x}_k \wedge \mathbf{x}_l) &= P_i(\mathbf{x}_k) \cdot P_j(\mathbf{x}_l) - P_i(\mathbf{x}_l) \cdot P_j(\mathbf{x}_k) \\ &= \mathbf{L}_{ik} \cdot \mathbf{L}_{jl} - \mathbf{L}_{il} \cdot \mathbf{L}_{jk}. \end{aligned}$$

3. (a) Show that the Riemann curvature bivector can be expressed directly in terms of the shape operator by

$$R(\mathbf{a} \wedge \mathbf{b}) = P_{\mathbf{a}}(\partial_{\mathbf{v}} \wedge P_{\mathbf{v}}(\mathbf{b})) = P_{\mathbf{a}}(L(\mathbf{b}))$$

   (b) $\partial_{\mathbf{v}} \cdot R(\mathbf{v} \wedge \mathbf{b}) = L^2(\mathbf{b})$.
4. (a) Prove part iii) of Theorem 15.5.2.
   (b) Show that the second Bianchi identity can be expressed in the form

$$R_{\not{\mathbf{c}}}(\mathbf{a} \wedge \mathbf{b}) + R_{\not{\mathbf{a}}}(\mathbf{b} \wedge \mathbf{c}) + R_{\not{\mathbf{b}}}(\mathbf{c} \wedge \mathbf{a}) = 0,$$

   where the directional coderivative was defined in (15.40).
5. (a) Show that $R(\mathbf{a} \wedge \mathbf{b}) = P_{\mathbf{x}}\Big(L(\mathbf{a}) \boxtimes L(\mathbf{b})\Big)$.
   (b) More generally, show that

$$L(\mathbf{a}) \boxtimes L(\mathbf{b}) = R(\mathbf{a} \wedge \mathbf{b}) - P_{\mathbf{a}}(\partial_{\mathbf{w}}) \wedge P_{\mathbf{b}}(\mathbf{w}).$$

   In [43, p.195], the quantity $L(\mathbf{a}) \boxtimes L(\mathbf{b})$ is referred to as the *total curvature*. The total curvature consists of the *intrinsic curvature* part $R(\mathbf{a} \wedge \mathbf{b})$ and the *extrinsic curvature* part $-P_{\mathbf{a}}(\partial_{\mathbf{w}}) \wedge P_{\mathbf{b}}(\mathbf{w})$.
6. Prove the *generalized Bianchi identity*

$$\Big(L(\mathbf{a}) \boxtimes L(\mathbf{b})\Big)_{\mathbf{c}} + \Big(L(\mathbf{b}) \boxtimes L(\mathbf{c})\Big)_{\mathbf{a}} + \Big(L(\mathbf{c}) \boxtimes L(\mathbf{a})\Big)_{\mathbf{b}} = 0,$$

   where

$$\Big(L(\mathbf{a}) \boxtimes L(\mathbf{b})\Big)_{\mathbf{c}} := \mathbf{c} \cdot \partial_{\mathbf{x}}\Big(L(\mathbf{a}) \boxtimes L(\mathbf{b})\Big) - \Big(L(\mathbf{c} \cdot \partial_{\mathbf{x}}\mathbf{a}) \boxtimes L(\mathbf{b})\Big) - \Big(L(\mathbf{a}) \boxtimes L(\mathbf{c} \cdot \partial_{\mathbf{x}}\mathbf{b})\Big).$$

# Chapter 16
# Mappings Between Surfaces

> *We could use up two Eternities in learning all that is to be learned about our own world and the thousands of nations that have arisen and flourshed and vanished from it. Mathematics alone would occupy me eight million years.*
>
> —Mark Twain

In this chapter we explore when a mapping will preserve geodesics, the shortest distances between two points on a surface, and when a mapping will preserve the angles between the intersection of two curves on the surface, known as *conformal mappings*. A conformal mapping of a pseudo-Euclidean space $\mathbb{R}^{p,q}$ generalizes the idea of an analytic function in the two-dimensional theory of complex variables.[1]

## 16.1 Mappings Between Surfaces

Let a regular $k$-surface $\mathscr{M}$ be given in $\mathbb{R}^n$ by $\mathbf{x} = \mathbf{x}(s^1,\ldots,s^k)$ and a second regular $k$-surface $\mathscr{M}'$ in $\mathbb{R}^n$ by $\mathbf{x}' = \mathbf{x}'(v^1,\ldots,v^k)$. Let $\mathbf{x}' = f(\mathbf{x})$ be a local bijective differentiable (in a coordinate patch) invertible function, $f : \mathscr{M} \to \mathscr{M}'$. The mapping $f$ induces a second mapping $\mathbf{x}' = f(\mathbf{x}) = f(\mathbf{x}(s^1,\ldots,s^k))$ on the surface $\mathscr{M}'$, which we naturally assume to be compatible with $\mathbf{x}' = \mathbf{x}'(v^1,\ldots,v^k)$. This means that each coordinate $v^i = v^i(s^1,\ldots,s^k)$ is a differentiable function of the coordinates of $\mathscr{M}$. The mapping $f$ is called a *diffeomorphism* between the surfaces $\mathscr{M}$ and $\mathscr{M}'$ (at least in some neighborhood of a point $\mathbf{x} \in \mathscr{M}$). Although we have specified that our surfaces are embedded in $\mathbb{R}^n$, all our results in this chapter remain generally valid for surfaces embedded in $\mathbb{R}^{p,q}$.

---

[1]This chapter is based upon an article by the author that appeared in the American Mathematical Mathematical Society Notices [89].

G. Sobczyk, *New Foundations in Mathematics: The Geometric Concept of Number*, DOI 10.1007/978-0-8176-8385-6_16,

The mapping $f$ between the surfaces $\mathscr{M}$ and $\mathscr{M}'$ induces a linear outermorphism $\underline{f} = \underline{f}_{\mathbf{x}}$ between the tangent spaces $\mathscr{T}_{\mathbf{x}}$ and $\mathscr{T}_{\mathbf{x}'}$, defined for each $\mathbf{a} \in \mathscr{T}_{\mathbf{x}}^1$ by

$$\underline{f}(\mathbf{a}) = \mathbf{a} \cdot \partial_{\mathbf{x}} f(\mathbf{x}) \in \mathscr{T}_{\mathbf{x}'}^1. \tag{16.1}$$

The $\mathbf{a}$-derivative $\underline{f}(\mathbf{a})$ is called the *differential* of $f$ at the point $\mathbf{x} \in \mathscr{M}$ and maps each tangent vector $\mathbf{a} \in \mathscr{T}_{\mathbf{x}}^1$ into a corresponding tangent vector $\mathbf{a}' = \underline{f}(\mathbf{a}) \in \mathscr{T}_{\mathbf{x}'}^1$. Indeed, $\underline{f}$ maps the whole tangent geometric algebra $\mathscr{T}_{\mathbf{x}}$ at $\mathbf{x}$ into the tangent geometric algebra $\mathscr{T}_{\mathbf{x}'}^{\prime}$ at $\mathbf{x}'$.

The mapping $\mathbf{x}' = f(\mathbf{x})$ induces a relationship between the basis vectors $(\mathbf{x})_{(k)}$ of $\mathscr{T}_{\mathbf{x}}^1$ and the corresponding basis vectors $(\mathbf{x}')_{(k)}$ of $\mathscr{T}_{\mathbf{x}'}^1$, defined by

$$\mathbf{x}_i' = \frac{\partial f(\mathbf{x})}{\partial s^i} = \mathbf{x}_i \cdot \partial_{\mathbf{x}} f(\mathbf{x}) = \underline{f}(\mathbf{x}_i). \tag{16.2}$$

The tangent vectors $(\mathbf{x})_{(k)}$ of $\mathscr{T}_{\mathbf{x}}^1$ are said to be *pushed forward* by the differential $\underline{f}$ into the tangent vectors $(\mathbf{x}')_{(k)} = \underline{f}(\mathbf{x})_{(k)}$ of the tangent space $\mathscr{T}_{\mathbf{x}'}^1$ at the point $\mathbf{x}' \in \mathscr{M}'$. The additional relationship

$$[1]_k = (\mathbf{x}')^{(k)} \cdot (\mathbf{x}')_{(k)} = (\mathbf{x}')^{(k)} \cdot \underline{f}(\mathbf{x})_{(k)} = \overline{f}(\mathbf{x}')^{(k)} \cdot (\mathbf{x})_{(k)}, \tag{16.3}$$

where $[1]_k$ is the identity $k \times k$-matrix, shows that the reciprocal basis $(\mathbf{x}')^{(k)}$ is *pulled back* by the adjoint mapping $\overline{f} = \overline{f}_{\mathbf{x}}$ into the reciprocal basis vectors $(\mathbf{x})^{(k)}$ of the tangent space $\mathscr{T}_{\mathbf{x}}$ of $\mathscr{M}$ at the point $\mathbf{x}$. Thus, we have

$$\overline{f}(\mathbf{x}')^{(k)} = (\mathbf{x})^{(k)},$$

and consequently,

$$\overline{f}(\partial_{\mathbf{x}'}') = \partial_{\mathbf{x}}. \tag{16.4}$$

Taking the differential of both sides of the mapping (16.1) with respect to $\mathbf{b} \in \mathscr{T}_{\mathbf{x}}^1$ gives

$$\mathbf{b} \cdot \partial_{\mathbf{x}} \underline{f}(\mathbf{a}) = \underline{f}_{\mathbf{b}}(\mathbf{a}) + \underline{f}(\mathbf{b} \cdot \partial_{\mathbf{x}} \mathbf{a}) = \mathbf{b} \cdot \partial_{\mathbf{x}} \mathbf{a} \cdot \partial_{\mathbf{x}} f(\mathbf{x}). \tag{16.5}$$

Using this relationship, the integrability condition (13.8), and the change of variables formula $\mathbf{a} \cdot \partial_{\mathbf{x}} = \underline{f}(\mathbf{a}) \cdot \partial_{\mathbf{x}'}$, implies that

$$[\mathbf{b}', \mathbf{a}'] = \mathbf{b}' \cdot \partial_{\mathbf{x}'} \underline{f}(\mathbf{a}) - \mathbf{a}' \cdot \partial_{\mathbf{x}'} \underline{f}(\mathbf{b}) = \underline{f}([\mathbf{a}, \mathbf{b}]), \tag{16.6}$$

and shows that the Lie bracket $[\mathbf{a}, \mathbf{b}]$ of the tangent vector fields $\mathbf{a}, \mathbf{b}$ on the surface $\mathscr{M}$ is pushed forward by the differential $\underline{f}$ into the corresponding Lie bracket $[\mathbf{a}', \mathbf{b}']$ of the corresponding tangent vector fields $\mathbf{a}', \mathbf{b}'$ on the surface $\mathscr{M}'$. The integrability condition also implies that

$$\underline{f}_{\mathbf{a}}(\mathbf{b}) = \underline{f}_{\mathbf{b}}(\mathbf{a}), \tag{16.7}$$

as can be easily verified by using (16.5) and (16.6). Since $\underline{f}_{\mathbf{a}}$ is a derivation satisfying the derivation rule (15.19), a more general version of (16.7) is

$$\underline{f}_{\mathbf{a}}(\mathbf{B}_r) = \underline{f}_{\mathbf{v}'}(\mathbf{a}) \wedge \left(\partial_{\mathbf{v}'} \cdot \mathbf{B}'_r\right), \tag{16.8}$$

where $\mathbf{B}'_r = \underline{f}(\mathbf{B}_r)$ is an $r$-vector in $\mathscr{T}_{\mathbf{x}'}$ for $r \geq 1$ and the tangent vector derivative $\partial_{\mathbf{v}'}$ operates to the left.

Let $\mathbf{a}, \mathbf{b}, \mathbf{c}$ be tangent vector fields in $\mathscr{T}_{\mathbf{x}}$. Defining the $\mathbf{b}$-differential of $\underline{f}_{\mathbf{a}}(\mathbf{c})$ by

$$\underline{f}_{\mathbf{a},\mathbf{b}}(\mathbf{c}) = \mathbf{b} \cdot \partial_{\mathbf{x}} \underline{f}_{\mathbf{a}}(\mathbf{c}) - \underline{f}_{\mathbf{b}\cdot\partial\mathbf{a}}(\mathbf{c}) - \underline{f}_{\mathbf{a}}(\mathbf{b} \cdot \partial_{\mathbf{x}}\mathbf{c}) \tag{16.9}$$

guarantees that $\underline{f}_{\mathbf{a},\mathbf{b}}(\mathbf{c})$ retains its tensor-like quality (linearity) in the arguements $\mathbf{a}, \mathbf{b}$, and $\mathbf{c}$. With this definition, another consequence of the integrability conditions (13.8) and (16.7) is

$$\underline{f}_{\mathbf{a},\mathbf{b}}(\mathbf{c}) = \underline{f}_{\mathbf{b},\mathbf{a}}(\mathbf{c}). \tag{16.10}$$

Just like we defined an intrinsic coderivative (15.40) of a vector field, we now define the intrinsic coderivatives of the differential $\underline{f}$ of the mapping $f$ between the surfaces $\mathscr{M}$ and $\mathscr{M}'$.

**Definition 16.1.1.** Let $f : \mathscr{M} \to \mathscr{M}'$ be a differentiable mapping between the $k$-surfaces $\mathscr{M}$ and $\mathscr{M}'$. The $\mathbf{a}$-coderivative of the differential $\underline{f}$ of $f$ at the point $\mathbf{x} \in \mathscr{M}$ is given by

$$\underline{f}_{/\!\!\!\mathbf{a}} = P' \underline{f}_{\mathbf{a}} P \tag{16.11}$$

The intrinsic second coderivative is given by

$$\underline{f}_{/\!\!\!\mathbf{a},/\!\!\!\mathbf{b}} = P'(\mathbf{b} \cdot \partial \underline{f}_{/\!\!\!\mathbf{a}})P - P' \underline{f}_{\mathbf{b}\cdot\nabla\mathbf{a}} P. \tag{16.12}$$

Note that the second term on the right side of (16.12) is necessary to guarantee that the second coderivative depends only upon the value of $\mathbf{a}$ at the point $\mathbf{x} \in \mathscr{M}$.

Calculating $\underline{f}_{/\!\!\!\mathbf{a},/\!\!\!\mathbf{b}}$, we find that

$$\begin{aligned}\underline{f}_{/\!\!\!\mathbf{a},/\!\!\!\mathbf{b}} &= P'\left(\mathbf{b} \cdot \partial_{\mathbf{x}} P' \underline{f}_{\mathbf{a}} P\right)P - P' \underline{f}_{\mathbf{b}\cdot\nabla\mathbf{a}} P = P'\left(P' \underline{f}_{\mathbf{a}} P\right)_{\mathbf{b}} P \\ &= P'\left(P'_{\mathbf{b}} \underline{f}_{\mathbf{a}} + \underline{f}_{\mathbf{a},\mathbf{b}} + P' \underline{f}_{\mathbf{b}\cdot\nabla\mathbf{a}}\right)P.\end{aligned}$$

Using (16.10) and the fact that $\underline{f} = P' \underline{f} P$, so that

$$\underline{f}_{\mathbf{a}} = P'_{\mathbf{a}} \underline{f} P + P' \underline{f}_{\mathbf{a}} P + P' \underline{f} P_{\mathbf{a}},$$

we can now calculate

$$\underline{f}_{\dot{a},\dot{b}} - \underline{f}_{\dot{b},\dot{a}} = P'\left(P'_{\mathbf{b}}\underline{f}_{\mathbf{a}} - P'_{\mathbf{a}}\underline{f}_{\mathbf{b}} + \underline{f}_{\mathbf{a}}P_{\mathbf{b}} - \underline{f}_{\mathbf{b}}P_{\mathbf{a}}\right)P$$

$$= \left(P'_{\mathbf{b}}P'_{\mathbf{a}} - P'_{\mathbf{a}}P'_{\mathbf{b}}\right)\underline{f} - \underline{f}\left(P_{\mathbf{b}}P_{\mathbf{a}} - P_{\mathbf{a}}P_{\mathbf{b}}\right) = [P'_{\mathbf{b}}, P'_{\mathbf{a}}]\underline{f} - \underline{f}[P_{\mathbf{b}}, P_{\mathbf{a}}]. \quad (16.13)$$

Applying both sides of this operator equation to the tangent vector $\mathbf{c}$ then gives

$$\underline{f}_{\dot{a}_{(2)}}(\mathbf{c}) := \left(\underline{f}_{\dot{a}_1,\dot{a}_2} - \underline{f}_{\dot{a}_2,\dot{a}_1}\right)(\mathbf{c}) = R'(\mathbf{a}'_{(2)})\cdot\mathbf{c}' - \underline{f}\left(R(\mathbf{a}_{(2)})\cdot\mathbf{c}\right), \quad (16.14)$$

which is the basic relationship between the curvature tensors of $\mathscr{M}$ and $\mathscr{M}'$.

A differentiable bijective mapping $f : \mathscr{M} \longrightarrow \mathscr{M}'$ is said to be an *isometry* between the surfaces $\mathscr{M}$ and $\mathscr{M}'$, if for all $\mathbf{a}, \mathbf{b} \in \mathscr{T}_{\mathbf{x}}^1$

$$\mathbf{a}\cdot\mathbf{b} = \underline{f}(\mathbf{a})\cdot\underline{f}(\mathbf{b}). \quad (16.15)$$

Let $\mathbf{x} = \mathbf{x}(t)$ be a regular curve in $\mathscr{M}$. Then $\mathbf{x}'(t) = f(\mathbf{x}(t))$ is the corresponding regular curve in $\mathscr{M}'$. Calculating the lengths of the corresponding curves in $\mathscr{M}$ and $\mathscr{M}'$, for $t_0 < t < t_1$, we find that

$$L = \int |\mathrm{d}\mathbf{x}'| = \int_{t_0}^{t_1}\left|\frac{\mathrm{d}\mathbf{x}'}{\mathrm{d}t}\right|\mathrm{d}t = \int_{t_0}^{t_1}\left|\frac{\mathrm{d}\mathbf{x}}{\mathrm{d}t}\cdot\partial_{\mathbf{x}}f(\mathbf{x})\right|\mathrm{d}t = \int_{t_0}^{t_1}\left|\underline{f}\left(\frac{\mathrm{d}\mathbf{x}}{\mathrm{d}t}\right)\right|\mathrm{d}t$$

$$= \int_{t_0}^{t_1}\left|\underline{f}\left(\frac{\mathrm{d}\mathbf{x}}{\mathrm{d}t}\right)\cdot\underline{f}\left(\frac{\mathrm{d}\mathbf{x}}{\mathrm{d}t}\right)\right|^{\frac{1}{2}}\mathrm{d}t = \int_{t_0}^{t_1}\left|\left(\frac{\mathrm{d}\mathbf{x}}{\mathrm{d}t}\right)\cdot\left(\frac{\mathrm{d}\mathbf{x}}{\mathrm{d}t}\right)\right|^{\frac{1}{2}}\mathrm{d}t = \int |\mathrm{d}\mathbf{x}|,$$

which shows that the lengths of the corresponding curves $\mathbf{x}(t)$ and $\mathbf{x}'(t) = f(\mathbf{x}(t))$ under the isometry $f$ are preserved.

Let $\mathscr{M}$ be defined by the coordinates $\mathbf{x} = \mathbf{x}(s^1,\dots,s^k)$ and $\mathscr{M}'$ by the coordinates $\mathbf{x}' = \mathbf{x}'(s^1,\dots,s^k) = f(\mathbf{x}(s^1,\dots,s^k))$, where $f$ is an isometry between them. Then we easily calculate

$$g'_{ij} = \mathbf{x}'_i\cdot\mathbf{x}'_j = \frac{\partial\mathbf{x}'}{\partial s^i}\cdot\frac{\partial\mathbf{x}'}{\partial s^j} = \underline{f}(\mathbf{x}_i)\cdot\underline{f}(\mathbf{x}_j) = \mathbf{x}_i\cdot\mathbf{x}_j = g_{ij},$$

showing that the corresponding metric tensors are preserved under an isometry.

A mapping $f : \mathbb{R}^n \longrightarrow \mathbb{R}^n$ is called a *rigid motion* if $f(\mathbf{x})$ is the composition of a rotation and a translation,

$$f(\mathbf{x}) = \exp\left(\frac{1}{2}\mathbf{B}\right)\mathbf{x}\exp\left(-\frac{1}{2}\mathbf{B}\right) + \mathbf{c}$$

where $\mathbf{B}$ is a constant bivector and $\mathbf{c}$ is a constant vector. Two surfaces $\mathscr{M}, \mathscr{M}'$ are *rigidly equivalent* if $f(\mathscr{M}) = \mathscr{M}'$ for some rigid motion of $\mathbb{R}^n$.

A *ruled surface* can be parameterized by

$$\mathbf{x}(s,t) = \mathbf{a}(s) + t\mathbf{b}(s),$$

where $\mathbf{a}(s)$ is a unit speed curve and $|\mathbf{b}(s)| = 1$, [56, p.139].

## Exercises

Let $f : \mathscr{M} \to \mathscr{M}'$ be a regular mapping between the $k$-surfaces $\mathscr{M}$ and $\mathscr{M}'$ at the point $\mathbf{x} \in \mathscr{M}$, and let $\mathbf{x}' = f(\mathbf{x}) \in \mathscr{M}'$. Let $\mathbf{a}, \mathbf{b} \in \mathscr{T}_{\mathbf{x}}^1$, and let $\mathbf{a}' = \underline{f}(\mathbf{a})$ and $\mathbf{b}' = \underline{f}(\mathbf{b})$.

1. Let $h(\mathbf{x}')$ be a differentiable geometric-valued function on $\mathscr{M}'$. Show that

$$\mathbf{a}' \cdot \partial_{\mathbf{x}'} h(\mathbf{x}') = \mathbf{a} \cdot \partial_{\mathbf{x}} h(\mathbf{x}'(\mathbf{x}))$$

   is a statement of the chain rule.
2. (a) Show that the integrability conditions (13.8) on $\mathscr{M}$ and $\mathscr{M}'$ are related by

$$(\mathbf{a}' \wedge \mathbf{b}') \cdot (\partial_{\mathbf{x}'} \wedge \partial_{\mathbf{x}'}) = (\mathbf{a} \wedge \mathbf{b}) \cdot \overline{f}(\partial_{\mathbf{x}'} \wedge \partial_{\mathbf{x}'}) = (\mathbf{a} \wedge \mathbf{b}) \cdot (\partial_{\mathbf{x}} \wedge \partial_{\mathbf{x}}) = 0.$$

   (b) Show that $\underline{f}_{\mathbf{a}'}(\mathbf{b}) = \underline{f}_{\mathbf{b}'}(\mathbf{a})$ for all $\mathbf{a}, \mathbf{b} \in \mathscr{T}_{\mathbf{x}}^1$. When $f(\mathbf{x}) = \mathbf{x}$ is the identity mapping of $\mathscr{M}$ onto itself, this reduces to the integrability condition that $P_{\mathbf{a}}(\mathbf{b}) = P_{\mathbf{b}}(\mathbf{a})$ for the projection $P_{\mathbf{x}}$.
3. Prove the more general integrability condition (16.8) follows from the integrability condition (16.7).
4. Show that a cylinder is a ruled surface.
5. (a) Show that the surface $z = x^2 - y^2$ is it doubly ruled in the sense that through each point on the surface, two straight lines can be drawn that are on the surface.
   (b) Show that the surface

$$\frac{x^2}{a^2} + \frac{y^2}{b^2} - \frac{z^2}{c^2} = 1$$

   is doubly ruled.
6. Show that a right circular cone is a ruled surface.

## 16.2 Projectively Related Surfaces

Let $\mathscr{M}$ and $\mathscr{M}'$ be $k$-surfaces in $\mathbb{R}^n$ related by the diffeomorphism $\mathbf{x}' = f(\mathbf{x})$, with the induced tangent outermorphism $\underline{f} : \mathscr{T}_{\mathbf{x}} \to \mathscr{T}_{\mathbf{x}'}$ as given in (16.1) of the previous section. Much information about how the geometries of the surfaces are related is contained in the *generalized shape operator* $L'(A)$, defined by

$$L'(A) = \dot{\partial}_{\mathbf{x}'}\underline{\dot{f}}(A), \tag{16.16}$$

where $A \in \mathbb{G}_n$. The generalized shape operator between the $k$-surfaces $\mathscr{M}$ and $\mathscr{M}'$ reduces to the shape operator (15.26) when $f(\mathbf{x})$ is the identity mapping, $\mathbf{x}' = f(\mathbf{x}) = \mathbf{x}$ for all $\mathbf{x} \in \mathscr{M}$. We shall be particularly interested in the generalized *shape divergence* when evaluated at tangent $r$-vector fields $A_r \in \mathscr{T}_{\mathbf{x}}^r$. We define

$$\phi(A_r) := \frac{1}{\mu}\dot{\partial}' \cdot \underline{\dot{f}}(A_r),$$

where $\mu$ is a normalizing constant chosen for convenience.

We wish to characterize when a mapping $\mathbf{x}' = f(\mathbf{x})$ between two surfaces preserves geodesics, that is, given a geodesic $\mathbf{x}(t)$ on $\mathscr{M}$, we require that the corresponding curve $\mathbf{x}'(t) = f(\mathbf{x}(t))$ be a geodesic on $\mathscr{M}'$. We calculate

$$\mathbf{v}' = \frac{\mathrm{d}\mathbf{x}'}{\mathrm{d}t} = \frac{\mathrm{d}\mathbf{x}}{\mathrm{d}t} \cdot \partial_{\mathbf{x}} f(\mathbf{x}) = \underline{f}(\mathbf{v}).$$

Taking the second derivative, we get

$$\frac{\mathrm{d}\mathbf{v}'}{\mathrm{d}t} = \frac{\mathrm{d}\underline{f}(\mathbf{v})}{\mathrm{d}t} = \underline{f}_{\mathbf{v}}(\mathbf{v}) + \underline{f}\left(\frac{\mathrm{d}\mathbf{v}}{\mathrm{d}t}\right).$$

Wedging this last equation with $\mathbf{v}' = \underline{f}(\mathbf{v})$, we get

$$\frac{\mathrm{d}\mathbf{v}'}{\mathrm{d}t} \wedge \mathbf{v}' = \underline{f}_{\mathbf{v}}(\mathbf{v}) \wedge \mathbf{v}' + \underline{f}\left(\frac{\mathrm{d}\mathbf{v}}{\mathrm{d}t} \wedge \mathbf{v}\right).$$

Recalling (15.38) that a non-unit speed curve $\mathbf{x}(t)$ is a geodesic iff $P_{\mathbf{x}}(\frac{\mathrm{d}\mathbf{v}}{\mathrm{d}t} \wedge \mathbf{v}) = 0$, we are led to the following:

**Definition 16.2.1.** The $f$-related surfaces $\mathscr{M}$ and $\mathscr{M}'$ are said to be projectively related by $f$ if for all tangent vectors $\mathbf{a} \in \mathscr{T}_{\mathbf{x}}^1$, and corresponding tangent vectors $\mathbf{a}' = \underline{f}(\mathbf{a})$ in $\mathscr{T}_{\mathbf{x}'}^1$,

$$P'(\underline{f}_{\mathbf{a}'}(\mathbf{a})) \wedge \mathbf{a}' = 0,$$

where

$$\underline{f}_{\mathbf{a}'}(\mathbf{b}) = \mathbf{a} \cdot \partial_{\mathbf{x}} \underline{f}(\mathbf{b}) - \underline{f}(\mathbf{a} \cdot \partial_{\mathbf{x}} \mathbf{b}) = \underline{f}_{\mathbf{b}'}(\mathbf{a}),$$

and $P' = P_{\mathbf{x}'}$ is the projection onto the tangent space $\mathscr{T}_{\mathbf{x}'}$ of $\mathscr{M}'$.

Saying that $\mathscr{M}$ and $\mathscr{M}'$ are projectively related by the regular mapping $f$ means that the mapping $f(\mathbf{x})$ preserves geodesics; if $\mathbf{x}(t)$ is a geodesic on $\mathscr{M}$, then $\mathbf{x}'(t) = f(\mathbf{x}(t))$ is a geodesic on $\mathscr{M}'$.

In studying projectively related $k$-surfaces, we define the normalizing factor $\mu = k+1$ in the definition of the shape divergence $\phi(\mathbf{a})$. Let $A_r \in \mathscr{T}_\mathbf{x}^r$ be any tangent $r$-vector field, then

$$\Phi'(A_r) = \frac{1}{k+1}\partial_{\mathbf{v}'} \cdot \underline{f}_{\mathbf{v}'}(A_r) = \underline{f}(\Phi(A_r)). \tag{16.17}$$

Suppose now that the $k$-surfaces $\mathscr{M}$ and $\mathscr{M}'$ are projectively related by the mapping $f$, so that for all $\mathbf{a} \in \mathscr{T}_\mathbf{a}$

$$P'(\underline{f}_{\mathbf{a}'}(\mathbf{a})) \wedge \mathbf{a}' = 0. \tag{16.18}$$

Treating $\mathbf{a}'$ as a variable in the tangent space $\mathscr{T}_{\mathbf{x}'}^1$, noting that $\mathbf{a} = \underline{f}^{-1}(\mathbf{a}')$, and taking the divergence of both sides of the equation (16.18) with $\partial_{\mathbf{a}'}$ gives

$$0 = \partial_{\mathbf{a}'} \cdot \left(P'(\underline{f}_{\mathbf{a}'}(\mathbf{a})) \wedge \mathbf{a}'\right) = P'\left(\partial_{\mathbf{a}'} \cdot (\underline{f}_{\mathbf{a}'}(\mathbf{a}) \wedge \mathbf{a}')\right).$$

Continuing the calculation, we get

$$\begin{aligned} 0 &= P'\left(\partial_{\mathbf{a}'} \cdot \underline{f}_{\mathbf{a}'}(\mathbf{a})\mathbf{a}' - \partial_{\mathbf{a}'} \cdot \mathbf{a}'\underline{f}_{\mathbf{a}'}(\mathbf{a})\right) \\ &= 2\partial_{\mathbf{a}'} \cdot \underline{f}_{\mathbf{a}'}(\mathbf{a})\mathbf{a}' + P'(f_{\mathbf{a}'}(\mathbf{a})) - 2P'(f_{\mathbf{a}'}(\mathbf{a})) - kP'(f_{\mathbf{a}'}(\mathbf{a})). \end{aligned}$$

Solving this last relationship for $P'(f_{\mathbf{a}'}(\mathbf{a}))$ gives

$$P'(\underline{f}_{\mathbf{a}'}(\mathbf{a})) = 2\Phi(\mathbf{a})\mathbf{a}' = 2\underline{f}(\Phi(\mathbf{a})\mathbf{a}). \tag{16.19}$$

It is not difficult to show that the relationship (16.19) can be equivalently expressed in each of the alternative forms

$$P'(\underline{f}_{\mathbf{a}'}(\mathbf{c})) = \underline{f}(\Phi(\mathbf{a})\mathbf{c} + \Phi(\mathbf{c})\mathbf{a}), \tag{16.20}$$

and

$$P'(\underline{f}_{\mathbf{a}'}(\mathbf{C}_r)) = \underline{f}(r\Phi(\mathbf{a})\mathbf{C}_r + \mathbf{a}\Phi(\mathbf{C}_r)) \tag{16.21}$$

where $\mathbf{C}_r \in \mathscr{T}_\mathbf{x}^r$.

## Exercises

1. Show that the relationship (16.19) is equivalent to (16.20).
2. Show that the relationship (16.19) is equivalent to (16.21).

3. Find the equations of the geodesic curves on a unit sphere centered at the origin.
4. Find the equations of the geodesics on a right circular cone through the origin.
5. Find the equations of the geodesics on a right circular cylinder centered at the origin.

## 16.3 Conformally Related Surfaces

We now study conformal mappings between surfaces.

**Definition 16.3.1.** A mapping $f : \mathscr{M} \rightarrow \mathscr{M}'$ between the $k$-surfaces in $\mathbb{R}^n$ is said to be a (proper) conformal transformation if for all tangent vectors $\mathbf{a}, \mathbf{b} \in \mathscr{T}_{\mathbf{x}}^1$,

$$\underline{f}(\mathbf{a}) \cdot \underline{f}(\mathbf{b}) = \mathrm{e}^{2\phi} \mathbf{a} \cdot \mathbf{b} \iff \underline{f}(\mathbf{a}) = \psi \mathbf{a} \psi^{\dagger}, \text{ and } \overline{f}(\mathbf{a}') = \psi^{\dagger} \mathbf{a}' \psi \tag{16.22}$$

where $\psi = \mathrm{e}^{\frac{\phi}{2}} U$ is an even multivector and $\psi\psi^{\dagger} = \mathrm{e}^{\phi}$ where $\phi = \phi(\mathbf{x}) \in \mathbb{R}$.

Although we have restricted our considerations to proper conformal mappings where $U$ is an even multivector (defining a rotation), little generality is lost since a *conformal reflection* can be made into a rotation by composing it with a reflection defined by a constant unit vector. A conformal mapping reduces to an isometry (16.15) when $\phi \equiv 0$.

We now calculate

$$\psi_{\mathfrak{b}} = \frac{1}{2} \mathrm{e}^{\frac{\phi}{2}} \phi_{\mathbf{b}} U + \frac{1}{2} \mathrm{e}^{\frac{\phi}{2}} U \mathbf{B}_{\mathfrak{b}} = \frac{1}{2} \psi (\phi_{\mathbf{b}} + \mathbf{B}_{\mathfrak{b}}),$$

where $\mathbf{B}_{\mathfrak{b}} = U^{\dagger} U_{\mathfrak{b}}$ is a bivector. It follows that $\psi_{\mathfrak{b}}^{\dagger} = \frac{1}{2}(\phi_{\mathbf{b}} - \mathbf{B}_{\mathfrak{b}})\psi^{\dagger}$ and

$$\begin{aligned} \underline{f}_{\mathfrak{b}}(\mathbf{a}) &= \psi_{\mathfrak{b}} \mathbf{a} \psi^{\dagger} + \psi \mathbf{a} \psi_{\mathfrak{b}}^{\dagger} = \frac{1}{2} \psi \Big( (\phi_{\mathbf{b}} + \mathbf{B}_{\mathfrak{b}}) \mathbf{a} + \mathbf{a} (\phi_{\mathbf{b}} - \mathbf{B}_{\mathfrak{b}}) \Big) \psi^{\dagger} \\ &= \psi \Big( \phi_{\mathbf{b}} \mathbf{a} + \mathbf{B}_{\mathfrak{b}} \cdot \mathbf{a} \Big) \psi^{\dagger} = \underline{f}(\phi_{\mathbf{b}} \mathbf{a} + \mathbf{B}_{\mathfrak{b}} \cdot \mathbf{a}). \end{aligned} \tag{16.23}$$

Similarly, using that $\overline{f}\underline{f}(\mathbf{a}) = \mathrm{e}^{2\phi}\mathbf{a}$, we calculate

$$\overline{f}_{\mathbf{b}}(\mathbf{a}') = \mathrm{e}^{2\phi} (\phi_{\mathbf{b}} \mathbf{a} + \mathbf{a} \cdot \mathbf{B}_{\mathfrak{b}}).$$

Taking the divergence of both sides of (16.23) with respect to $\mathbf{b}'$, and using (16.4), we find that

$$\begin{aligned} \partial_{\mathbf{b}'}' \cdot \underline{f}_{\mathfrak{b}}(\mathbf{a}) &= \partial_{\mathbf{b}'}' \cdot \underline{f}(\phi_{\mathbf{b}} \mathbf{a} + \mathbf{B}_{\mathfrak{b}} \cdot \mathbf{a}) \\ &= \partial_{\mathbf{b}} \cdot (\phi_{\mathbf{b}} \mathbf{a} + \mathbf{B}_{\mathfrak{b}} \cdot \mathbf{a}) = \phi_a + (k-1) \mathbf{a} \cdot \partial_{\mathbf{x}} \phi = k \phi_{\mathbf{a}}, \end{aligned}$$

from which it follows that $\phi_{\mathbf{a}} = \frac{1}{k}\partial'_{\mathbf{b}'} \cdot \underline{f}_{\mathbf{b}}(\mathbf{a})$ and also that $\partial\phi = \frac{1}{k-1}\partial \cdot \mathbf{B}$. This last relationship implies that

$$\mathbf{B}_{\grave{a}} = \mathbf{a} \wedge \partial_{\mathbf{x}}\phi \tag{16.24}$$

provided that $\nabla_{\mathbf{x}} \wedge \mathbf{B} = 0$.

Note that

$$\underline{f}_{\mathbf{a}}(\mathbf{b}) = \underline{f}_{\mathbf{b}}(\mathbf{a}) \iff \partial_{\mathbf{a}} \wedge \partial_{\mathbf{b}} \underline{f}_{\grave{a}'}(\mathbf{b}) = 0.$$

Since

$$\overline{f}_{\mathbf{c}}(\mathbf{b}') = \partial_{\mathbf{a}} \underline{f}_{\mathbf{c}}(\mathbf{a}) \cdot \mathbf{b}',$$

it follows that

$$\dot{\nabla} \wedge \dot{\overline{f}}(\mathbf{b}') = \partial_{\mathbf{c}} \wedge \partial_{\mathbf{a}} \underline{f}_{\mathbf{c}}(\mathbf{a}) \cdot \mathbf{b}' = 0.$$

Let us directly calculate $\partial_{\mathbf{a}} \wedge B_{\grave{a}} = \nabla \wedge B$, where $\mathbf{B}_{\grave{a}} = U^{\dagger} U_{\grave{a}}$. Equation (16.23), together with the integrability condition $\underline{f}_{\mathbf{a}}(\mathbf{b}) = \underline{f}_{\mathbf{b}}(\mathbf{a})$, implies that

$$\phi_{\mathbf{b}}\mathbf{a} - \phi_{\mathbf{a}}\mathbf{b} = \mathbf{B}_{\grave{a}} \cdot \mathbf{b} - \mathbf{B}_{\grave{b}} \cdot \mathbf{a} = \mathbf{a} \cdot \mathbf{B}_{\grave{b}} - \mathbf{b} \cdot \mathbf{B}_{\grave{a}}.$$

Taking the outer product of both sides of this equation with $\partial_{\mathbf{b}}$ gives

$$\mathbf{a} \wedge \partial\phi = 2\mathbf{B}_{\grave{a}} + \partial_{\mathbf{b}} \wedge (\mathbf{B}_{\grave{b}} \cdot \mathbf{a}) = \mathbf{B}_{\grave{a}} + (\nabla \wedge \mathbf{B}) \cdot \mathbf{a}.$$

But taking the outer product of both sides of this equation with $\partial_{\mathbf{a}}$ shows that

$$0 = \partial_{\mathbf{a}} \wedge \mathbf{B}_{\grave{a}} - 2\nabla \wedge \mathbf{B} = -\nabla \wedge \mathbf{B},$$

so that $\mathbf{a} \wedge \partial\phi = \mathbf{B}_{\grave{a}}$ as given in (16.24).

Equation (16.24) shows that a conformal mapping between surfaces is completely determined by dilation factor $e^{\phi}$. Let us explore this interesting property further. Note first that from $\mathbf{B}_{\grave{a}} = U^{\dagger} U_{\grave{a}}$, we immediately get

$$U_{\grave{a}} = U\mathbf{B}_{\grave{a}} = U\mathbf{B}_{\grave{a}}U^{\dagger}U = \mathbf{B}'_{\grave{a}'}U,$$

where

$$\mathbf{B}'_{\grave{a}'} = U\mathbf{B}_{\grave{a}}U^{\dagger} = e^{-\phi}\underline{f}(\mathbf{B}_{\mathbf{a}}) = e^{-\phi}\underline{f}(\mathbf{a} \wedge \partial\phi) = e^{\phi}\mathbf{a}' \wedge \partial'\phi' \tag{16.25}$$

is a bivector in the tangent space $\mathcal{T}^2_{\mathbf{x}'}$. In deriving the last step in (16.25), we have used the fact that $\mathbf{a}' = \underline{f}(\mathbf{a})$ and $\underline{f}(\partial_{\mathbf{v}}) = \underline{f}\overline{f}(\partial'_{\mathbf{v}'}) = e^{2\phi}\partial'_{\mathbf{v}'}$.

We wish now to determine how the Riemann curvature bivector transforms under a conformal mapping. First, note that $P(\partial \wedge \partial\phi) = 0$ implies that for any tangent vector $\mathbf{a}$,

$$P(\mathbf{a} \cdot \partial\partial\phi) = \partial\phi_{\mathbf{a}} = \partial_{\mathbf{v}}\phi_{\mathbf{v},\mathbf{a}}.$$

Now, recalling (16.23) and (16.24),

$$\underline{f}_{\dot{a}}(\mathbf{c}) = \underline{f}(\phi_{\mathbf{a}}\mathbf{c} + \mathbf{B}_{\dot{a}} \cdot \mathbf{c}) = \underline{f}(\phi_{\mathbf{a}}\mathbf{c} + \phi_{\mathbf{c}}\mathbf{a} - \mathbf{a} \cdot \mathbf{c}\,\partial\phi) = \frac{1}{2}\underline{f}(\mathbf{a}\nabla\phi\mathbf{c} + \mathbf{c}\nabla\phi\mathbf{a}).$$

We then somewhat tediously calculate

$$\begin{aligned}
\underline{f}_{\dot{a},\dot{b}}(\mathbf{c}) &= \frac{1}{2}\underline{f}_{\dot{b}}(\mathbf{a}\nabla\phi\mathbf{c} + \mathbf{c}\nabla\phi\mathbf{a}) + \frac{1}{2}\underline{f}(\mathbf{a}\nabla\phi_{\mathbf{b}}\mathbf{c} + \mathbf{c}\nabla\phi_{\mathbf{b}}\mathbf{a}) \\
&= \frac{1}{4}\underline{f}[(\mathbf{a}\nabla\phi\mathbf{c} + \mathbf{c}\nabla\phi\mathbf{a})\nabla\phi\mathbf{b} + \mathbf{b}\nabla\phi(\mathbf{a}\nabla\phi\mathbf{c} + \mathbf{c}\nabla\phi\mathbf{a}) + 2\mathbf{a}\nabla\phi_{\mathbf{b}}\mathbf{c} + 2\mathbf{c}\nabla\phi_{\mathbf{b}}\mathbf{a}],
\end{aligned}$$

which we use to get

$$\begin{aligned}
\left(\underline{f}_{\dot{a},\dot{b}} - \underline{f}_{\dot{b},\dot{a}}\right)(\mathbf{c}) &= \frac{1}{4}\underline{f}\Big(\mathbf{c}\nabla\phi(\mathbf{a}\nabla\phi\mathbf{b} - \mathbf{b}\nabla\phi\mathbf{a}) - (\mathbf{a}\nabla\phi\mathbf{b} - \mathbf{b}\nabla\phi\mathbf{a})\nabla\phi\mathbf{c} \\
&\quad + 2\mathbf{c}\nabla(\phi_{\mathbf{b}}\mathbf{a} - \phi_{\mathbf{a}}\mathbf{b}) - 2(\mathbf{b}\nabla\phi_{\mathbf{a}} - \mathbf{a}\nabla\phi_{\mathbf{b}})\mathbf{c}\Big) \\
&= \frac{1}{4}\underline{f}\Big(\mathbf{c}\nabla\phi(\mathbf{a}\wedge\nabla\phi\mathbf{b} + \mathbf{b}\mathbf{a}\wedge\nabla\phi) - (\mathbf{a}\wedge\nabla\phi\mathbf{b} + \mathbf{b}\mathbf{a}\wedge\nabla\phi)\nabla\phi\mathbf{c} \\
&\quad + 2\mathbf{c}\nabla(\phi_{\mathbf{b}}\mathbf{a} - \phi_{\mathbf{a}}\mathbf{b}) - 2(\mathbf{b}\wedge\nabla\phi_{\mathbf{a}} - \mathbf{a}\wedge\nabla\phi_{\mathbf{b}})\mathbf{c}\Big) \\
&= \frac{1}{2}\underline{f}\Big(\mathbf{c}\nabla\phi(\mathbf{b}\wedge\mathbf{a}\wedge\nabla\phi - (\mathbf{b}\wedge\mathbf{a}\wedge\nabla)\nabla\phi\mathbf{c} + \mathbf{c}\nabla(\phi_{\mathbf{b}}\mathbf{a} - \phi_{\mathbf{a}}\mathbf{b}) \\
&\quad - (\mathbf{b}\wedge\nabla\phi_{\mathbf{a}} - \mathbf{a}\wedge\nabla\phi_{\mathbf{b}})\mathbf{c}\Big) \\
&= \underline{f}\Big([\nabla\phi\cdot(\mathbf{a}\wedge\mathbf{b}\wedge\nabla\phi)]\cdot\mathbf{c} + ([(\mathbf{a}\wedge\mathbf{b})\cdot\partial_{\mathbf{u}}]\wedge\partial_{\mathbf{v}}\phi_{\mathbf{u},\mathbf{v}})\cdot\mathbf{c}\Big) = \underline{f}(\Omega_{\mathbf{c}}\cdot\mathbf{c}),
\end{aligned}$$

where

$$\begin{aligned}
\Omega = \Omega(\mathbf{a}\wedge\mathbf{b}) &= \nabla\phi\cdot(\mathbf{a}\wedge\mathbf{b}\wedge\nabla\phi) - [(\mathbf{a}\wedge\mathbf{b})\cdot\partial_{\mathbf{u}}]\wedge\partial_{\mathbf{v}}\phi_{\mathbf{u},\mathbf{v}} \\
&= \mathbf{a}\wedge\mathbf{b}\wedge\partial\phi\;\partial\phi - [(\mathbf{a}\wedge\mathbf{b})\cdot\partial_{\mathbf{u}}]\partial_{\mathbf{v}}\phi_{\mathbf{u},\mathbf{v}}.
\end{aligned} \tag{16.26}$$

Defining $\mathbf{a}_{(2)} = \mathbf{a}_1 \wedge \mathbf{a}_2$, and $\mathbf{a}'_{(2)} = \underline{f}(\mathbf{a}_{(2)})$, $\mathbf{c}' = \underline{f}(\mathbf{c})$, and using (16.14), we get a direct relationship between the Riemann curvature bivectors on $\mathscr{M}$ and $\mathscr{M}'$. Letting $\underline{f}_{\dot{a}_{(2)}}(\mathbf{c}) = (\underline{f}_{\dot{a}_1,\dot{a}_2} - \underline{f}_{\dot{a}_2,\dot{a}_1})(\mathbf{c})$, we find that

$$\underline{f}_{\dot{a}_{(2)}}(\mathbf{c}) = R'(\mathbf{a}'_{(2)})\cdot\mathbf{c}' - \underline{f}\Big(R(\mathbf{a}_{(2)})\cdot\mathbf{c}\Big) = \underline{f}(\Omega\cdot\mathbf{c}).$$

Taking the outer product of both sides of this equation with $\underline{f}(\partial_{\mathbf{c}}) = \mathrm{e}^{2\phi}\partial'_{\mathbf{c}'}$ we find the important relationship

$$\underline{f}(\Omega) = \mathrm{e}^{2\phi}R'(\mathbf{a}'_{(2)}) - \underline{f}(R(\mathbf{a}_{(2)})). \tag{16.27}$$

Wedging both sides of (16.27) with the bivector $\mathbf{a}'_{(2)}$ and noting that because of (16.26) $\underline{f}(\Omega) \wedge \mathbf{a}'_{(2)} = 0$, we find that

$$\underline{f}\Big(R(\mathbf{a}_{(2)}) \wedge \mathbf{a}_{(2)}\Big) = e^{2\phi} R'(\mathbf{a}'_{(2)}) \wedge \mathbf{a}'_{(2)}. \tag{16.28}$$

We call $\mathcal{W}_4(\mathbf{a}_{(2)}) := R(\mathbf{a}_{(2)}) \wedge \mathbf{a}_{(2)}$ the *conformal Weyl 4-vector* because it is closely related to the classical conformal Weyl tensor. We see from (16.28) that $\underline{f}(\mathcal{W}_4(\mathbf{a}_{(2)})) = e^{2\phi} \mathcal{W}'_4(\mathbf{a}'_{(2)})$. Obviously, if $k = 3$, the conformal Weyl 4-vector $\mathcal{W}_4(\mathbf{a}_{(2)}) \equiv 0$.

More closely related to the classical conformal Weyl tensor is the quantity

$$\mathcal{W}_C(\mathbf{a}_{(2)}) = \frac{1}{(k-1)(k-2)} \partial_{\mathbf{a}_2} \cdot [\partial_{\mathbf{a}_1} \cdot \mathcal{W}_4(\mathbf{a}_{(2)})].$$

Calculating $\mathcal{W}_C(\mathbf{a}_{(2)})$, we find that

$$\mathcal{W}_C(\mathbf{a}_{(2)}) = R(\mathbf{a}_{(2)}) - \frac{1}{k-2}[R(\mathbf{a}_1) \wedge \mathbf{a}_2 + \mathbf{a}_1 \wedge R(\mathbf{a}_2)] + \frac{R}{(k-1)(k-2)} \mathbf{a}_{(2)}. \tag{16.29}$$

Taking the divergence of both sides of (16.28) with $\partial_{\mathbf{a}'_2} \wedge \partial_{\mathbf{a}'_1}$ shows that

$$\underline{f}(\mathcal{W}_C(\mathbf{a}_{(2)})) = e^{2\phi} \mathcal{W}'_C(\mathbf{a}'_{(2)}). \tag{16.30}$$

Furthermore, by dotting both sides of this equation on the right by $\mathbf{c}'$, and using the fact that $\mathbf{c}' = e^{-2\phi} \overline{f} \underline{f}(\mathbf{c})$, we get that

$$\underline{f}\Big(\mathcal{W}_C(\mathbf{a}_{(2)}) \cdot \mathbf{c}\Big) = \mathcal{W}'_C(\mathbf{a}'_{(2)}) \cdot \mathbf{c}'. \tag{16.31}$$

## Exercises

1. Let

$$\mathbf{A} = [R'(\mathbf{a}'_{(2)}) \cdot \mathbf{c}'] \wedge \mathbf{a}'_{(2)} - \underline{f}\Big([R(\mathbf{a}_{(2)}) \cdot \mathbf{c}] \wedge \mathbf{a}_{(2)}\Big)$$

and

$$\mathbf{B} = R'(\mathbf{v}) \wedge \mathbf{a}'_{(2)} - \underline{f}[R(\mathbf{v}) \wedge \mathbf{a}_{(2)}],$$

where $\mathbf{v} = \mathbf{a}_{(2)} \cdot \mathbf{c}$. Using the property (16.14), show that

$$\mathbf{A} - \frac{1}{k-2}\mathbf{B} = 0.$$

2. Using the Bianchi identities (15.5.2), show that

$$[R(\mathbf{a}_1) \wedge \mathbf{a}_2 + \mathbf{a}_1 \wedge R(\mathbf{a}_2)] \cdot \mathbf{c} = \mathbf{a}_{(2)} \cdot R(\mathbf{c}) + R(\mathbf{a}_{(2)} \cdot \mathbf{c}).$$

3. Calculate $\mathscr{W}_C$ given in (16.29) by taking the second divergence of the Weyl 4-vector.
4. Use (16.30) to show that $\mathscr{W}_C^2 = \mathscr{W}_C'^2$.

## 16.4 Conformal Mapping in $\mathbb{R}^{p,q}$

When we make the assumption that our conformal mapping $\mathbf{x}' = f(\mathbf{x})$ is between the flat space $\mathscr{M} = \mathbb{R}_{p,q} = \mathscr{M}'$ and itself, for which the curvature bivectors $R(\mathbf{a}\wedge\mathbf{b})$ and $R'(\mathbf{a}'\wedge\mathbf{b}')$ vanish, (16.27) simplifies to the identity $\Omega(\mathbf{a}\wedge\mathbf{b}) \equiv 0$. Taking the contraction of this equation with respect to the bivector variable $\mathbf{B} = \mathbf{a}\wedge\mathbf{b}$ gives the relationship

$$\nabla\cdot\mathbf{w} = -\frac{k-2}{2}\mathbf{w}^2$$

for all values of $k = n > 2$. Calculating $\partial_{\mathbf{b}}\cdot\Omega = 0$ and eliminating $\nabla\cdot\mathbf{w}$ from this equation leads to the surprisingly simple differential equation

$$\mathbf{w_a} = \mathbf{a}\cdot\nabla\mathbf{w} = \frac{1}{2}\mathbf{waw}. \tag{16.32}$$

Equation (16.32) specifies the extra condition that $\mathbf{w} = \nabla\phi$ must satisfy in order that $f(\mathbf{x})$ is a nondegenerate conformal mapping of the pseudo-Euclidean space $\mathbb{R}^{p,q}$ onto itself, where $n = p+q > 2$.

Trivial solutions of (16.32) satisfying (16.22) consist of i) $\nabla\phi = 0$ so that $\psi$ is a constant dilation factor and ii) $\psi = U$ where $U$ is a constant rotation in the plane of some constant bivector $\mathbf{B}$.

Let $\mathbf{c}$ be a constant non-null vector in $\mathbb{R}^{p,q}$. A nontrivial solution to (16.32) is

$$f(x) = \psi\mathbf{x} = \mathbf{x}(1-\mathbf{cx})^{-1} = \frac{1}{2}\mathbf{c}^{-1}\mathbf{wx} \tag{16.33}$$

where $\mathbf{w} = \nabla\phi = 2(1-\mathbf{cx})^{-1}\mathbf{c}$ and

$$\mathrm{e}^{-\phi} = (1-\mathbf{xc})(1-\mathbf{cx}) = 4\mathbf{c}^2\mathbf{w}^2. \tag{16.34}$$

Equivalently, we can write (16.33) in the form

$$f(\mathbf{x}) = \frac{\mathbf{x}-\mathbf{x}^2\mathbf{c}}{1-2\mathbf{c}\cdot\mathbf{x}+\mathbf{c}^2\mathbf{x}^2}.$$

The mapping $f(\mathbf{x})$ is called a *transversion* in the identity component of the conformal or *Möbius group* of $\mathbb{R}^{p,q}$ and is well defined for all values of $\mathbf{x}\in\mathbb{R}^{p,q}$ for which

$$1-2\mathbf{c}\cdot\mathbf{x}+\mathbf{c}^2\mathbf{x}^2 = 4\mathbf{c}^2\mathbf{w}^2 > 0.$$

Of course, for a Euclidean signature, this is always true. In addition to transversions, the sense-preserving conformal group is generated by rotations, translations, and dilations. A related derivation of the sense-preserving conformal group can be found in [43, p.210–19].

## Exercises

1. Given the transversion $\mathbf{y} = f(\mathbf{x}) = \mathbf{x}(1-\mathbf{c}\mathbf{x})^{-1}$, show that

$$\mathbf{x} = f^{-1}(\mathbf{y}) = \mathbf{y}(1+\mathbf{c}\mathbf{y})^{-1}.$$

2. (a) Show that the transversion $\mathbf{y} = f(\mathbf{x}) = \mathbf{x}(1-\mathbf{c}\mathbf{x})^{-1}$, with $\mathbf{c} = (0,0,-1)$, maps the $xy$-plane in $\mathbb{R}^3$ into the unit sphere lying above the $xy$-plane with the south pole at the origin.
   (b) What happens to the mapping when the constant vector $\mathbf{c} = (0,0,-1/4)$?
   (c) Show that if we choose $\mathbf{c} = \mathbf{e}_1$, the transversion $f(\mathbf{x})$ maps the $xy$-plane into itself and that this mapping is equivalent to the complex mapping $w = \frac{z}{1-z}$ mapping the complex number plane conformally onto itself, where $z = \mathbf{e}_1\mathbf{x}$ and $w = \mathbf{e}_1\mathbf{y}$.

## 16.5 Möbius Transformations and Ahlfors–Vahlen Matrices

We must take a step back to see the whole picture of the structure of conformal transformations of $\mathbb{R}^{p,q}$. Following [55, 248–51], by the *Ahlfors–Vahlen matrix* $[f]$ of a conformal transformation of the form $f(\mathbf{x}) = (a\mathbf{x}+b)(c\mathbf{x}+d)^{-1}$ where $a,b,c,d \in \mathbb{G}_{p,q}$, we mean

$$[f] = \begin{pmatrix} a & b \\ c & d \end{pmatrix}. \tag{16.35}$$

Note that when the Ahlfors–Vahlen matrix operates on the column matrix $\begin{pmatrix} \mathbf{x} \\ 1 \end{pmatrix}$, we get

$$\begin{pmatrix} a & b \\ c & d \end{pmatrix}\begin{pmatrix} \mathbf{x} \\ 1 \end{pmatrix} = \begin{pmatrix} a\mathbf{x}+b \\ c\mathbf{x}+d \end{pmatrix}. \tag{16.36}$$

We will see what is behind this unusual circumstance in the next chapter. The Ahlfors–Vahlen matrix $[f]$ of the identity transformation $f(\mathbf{x}) = \mathbf{x}$ is easily seen to be the identity matrix $[f] = \begin{pmatrix} 1 & 0 \\ 0 & 1 \end{pmatrix}$.

The Ahlfors–Vahlen matrices of conformal transformations simplify the study of the conformal group because the group action of composition is reduced to the

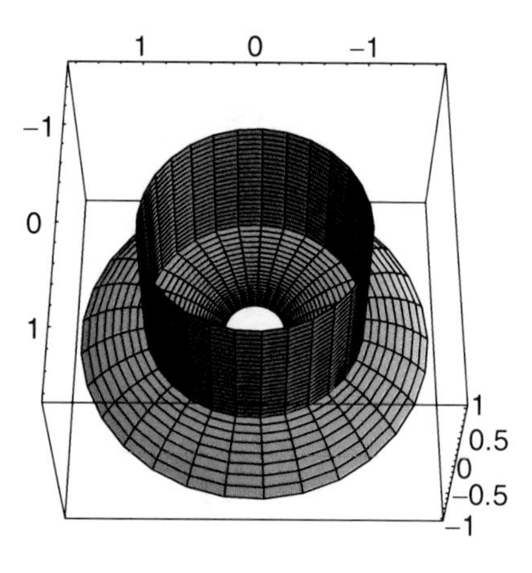


**Fig. 16.1** The transversion $f(\mathbf{x}) = \mathbf{x}(1-\mathbf{c}\mathbf{x})^{-1}$, with $\mathbf{c} = (0, 0, 3/4)$, maps the cylinder in $\mathbb{R}^3$ conformally onto the figure shown

matrix product of the corresponding Ahlfors–Vahlen matrices. If $f(\mathbf{x})$ and $g(\mathbf{x})$ are conformal transformations, then $h(\mathbf{x}) = g(f(\mathbf{x}))$ is also a conformal transformation of $\mathbb{R}^{p,q}$ and their Ahlfors–Vahlen matrices satisfy the rule

$$[h] = [g \circ f] = [g][f].$$

By the *pseudodeterminant* of a Ahlfors–Vahlen matrix, we mean

$$\text{pdet}\begin{pmatrix} a & b \\ c & d \end{pmatrix} = ad^\dagger - bc^\dagger.$$

A conformal transformation $f$ is a *sense preserving* (rotation) if $\text{pdet}[f] > 0$ and *sense reversing* (reflection) if $\text{pdet}[f] < 0$. By the *normalized identity component* of the conformal group, we mean all conformal transformations $f$ with the property that $\text{pdet}[f] = 1$.

We can now talk in greater detail about the conformal Möbius transformations shown in Figs. 16.1–16.5. Figures 16.1–16.3 all have Ahlfors–Vahlen matrices of the form $\begin{pmatrix} 1 & 0 \\ -\mathbf{c} & 1 \end{pmatrix}$ with the pseudodeterminant

$$\text{pdet}\begin{pmatrix} 1 & 0 \\ -\mathbf{c} & 1 \end{pmatrix} = 1.$$

This means that these transformations are sense-preserving conformal transformations of $\mathbb{R}^3$ onto itself which are continuously connected to the identity component, represented by the identity $2 \times 2$ matrix.

Figures 16.4 and 16.5 have Ahlfors–Vahlen matrices of the respective forms

$$\begin{pmatrix} \mathbf{c} & 1 \\ 1 & -\mathbf{c} \end{pmatrix} \text{ and } \begin{pmatrix} 1 & -\mathbf{c} \\ \mathbf{c} & 1 \end{pmatrix}$$

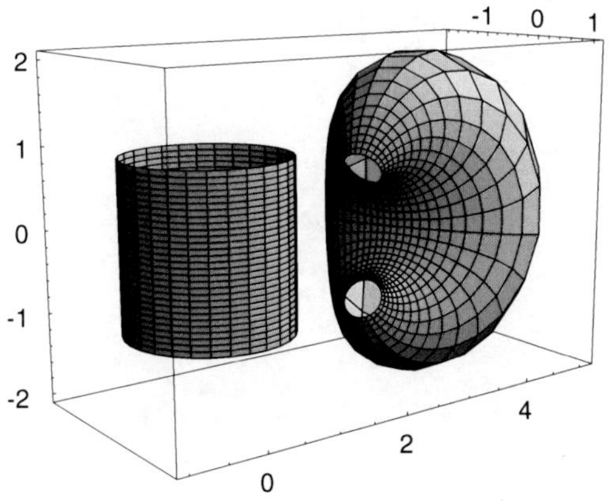

**Fig. 16.2** The transversion $f(\mathbf{x}) = \mathbf{x}(1 - \mathbf{c}\mathbf{x})^{-1}$ with $\mathbf{c} = (-1/4, 1/2, 0)$ maps the cylinder in $\mathbb{R}^3$ conformally onto the figure shown. The image is translated by 3 units on the $y$-axis for clarity

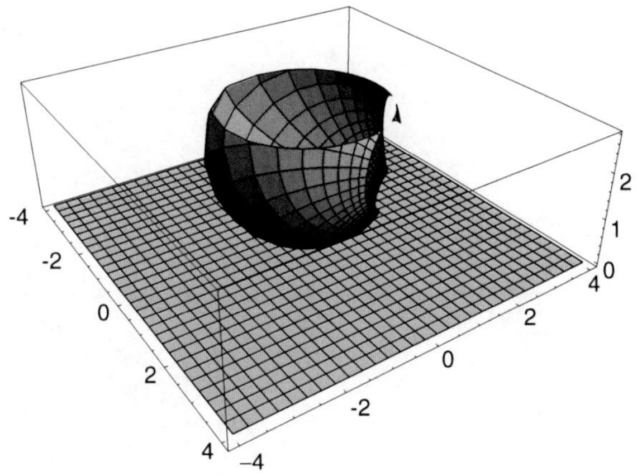

**Fig. 16.3** The transversion $f(\mathbf{x}) = \mathbf{x}(1 - \mathbf{c}\mathbf{x})^{-1}$ with $\mathbf{c} = (-1/4, -1/4, -1/4)$ maps the $xy$-plane conformally onto the unit sphere in $\mathbb{R}^3$ above plane as shown

with pseudodeterminants $-2$ and 2, respectively. These *global conformal transformations* on $\mathbb{R}^3$ and $\mathbb{R}^{2,1}$ are the extensions of the standard *stereographic projection* from the Euclidean plane in $\mathbb{R}^2$ to the unit sphere in $\mathbb{R}^3$ in the case of Fig. 16.4 and from the hyperbolic plane in $\mathbb{R}^{1,1}$ to the unit hyperboloid in $\mathbb{R}^{2,1}$ in the case of Fig. 16.5. Whereas Fig. 16.5 is continuously connected to the identity component as $\mathbf{c} \to 0$, this is not the case for Fig. 16.4.

We wish to call one peculiarity to the attention of the reader. The standard stereographic transformation from the Euclidean plane in $\mathbb{R}^2$ to the unit sphere in $\mathbb{R}^3$, with the north pole at the unit vector $\mathbf{e}_3$ on the $z$-axis, can be represented either by $f(\mathbf{x}) = (\mathbf{e}_3\mathbf{x} + 1)(\mathbf{x} - \mathbf{e}_3)^{-1}$ or by $g(\mathbf{x}) = (\mathbf{x} - \mathbf{e}_3)(\mathbf{e}_3\mathbf{x} + 1)^{-1}$. Both of these transformations are identical when restricted to the $xy$-plane, but are globally distinct on $\mathbb{R}^3$. One of these conformal transformations is sense preserving and continuously connected to the identity, while the other one is not. How is this possible?

I highly recommend the Clifford algebra calculator software [54], which can be downloaded, for checking calculations. Also, transversions (16.34) can be easily plotted with the help of software graphics programs such as Mathematica or

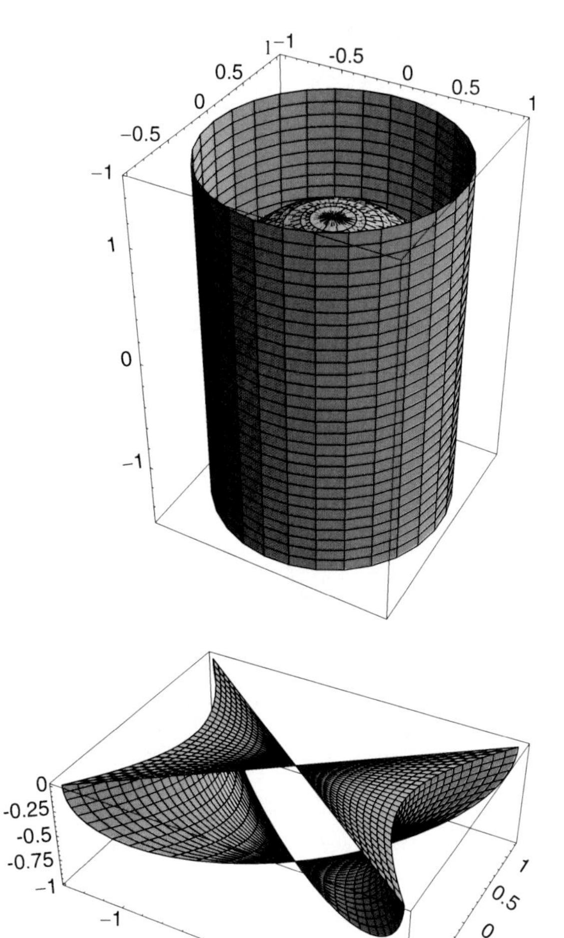

**Fig. 16.4** The stereographic projection $f(\mathbf{x}) = (\mathbf{c}\mathbf{x}+1)(\mathbf{x}-\mathbf{c})^{-1}$ in $\mathbb{R}^3$, with $\mathbf{c} = (0,0,1)$, wraps the cylinder conformally around the unit sphere as shown. For a cylinder four times as long as the radius of the sphere, the sphere is covered twice

**Fig. 16.5** The hyperbolic stereographic projection $f(\mathbf{x}) = (\mathbf{x}-\mathbf{c})(1+\mathbf{c}\mathbf{x})^{-1}$ with $\mathbf{c} = (0,0,1)$ maps the hyperbolic $xy$-plane conformally onto the unit hyperboloid. The metric $g$ has the Lorentzian signature $(+,-,+)$

Maple. The following website gives the graphs of some more exotic conformal transformations, and other information:

```
http://www.garretstar.com/algebra
```

# Exercises

1. Explain the "peculiarity" mentioned above regarding the mappings
   $f(\mathbf{x}) = (\mathbf{e}_3\mathbf{x}+1)(\mathbf{x}-\mathbf{e}_3)^{-1}$ and $g(\mathbf{x}) = (\mathbf{x}-\mathbf{e}_3)(\mathbf{e}_3\mathbf{x}+1)^{-1}$.
2. In $\mathbb{R}^3$, let

$$f(\mathbf{x}) = \mathbf{x}(1-\mathbf{a}\mathbf{x})^{-1} \quad \text{and} \quad g(\mathbf{x}) = \mathbf{x}(1-\mathbf{b}\mathbf{x})^{-1}.$$

Show that

$$f(g(\mathbf{x})) = g(f(\mathbf{x})) = \mathbf{x}\Big(1-(\mathbf{a}+\mathbf{b})\mathbf{x}\Big)^{-1}.$$

3. (a) Let $\mathbf{b} \in \mathbb{R}^3$ be a constant nonzero vector. Define the reflection $h(\mathbf{x}) = -\mathbf{b}\mathbf{x}\mathbf{b}^{-1}$. Find the Ahlfors–Vahlen matrix $[h]$ of the reflection $h(\mathbf{x})$.
   (b) Let $\mathbf{a} \in \mathbb{R}^3$ be a constant vector and let $i = \mathbf{e}_{123} \in \mathbb{G}_3$. Define the rotation in $\mathbb{R}^3$ by

$$\mathrm{rot}(\mathbf{x}) = \mathrm{e}^{\frac{1}{2}i\mathbf{a}}\mathbf{x}\mathrm{e}^{\frac{-1}{2}i\mathbf{a}}.$$

   Find the Ahlfors–Vahlen matrix $[\mathrm{rot}]$ of the rotation $\mathrm{rot}(\mathbf{x})$.
   (c) Let $f(\mathbf{x}) = \mathbf{x}(1-\mathbf{c}\mathbf{x})^{-1}$ be a transversion. For the reflection $h(\mathbf{x})$ defined in part a), find the composition $f(h(\mathbf{x}))$, and verify that $[f \circ h] = [f][h]$ for their corresponding Ahlfors–Vahlen matrices.
4. (a) Let $\mathbf{d}$ be a constant vector. Define the translation $t(\mathbf{x}) = \mathbf{x}+\mathbf{d}$ for all $\mathbf{x} \in \mathbb{R}^n$. Find the Ahlfors–Vahlen matrix $[t]$ of the translation $t(\mathbf{x})$.
   (b) Let $f(\mathbf{x})$ be the transversion given in Problem 3 (c). Find the composition $f(t(\mathbf{x}))$, and verify that their Ahlfors–Vahlen matrices satisfy $[f \circ t] = [f][t]$.
5. Given the transversion $\mathbf{y} = f(\mathbf{x}) = \mathbf{x}(1-\mathbf{c}\mathbf{x})^{-1}$, show that

$$\mathbf{x} = f^{-1}(\mathbf{y}) = \mathbf{y}(1+\mathbf{c}\mathbf{y})^{-1}$$

   by finding the inverse of the Ahlfors–Vahlen matrix $[f]$ of the transversion $f(\mathbf{x})$.

## *16.6 Affine Connections

The concept of a coderivative (15.39), and directional coderivative (15.40), on a $k$-surface embedded in $\mathbb{R}^n$ can be generalized by the concept of an *affine connection*. The concept of an affine connection has been used for an interesting new formulation of Einstein's Theory of General Relativity (GR) in the flat spacetime algebra $G_{1,3} = G(R^{1,3})$ of the pseudoEuclidean space $R^{1,3}$ of special relativity, [39, 51]. We should also like to mention another interesting model of GR in flat spacetime which shows great promise in-so-far as that it is a much simpler theory but makes many of the same famous predictions of GR, [87, 98]. Unfortunately for science fiction fans, if the latter model is correct, there will be no black holes for unsuspecting space travellers to fall into. A powerpoint talk "Do Black Holes Really Exist" can be found on my website: http://www.garretstar.com/blackholestalk.ppt

Following [43, p.220], we let

$$\underline{h}_{\mathbf{x}} : \mathscr{T}_{\mathbf{x}} \longrightarrow \mathscr{T}'_{\mathbf{x}}$$

be an invertible outermorphism of the tangent algebra $\mathscr{T}_{\mathbf{x}}$ to a $k$-surface $\mathscr{M}$ in $\mathbb{R}^n$ at the point $\mathbf{x} \in \mathscr{M}$ to a geometric algebra $\mathscr{T}'_{\mathbf{x}}$. We assume that the range of $\underline{h}_{\mathbf{x}}$ is a

$2^k$-dimensional geometric algebra $\mathscr{T}'_{\mathbf{x}}$, and we further suppose that $\underline{h}_{\mathbf{x}}$ is smoothly differentiable on $\mathscr{M}$. For example, $\underline{h}_{\mathbf{x}}$ could be the differential $\underline{f}_{\mathbf{x}}$ of a mapping $f: \mathscr{M} \to \mathscr{M}'$, in which case $\mathscr{T}'_{\mathbf{x}}$ is the tangent algebra to $\mathscr{M}'$ at the point $\mathbf{x}' = f(\mathbf{x})$. In the case that $f(\mathbf{x}) = \mathbf{x}$, the identity mapping on $\mathscr{M}$, $h_{\mathbf{x}} = P_{\mathbf{x}}$ is the projection operator onto the tangent algebra $\mathscr{T}_{\mathbf{x}}$ of $\mathscr{M}$ at the point $\mathbf{x}$.

The induced *affine directional h-coderivative* $\mathbf{a} \cdot \delta_{\mathbf{x}}$ is defined by

$$\mathbf{a} \cdot \delta_{\mathbf{x}} = \underline{h}_{\mathbf{x}}^{-1} \mathbf{a} \cdot \partial_{\mathbf{x}} \underline{h}_{\mathbf{x}} = \mathbf{a} \cdot \nabla_{\mathbf{x}} + \underline{h}_{\mathbf{x}}^{-1} \underline{h}_{\mathbf{a}} \tag{16.37}$$

and the induced *affine h-coderivative* is defined by

$$\delta_{\mathbf{x}} = \partial_{\mathbf{a}} \underline{h}_{\mathbf{x}}^{-1} \mathbf{a} \cdot \partial_{\mathbf{x}} \underline{h}_{\mathbf{x}} = \nabla_{\mathbf{x}} + \partial_{\mathbf{a}} \underline{h}_{\mathbf{x}}^{-1} \underline{h}_{\mathbf{a}}. \tag{16.38}$$

It is understood that the affine h-coderivative and affine directional h-coderivative only operate on fields in the tangent algebra $\mathscr{T}_{\mathbf{x}}$. This becomes explicit when we assume that

$$\underline{h} = P'_{\mathbf{x}} \underline{h}_{\mathbf{x}} P_{\mathbf{x}}, \tag{16.39}$$

where $P_{\mathbf{x}}$ is the projection onto the tangent algebra $\mathscr{T}_{\mathbf{x}}$ and $P'_{\mathbf{x}}$ is the projection onto the range space $\mathscr{T}'_{\mathbf{x}}$.

If $\underline{h}_{\mathbf{x}} = P_{\mathbf{x}}$, the projection onto the tangent algebra $\mathscr{T}_{\mathbf{x}}$, then $h_{\mathbf{x}}^{-1} = P_{\mathbf{x}}$ and

$$\mathbf{a} \cdot \delta_{\mathbf{x}} = \mathbf{a} \cdot \nabla_{\mathbf{x}}, \quad \text{and} \quad \delta_{\mathbf{x}} = \nabla_{\mathbf{x}},$$

so the affine directional and affine h-coderivative reduce to the directional and coderivatives defined in (15.39) and (15.40), respectively. We see that the directional and h-coderivatives distort the geometry of the natural coderivative inherited from $\mathbb{R}^n$. We will see what the consequences of this distortion is below.

As a measure of this distortion, we define the *affine tensor*

$$H_{\grave{\mathbf{a}}} = \underline{h}^{-1} \underline{h}_{\mathbf{a}} = \underline{h}^{-1} \underline{h}_{\grave{\mathbf{a}}}. \tag{16.40}$$

For a scalar field $\phi = \phi_{\mathbf{x}}$,

$$\delta_{\mathbf{x}} \phi = \nabla_{\mathbf{x}} \phi \iff \mathbf{a} \cdot \delta_{\mathbf{x}} \phi = \mathbf{a} \cdot \nabla_{\mathbf{x}} \phi,$$

since $H(\phi) = \phi$. More generally, in terms of the affine tensor $H_{\grave{\mathbf{a}}}$, we see that for any multivector field $F \in \mathscr{T}_{\mathbf{x}}$,

$$\mathbf{a} \cdot \delta_{\mathbf{x}} F = \mathbf{a} \cdot \nabla_{\mathbf{x}} F + H_{\grave{\mathbf{a}}}(F). \tag{16.41}$$

Taking the $\mathbf{a}$-coderivative of

$$\underline{h}_{\mathbf{x}}^{-1} \underline{h}_{\mathbf{x}} = P_{\mathbf{x}},$$

and simplifying by using the fact that $P_{\dot{a}} = 0$, gives

$$\underline{h}_{\dot{a}}^{-1} = -\underline{h}^{-1}\underline{h}_{\dot{a}}\underline{h}^{-1} = -H_{\dot{a}}\underline{h}^{-1}. \tag{16.42}$$

We also have for vector fields $\mathbf{b}, \mathbf{c} \in \mathscr{T}_{\mathbf{x}}^{1}$

$$H_{\dot{a}}(\mathbf{b} \wedge \mathbf{c}) = H_{\dot{a}}(\mathbf{b}) \wedge \mathbf{c} + \mathbf{b} \wedge H_{\dot{a}}(\mathbf{c}), \tag{16.43}$$

just as we had the similar property (15.19) for the projection $P_{\mathbf{x}}$.

Decomposing the affine tensor $H_{\dot{a}}(\mathbf{b})$ into symmetric and skew-symmetric parts,

$$H_{\dot{a}}(\mathbf{b}) = \mathbf{K}_{\dot{a}}(\mathbf{b}) + \mathbf{T}_{\dot{a}}(\mathbf{b}) \tag{16.44}$$

where

$$\mathbf{K}_{\dot{a}}(\mathbf{b}) = \frac{1}{2}(H_{\dot{a}}(\mathbf{b}) + H_{\dot{b}}(\mathbf{a})), \quad \text{and} \quad \mathbf{T}_{\dot{a}}(\mathbf{b}) = \frac{1}{2}(H_{\dot{a}}(\mathbf{b}) - H_{\dot{b}}(\mathbf{a})).$$

The symmetric part $\mathbf{K}_{\dot{a}}$ is the *Christoffel tensor* of the affine tensor $H$, and the skew-symmetric part $\mathbf{T}$ is called the *torsion tensor* of the affine tensor $H$. If $\underline{h} = \underline{f}$ is the differential of a mapping $f : \mathscr{M} \to \mathscr{M}'$, then by the integrability condition (16.7) gives $\underline{h}_{\dot{a})}(\mathbf{b}) = \underline{h}_{\dot{b}}(\mathbf{a})$, so that

$$\mathbf{T}_{\dot{a}}(\mathbf{b}) = \frac{1}{2}(H_{\dot{a}}(\mathbf{b}) - H_{\dot{b}}(\mathbf{a})) = \frac{1}{2}\underline{h}^{-1}(\underline{h}_{\dot{a}}(\mathbf{b}) - \underline{h}_{\dot{b}}(\mathbf{a})) = 0.$$

It follows that the torsion tensor $\mathbf{T}_{\dot{a}}(\mathbf{b})$, in the affine geometry induced by a mapping between surfaces, always vanishes.

The affine geometry induced by $\underline{h}$ can be used to define a new metric tensor $g$ on $\mathscr{M}$, given by

$$\underline{h}(\mathbf{a}) \cdot \underline{h}(\mathbf{b}) \quad \Longleftrightarrow \quad \mathbf{a} \cdot \overline{h}(\underline{h}(\mathbf{b})) = \mathbf{a} \cdot g(\mathbf{b}), \tag{16.45}$$

where $g(\mathbf{a}) := \overline{h}(\underline{h}(\mathbf{a}))$. Since $H_{\dot{a}}(\mathbf{b}) = \underline{h}^{-1}\underline{h}_{\dot{a}}(\mathbf{b})$, it follows that $\overline{H}_{\dot{a}}(\mathbf{b}) = \overline{h}_{\dot{a}}\overline{h}^{-1}(\mathbf{b})$. Using this, we calculate the differential of $g$,

$$g_{\dot{b}}(\mathbf{a}) = \overline{h}_{\dot{b}}\overline{h}^{-1}\overline{h}\underline{h}(\mathbf{a}) + \overline{h}\underline{h}\underline{h}^{-1}\underline{h}_{\dot{b}}(\mathbf{a}) = \overline{H}_{\dot{b}}g(\mathbf{a}) + gH_{\dot{b}}(\mathbf{a}).$$

Assuming again $\underline{h} = \underline{f}$, so that $\underline{h}$ is the differential of a mapping $f$ between surfaces, it follows that $H_{\dot{b}}(\mathbf{a}) = H_{\dot{a}}(\mathbf{b})$. In this case, we further calculate that

$$\dot{\partial}\dot{g}(\mathbf{a}) \cdot \mathbf{b} = \overline{H}_{\dot{a}}g(\mathbf{b}) + \overline{H}_{\dot{b}}g(\mathbf{a}).$$

With the above calculations in hand, in the case that $\underline{h}$ is the differential of a mapping, we find that

$$g_{\dot{a}}(\mathbf{b}) + g_{\dot{b}}(\mathbf{a}) - \dot{\partial}\mathbf{a} \cdot \dot{g}(\mathbf{b}) = 2g(H_{\dot{a}}(\mathbf{b})),$$

which gives the Christoffel tensor

$$\mathbf{K}_{\dot{\mathbf{a}}}(\mathbf{b}) = \frac{1}{2}(H_{\dot{\mathbf{a}}}(\mathbf{b}) + H_{\dot{\mathbf{b}}}(\mathbf{a})) = \frac{1}{2}g^{-1}\Big(g_{\dot{\mathbf{a}}}(\mathbf{b}) + g_{\dot{\mathbf{b}}}(\mathbf{a}) - \dot{\partial}\mathbf{a}\cdot\dot{g}(\mathbf{b})\Big). \tag{16.46}$$

In the special case that $\underline{h}$ is the differential of the mapping of a coordinate patch in $\mathbb{R}^n$ to a $n$-surface $\mathscr{M}$ in $\mathbb{R}^n$, the formula for (16.46) becomes equivalent to (15.25), which we found in Chap. 15. We see that

$$\mathbf{e}_k\cdot g(\mathbf{K}_{ij}) = \frac{1}{2}\left(\partial_i g_{kj} + \partial_j g_{ki} - \partial_k g_{ij}\right)$$

or

$$g(\mathbf{K}_{ij}) = \frac{1}{2}\sum \mathbf{e}^k\left(\partial_i g_{kj} + \partial_j g_{ki} - \partial_k g_{ij}\right).$$

Using (16.45), we are thus able to *pull back* the geometry on $\mathscr{M}$ to the geometry on a flat coordinate patch in $\mathbb{R}^n$ by introducing the new affine metric tensor $g$ on $\mathbb{R}^n$.

In (16.41), we saw how the affine derivative is related to the coderivative on $\mathscr{M}$. We now calculate how the affine curvature is related to the Riemann curvature, assuming that the torsion tensor $\mathbf{T}_{\dot{\mathbf{a}}}(\mathbf{b}) = 0$. Using (16.41), we find

$$\begin{aligned}
Q(\mathbf{a}\wedge\mathbf{b})\cdot\mathbf{c} :&= (\mathbf{a}\wedge\mathbf{b})\cdot(\delta_{\mathbf{x}}\wedge\delta_{\mathbf{x}})\mathbf{c} = \Big([\mathbf{a}\cdot\delta_{\mathbf{x}},\mathbf{b}\cdot\delta_{\mathbf{x}}] - [\mathbf{a},\mathbf{b}]\cdot\delta_{\mathbf{x}}\Big)\mathbf{c}\\
&= \Big((\mathbf{a}\cdot\nabla + H_{\dot{\mathbf{a}}})(\mathbf{b}\cdot\nabla + H_{\dot{\mathbf{b}}}) - [\mathbf{a},\mathbf{b}]\cdot\nabla - H_{[\mathbf{a},\mathbf{b}]}\Big)\mathbf{c}\\
&= \Big([\mathbf{a}\cdot\nabla,\mathbf{b}\cdot\nabla] - [\mathbf{a},\mathbf{b}]\cdot\nabla + (\mathbf{a}\cdot\nabla H_{\dot{\mathbf{b}}} - H_{\dot{\mathbf{b}}}\mathbf{a}\cdot\nabla) - (\mathbf{b}\cdot\nabla H_{\dot{\mathbf{a}}} - H_{\dot{\mathbf{a}}}\mathbf{b}\cdot\nabla)\\
&\quad - H_{[\mathbf{a},\mathbf{b}]} + [H_{\dot{\mathbf{a}}},H_{\dot{\mathbf{b}}}]\Big)\mathbf{c}\\
&= R(\mathbf{a}\wedge\mathbf{b})\cdot\mathbf{c} + \Big(H_{\dot{\mathbf{b}},\dot{\mathbf{a}}} - H_{\dot{\mathbf{a}},\dot{\mathbf{b}}} + [H_{\dot{\mathbf{a}}},H_{\dot{\mathbf{b}}}]\Big)\mathbf{c}.
\end{aligned} \tag{16.47}$$

Recalling that $H_{\dot{\mathbf{a}}} = \underline{h}^{-1}\underline{h}_{\dot{\mathbf{a}}}$, from $\underline{h}^{-1}\underline{h} = P$, we easily find that

$$\underline{h}_{\dot{\mathbf{a}}}^{-1}\underline{h} + \underline{h}^{-1}\underline{h}_{\dot{\mathbf{a}}} = 0 \iff \underline{h}_{\dot{\mathbf{a}}}^{-1}\underline{h} = -H_{\dot{\mathbf{a}}}.$$

Using this, we calculate

$$\begin{aligned}
H_{\dot{\mathbf{b}},\dot{\mathbf{a}}} - H_{\dot{\mathbf{a}},\dot{\mathbf{b}}} &= \underline{h}_{\dot{\mathbf{a}}}^{-1}\underline{h}_{\dot{\mathbf{b}}} - \underline{h}_{\dot{\mathbf{b}}}^{-1}\underline{h}_{\dot{\mathbf{a}}} + \underline{h}^{-1}(\underline{h}_{\dot{\mathbf{b}},\dot{\mathbf{a}}} - \underline{h}_{\dot{\mathbf{a}},\dot{\mathbf{b}}})\\
&= -[H_{\dot{\mathbf{a}}},H_{\dot{\mathbf{b}}}] + \underline{h}^{-1}(\underline{h}_{\dot{\mathbf{b}},\dot{\mathbf{a}}} - \underline{h}_{\dot{\mathbf{a}},\dot{\mathbf{b}}}),
\end{aligned}$$

giving the simple expression

$$Q(\mathbf{a}\wedge\mathbf{b})\cdot\mathbf{c} - R(\mathbf{a}\wedge\mathbf{b})\cdot\mathbf{c} = \underline{h}^{-1}(\underline{h}_{\dot{\mathbf{b}},\dot{\mathbf{a}}} - \underline{h}_{\dot{\mathbf{a}},\dot{\mathbf{b}}})(\mathbf{c}). \tag{16.48}$$

## Exercises

1. Compare the results found for conformal mappings for the affine geometry obtained by letting $\underline{h} = \underline{f}$, where $\mathbf{x}' = f(\mathbf{x})$ is a conformal mapping satisfying (16.22).

# Chapter 17
# Non-euclidean and Projective Geometries

*The secret of science is to ask the right questions, and it is the choice of problem more than anything else that marks the man of genius in the scientific world.*

—Sir Henry Tizard

We investigate the relationship between conformal transformations in $\mathbb{R}^{p,q}$, studied in the previous chapter, and orthogonal transformations acting on the *horosphere* in $\mathbb{R}^{p+1,q+1}$. The horosphere is a nonlinear model of a pseudo-Euclidean space in a pseudo-Euclidean space of two dimensions higher. In the process, we are led to consider the relationships between non-Euclidean and projective geometries, beginning with the definition of the *affne plane* and the *meet* and *join* operations. One of the first to consider the problems of projective geometry was Leonardo da Vinci (1452–1519). But it was only much later that the French mathematician Jean-Victor Poncelet (1788–1867) developed projective geometry into a well-defined subject. The generality and simplicity of projective geometry led the English mathematician Arthur Cayley (1821–1895) to proclaim "Projective Geometry is all of geometry". Whereas we cannot delve deeply into this beautiful subject, we can lay down just enough of the basic ideas to understand what led Cayley to make his famous proclamation.

## 17.1 The Affine $n$-Plane $\mathscr{A}_h^n$

Following [35, 44], we begin by extending the geometric algebras $\mathbb{G}_n$ of $\mathbb{R}^n$ to the geometric algebra $\mathbb{G}_{n+1}$ of $\mathbb{R}^{n+1}$. Recall (10.2) that the standard orthonormal basis of $\mathbb{R}^n$ is given by

$$(\mathbf{e})_{(n)} = (\mathbf{e}_1, \dots, \mathbf{e}_n),$$

G. Sobczyk, *New Foundations in Mathematics: The Geometric Concept of Number*, DOI 10.1007/978-0-8176-8385-6_17,

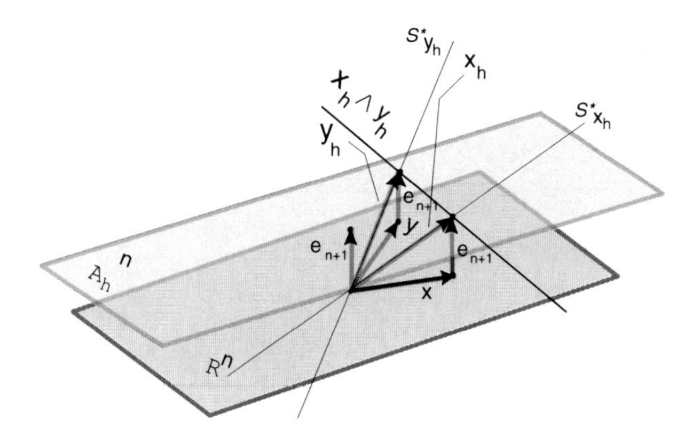


**Fig. 17.1** The affine $n$-plane $\mathscr{A}_h^n$. Each point $\mathbf{x} \in \mathbb{R}^n$ determines a unique point $\mathbf{x}_h = \mathbf{x} + \mathbf{e}_{n+1} \in \mathscr{A}_h^n$ which in turn determines the unique ray $S^*_{\mathbf{x}_h}$. Two distinct points $\mathbf{x}_h$ and $\mathbf{y}_h$ determine the the unique line $\mathbf{x}_h \wedge \mathbf{y}_h$

where $\mathbf{e}_i^2 = 1$ for $1 \le i \le n$. For the standard basis of $\mathbb{R}^{n+1}$, we naturally extend the standard basis of $\mathbb{R}^n$ to include the new orthonormal vector $\mathbf{e}_{n+1}$, so the standard basis of $\mathbb{R}^{n+1}$ becomes $(\mathbf{e})_{(n+1)} = (\mathbf{e})_{(n)} \cup \{\mathbf{e}_{n+1}\}$.

**Definition 17.1.1.** By the affine $n$-plane $\mathscr{A}_h^n$ of the Euclidean space $\mathbb{R}^n$, we mean

$$\mathscr{A}_h^n = \{\mathbf{x}_h = \mathbf{x} + \mathbf{e}_{n+1} |\ \mathbf{x} \in \mathbb{R}^n\}.$$

The affine $n$-plane $\mathscr{A}_h^n$ consists of all of the points $\mathbf{x} \in \mathbb{R}^n$ displaced or translated by the unit vector $\mathbf{e}_{n+1} \in \mathbb{R}^{n+1}$. Since any point $\mathbf{y} \in \mathbb{R}^{n+1}$ is of the form $\mathbf{y} = \mathbf{x} + \alpha \mathbf{e}_{n+1}$ for $\mathbf{x} \in \mathbb{R}^n$ and $\alpha \in \mathbb{R}$, it follows that the affine plane $\mathscr{A}_h^n$ is the hyperplane in $\mathbb{R}^{n+1}$ consisting of all points $\mathbf{y} \in \mathbb{R}^{n+1}$ satisfying the equation $\mathbf{y} \cdot \mathbf{e}_{n+1} = 1$. The affine $n$-plane $\mathscr{A}_h^n$ is pictured in Fig. 17.1.

Notice that each point $\mathbf{x}_h \in \mathscr{A}_h^n$ is uniquely determined by the direction of a nonzero vector or *ray*

$$S^*_{\mathbf{x}_h} = \{\mathbf{y} \in \mathbb{R}^{n+1} |\ \mathbf{y} \wedge \mathbf{x}_h = 0, \text{ and } \mathbf{y} \neq 0\} \tag{17.1}$$

through the origin of $\mathbb{R}^{n+1}$. Each ray $S^*_{\mathbf{x}_h}$ through the origin of $\mathbb{R}^{n+1}$ meets the affine $n$-plane $\mathscr{A}_h^n$ in precisely the point $\mathbf{x}_h$. The point $\mathbf{x}_h \in \mathscr{A}_h^n$ is called the *homogeneous representant* of the ray $S^*_{\mathbf{x}_h}$ and can be easily recovered from any other representant $\mathbf{y} \in S^*_{\mathbf{x}_h}$ by the simple formula $\mathbf{x}_h = \frac{\mathbf{y}}{\mathbf{y} \cdot \mathbf{e}_{n+1}}$. The ray $S^*_{\mathbf{x}_h}$ becomes a one-dimensional subspace $S_{\mathbf{x}_h}$ of $\mathbb{R}^{n+1}$ if we add in the missing origin $\mathbf{y} = 0$, i.e., so that

$$S_{\mathbf{x}_h} := \{\mathbf{y} \in \mathbb{R}^{n+1} |\ \mathbf{y} \wedge \mathbf{x}_h = 0\} = S^*_{\mathbf{x}_h} \cup \{0\}. \tag{17.2}$$

Definition 17.1 brings us closer to a *projective geometry* of rays through the origin.

## Exercises

Consider the affine plane $\mathscr{A}_h^2 \subset \mathbb{R}^3$.

1. Find the ray $S^*_{\mathbf{x}_h}$ and one-dimensional subspace $S_{\mathbf{x}_h}$ of $\mathbb{R}^3$ corresponding to each of the given homogeneous representants $\mathbf{x}_h \in \mathscr{A}_h^2$:

   (a) $\mathbf{x}_h = (1,2,1)$. b) $\mathbf{x}_h = (0,0,1)$. c) $\mathbf{x}_h = (2,-1,1)$.

2. Find the unique line $\mathbf{x}_h \wedge \mathbf{y}_h \in \mathscr{A}_h^2$ determined by the points $\mathbf{x}_h = (1,2,1)$ and $\mathbf{y}_h = (2,-1,1)$ in $\mathscr{A}_h^2$. Give a figure showing the affine plane $\mathscr{A}_h^2$, the points $\mathbf{x}_h, \mathbf{y}_h$, the ray $S^*_{\mathbf{x}_h}$, the subspace $S_{\mathbf{x}_h}$, and the line $\mathbf{x}_h \wedge \mathbf{y}_h$.

## 17.2 The Meet and Joint Operations

Not only can rays $S^*_{\mathbf{x}_h}$ through the origin of $\mathbb{R}^{n+1}$, and their associated 1-dimensional subspaces $S_{\mathbf{x}_h}$, be identified with the points $\mathbf{x}_h \in \mathscr{A}_h^n$, but higher dimensional $(k+1)$-rays, for $0 < k \le n$, can be identified with $k$-planes lying in the affine $n$-plane $\mathscr{A}_h^n$. For example, two distinct points $\mathbf{x}_h, \mathbf{y}_h \in \mathscr{A}_h^n$ determine the unique line $\mathbf{x}_h \wedge \mathbf{y}_h$ in $\mathscr{A}_h^n$ which passes through these points. The line $\mathbf{x}_h \wedge \mathbf{y}_h$ is the intersection of the plane of the bivector $\mathbf{x}_h \wedge \mathbf{y}_h$ passing through the origin of $\mathbb{R}^{n+1}$ and the affine $n$-plane $\mathscr{A}_h^n$. The plane of the bivector $\mathbf{x}_h \wedge \mathbf{y}_h$ passing through the origin of $\mathbb{R}^{n+1}$ uniquely determines the 2-ray $S^*_{\mathbf{x}_h \wedge \mathbf{y}_h}$ and the corresponding 2-dimensional subspace $S_{\mathbf{x}_h \wedge \mathbf{y}_h}$ of $\mathbb{R}^{n+1}$, as shown in Fig. 17.1.

More generally, the $k$-plane $\mathbf{x}^h_{(k+1)} = \mathbf{x}^h_1 \wedge \cdots \wedge \mathbf{x}^h_k$ in $\mathscr{A}_h^n$ is uniquely determined by both the $(k+1)$-ray

$$S^*_{\mathbf{x}^h_{(k+1)}} = \{\mathbf{y} \in \mathbb{R}^{n+1} |\ \mathbf{y} \wedge \mathbf{x}^h_{(k+1)} = 0, \text{ and } \mathbf{y} \cdot \mathbf{e}_{n+1} \neq 0\}$$

and its corresponding $(k+1)$-dimensional subspace

$$S_{\mathbf{x}^h_{(k+1)}} = \{\mathbf{y} \in \mathbb{R}^{n+1} |\ \mathbf{y} \wedge \mathbf{x}^h_{(k+1)} = 0\} = S^*_{\mathbf{x}^h_{(k+1)}} \cup \{0\}.$$

Just as a *point* in $\mathscr{A}_h^n$ is identified by the *vector* $\mathbf{x}_h \in \mathbb{R}^{n+1}$, a *k-plane* in $\mathscr{A}_h^n$ is identified with the $(k+1)$-vector $\mathbf{x}^h_{(k+1)} \in \mathbb{G}_{n+1}$. We write

$$\mathbf{x}^h_{(k+1)} \doteq S^*_{\mathbf{x}^h_{(k+1)}} \cap \mathscr{A}_h^n$$

to formally express this idea, the dot over the equal sign to signify that $\mathbf{x}^h_{(k+1)}$ is only determined up to a nonzero scalar. As such, the affine $n$-plane $\mathscr{A}_h^n$ itself is identified with $\mathbf{x}^h_{(n+1)} \neq 0$, the outer product of any $(n+1)$ points $\mathbf{x}^h_i$ of $\mathscr{A}_h^n$ which are linearly independent as vectors in $\mathbb{R}^{n+1}$, so that

$$\mathbf{x}^h_{(n+1)} \doteq \mathscr{A}_h^n = S_{(\mathbf{x}^h)_{(n+1)}} \cap \mathscr{A}_h^n = \cup_{\mathbf{x}_h \in \mathscr{A}_h^n} S^*_{\mathbf{x}_h}.$$

We see that the objects of points (0-planes), lines (1-planes), and higher dimensional $k$-planes in $\mathscr{A}_h^n$ can all be identified with their corresponding $(k+1)$-rays $S^*_{\mathbf{x}^h_{k+1}}$ and $(k+1)$-dimensional subspaces $S_{\mathbf{x}^h_{k+1}}$, respectively. We now introduce the the *meet* and *join* operations on these subspaces to characterize the *incidence relationship* between these objects, which are also the basic objects of projective geometry. As we now show, the meet and join operations of these objects become, respectively, the *intersection* and *union* of the corresponding subspaces which name these objects in $\mathscr{A}_h^n$.

The direct correspondence between simple nonzero $r$-blades $A_r$ in $\mathbb{G}_{n+1}$ and subspaces of $\mathbb{R}^{n+1}$ is given by

$$S_{A_r} = \{\mathbf{x} \in \mathbb{R}^{n+1} | \ \mathbf{x} \wedge A_r = 0\}.$$

We say that the nonzero $r$-blade $A_r$ *represents*, or is a *representant* of, the $r$-subspace $S_{A_r}$. However, this correspondence is not unique since any nonzero multiple $kA_r$ of the nonzero $r$-vector $A_r$ will represent the same subspace $S_{A_r}$, i.e.,

$$S_{A_r} = S_{kA_r} \quad \text{for all } k \in \mathbb{R}^*.$$

Let $A_r$, $B_s$, and $C_t$ be nonzero blades representing three subspaces $S_{A_r}$, $S_{B_s}$, and $S_{C_t}$, respectively. We say that

**Definition 17.2.1.** The $t$-blade $C_t \doteq A_r \cap B_s$ is the *meet* of $A_r$ and $B_s$ if $S_{C_t}$ is the intersection of the subspaces $S_{A_r}$ and $S_{B_s}$,

$$S_{C_s} = S_{A_r} \cap S_{B_s}. \tag{17.3}$$

We say that

**Definition 17.2.2.** The $t$-blade $C_t \doteq A_r \cup B_s$ is the *join* of $A_r$ and $B_s$ if $S_{C_t}$ is the union of the subspaces $S_{A_r}$ and $S_{B_s}$,

$$S_{C_t} = S_{A_r} \cup S_{B_s}. \tag{17.4}$$

For $r, s \geq 1$, suppose that an $(r-1)$-plane $S_{A_r}$ in $\mathscr{A}_h^n$ is represented by the $r$-blade

$$A_r = \mathbf{a}_1 \wedge \mathbf{a}_2 \wedge \cdots \wedge \mathbf{a}_r$$

and an $(s-1)$-plane $S_{B_s}$ by

$$B_s = \mathbf{b}_1 \wedge \mathbf{b}_2 \wedge \cdots \wedge \mathbf{b}_s.$$

Considering the $\mathbf{a}$'s and $\mathbf{b}$'s to be the basis elements spanning the respective subspaces $S_{A_r}$ and $S_{B_s}$, they can be sorted in such a way that

$$S_{A_r} \cup S_{B_s} = span\{\mathbf{a}_1, \mathbf{a}_2, \ldots \mathbf{a}_s, \mathbf{b}_{\lambda_1}, \ldots, \mathbf{b}_{\lambda_k}\},$$

where the $\lambda$'s are chosen as small as possible and are ordered to satisfy $1 \le \lambda_1 < \lambda_2 < \cdots < \lambda_k \le s$. It follows that the join

$$A_r \cup B_s \doteq A_r \wedge \mathbf{b}_{\lambda_1} \wedge \cdots \wedge \mathbf{b}_{\lambda_k} \neq 0. \tag{17.5}$$

Once the join $A_r \cup B_s$ of the subspaces $S_{A_r}$ and $S_{B_s}$ has been determined, it can be used to determine the meet $A_r \cap B_s$ of those subspaces. Letting $\mathrm{i} \doteq A_r \cup B_s$, then the meet is specified by

$$A_r \cap B_s \doteq (\mathrm{i}A_r) \cdot B_s \doteq A_r \cdot (\mathrm{i}B_s) \neq 0. \tag{17.6}$$

The problem of "meet" and "join" has thus been solved by finding the union and intersection of linear subspaces and their equivalent $(s+k)$-blade and $(s-k)$-blade representants.

It is important to note that it is only in the special case when $A_r \cap B_s = 0$, which is equivalent to writing $S_{A_r} \cap S_{B_s} = \{0\}$, that the join reduces to the outer product. That is,

$$A_r \cap B_s = 0 \iff A_r \cup B_s \doteq A_r \wedge B_s$$

After either the join $C_{s+k} \doteq A_r \cup B_s$ or the meet $D_{s-k} \doteq A_r \cap B_s$ has been found, it can be used to find the meet $D_{s-k}$ or the join $C_{s+k}$, respectively. We have

$$A_r \cap B_s \doteq [(A_r \cup B_s) \cdot A_r] \cdot B_s, \iff A_r \cup B_s \doteq A_r \wedge [B_s(A_r \cap B_s)]. \tag{17.7}$$

This is a reflection of the powerful *principle of duality* in projective geometry about which we will have more to say later.

With the basic definitions of meet and join defined on subspaces, and the objects in the affine plane that they represent, we can now give the basic definition regarding the incidence relationships among the objects in the affine plane.

**Definition 17.2.3.** A point $\mathbf{x}_h \in \mathscr{A}_h^n$ is said to be incident to, or on an $r$-plane $\mathbf{c}^h_{(r+1)} = \wedge_{i=1}^{r+1} \mathbf{c}_i^h$ in $\mathscr{A}_h^n$ if

$$\mathbf{x}_h \wedge \mathbf{c}^h_{(r+1)} = 0 \iff \mathbf{x}_h \cap \mathbf{c}^h_{(r+1)} \doteq \mathbf{x}_h.$$

If $\mathbf{x}_h \wedge \mathbf{c}^h_{(r+1)} \neq 0$, the point $\mathbf{x}_h$ is not on the $r$-plane $\mathbf{c}^h_{(r+1)}$. More generally, a $t$-plane $\mathbf{x}^h_{(t+1)}$ in $\mathscr{A}_h^{p,q}$ is said to be the meet of the $r$- and $s$-planes $\mathbf{c}^h_{(r+1)}$ and $\mathbf{d}^h_{(s+1)}$ in $\mathscr{A}_h^n$ if

$$\mathbf{x}^h_{(t+1)} \doteq \mathbf{c}^h_{(r+1)} \cap \mathbf{d}^h_{(s+1)},$$

and there is no larger $t$-plane in $\mathscr{A}_h^n$ with this property.

For example, consider two lines $l_1 \doteq \mathbf{a}_h \wedge \mathbf{b}_h$ and $l_2 \doteq \mathbf{c}_h \wedge \mathbf{d}_h$ on the affine 2-plane $\mathscr{A}_h^2$ embedded in $\mathbb{R}^3$, where

$$\mathbf{a}_h = (a_1, a_2, 1),\ \mathbf{b}_h = (b_1, b_2, 1),\ \mathbf{c}_h = (c_1, c_2, 1),\ \mathbf{d}_h = (d_1, d_2, 1).$$

Recall that the pseudoscalar trivector in $\mathbb{G}_3$ is $\mathrm{i} = \mathbf{e}_{123}$. If no three of the four points are coplanar, then

$$(\mathbf{a}_h \wedge \mathbf{b}_h) \cup (\mathbf{c}_h \wedge \mathbf{d}_h) \doteq \mathrm{i},$$

and using (17.7), we find that

$$(\mathbf{a}_h \wedge \mathbf{b}_h) \cap (\mathbf{c}_h \wedge \mathbf{d}_h) \doteq (\mathrm{i}(\mathbf{a}_h \wedge \mathbf{b}_h)) \cdot (\mathbf{c}_h \wedge \mathbf{d}_h) = [(\mathrm{i}\mathbf{a}_h) \cdot \mathbf{b}_h] \cdot (\mathbf{c}_h \wedge \mathbf{d}_h)$$

$$= \Big([(\mathrm{i}\mathbf{a}_h) \cdot \mathbf{b}_h] \cdot \mathbf{c}_h\Big)\mathbf{c}_d - \Big([(\mathrm{i}\mathbf{a}_h) \cdot \mathbf{b}_h] \cdot \mathbf{d}_h\Big)\mathbf{c}_h = \mathrm{i}\ \mathbf{a}_h \wedge \mathbf{b}_h \wedge \mathbf{c}_h\ \mathbf{d}_h - \mathrm{i}\ \mathbf{a}_h \wedge \mathbf{b}_h \wedge \mathbf{d}_h\ \mathbf{c}_h$$

$$= -\det\begin{pmatrix} a_1 & a_2 & 1 \\ b_1 & b_2 & 1 \\ c_1 & c_2 & 1 \end{pmatrix} \mathbf{d}_h + \det\begin{pmatrix} a_1 & a_2 & 1 \\ b_1 & b_2 & 1 \\ d_1 & d_2 & 1 \end{pmatrix} \mathbf{c}_h. \tag{17.8}$$

Formula (17.8) gives the ray of the intersection. This ray is easily normalized to give the point $\mathbf{x}_h$ of intersecion on the affine plane $\mathscr{A}_h^2$. We find that

$$\mathbf{x}_h = \frac{(\mathbf{a}_h \wedge \mathbf{b}_h) \cap (\mathbf{c}_h \wedge \mathbf{d}_h)}{\Big((\mathbf{a}_h \wedge \mathbf{b}_h) \cap (\mathbf{c}_h \wedge \mathbf{d}_h)\Big) \cdot \mathbf{e}_3} = \frac{-\det\begin{pmatrix} a_1 & a_2 & 1 \\ b_1 & b_2 & 1 \\ c_1 & c_2 & 1 \end{pmatrix} \mathbf{d}_h + \det\begin{pmatrix} a_1 & a_2 & 1 \\ b_1 & b_2 & 1 \\ d_1 & d_2 & 1 \end{pmatrix} \mathbf{c}_h}{\det\begin{pmatrix} a_1 & a_2 & 1 \\ b_1 & b_2 & 1 \\ d_1 - c_1 & d_2 - c_2 & 0 \end{pmatrix}}.$$

Suppose that we are given two noncollinear points $\mathbf{a}_h, \mathbf{b}_h \in \mathscr{A}_h^2$, then as we have mentioned previously, the line $l_{ab}$ passing through the points $\mathbf{a}_h, \mathbf{b}_h \in \mathscr{A}_h^2$ is uniquely determined by the bivector $\mathbf{a}_h \wedge \mathbf{b}_h$. Suppose that we are given a third point $\mathbf{d}_h \in \mathscr{A}_h^2$, as in Fig. 17.2a, and we are asked to find the point $\mathbf{p}_h$ on the line $l_{ab}$ such that $l_{ab}$ is perpendicular to $l_{pd}$. The point $\mathbf{p}_h$ we are looking for is of the form $\mathbf{p}_h = \mathbf{d} + s\mathrm{i}(\mathbf{a}_h - \mathbf{b}_h)$ for some $s \in \mathbb{R}$ and lies on the line $l_{pd}$ which is specified by

$$\mathbf{p}_h \wedge \mathbf{d}_h = [\mathbf{d}_h + s\mathrm{i}(\mathbf{a}_h - \mathbf{b}_h)] \wedge \mathbf{d}_h = s[\mathrm{i}(\mathbf{a}_h - \mathbf{b}_h)] \wedge \mathbf{d}_h,$$

where $\mathrm{i} = \mathbf{e}_1\mathbf{e}_2$ is the bivector tangent to the affine plane $\mathscr{A}_h^2$. The scalar $s \neq 0$ is unimportant since the line is uniquely determined by the planar direction of the bivector $\mathbf{p}_h \wedge \mathbf{d}_h$ and not by its magnitude. The point $\mathbf{p}$ is therefore uniquely specified by

$$\mathbf{p} \doteq (\mathbf{a} \wedge \mathbf{b}) \cap \{[\mathrm{i}(\mathbf{a}_h - \mathbf{b}_h)] \wedge \mathbf{d}_h\}. \tag{17.9}$$

Evaluating (17.9) for the point $\mathbf{p}$, we find

$$\mathbf{p} \doteq (\mathbf{a}_h \wedge \mathbf{b}_h) \cap \{[\mathrm{i}(\mathbf{a}_h - \mathbf{b}_h)] \wedge \mathbf{d}_h\} \doteq \{[\mathbf{e}_3 \wedge (\mathbf{a}_h - \mathbf{b}_h)] \cdot \mathbf{d}_h\} \cdot (\mathbf{a}_h \wedge \mathbf{b}_h)$$

$$= [\mathbf{e}_3 \wedge (\mathbf{a}_h - \mathbf{b}_h)] \cdot (\mathbf{d}_h \wedge \mathbf{a}_h)\ \mathbf{b}_h - [\mathbf{e}_3 \wedge (\mathbf{a}_h - \mathbf{b}_h)] \cdot (\mathbf{d}_h \wedge \mathbf{b}_h)\ \mathbf{a}_h$$

$$= (\mathbf{a}_h - \mathbf{b}_h) \cdot (\mathbf{b}_h - \mathbf{d}_h)\ \mathbf{a}_h - (\mathbf{a}_h - \mathbf{b}_h) \cdot (\mathbf{a}_h - \mathbf{d}_h)\ \mathbf{b}_h$$

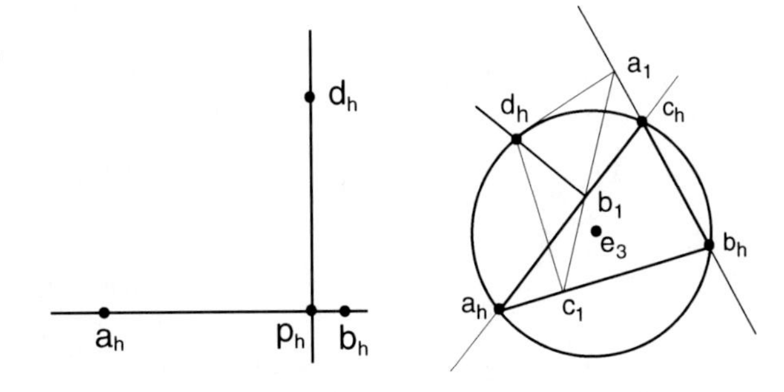


**Fig. 17.2** (**a**) Perpendicular point on a line. (**b**) Simpson's theorem for a circle

The *normalized point*

$$\mathbf{p}_h = \frac{\mathbf{p}}{\mathbf{p}\cdot\mathbf{e}_3} = -\frac{1}{(\mathbf{a}_h-\mathbf{b}_h)^2}[(\mathbf{a}_h-\mathbf{b}_h)\cdot(\mathbf{b}_h-\mathbf{d}_h)\,\mathbf{a}_h-(\mathbf{a}_h-\mathbf{b}_h)\cdot(\mathbf{a}_h-\mathbf{d}_h)\,\mathbf{b}_h] \tag{17.10}$$

will be in the affine plane $\mathscr{A}_h^2$.

We now prove Simpson's theorem for a circle in the affine plane of rays $\mathscr{A}_h^2$ using the operations of meet and join. Referring to the Fig. 17.2b, we have

**Theorem 17.2.4.** *(Simpson's theorem for the circle.) Three noncollinear points $\mathbf{a}_h,\mathbf{b}_h,\mathbf{c}_h \in \mathscr{A}_h^2$ define a unique circle. A fourth point $\mathbf{d}_h \in \mathscr{A}_h^2$ will lie on this circle iff $\mathbf{a}_1\wedge\mathbf{b}_1\wedge\mathbf{c}_1 = 0$ where*

$$\begin{aligned}\mathbf{a}_1 &\doteq (\mathbf{b}_h\wedge\mathbf{c}_h)\cap\{[i(\mathbf{b}_h-\mathbf{c}_h)]\wedge\mathbf{d}_h\}\\ \mathbf{b}_1 &\doteq (\mathbf{c}_h\wedge\mathbf{a}_h)\cap\{[i(\mathbf{c}_h-\mathbf{a}_h)]\wedge\mathbf{d}_h\}\\ \mathbf{c}_1 &\doteq (\mathbf{a}_h\wedge\mathbf{b}_h)\cap\{[i(\mathbf{a}_h-\mathbf{b}_h)]\wedge\mathbf{d}_h\}\end{aligned}$$

*Proof.* Using the above formula above for evaluating the meets, we find that

$$\mathbf{a}_1 \doteq (\mathbf{b}_h-\mathbf{c}_h)\cdot(\mathbf{d}_h-\mathbf{c}_h)\,\mathbf{b}_h-(\mathbf{b}_h-\mathbf{c}_h)\cdot(\mathbf{d}_h-\mathbf{b}_h)\,\mathbf{c}_h,$$

$$\mathbf{b}_1 \doteq (\mathbf{c}_h-\mathbf{a}_h)\cdot(\mathbf{d}_h-\mathbf{a}_h)\,\mathbf{c}_h-(\mathbf{c}_h-\mathbf{a}_h)\cdot(\mathbf{d}_h-\mathbf{c}_h)\,\mathbf{a}_h,$$

and

$$\mathbf{c}_1 \doteq (\mathbf{a}_h-\mathbf{b}_h)\cdot(\mathbf{d}_h-\mathbf{b}_h)\,\mathbf{a}_h-(\mathbf{a}_h-\mathbf{b}_h)\cdot(\mathbf{d}_h-\mathbf{a}_h)\,\mathbf{b}_h.$$

These points will be collinear if and only if $\mathbf{a}_1\wedge\mathbf{b}_1\wedge\mathbf{c}_1 = 0$, but

$$\begin{aligned}\mathbf{a}_1\wedge\mathbf{b}_1\wedge\mathbf{c}_1 \doteq \{&[(\mathbf{b}_h-\mathbf{c}_h)\cdot(\mathbf{d}_h-\mathbf{c}_h)][(\mathbf{c}_h-\mathbf{a}_h)\cdot(\mathbf{d}_h-\mathbf{a}_h)][(\mathbf{a}_h-\mathbf{b}_h)\cdot(\mathbf{d}_h-\mathbf{b}_h)]\\ &-[(\mathbf{c}_h-\mathbf{a}_h)\cdot(\mathbf{d}_h-\mathbf{c}_h)][(\mathbf{a}_h-\mathbf{b}_h)\cdot(\mathbf{d}_h-\mathbf{a}_h)][(\mathbf{b}_h-\mathbf{c}_h)\cdot(\mathbf{d}_h-\mathbf{b}_h)]\}\mathbf{a}_h\wedge\mathbf{b}_h\wedge\mathbf{c}_h\end{aligned}$$

Note that the right-hand side of the last equation only involves *differences* of the points $\mathbf{a}_h,\mathbf{b}_h,\mathbf{c}_h,\mathbf{d}_h$ and that these differences lie in the tangent plane $\mathbb{R}^2$ of $\mathscr{A}_h^2$.

Without loss of generality, we can assume that the center of the circle is the point $\mathbf{e}_3$ and that the circle has radius $\rho$. Using normalized points (17.10), it is not difficult to show that

$$\mathbf{a}_\mathrm{l} \wedge \mathbf{b}_\mathrm{l} \wedge \mathbf{c}_\mathrm{l} = \frac{\rho^2 - \mathbf{d}_h^2 - 1}{4\rho^2}(\mathbf{a}_h \wedge \mathbf{b}_h \wedge \mathbf{c}_h)$$

$$\iff (\mathbf{b}_\mathrm{l} - \mathbf{a}_\mathrm{l}) \wedge (\mathbf{c}_\mathrm{l} - \mathbf{b}_\mathrm{l}) = \frac{\rho^2 - \mathbf{d}^2 - 1}{4\rho^2}(\mathbf{b}_h - \mathbf{a}_h) \wedge (\mathbf{c}_h - \mathbf{a}_h)$$

Since the points $\mathbf{a}_h, \mathbf{b}_h, \mathbf{c}_h$ are not collinear, $\mathbf{a}_h \wedge \mathbf{b}_h \wedge \mathbf{c}_h \neq 0$, from which it follows that $\mathbf{a}_\mathrm{l} \wedge \mathbf{b}_\mathrm{l} \wedge \mathbf{c}_\mathrm{l} = 0$ iff $\mathbf{d}_h^2 - 1 = \rho^2$. □

Note in the proof that the distance of the point $\mathbf{d}_h$ on the circle of radius $\rho$ in the affine plane $\mathscr{A}_h^2$, centered at the point $\mathbf{e}_3$, is defined by $\rho^2 = (\mathbf{d}_h - \mathbf{e}_3)^2 = \mathbf{d}_h^2 - 1$.

## Exercises

1. For $\mathscr{A}_h^n \subset \mathbb{R}^{n+1}$, show that

$$A_r \cap B_s \doteq [(A_r \cup B_s) \cdot B_s] \cdot A_r.$$

2. For $\mathscr{A}_h^n \subset \mathbb{R}^{n+1}$, show that

$$A_r \cup B_s \doteq A_r \wedge [A_r(A_r \cap B_s)].$$

3. Given $\mathbf{a} = (1,2,1)$, $\mathbf{b} = (-1,2,3), \mathbf{c} = (-2,1,-1)$, and $\mathbf{d} = (1,1,-2) \in \mathbb{R}^3$:

   (a) Find $(\mathbf{a} \wedge \mathbf{b}) \cup (\mathbf{c} \wedge \mathbf{d})$. (b) Find $(\mathbf{a} \wedge \mathbf{b}) \cap (\mathbf{c} \wedge \mathbf{d})$.
   (b) Find $(\mathbf{a} \wedge \mathbf{c}) \cup (\mathbf{d} \wedge \mathbf{d})$. (b) Find $(\mathbf{a} \wedge \mathbf{c}) \cap (\mathbf{b} \wedge \mathbf{d})$.

4. Given $\mathbf{a}_h = (1,2,1), \mathbf{b}_h = (-1,2,1)$, and $\mathbf{c}_h = (1,-2,1) \in \mathscr{A}_h^2 \subset \mathbb{R}^3$. Show by using Simpson's theorem that the point $\mathbf{d}_h = (\sqrt{5},0,1)$ lies on the circle through the points $\mathbf{a}_h, \mathbf{b}_h, \mathbf{c}_h$.

## 17.3 Projective Geometry

The *projective n-plane* $\Pi^n$ is closely related to the affine $n$-plane $\mathscr{A}_h^n \subset \mathbb{R}^{n+1}$. It is the purpose of this section, and the next two sections, to explain this relationship and what led Cayley to make his famous proclimation that "projective geometry is all of geometry." The points of projective geometry include all of the points (*rays*) $\mathbf{x}_h = \mathbf{x} + \mathbf{e}_{n+1}$ of the affine plane $\mathscr{A}_h^n$ and, in addition, the points on an *ideal* $(n-1)$-*hyperplane at infinity* which are called *ideal points*. Each line $\mathbf{x}_h \wedge \mathbf{e}_{n+1}$ in $\mathscr{A}^n$ for $\mathbf{x} \in \mathbb{R}^n$, $\mathbf{x} \neq 0$ meets the ideal $(n-1)$-hyperplane $\mathrm{i} = \mathbf{e}_{1\cdots n}$ in the unique ideal point $\{\mathbf{x}\}_{\mathrm{ray}}$, i.e., $\mathbf{x}_{\mathrm{ray}} \cup i \doteq \mathbf{x} \wedge i = 0$. Formally,

$$\Pi^n := \mathscr{A}_h^n \cup \{\mathbf{x}_{\text{ray}} | \mathbf{x} \in \mathbb{R}^n \text{ and } \mathbf{x} \neq 0\} = \{\mathbf{y}_{\text{ray}} | \mathbf{y} \in \mathbb{R}^{n+1}, \text{ and } \mathbf{y} \neq 0\}. \quad (17.11)$$

The association of points in the projective $n$-plane with rays, or directions from the origin, is quite natural. If we think about a point of light being located at the origin, then the rays of light will radiate outward in all possible directions— each direction represented by the direction of a nonzero vector from the origin. In this association, the orientation of the vector is unimportant; the nonzero vector $\mathbf{v} \in \mathbb{R}^{n+1}$ represents the same point in $\Pi^n$ as the nonzero vector $-\mathbf{v} \in \mathbb{R}^{n+1}$. In order to express this idea more precisely, we write

$$\Pi^n \doteq \frac{\mathbb{R}^{n+1}}{\mathbb{R}^*}. \quad (17.12)$$

We say that a set of $n+1$ points $(\mathbf{a})_{(n+1)} \in \Pi^n$ is in *general position* if the vectors $(\mathbf{a})_{(n+1)} \in \mathbb{R}^{n+1}$ are linearly independent, i.e., $\mathbf{a}_{(n+1)} \neq 0$.

Linear transformations are a natural tool and play a crucial role in the study of the projective $n$-plane $\Pi^n$ embedded in $\mathbb{R}^{n+1}$. We have the following important

**Definition 17.3.1.** By a *projective transformation* $f$ between two sets of $(n+1)$ points in $\Pi^n$ in general position, we mean a nonsingular linear transformation $f$ with the property that

$$f(\mathbf{a})_{(n+1)} = (\mathbf{a})_{(n+1)}[f] = (\mathbf{b})_{(n+1)},$$

where $[f]$ is the real $(n+1)$-square matrix of the transformation with $\det(f) \neq 0$.

The theory of linear transformations developed in earlier chapters can be immediately applied to give key results in projective geometry. Any theorem about linear transformations on $\mathbb{R}^{n+1}$ translates into a corresponding theorem in the projective geometry of $\Pi^n$. We will present a simple proof of the classical *Desargues' configuration*, a basic result of projective geometry, in the projective plane $\Pi^2$ embedded in $\mathbb{R}^3$. We do so to emphasize the point that even though geometric algebra is endowed with a metric, there is no reason why we cannot use the tools of this structure to give a proof of this metric independent result, which is also valid in the affine plane $\mathscr{A}_h^2$ and in the Euclidean plane $\mathbb{R}^2$.

Two distinct points $\mathbf{a}, \mathbf{b} \in \Pi^2$ define the projective line $\mathbf{a} \wedge \mathbf{b} \neq 0$, and a third point $\mathbf{c} \in \Pi^2$ lies on this line (is collinear) if and only if $\mathbf{a} \wedge \mathbf{b} \wedge \mathbf{c} = 0$. Suppose now that $\mathbf{a}, \mathbf{b}, \mathbf{c}, \mathbf{d} \in \Pi^2$ are such that no 3 of them are collinear. Then the meet of the projectives lines $\mathbf{a} \wedge \mathbf{b}$ and $\mathbf{c} \wedge \mathbf{d}$ is the unique point $\mathbf{d} \in \Pi^3$ defined by

$$\mathbf{d} \doteq (\mathbf{a} \wedge \mathbf{b}) \cap (\mathbf{c} \wedge \mathbf{d}) \doteq (\mathbf{a} \wedge \mathbf{b}) \cdot [(\mathbf{a} \wedge \mathbf{b} \wedge \mathbf{c}) \cdot (\mathbf{c} \wedge \mathbf{d})] \doteq (\mathbf{a} \times \mathbf{b}) \times (\mathbf{c} \times \mathbf{d}). \quad (17.13)$$

This formula is considerably simpler than the corresponding formula (17.8) that we derived for the meet in the affine plane.

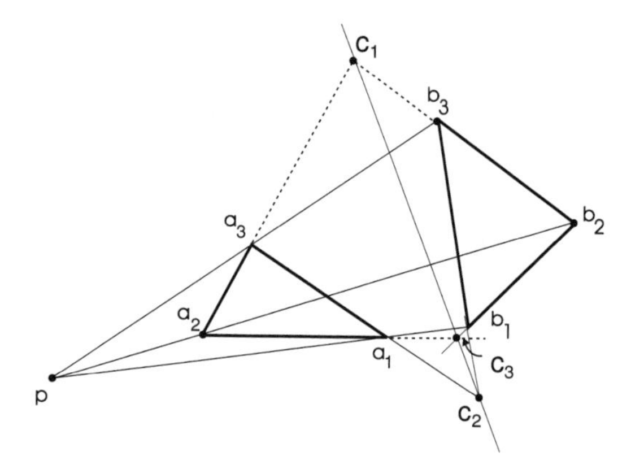


**Fig. 17.3** Desargue's configuration

Refering to Fig. 17.3, we are now ready to prove

**Theorem: Desargues' Configuration 17.3.2.** *Let* $\mathbf{a}_1, \mathbf{a}_2, \mathbf{a}_3$ *and* $\mathbf{b}_1, \mathbf{b}_2, \mathbf{b}_3$ *be the verticies of two triangles in* $\Pi^2$*, and suppose that*

$$(\mathbf{a}_1 \wedge \mathbf{a}_2) \cap (\mathbf{b}_1 \wedge \mathbf{b}_2) \doteq \mathbf{c}_3, \; (\mathbf{a}_2 \wedge \mathbf{a}_3) \cap (\mathbf{b}_2 \wedge \mathbf{b}_3) \doteq \mathbf{c}_1, \; (\mathbf{a}_3 \wedge \mathbf{a}_1) \cap (\mathbf{b}_3 \wedge \mathbf{b}_1) \doteq \mathbf{c}_2.$$

*Then* $\mathbf{c}_1 \wedge \mathbf{c}_2 \wedge \mathbf{c}_3 = 0$ *if and only if there is a point* $\mathbf{p}$ *such that*

$$\mathbf{a}_1 \wedge \mathbf{b}_1 \wedge \mathbf{p} = \mathbf{a}_2 \wedge \mathbf{b}_2 \wedge \mathbf{p} = \mathbf{a}_3 \wedge \mathbf{b}_3 \wedge \mathbf{p} = 0.$$

*Proof.*

$$\left.\begin{array}{l} \mathbf{a}_1 \wedge \mathbf{b}_1 \wedge \mathbf{p} = 0 \\ \mathbf{a}_2 \wedge \mathbf{b}_2 \wedge \mathbf{p} = 0 \\ \mathbf{a}_3 \wedge \mathbf{b}_3 \wedge \mathbf{p} = 0 \end{array}\right\} \iff \left\{\begin{array}{l} \mathbf{p} = \alpha_1 \mathbf{a}_1 + \beta_1 \mathbf{b}_1 \\ \mathbf{p} = \alpha_2 \mathbf{a}_2 + \beta_2 \mathbf{b}_2 \\ \mathbf{p} = \alpha_3 \mathbf{a}_3 + \beta_3 \mathbf{b}_3 \end{array}\right.$$

but this, in turn, implies that

$$\begin{aligned} \alpha_1 \mathbf{a}_1 - \alpha_2 \mathbf{a}_2 &= -(\beta_1 \mathbf{b}_1 - \beta_2 \mathbf{b}_2) = \mathbf{c}_3, \\ \alpha_2 \mathbf{a}_2 - \alpha_3 \mathbf{a}_3 &= -(\beta_2 \mathbf{b}_2 - \beta_3 \mathbf{b}_3) = \mathbf{c}_1, \text{ and} \\ \alpha_3 \mathbf{a}_3 - \alpha_1 \mathbf{a}_1 &= -(\beta_3 \mathbf{b}_3 - \beta_1 \mathbf{b}_1) = \mathbf{c}_2. \end{aligned}$$

Taking the sum of the last three equalities gives $\mathbf{c}_1 + \mathbf{c}_2 + \mathbf{c}_3 = 0$, which implies that $\mathbf{c}_1 \wedge \mathbf{c}_2 \wedge \mathbf{c}_3 = 0$. Since the above steps can be reversed, the other half of the proof follows. □

We mentioned earlier, when defining the meet and join operation, the powerful *principle of duality* that exists in projective geometry. This principle is most simply illustrated by considering points and lines which are dual objects in the projective plane $\Pi^2$, although the same principle applies relating points and $(n-1)$-hyperplanes in $\Pi^n$. By the principle of duality, any valid statement about points and lines in the projective plane $\Pi^2$ can be changed into an equally valid statement about lines and points in $\Pi^2$. It follows that it is only necessary to prove one of the statements, the proof of the dual statement following automatically by the principle of duality.

For example, given three distinct points $\mathbf{a},\mathbf{b},\mathbf{c} \in \Pi^2$, the statement $(\mathbf{a}\wedge\mathbf{b})\wedge\mathbf{c}=0$ which means that the projective point $\mathbf{c}$ is on the projective line $\mathbf{a}\wedge\mathbf{b}$, is equivalent to the dual statement that $[i(\mathbf{a}\wedge\mathbf{b})]\wedge(i\mathbf{c})=0$ which means that the point $i(\mathbf{a}\wedge\mathbf{b})$ is on the line $i\mathbf{c}$, where $i=\mathbf{e}_{123}$ is the unit pseudoscalar element in $\mathbb{G}_3$. The equivalence of these statements follows from the trivial algebraic identity

$$(\mathbf{a}\wedge\mathbf{b})\wedge\mathbf{c} \doteq <(\mathbf{a}\wedge\mathbf{b})\mathbf{c}>_3 \doteq <[i(\mathbf{a}\wedge\mathbf{b})](i\mathbf{c})>_3 \doteq [i(\mathbf{a}\wedge\mathbf{b})]\wedge(i\mathbf{c}).$$

A more substantive example is the classical theorem of Pappus relating collinear points and the corresponding dual theorem of Pappus which relates concurrent lines:

**Theorem of Pappus 17.3.3.** *Given triplets of points* $\mathbf{a},\mathbf{b},\mathbf{c}$ *and* $\mathbf{a}',\mathbf{b}',\mathbf{c}'$ *respectively on two distinct lines in* $\Pi^2$ *and letting*

$$\begin{aligned}\mathbf{p} &\doteq (\mathbf{a}\wedge\mathbf{b}')\cap(\mathbf{a}'\wedge\mathbf{b})\\ \mathbf{q} &\doteq (\mathbf{a}\wedge\mathbf{c}')\cap(\mathbf{a}'\wedge\mathbf{c})\\ \mathbf{r} &\doteq (\mathbf{b}\wedge\mathbf{c}')\cap(\mathbf{b}'\wedge\mathbf{c}),\end{aligned}$$

*then* $\mathbf{p}\wedge\mathbf{q}\wedge\mathbf{r}=0$.

*Proof.* Since $\mathbf{c}$ is on the line $\mathbf{a}\wedge\mathbf{b}$, $\mathbf{c}=\alpha\mathbf{a}+\beta\mathbf{b}$ for $\alpha,\beta\in\mathbb{R}$. Similarly,

$$\mathbf{c}'=\alpha'\mathbf{a}'+\beta'\mathbf{b}'$$

for $\alpha',\beta'\in\mathbb{R}$. Using (17.13), we find that

$$\mathbf{p}\doteq(\mathbf{a}\times\mathbf{b}')\times(\mathbf{a}'\times\mathbf{b}),\quad \mathbf{q}\doteq(\mathbf{a}\times\mathbf{c}')\times(\mathbf{a}'\times\mathbf{c})$$

and

$$\mathbf{r}\doteq(\mathbf{b}\times\mathbf{c}')\times(\mathbf{b}'\times\mathbf{c}).$$

The remainder of the proof consists in establishing the vector analysis identity that

$$\mathbf{p}\wedge\mathbf{q}\wedge\mathbf{r}=\mathbf{e}_{123}(\mathbf{p}\times\mathbf{q})\cdot\mathbf{r}=0.$$

□

See Fig. 17.4.

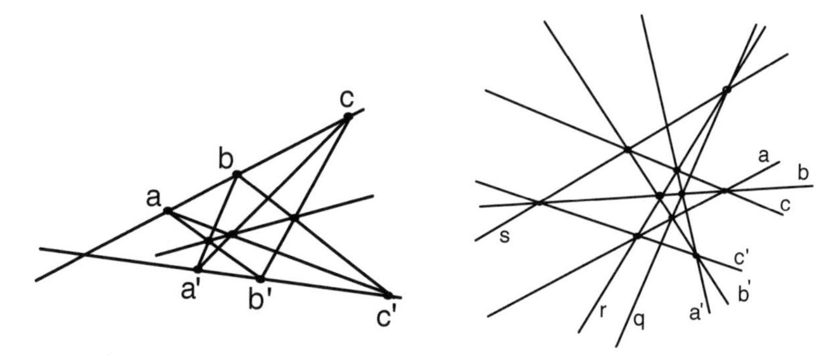


**Fig. 17.4** Theorem of Pappus for points and dual theorem of Pappus for lines

We now *dualize* the *theorem of Pappus* to obtain

**Dual Theorem of Pappus 17.3.4.** *Given triplets of concurrent lines* $\mathbf{ia}, \mathbf{ib}, \mathbf{ic}$ *and* $\mathbf{ia}', \mathbf{ib}', \mathbf{ic}'$ *respectively on two distinct points in* $\Pi^2$ *and letting*

$$\begin{aligned}\mathbf{ip} &\doteq (\mathbf{ia} \wedge \mathbf{b}') \cap (\mathbf{ia}' \wedge \mathbf{b})\\ \mathbf{iq} &\doteq (\mathbf{ia} \wedge \mathbf{c}') \cap (\mathbf{ia}' \wedge \mathbf{c})\\ \mathbf{ir} &\doteq (\mathbf{ib} \wedge \mathbf{c}') \cap (\mathbf{ib}' \wedge \mathbf{c}),\end{aligned}$$

*then* $(\mathbf{ip}) \cap (\mathbf{iq}) \doteq (\mathbf{iq}) \cap (\mathbf{ir})$.

*Proof.* Since the line $\mathbf{ic}$ is concurrent with the lines $\mathbf{ia}$ and $\mathbf{ib}$ and $\mathbf{ic}'$ is concurrent with the lines $\mathbf{ia}'$ and $\mathbf{ib}'$, it follows that $\mathbf{ic} = \alpha \mathbf{ia} + \beta \mathbf{ib}$ for $\alpha, \beta \in \mathbb{R}$. Similarly,

$$\mathbf{ic}' = \alpha' \mathbf{ia}' + \beta' \mathbf{ib}'$$

for $\alpha', \beta' \in \mathbb{R}$. Using (17.13), we calculate

$$\begin{aligned}\mathbf{ip} &\doteq (\mathbf{ia} \wedge \mathbf{b}') \wedge (\mathbf{ia}' \wedge \mathbf{b}) \doteq (\mathbf{a} \times \mathbf{b}') \wedge (\mathbf{a}' \times \mathbf{b}),\\ \mathbf{iq} &\doteq (\mathbf{ia} \wedge \mathbf{c}') \wedge (\mathbf{ia}' \wedge \mathbf{c}) \doteq (\mathbf{a} \times \mathbf{c}') \wedge (\mathbf{a}' \times \mathbf{c}),\end{aligned}$$

and

$$\mathbf{ir} \doteq (\mathbf{ib} \wedge \mathbf{c}') \wedge (\mathbf{ib}' \wedge \mathbf{c}) \doteq (\mathbf{b} \times \mathbf{c}') \wedge (\mathbf{b}' \times \mathbf{c}).$$

The remainder of the proof consists in showing that

$$(\mathbf{ip}) \cap (\mathbf{iq}) \doteq (\mathbf{iq}) \cap (\mathbf{ir}) \quad \iff \quad \mathbf{p} \times \mathbf{q} \doteq \mathbf{q} \times \mathbf{r},$$

which is equivalent to the vector analysis identity

$$(\mathbf{p} \times \mathbf{q}) \times (\mathbf{q} \times \mathbf{r}) = [(\mathbf{p} \times \mathbf{q}) \cdot \mathbf{r}]\mathbf{q} = 0.$$

□

See Fig. 17.4.

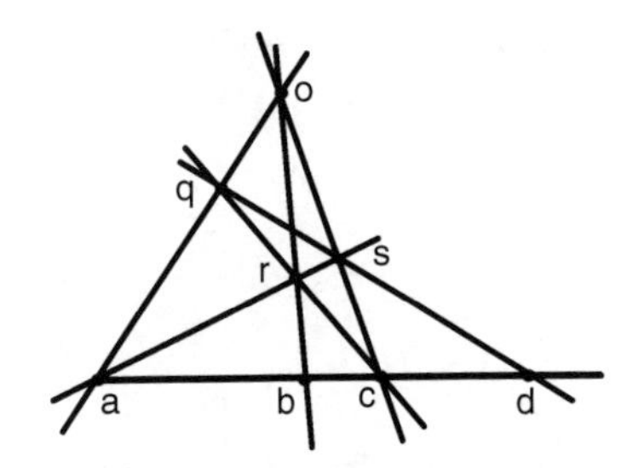


**Fig. 17.5** Given 3 points $\mathbf{a}, \mathbf{b}, \mathbf{c}$ on a line in $\Pi^n$, a fourth point $\mathbf{d}$ can always be constructed that is harmonically related to the given three points

Two useful tools of projective geometry are the concept of *harmonically related* points on a line and the *cross ratio* of points on a line. Suppose that we are given three distinct points $\mathbf{a}, \mathbf{b}, \mathbf{c}$ on a line in the projective plane $\Pi^n$. This means that $\mathbf{a} \wedge \mathbf{b} \wedge \mathbf{c} = 0$, but $\mathbf{a} \wedge \mathbf{b}, \mathbf{b} \wedge \mathbf{c}$ and $\mathbf{a} \wedge \mathbf{c}$ are all nonzero. We will now construct a fourth point $\mathbf{d} \in \Pi^n$ on the line $\mathbf{a} \wedge \mathbf{b}$ *harmonically related* to $\mathbf{a}, \mathbf{b}, \mathbf{c}$ in that order. The construction is shown in Fig. 17.5.

1. Choose any point $\mathbf{o} \in \Pi^n$ such that $\mathbf{o} \wedge \mathbf{a} \wedge \mathbf{b} \neq 0$ and draw the lines $\mathbf{a} \wedge \mathbf{o}$, $\mathbf{b} \wedge \mathbf{o}$ and $\mathbf{c} \wedge \mathbf{o}$.
2. Choose any point $\mathbf{q}$ on the line $\mathbf{a} \wedge \mathbf{o}$, which means that $\mathbf{q} \wedge \mathbf{a} \wedge \mathbf{b} = 0$, and construct the line $\mathbf{q} \wedge \mathbf{c}$. Now let $\mathbf{r} \doteq (\mathbf{q} \wedge \mathbf{c}) \cap (\mathbf{o} \wedge \mathbf{b})$ and construct the line $\mathbf{a} \wedge \mathbf{r}$.
3. Let $\mathbf{s} \doteq (\mathbf{a} \wedge \mathbf{r}) \cap (\mathbf{o} \wedge \mathbf{c})$. The point $\mathbf{d} \doteq (\mathbf{q} \wedge \mathbf{s}) \cap (\mathbf{a} \wedge \mathbf{b})$.

The point $\mathbf{d}$ is said to be *harmonically related* to the points $\mathbf{a}, \mathbf{b}, \mathbf{c}$ in that order. Symbolically, we write $\mathscr{H}(\mathbf{a}, \mathbf{b}, \mathbf{c}, \mathbf{d})$ to express this relationship.

To see analytically that the point $\mathbf{d}$ on the line $\mathbf{a} \wedge \mathbf{b}$ is uniquely determined by the points $\mathbf{a}, \mathbf{b}, \mathbf{c}$ and independent of the choice of the points $\mathbf{o}$ and $\mathbf{q}$, we carry out the following calculations in $\Pi^2$. Let $\mathbf{a}, \mathbf{b} \in \Pi^2$ such that $\mathbf{a} \wedge \mathbf{b} \neq 0$, and let $\mathbf{c} = \alpha \mathbf{a} + \beta \mathbf{b}$ be any point on the line joining $\mathbf{a}$ and $\mathbf{b}$. Now choose any point $\mathbf{o} \in \Pi^2$ such that $\mathbf{a} \wedge \mathbf{b} \wedge \mathbf{c} \neq 0$ so that the points $\mathbf{a}, \mathbf{b}, \mathbf{o}$ are not collinear, and define $\mathbf{q} = s\mathbf{a} + t\mathbf{o}$. Using (17.13), we calculate successively

$$\mathbf{q} \wedge \mathbf{c} = (s\mathbf{a} + t\mathbf{c}) \wedge (\alpha \mathbf{a} + \beta \mathbf{b}) = s\beta \mathbf{a} \wedge \mathbf{b} + \alpha t \mathbf{c} \wedge \mathbf{a} + t\beta \mathbf{c} \wedge \mathbf{b},$$

$$\mathbf{r} \doteq (\mathbf{q} \wedge \mathbf{c}) \cap (\mathbf{o} \wedge \mathbf{b}), \ \mathbf{s} \doteq (\mathbf{a} \wedge \mathbf{r}) \cap (\mathbf{o} \wedge \mathbf{c}),$$

and finally

$$\mathbf{d} \doteq (\mathbf{q} \wedge \mathbf{s}) \cap (\mathbf{a} \wedge \mathbf{b}) \doteq 2\alpha \mathbf{a} + \beta \mathbf{b}. \tag{17.14}$$

Note that the final result depends only on the choice of values $\alpha$ and $\beta$, which determine the position of the point $\mathbf{c}$ on the line $\mathbf{a} \wedge \mathbf{b}$, but not upon the values of $s$ and $t$ which determine the position of the point $\mathbf{q}$ on the line $\mathbf{o} \wedge \mathbf{a}$.

Given four points $\mathbf{a}, \mathbf{b}, \mathbf{c}, \mathbf{d}$ on a line in $\Pi^n$, at least three of which are distinct, another important invarient is the *cross ratio* $r(\mathbf{a}, \mathbf{b}, \mathbf{c}, \mathbf{d})$ of those points, defined by

$$r(\mathbf{a}, \mathbf{b}, \mathbf{c}, \mathbf{d}) = \frac{(\mathbf{a} \wedge \mathbf{b})(\mathbf{c} \wedge \mathbf{d})}{(\mathbf{c} \wedge \mathbf{b})(\mathbf{a} \wedge \mathbf{d})} = \frac{\det \begin{pmatrix} \mathbf{b} \cdot \mathbf{c} & \mathbf{b} \cdot \mathbf{d} \\ \mathbf{a} \cdot \mathbf{c} & \mathbf{a} \cdot \mathbf{d} \end{pmatrix}}{\det \begin{pmatrix} \mathbf{b} \cdot \mathbf{a} & \mathbf{b} \cdot \mathbf{d} \\ \mathbf{c} \cdot \mathbf{a} & \mathbf{c} \cdot \mathbf{d} \end{pmatrix}}. \tag{17.15}$$

The points $\mathbf{a}, \mathbf{b}, \mathbf{c}, \mathbf{d}$ are harmonically related iff $r(\mathbf{a}, \mathbf{b}, \mathbf{c}, \mathbf{d}) = -1$. In symbols,

$$\mathscr{H}(\mathbf{a}, \mathbf{b}, \mathbf{c}, \mathbf{d}) \iff r(\mathbf{a}, \mathbf{b}, \mathbf{c}, \mathbf{d}) = -1. \tag{17.16}$$

This relationship can be checked analytically simply by substituting in the harmonically related points $\mathbf{a}, \mathbf{b}, \mathbf{c}, \mathbf{d}$ just found by using (17.14). Any variation of the position of the point $\mathbf{d}$ on the line $\mathbf{a} \wedge \mathbf{b}$ will give a value of the cross ratio different than $-1$, as can also be easily checked analytically.

Consider now the case of a projective transformation $f$ on the projective line $\Pi^1$ embedded in $\mathbb{R}^2$. From (8.9), the spectral decomposition of a nonsingular linear transformation $f$ on $\mathbb{R}^2$ can only be of the two possible types:

$$f = \lambda_1 p_1 + \lambda_2 p_2 \quad \text{or} \quad f = \lambda_1 p_1 + q_1, \tag{17.17}$$

where the eigenvalues $\lambda_1$ and $\lambda_2$ are both nonzero and $q_1^2 = 0$.

**Definition 17.3.5.** A nonidentity projectivity of a line onto itself is elliptic, parabolic, or hyperbolic, if it has no, one, or two fixed points.

It follows from (17.17) that $f$ is elliptic if $\lambda_2 = \overline{\lambda}_1$ are complex conjugate (no real eigenvalues), parabolic if $f = \lambda_1 p_1 + q_1$, and hyperbolic if $\lambda_1 \neq \lambda_2$ and are both real.

Using the outermorphism rule 7.1.1,

$$\begin{aligned} r(f(\mathbf{a}), f(\mathbf{b}), f(\mathbf{c}), f(\mathbf{d})) &= \frac{f[(\mathbf{a} \wedge \mathbf{b})] f[(\mathbf{c} \wedge \mathbf{d})]}{f[(\mathbf{c} \wedge \mathbf{b})] f[(\mathbf{a} \wedge \mathbf{d})]} \\ &= \frac{(\mathbf{a} \wedge \mathbf{b})(\mathbf{c} \wedge \mathbf{d})}{(\mathbf{c} \wedge \mathbf{b})(\mathbf{a} \wedge \mathbf{d})} = r(\mathbf{a}, \mathbf{b}, \mathbf{c}, \mathbf{d}), \end{aligned}$$

which shows that a projectivity of a line preserves the cross ratio $r(\mathbf{a}, \mathbf{b}, \mathbf{c}, \mathbf{d})$ of four points on that line. Consequently, by (17.16), the harmonic property of four points on a projective line is also preserved under a projective transformation.

By an *involution* $f$ on the projective line $\Pi^1$, we mean a nonsingular linear transformation, which is not of the form $f = \lambda \mathbf{1}$, i.e., $f$ is not a multiple of the identity operator $\mathbf{1}$, and $f^2 = \lambda^2$. It immediately follows that the involution $f$ must be one of the forms

$$f = \lambda p_1 - \lambda p_2 \quad \text{or} \quad f = \lambda p_1 + q_1$$

for some nonzero $\lambda \in \mathbb{R}^*$. In other words, an involution on a projective line must be either elliptic or hyperbolic, respectively.

## Exercises

1. Prove the "only if" part of Desargue's theorem by reversing the steps of the "if" part.
2. In order for $\mathbf{d}$ to be well defined in (17.14), for $\mathbf{b} = (b_1, b_2, b_3)$ and $\mathbf{s} = (s_1, s_2, s_3)$, the denominator $s_1 s_3 (b_1 - 2b_2 - b_3 + b_2 s_2)^3$ must not be zero. Explain what conditions this places on the choices of the points in the construction of the harmonically related points $\mathscr{H}(\mathbf{a}, \mathbf{b}, \mathbf{c}, \mathbf{d})$.
3. Give a geometric proof, based on Desargues' theorem, of

$$\mathscr{H}(\mathbf{a}, \mathbf{b}, \mathbf{c}, \mathbf{d}) \iff r(\mathbf{a}, \mathbf{b}, \mathbf{c}, \mathbf{d}) = -1.$$

4. Let $f(\mathbf{x}) = \mathbf{axa}$ be defined for $\mathbf{x} \in \mathbb{R}^2$, where $\mathbf{a} \in \mathbb{R}^2$ is nonzero. The linear transformation $f$ defines a projectivity on the projective line $\Pi^1$.
   (a) Is $f$ elliptic, parabolic, or hyperbolic?
   (b) What about $f = -\mathbf{axa}$?
5. (a) Suppose $[f] = \begin{pmatrix} 4 & 1 \\ -1 & 2 \end{pmatrix}$. Is $f$ elliptic, parabolic, or hyperbolic?
   (b) Suppose $[f] = \begin{pmatrix} 4 & 1 \\ 1 & 2 \end{pmatrix}$. Is $f$ elliptic, parabolic, or hyperbolic?
   (c) Suppose $[f] = \begin{pmatrix} 1 & -4 \\ 2 & 1 \end{pmatrix}$. Is $f$ elliptic, parabolic, or hyperbolic?
6. (a) Determine the cross ratio of the points a) $\mathbf{a} = (2,1), \mathbf{b} = (3,2), \mathbf{c} = (-1,4)$, $\mathbf{d} = (1,1)$.
   (b) Construct a point $\mathbf{d}$ that is harmonically related to $\mathbf{a} = (2,1), \mathbf{b} = (3,2)$, $\mathbf{c} = (-1,4)$.
   (c) Determine the cross ratio of the points a) $\mathbf{a} = (3,5), \mathbf{b} = (4,1), \mathbf{c} = (-2,7)$, $\mathbf{d} = (2,3)$.
   (d) Construct a point $\mathbf{d}$ that is harmonically related to $\mathbf{a} = (3,5), \mathbf{b} = (4,1)$, $\mathbf{c} = (-2,7)$.
7. (a) Let $\mathbf{a}, \mathbf{b}, \mathbf{c} \in \Pi^2$ and suppose that $\mathbf{a} \wedge \mathbf{b} \neq 0$. Show that the projective point $\mathbf{c}$ lies on the projective line $\mathbf{a} \wedge \mathbf{b}$ if and only if $(\mathbf{a} \wedge \mathbf{b}) \wedge \mathbf{c} = 0$.
   (b) By a nondegenerate *projective r-simplex* in $\Pi^n$, where $r < n$, we mean a set of points $(\mathbf{p})_{(r+1)}$ with the property that $\mathbf{p}_{(r+1)} = \wedge(\mathbf{p})_{(r+1)} \neq 0$. Show that a point $\mathbf{b}$ lies on the nondegenerate $r$-simplex $\mathbf{p}_{(r+1)}$ if and only if $\mathbf{p}_{(r+1)} \wedge \mathbf{b} = 0$.

8. (a) Given the projective triangle defined by the points $(1,1,2),(-1,1,0)$ and $(2,1,1)$ in $\Pi^2$, find the matrix $[f]$ of the projective transformation $f$ which maps this triangle into an arbitrary projective triangle $(\mathbf{a})_{(3)}$ defined by $\mathbf{a}_i = (a_{i1}, a_{i2}, a_{i3})$ for $i = 1,2,3$, where $\det(\mathbf{a})_{(3)} \neq 0$.
   (b) Give a geometric interpretation of this result in terms of the shadow a cutout of the given triangle would cast when held in various positions between a point source of light and a screen.
9. Find the 2-ray $S^*_{\mathbf{x}\wedge\mathbf{y}} \in \Pi^2$ that corresponds to the projective line through the projective points $\mathbf{x} = (1,2,1)$ and $\mathbf{y} = (2,-1,2)$ in $\Pi^2$. Is the projective point $\mathbf{w} = (3,1,3)$ on this projective line?

## 17.4 Conics

We now turn our attention to the study of projective transformations on $\Pi^2$. A projective transformation on $\Pi^2$, which is embedded in $\mathbb{R}^3$, is simply a nonsingular linear transformation $f$ on $\mathbb{R}^3$. Each projective transformation on $\Pi^2$ defines a unique conic, as we now explain.

Let $f$ be a projectivity on $\Pi^2$. Given three noncollinear points $\mathbf{a}, \mathbf{x}_1, \mathbf{x}_2 \in \Pi^2$, let $\mathbf{b} = f(\mathbf{a}), \mathbf{y}_1 = f(\mathbf{x}_1)$, and $\mathbf{y}_2 = f(\mathbf{x}_1)$. Given that the points are noncollinear means that $\mathbf{a} \wedge \mathbf{x}_1 \wedge \mathbf{x}_2 \neq 0$, so they form a basis of $\mathbb{R}^3$. Thus, any nonzero vector $\mathbf{x}_s \in \mathbb{R}^3$, which represents a point in $\Pi^2$, is of the form

$$\mathbf{x}_s = s_a \mathbf{a} + s_1 \mathbf{x}_1 + s_2 \mathbf{x}_2, \quad \text{for} \quad s_a, s_1, s_2 \in \mathbb{R}. \tag{17.18}$$

The projectivity $f$ defines a relationship between two *pencils of lines* through the points $\mathbf{a}$ and $\mathbf{b}$. The lines $\mathbf{a} \wedge \mathbf{x}_1$, $\mathbf{a} \wedge \mathbf{x}_2$, and $\mathbf{a} \wedge \mathbf{x}_s$ correspond to the lines $f(\mathbf{a} \wedge \mathbf{x}_1) = \mathbf{b} \wedge \mathbf{y}_1$, $f(\mathbf{a} \wedge \mathbf{x}_2) = \mathbf{b} \wedge \mathbf{y}_2$, and $f(\mathbf{a} \wedge \mathbf{x}_s) = s_1 \mathbf{b} \wedge \mathbf{y}_1 + s_2 \mathbf{b} \wedge \mathbf{y}_2$, respectively, as shown in Fig. 17.6. The property of interest here is to note that the intersections of the correlated lines of the two pencils define a unique *conic* in $\Pi^2$.

Given the projective relationship $f$ between the pencils of lines through the points $\mathbf{a}$ and $\mathbf{b} = f(\mathbf{a})$, any point $\mathbf{x}$ on the conic defined by $f$ must satisfy $\mathbf{x} = (\mathbf{a} \wedge \mathbf{x}_s) \cap (\mathbf{b} \wedge \mathbf{y}_s)$. In order that the point $\mathbf{x}$ is incident to the line $\mathbf{a} \wedge \mathbf{x}_s$, we must have $\mathbf{x} \wedge \mathbf{a} \wedge \mathbf{x}_s = 0$.

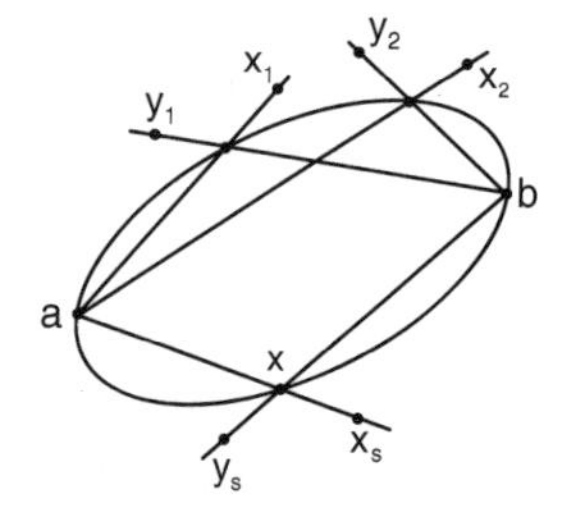


**Fig. 17.6** Two projectively related pencils of lines located at the points $\mathbf{a}$ and $\mathbf{b} = f(\mathbf{a})$ in $\Pi^2$ define a conic, shown here as an ellipse

Similarly, in order that $\mathbf{x}$ is incident to the line $\mathbf{b}\wedge\mathbf{y}_s = f(\mathbf{a}\wedge\mathbf{x}_s)$, we must have $\mathbf{x}\wedge\mathbf{b}\wedge\mathbf{y}_s = 0$. From these two equations, it follows that

$$s_1\mathbf{x}\wedge\mathbf{a}\wedge\mathbf{x}_1 + s_2\mathbf{x}\wedge\mathbf{a}\wedge\mathbf{x}_2 = 0 = s_1\mathbf{x}\wedge\mathbf{b}\wedge\mathbf{y}_1 + s_2\mathbf{x}\wedge\mathbf{b}\wedge\mathbf{y}_2$$

or, equivalently,

$$-\frac{s_1}{s_2} = \frac{\mathbf{x}\wedge\mathbf{a}\wedge\mathbf{x}_2}{\mathbf{x}\wedge\mathbf{a}\wedge\mathbf{x}_1} = \frac{\mathbf{x}\wedge\mathbf{b}\wedge\mathbf{y}_2}{\mathbf{x}\wedge\mathbf{b}\wedge\mathbf{y}_1},$$

which by cross multiplying gives the desired relationship

$$(\mathbf{x}\wedge\mathbf{a}\wedge\mathbf{x}_2)(\mathbf{x}\wedge\mathbf{b}\wedge\mathbf{y}_1) - (\mathbf{x}\wedge\mathbf{a}\wedge\mathbf{x}_1)(\mathbf{x}\wedge\mathbf{b}\wedge\mathbf{y}_2) = 0. \tag{17.19}$$

The expression (17.19) is quadratic in $\mathbf{x}$ and can be used to define an associated *quadratic form*

$$\begin{aligned} q(\mathbf{x}) &= (\mathbf{x}\wedge\mathbf{a}\wedge\mathbf{x}_2)(\mathbf{x}\wedge\mathbf{b}\wedge\mathbf{y}_1) - (\mathbf{x}\wedge\mathbf{a}\wedge\mathbf{x}_1)(\mathbf{x}\wedge\mathbf{b}\wedge\mathbf{y}_2) \\ &= \det\begin{pmatrix} \mathbf{a}\cdot\mathbf{x} & \mathbf{a}\cdot\mathbf{b} & \mathbf{a}\cdot\mathbf{y}_2 \\ \mathbf{x}_1\cdot\mathbf{x} & \mathbf{x}_1\cdot\mathbf{b} & \mathbf{x}_1\cdot\mathbf{y}_2 \\ \mathbf{x}\cdot\mathbf{x} & \mathbf{x}\cdot\mathbf{b} & \mathbf{x}\cdot\mathbf{y}_2 \end{pmatrix} \\ &\quad -\det\begin{pmatrix} \mathbf{a}\cdot\mathbf{x} & \mathbf{a}\cdot\mathbf{b} & \mathbf{a}\cdot\mathbf{y}_1 \\ \mathbf{x}_2\cdot\mathbf{x} & \mathbf{x}_2\cdot\mathbf{b} & \mathbf{x}_2\cdot\mathbf{y}_1 \\ \mathbf{x}\cdot\mathbf{x} & \mathbf{x}\cdot\mathbf{b} & \mathbf{x}\cdot\mathbf{y}_1 \end{pmatrix} \\ &= (x)_{(3)}S(x)^{\mathrm{T}}_{(3)}, \end{aligned} \tag{17.20}$$

where $S$ is the matrix of the quadratic form $q$ with respect to the standard basis $(\mathbf{e})_{(3)}$ of $\mathbb{R}^3$. The quadratic form $q(\mathbf{x})$ also defines a symmetric linear transformation

$$g(\mathbf{x}) = \frac{1}{2}\partial_{\mathbf{x}}q(\mathbf{x}), \tag{17.21}$$

which, in turn, defines the *bilinear form*

$$S(\mathbf{x},\mathbf{y}) = \mathbf{x}\cdot g(\mathbf{y}). \tag{17.22}$$

Before going further, it is worthwhile to consider an example. Let

$$\mathbf{a} = \begin{pmatrix}1 & 0 & 1\end{pmatrix}, \ \mathbf{x}_1 = \begin{pmatrix}0 & 2 & 1\end{pmatrix},$$
$$\mathbf{x}_2 = \begin{pmatrix}1 & 2 & 1\end{pmatrix}, \ \mathbf{y}_1 = \begin{pmatrix}-2 & -1 & 1\end{pmatrix}, \ \mathbf{y}_2 = \begin{pmatrix}-2 & 3 & 1\end{pmatrix}$$

be points in the projective plane $\Pi^2$. Since

$$\mathbf{a}\wedge\mathbf{x}_1\wedge\mathbf{x}_2 \neq 0 \ \text{ and } \ \mathbf{b}\wedge\mathbf{y}_1\wedge\mathbf{y}_2 \neq 0,$$

the points are noncollinear in $\Pi^2$. The linear mapping $f : \mathbb{R}^3 \to \mathbb{R}^3$, such that

$$f\left(\mathbf{a}\ \mathbf{x}_1\ \mathbf{x}_2\right) = \left(\mathbf{a}\ \mathbf{x}_1\ \mathbf{x}_2\right)[f] = \left(\mathbf{b}\ \mathbf{y}_1\ \mathbf{y}_2\right)$$

is specified by the matrix

$$[f] = \begin{pmatrix} 0 & 4 & 0 \\ -2 & \frac{5}{2} & 0 \\ 2 & -6 & 1 \end{pmatrix}.$$

Using (17.18)–(17.20), we then find the quadratic form

$$q(\mathbf{x}) = \det\begin{pmatrix} \mathbf{a}\cdot\mathbf{x} & \mathbf{a}\cdot\mathbf{b} & \mathbf{a}\cdot\mathbf{y}_2 \\ \mathbf{x}_1\cdot\mathbf{x} & \mathbf{x}_1\cdot\mathbf{b} & \mathbf{x}_1\cdot\mathbf{y}_2 \\ \mathbf{x}\cdot\mathbf{x} & \mathbf{x}\cdot\mathbf{b} & \mathbf{x}\cdot\mathbf{y}_2 \end{pmatrix} - \det\begin{pmatrix} \mathbf{a}\cdot\mathbf{x} & \mathbf{a}\cdot\mathbf{b} & \mathbf{a}\cdot\mathbf{y}_1 \\ \mathbf{x}_2\cdot\mathbf{x} & \mathbf{x}_2\cdot\mathbf{b} & \mathbf{x}_2\cdot\mathbf{y}_1 \\ \mathbf{x}\cdot\mathbf{x} & \mathbf{x}\cdot\mathbf{b} & \mathbf{x}\cdot\mathbf{y}_1 \end{pmatrix}$$

$$= \det\begin{pmatrix} x_1+x_3 & 3 & -1 \\ 2x_2+x_3 & -3 & 7 \\ x_1^2+x_2^2+x_3^2 & 2x_1-2x_2+x & -2x_1+3x_2+x_3 \end{pmatrix}$$

$$-\det\begin{pmatrix} x_1+x_3 & 3 & -1 \\ x_1+2x_2+x_3 & -1 & -3 \\ x_1^2+x_2^2+x_3^2 & 2x_1-2x_2+x_3 & -2x_1-x_2+x_3 \end{pmatrix}$$

$$= 8x_1^2+5x_1x_2+4x_2^2-24x_1x_3-2x_2x_3+16x_3^2.$$

Using (17.21) and (17.22), we also find the associated linear mapping

$$g(\mathbf{x}) = \frac{1}{2}\partial_{\mathbf{x}}q(\mathbf{x}) = \left(8x_1+\frac{5}{2}x_2-12x_3\ \ \frac{5}{2}x_1+4x_2-x_3\ \ -12x_1-x_2+16x_3\right)$$

$$= \begin{pmatrix} 8 & \frac{5}{2} & -12 \\ \frac{5}{2} & 4 & -1 \\ -12 & -1 & 16 \end{pmatrix}\begin{pmatrix} x_1 \\ x_2 \\ x_3 \end{pmatrix}$$

and the bilinear mapping $S(\mathbf{x},\mathbf{y})$ for which

$$S(\mathbf{x},\mathbf{y}) = \mathbf{x}\cdot g(\mathbf{y}) = \left(x_1\ x_2\ x_3\right)\begin{pmatrix} 8 & \frac{5}{2} & -12 \\ \frac{5}{2} & 4 & -1 \\ -12 & -1 & 16 \end{pmatrix}\begin{pmatrix} y_1 \\ y_2 \\ y_3 \end{pmatrix}$$

$$= 8x_1y_1+\frac{5}{2}x_2y_1-12x_3y_1+\frac{5}{2}x_1y_2+4x_2y_2-x_3y_2-12x_1y_3-x_2y_3+16x_3y_3. \tag{17.23}$$

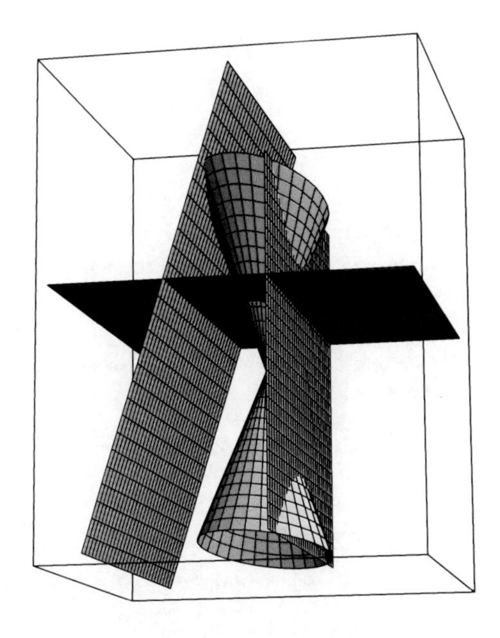

**Fig. 17.7** The quadratic form $q(\mathbf{x}) = 0$ defines a projective cone in $\mathbb{R}^3$. Three distinct affine planes cut the cone in an ellipse, hyperpola, and parabola

Using the methods of Sect. 9.4, we can complete the square for (17.23), getting

$$S(\mathbf{x}, \mathbf{y}) = (x)_{(3)} S(y)^{\mathrm{T}}_{(3)} = (x)_{(3)} (A^{-1})^{\mathrm{T}} A^{\mathrm{T}} S A A^{-1} (y)^{\mathrm{T}}_{(3)}$$
$$= (x)^A_{(3)} \begin{pmatrix} 1 & 0 & 0 \\ 0 & 1 & 0 \\ 0 & 0 & -1 \end{pmatrix} ((y)^A_{(3)})^{\mathrm{T}} = x^A_1 y^A_1 + x^A_2 y^A_2 - x^A_3 y^A_3.$$

By intersecting this projective cone with the affine planes, defined by the equations $x^A_3 = 1, x^A_2 = 1$, and $x^A_2 + x^A_3 = 1$, we get the equations of an ellipse, $(x^A_1)^2 + (x^A_2)^2 = 1$; hyperbola, $(x^A_3)^2 - (x^A_1)^2 = 1$; and parabola, $(x^A_1)^2 + 2x^A_2 = 1$, in the respective affine planes defined by these equations, shown in Fig. 17.7. The points on the projective cone in $\Pi^2$ are all the rays (lines) on the cone passing through, but not including the origin. The ideal line for the affine plane $x^A_3 = 1$ is $x^A_3 = 0$. The ideal line for the affine plane $x^A_2 = 1$ is $x^A_2 = 0$, and the ideal line for the affine plane $x^A_2 + x^A_3 = 1$ is $x^A_2 + x^A_3 = 0$.

By the theory of quadratic forms, worked out in Sect. 9.4 of Chap. 9, we can find a coordinate system of $\mathbb{R}^3$ in which the general quadratic form $q(\mathbf{x})$, given in (17.20), is reduced to a sum of squares of the form

$$q(\mathbf{x}) = (\kappa)_{(3)} (x)^{\mathrm{T}}_{(3)} = \kappa_1 x_1^2 + \kappa_2 x_2^2 + \kappa_3 x_3^2,$$

where each $\kappa_i = -1, 0, 1$ for $i = 1, 2, 3$. This reduces to the 6 distinct cases

$$(\kappa)_{(3)} = \{(1\ 1\ 1),\ (1\ 1\ -1),\ (1\ 1\ 0),\ (1\ -1\ 0),\ (1\ 0\ 0),\ (0\ 0\ 0)\} \tag{17.24}$$

in the projective plane $\Pi^2$, [22, p.166].

Other concepts that arise in the study of conics in $\Pi^2$ are the idea of *conjugate points* and the concept of *polarity* with respect to a given nondegenerate conic. We can hardly give more than definitions of these beautiful ideas.

For $\mathbf{x}, \mathbf{y} \in \Pi^2$, let $\mathbf{x} \cdot g(\mathbf{y})$ be the associated bilinear form of a given nondegenerate conic $q(\mathbf{x}) = 0$ in $\Pi^2$.

**Definition 17.4.1.** Two points $\mathbf{x}, \mathbf{y} \in \Pi^2$ are said to be conjugate with respect to the conic $q(\mathbf{x}) = 0$ if $\mathbf{x} \cdot g(\mathbf{y}) = 0$.

**Definition 17.4.2.** Given a point $\mathbf{y} \in \Pi^2$ and two distinct points $\mathbf{y}_1, \mathbf{y}_2 \in \Pi^2$, the projective line $\mathbf{y}_1 \wedge \mathbf{y}_2$ is called the polar line of the point $\mathbf{y}$ if $g(\mathbf{y}) \cdot (\mathbf{y}_1 \wedge \mathbf{y}_2) = 0$. The point $\mathbf{y}$ is said to be the pole of its polar line $\mathbf{y}_1 \wedge \mathbf{y}_2$.

Clearly, if $\mathbf{y}_1 \wedge \mathbf{y}_2$ is the polar line of the point $\mathbf{y}$, then $\mathbf{y}_1 \wedge \mathbf{y}_2 \doteq ig(\mathbf{y})$ where $i = \mathbf{e}_{123} \in \mathbb{G}_3$, since

$$g(\mathbf{y}) \cdot [ig(\mathbf{y})] = [g(\mathbf{y}) \wedge g(\mathbf{y})]i = 0.$$

Also if $\mathbf{y}_1 \wedge \mathbf{y}_2 \doteq ig(\mathbf{y})$, using the fact that $g(\mathbf{y})$ is a symmetric nonsingular linear operator on $\mathbb{R}^3$, we can with the help of (7.20) solve this equation for the pole $\mathbf{y}$ in terms of its polar line $\mathbf{y}_1 \wedge \mathbf{y}_2$, getting

$$\mathbf{y} \doteq g^{-1}(\mathrm{i}\, \mathbf{y}_1 \wedge \mathbf{y}_2) \doteq \underline{g}(\mathbf{y}_1 \wedge \mathbf{y}_2)i = g(\mathbf{y}_1) \times g(\mathbf{y}_2), \tag{17.25}$$

where $\underline{g}$ is the outermorphism extension of $g$ to the geometric algebra $\mathbb{G}_3$. Thus, there is a one-to-one correspondence between polar lines $p_{\mathbf{y}}$ and their corresponding poles $\mathbf{y}$ with respect to a given nondegenerate conic $q(\mathbf{x}) = 0$.

The following theorem gives the relationship between a given nondegenerate conic $q(\mathbf{x}) = 0$ in $\Pi^2$ and the polar line $p_{\mathbf{y}}$ with its pole $\mathbf{y}$ for any point $\mathbf{y} \in \Pi^2$. The theorem is in three parts, depending upon whether the polar line $p_{\mathbf{y}}$ intersects the the conic in two points, is a tangent line to the conic, or does not intersect the conic.

**Theorem 17.4.3.** *Given the nondegenerate conic $q(\mathbf{x}) = 0$ and a point $\mathbf{y} \in \Pi^2$:*

*(a) The polar line $p_{\mathbf{x}} \doteq ig(\mathbf{x})$ to a point $\mathbf{x}$ on the conic $q(\mathbf{x}) = 0$ is the tangent line to the conic at the point $\mathbf{x}$.*
*(b) If the polar line $p_{\mathbf{y}}$ in $\Pi^2$ intersects the conic at two distinct points $\mathbf{x}_1$ and $\mathbf{x}_2$, then $p_{\mathbf{y}} \doteq \mathbf{x}_1 \wedge \mathbf{x}_2$ and*

$$\mathbf{y} \doteq [ig(\mathbf{x}_1)] \cap [ig(\mathbf{x}_2)] \doteq g(\mathbf{x}_1) \times g(\mathbf{x}_2).$$

(c) *If the polar line* $p_{\mathbf{y}} = ig(\mathbf{y})$ *to the point* $\mathbf{y}$ *does not intersect the conic, then each line through the point* $\mathbf{y}$ *will meet the conic at two distinct points, say,* $\mathbf{x}_1$ *and* $\mathbf{x}_2$. *If* $l_{\mathbf{x}_1}$ *and* $l_{\mathbf{x}_2}$ *are the tangent lines to the conic at the points* $\mathbf{x}_1$ *and* $\mathbf{x}_2$, *then*

$$p_{\mathbf{y}} \wedge [l_{\mathbf{x}_1} \cap l_{\mathbf{x}_2}] = p_{\mathbf{y}} \wedge (g(\mathbf{x}_1) \times g(\mathbf{x}_2)) = 0.$$

*Proof.* (a) Suppose that $\mathbf{x}_1 \cdot g(\mathbf{x}_1) = 0$ so that $\mathbf{x}_1$ is on the conic. Since $\mathbf{x}_1 \wedge (ig(\mathbf{x}_1)) = (\mathbf{x}_1 \cdot g(\mathbf{x}_1))i = 0$, it follows that the polar line $p_{\mathbf{x}_1} = ig(\mathbf{x}_1)$ passes through the point $\mathbf{x}_1$. Now suppose that $p_{\mathbf{x}_1} = \mathbf{x}_1 \wedge \mathbf{x}_2 = 0$ for a second point $\mathbf{x}_2$ on the conic which is distinct from $\mathbf{x}_1$. But then, any point $\mathbf{y} = \alpha\mathbf{x}_1 + \beta\mathbf{x}_2$ would also be on the conic since

$$g(\mathbf{y}) \cdot \mathbf{y} = \alpha^2 g(\mathbf{x}_1) \cdot \mathbf{x}_1 + 2\alpha\beta g(\mathbf{x}_1) \cdot \mathbf{x}_2 + \beta^2 g(\mathbf{x}_2) \cdot \mathbf{x}_2 = 2\alpha\beta g(\mathbf{x}_1) \cdot \mathbf{x}_2 = 0. \tag{17.26}$$

However, this is impossible since the line $\mathbf{x}_1 \wedge \mathbf{x}_2$ will cut the conic in at most two points. The last equality in (17.26) follows from

$$0 = g(\mathbf{x}_1) \cdot p_{\mathbf{x}_1} \doteq g(\mathbf{x}_1) \cdot (\mathbf{x}_1 \wedge \mathbf{x}_2) = (g(\mathbf{x}_1) \cdot \mathbf{x}_1)\mathbf{x}_2 - (g(\mathbf{x}_1) \cdot \mathbf{x}_2)\mathbf{x}_1 \doteq -(g(\mathbf{x}_1) \cdot \mathbf{x}_2)\mathbf{x}_1.$$

(b) Suppose that $p_{\mathbf{y}} \doteq ig(\mathbf{y}) = \mathbf{x}_1 \wedge \mathbf{x}_2$ where $\mathbf{x}_1$ and $\mathbf{x}_2$ are distinct points on the conic. Then by part a) of the theorem, $l_{\mathbf{x}_1} = ig(\mathbf{x}_1)$ and $l_{\mathbf{x}_2} = ig(\mathbf{x}_2)$ are the tangent lines through the conic at the points $\mathbf{x}_1$ and $\mathbf{x}_2$, respectively. But by (17.25),

$$\mathbf{y} \doteq g(\mathbf{x}_1) \times g(\mathbf{x}_2) \doteq [ig(\mathbf{x}_1)] \cap [ig(\mathbf{x}_2)] = l_{\mathbf{x}_1} \cap l_{\mathbf{x}_2}.$$

(c) Let $p_{\mathbf{y}} = ig(\mathbf{y})$ be any polar line which does not meet the conic, and let $\mathbf{x}_1$ be any point on the conic. Then

$$\mathbf{x}_1 \wedge (ig(\mathbf{y})) = i(\mathbf{x}_1 \cdot g(\mathbf{y})) = i(g(\mathbf{x}_1) \cdot \mathbf{y}) = \mathbf{y} \wedge (ig(\mathbf{x}_1)) \neq 0$$

for any point $\mathbf{x}_1$ on the conic implies that $\mathbf{y}$ is not on any tangent line to the conic. This means that every line that meets the point $\mathbf{y}$ will meet the conic in two distinct points, say, $\mathbf{x}_1$ and $\mathbf{x}_2$, so that $\mathbf{y} \wedge \mathbf{x}_1 \wedge \mathbf{x}_2 = 0$.
Since $\mathbf{y} \wedge \mathbf{x}_1 \wedge \mathbf{x}_2 = 0$, it follows that

$$\underline{g}(\mathbf{y} \wedge \mathbf{x}_1 \wedge \mathbf{x}_2) = g(\mathbf{y}) \wedge g(\mathbf{x}_1) \wedge g(\mathbf{x}_2) = [ig(\mathbf{y})] \wedge [g(\mathbf{x}_1) \times g(\mathbf{x}_2)] = 0.$$

Letting $l_{\mathbf{x}_1} = ig(\mathbf{x}_1)$ and $l_{\mathbf{x}_2} = ig(\mathbf{x}_2)$ be the tangent lines to the conic at the points $\mathbf{x}_1$ and $\mathbf{x}_2$, it then follows that

$$p_{\mathbf{y}} \wedge [l_{\mathbf{x}_1} \cap l_{\mathbf{x}_2}] = p_{\mathbf{y}} \wedge (g(\mathbf{x}_1) \times g(\mathbf{x}_2)) = 0.$$

□

Figure 17.8 constructs the polarity for the cases b) and c) of the theorem.

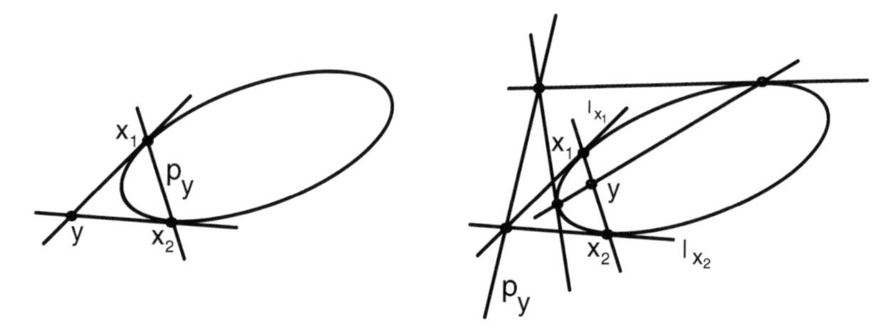


**Fig. 17.8** The polarity between a pole $\mathbf{y}$ and its polar line $p_\mathbf{y} = ig(\mathbf{y})$ is shown for the cases (b) and (c) of the theorem

## Exercises

1. For $\mathbf{x} = (x_1, x_2, x_3) \in \mathbb{R}^3$, define the quadratic form $q(\mathbf{x}) = x_1^2 - 4x_1x_3 + 3x_2^2 - x_3^2$.

   (a) Find the associated symmetric linear tramsformation $g(\mathbf{x})$, defined in (17.21), and the matrix $S$ of the bilinear form $\mathbf{x} \cdot g(\mathbf{y})$.
   (b) Show that the points $\mathbf{x} = (1, 0, -2 \pm \sqrt{5})$ satisfy the equation $q(\mathbf{x}) = 0$, and find the polar lines tangent to the conic at these points. Find the projective line that passes through these two points and the pole of this line.
   (c) Find a point $\mathbf{y} = (y_1, y_2, y_3) \in \Pi^2$ which is conjugate to the point $\mathbf{x} \doteq (1, 2, 3) \in \Pi^2$.
   (d) Given the two points $\mathbf{x} = (1, -1, 1)$ and $\mathbf{y} = (2, 2, 2)$ in $\Pi^2$, find the projective line $\mathbf{x} \wedge \mathbf{y}$ which passes through these points. Find the projective point $\mathbf{p} = (p_1, p_2, p_3)$ which is the pole of the polar line $\mathbf{x} \wedge \mathbf{y}$.

2. For $\mathbf{x} = (x_1, x_2, x_3) \in \mathbb{R}^3$, define the quadratic form $q(\mathbf{x}) = x_1^2 + 2x_1x_2 + x_2^2 + 4x_2x_3 + 4x_3^2$.

   (a) Find the associated symmetric linear tramsformation $g(\mathbf{x})$, defined in (17.21), and the matrix $S$ of the bilinear form $\mathbf{x} \cdot g(\mathbf{y})$.
   (b) Show that the point $\mathbf{x} = (1, -1, 0)$ satisfies the equation $q(\mathbf{x}) = 0$, and find the polar line tangent to the conic at this point.
   (c) Find a point $\mathbf{y} = (y_1, y_2, y_3) \in \Pi^2$ which is conjugate to the point $\mathbf{x} \doteq (1, 2, 3) \in \Pi^2$.
   (d) Given the two points $\mathbf{x} = (1, -1, 1)$ and $\mathbf{y} = (2, 0, 2)$ in $\Pi^2$, find the projective line $\mathbf{x} \wedge \mathbf{y}$ which passes through these points. Find the projective point $\mathbf{p} = (p_1, p_2, p_3)$ which is the pole of the polar line $\mathbf{x} \wedge \mathbf{y}$.

## 17.5 Projective Geometry Is All of Geometry

We are now in a position to explain what Cauchy meant by his famous exclamation that "projective geometry is all of geometry." We shall only discuss the projective plane $\Pi^2$ and some of the geometries for which $\Pi^2$ is the *precusor* or *parent* geometry, although the discussion could be broadened to the higher dimensional geometries for which $\Pi^n$ is the precursor geometry.

Projective geometry of the real projective plane $\Pi^2$ is the study of all those properties that remain invariant or unchanged under a projective transformation or the composition of projective transformations. In our study of the analytic projective geometry of $\Pi^2$, all projective transformations are represented by nonsingular linear transformations on $\mathbb{R}^3$. Thus, the *group of projective transformations* $GL_3(\mathbb{R})$ consists of all linear transformations on $\mathbb{R}^3$ which have nonzero determinant. The group $GL_3(\mathbb{R})$ is also known as the *general linear group*. We will have a lot more to say about the general linear group in the next chapter.

The group of projective transformations $GL_3(\mathbb{R})$ is the parent geometry of a host of other geometries, only some of which we will now discuss. The affine plane $\mathscr{A}_h^2$ is the *subgeometry* obtained from the projective plane $\Pi^2$, by picking out the projective line $\mathbf{e}_{12}$ in $\Pi^2$ and calling it the *ideal line at infinity*. Recalling (17.24), the equation of this ideal line is the degenerate conic $x_3^2 = 0$. The *ordinary points* of the affine plane $\mathscr{A}_h^2$ are all those points of $\Pi^2$ which do not lie on the ideal line at infinity. This definition of $\mathscr{A}_h^2$ agrees with Definition 17.1.1 and (17.1) given earlier. The affine group is the subgroup of the projective transformations $GL_3(\mathbb{R})$ of $\Pi^2$ which leave the ideal line fixed and map ordinary points into ordinary points.

Let $\mathbf{x} = (x_1, x_2, 1) \in \mathscr{A}_h^2$, and suppose $\mathbf{y} = (y_1, y_2, 1)(\mathbf{e})_{(3)}^{\mathrm{T}}$ is given such that

$$\mathbf{y} = f(\mathbf{x}) = (\mathbf{e})_{(3)}[f]\begin{pmatrix} x_1 \\ x_2 \\ 1 \end{pmatrix}.$$

Then the matrix $[f]$ of the projective transformation $f$ must be of the form

$$[f] = \begin{pmatrix} f_{11} & f_{12} & b_1 \\ f_{21} & f_{22} & b_2 \\ 0 & 0 & 1 \end{pmatrix}, \tag{17.27}$$

where $\det(f) = \det[f] = f_{11}f_{22} - f_{12}f_{21} \neq 0$. This is the most general projective transformation that has the required property. We see that

$$f(\mathbf{x}) = (\mathbf{e})_{(3)}[f]\begin{pmatrix} x_1 \\ x_2 \\ 0 \end{pmatrix} + (\mathbf{e})_{(3)}\begin{pmatrix} b_1 \\ b_2 \\ 1 \end{pmatrix},$$

which has the interpretation a projective transformation of $GL_2(\mathbb{R})$ in the first two components plus a translation in $\mathscr{A}_h^2$. An affine transformation does not preserve the angles or the area of a triangle.

If we further restrict the affine group by requiring that it leaves a given involution, called the *absolute involution* invariant, we obtain a *similarity transformation*. The group of similarity transformations on the affine plane $\mathscr{A}_h^2$ is called *similarity geometry*. The matrix $[f]$ of such a transformation will have the form

$$[f] = \begin{pmatrix} r\cos\theta & -r\sin\theta & b_1 \\ rs\sin\theta & rs\cos\theta & b_2 \\ 0 & 0 & 1 \end{pmatrix}, \tag{17.28}$$

where $r > 0$ and $s^2 = 1$.

It is only when we further restrict the similarity transformations (17.28) of $\mathscr{A}_h^2$ by the requirement that $\det[f] = s$, or $r = 1$ and $s = \pm 1$, that we obtain the familiar *Euclidean geometry* $\mathbb{R}^2$ on the plane. Thus, a Euclidean transformation, or *motion* in $\mathscr{A}_h^2$, consists of reflections when $r = 1$ and $s = -1$, rotations when $r = 1 = s$, and translations.

The path

$$\Pi^2 \longrightarrow \mathscr{A}_h^2 \longrightarrow \text{similarity} \longrightarrow \text{Euclidean}$$

to Euclidean geometry is only one of a number of different paths that are available to us. There are other paths that lead us instead to *hyperbolic geometry* and *elliptic geometry*. To get from $\Pi^2$ to hyperbolic geometry $\mathscr{H}^2$, we pick out a nondegenerate conic of the form $x_1^2 + x_2^2 = x_3^2$, which we call the *absolute conic*. The points of $\mathscr{H}^2$ are those points of the form $x_1^2 + x_2^2 < x_3^2$, called *ordinary points*. The points $x_1^2 + x_2^2 = x_3^2$ on the conic are called *ideal points*, and those satisfying $x_1^2 + x_2^2 > x_3^2$ are called *ultra ideal*. The reader is encouraged to further study these fascinating geometries. A good elementary account can be found in [22, p.229–260].

http://en.wikipedia.org/wiki/Hyperbolic_geometry

There is still another path that is available to us, and that is the path to *elliptic geometry* obtained by selecting a nondegenerate *absolute conic* of the form $x_1^2 + x_2^2 + x_3^2 = 0$. Of course, there are no real points on this conic so all the points of $\Pi^2$ are points of the elliptic plane. For a brief introduction to elliptic geometry, see [22, p.261-285].

http://en.wikipedia.org/wiki/Elliptic_geometry

## Exercises

1. (a) Show that the similarity transformation $f(\mathbf{x})$, with the matrix given in (17.28) with $s = 1$, can be equivalently expressed in the form

$$f(\mathbf{x}) = re^{\frac{1}{2}\theta\mathbf{e}_{21}}\mathbf{x}re^{-1/2\theta\mathbf{e}_{21}} + \mathbf{b},$$

where $\mathbf{x} = (x_1, x_2, 0)(\mathbf{e})_{(3)}^{\mathrm{T}}$ and $\mathbf{b} = (b_1, b_2, 1)(\mathbf{e})_{(3)}^{\mathrm{T}}$. Explain your result.

(b) Show that the similarity transformation $f(\mathbf{x})$, with the matrix given in (17.28) with $s=-1$, can be equivalently expressed in the form

$$f(\mathbf{x}) = -re^{\frac{1}{2}\theta\mathbf{e}_{21}}\mathbf{e}_2\mathbf{x}\mathbf{e}_2 re^{-\frac{1}{2}\theta\mathbf{e}_{21}} + \mathbf{b},$$

where $\mathbf{x} = (x_1, x_2, 0)(\mathbf{e})^{\mathrm{T}}_{(3)}$ and $\mathbf{b} = (b_1, b_2, 1)(\mathbf{e})^{\mathrm{T}}_{(3)}$. Explain your result.

2. Verify that a projective transformation $\mathbf{y} = f(\mathbf{x})$, whose matrix $[f]$ is of the form (17.27), leaves the projective line $\mathbf{e}_{12}$ invariant, i.e., $\underline{f}(\mathbf{e}_{12}) \doteq \mathbf{e}_{12}$.

## 17.6 The Horosphere $\mathscr{H}^{p,q}$

We began this chapter by defining the affine $n$-plane $\mathscr{A}_h^n$ as a translation of $\mathbb{R}^n$ in the larger Euclidean space $\mathbb{R}^{n+1}$, and we saw how the concept of points in $\mathscr{A}_h^n$ as rays led us to the study of projective geometry. We take this idea a step further in this section by embedding the pseudo-Euclidean space $\mathbb{R}^{p,q}$, and its corresponding geometric algebra $\mathbb{G}_{p,q}$, in the larger pseudo-Euclidean space $\mathbb{R}^{p+1,q+1}$ and its corresponding geometric algebra $\mathbb{G}_{p+1,q+1}$. In order to better understand the nature of Ahlfors–Vahlen matrices and how they arise, we introduce the *horosphere* $\mathscr{H}^{p,q}$ in $\mathbb{R}^{p+1,q+1}$. The horosphere is a nonlinear model of the pseudo-Euclidean space $\mathbb{R}^{p,q}$. It was first introduced for the Euclidean space $\mathbb{R}^n$ by F. A. Wachter, a student of Gauss, and has been recently finding diverse applications [6, 20, 31].

Recall (10.2) that the standard orthonormal basis of $\mathbb{R}^{p,q}$ is given by

$$(\mathbf{e})_{(p+q)} = (\mathbf{e}_1, \ldots, \mathbf{e}_p, \mathbf{e}_{p+1}, \ldots, \mathbf{e}_{p+q}),$$

where $\mathbf{e}_i^2 = 1$ for $1 \le i \le p$ and $\mathbf{e}_i^2 = -1$ for $p < i \le p+q$. For the standard basis of $\mathbb{R}^{p+1,q+1}$, we extend the standard basis of $\mathbb{R}^{p,q}$ to include two new orthonormal (anticommuting) vectors $\mathbf{e}$ and $\mathbf{f}$ with the properties that $\mathbf{e}^2 = 1$ and $\mathbf{f}^2 = -1$. The standard basis of $\mathbb{R}^{p+1,q+1}$ then becomes

$$(\mathbf{e})_{(n)} = (\mathbf{e})_{(p+q)} \cup (\mathbf{e}, \mathbf{f}),$$

where $n = p+q+2$. We also introduce two *reciprocal null vectors*

$$\mathbf{a} = \frac{1}{2}(\mathbf{e}+\mathbf{f}) \quad \text{and} \quad \mathbf{b} = \mathbf{e} - \mathbf{f}, \tag{17.29}$$

which play a prominant role in what follows. The null vectors $\mathbf{a}$ and $\mathbf{b}$ satisfy the properties

$$\mathbf{a}^2 = \mathbf{b}^2 = 0 \quad \text{and} \quad \mathbf{a} \cdot \mathbf{b} = 1.$$

Now define the bivector $\mathrm{u} = \mathbf{ef} = \mathbf{b} \wedge \mathbf{a}$ and idempotents $\mathrm{u}_\pm = \frac{1}{2}(1 \pm \mathrm{u})$. A general geometric number $G \in \mathbb{G}_{p+1,q+1}$ in this basis can be written

$$G = (g_1 + g_2\mathrm{u}) + (g_3 + g_4\mathrm{u})\mathbf{e} \quad \text{for} \quad g_1, g_2, g_3, g_4 \in \mathbb{G}_{p,q}, \tag{17.30}$$

with the three operations of conjugation given by

$$G^{\dagger}=(g_1^{\dagger}-g_2^{\dagger}\mathrm{u})+(\widetilde{g}_3+\widetilde{g}_4\mathrm{u})\mathbf{e},\quad G^{-}=(g_1^{-}+g_2^{-}\mathrm{u})-(g_3^{-}+g_4^{-}\mathrm{u})\mathbf{e},\ \text{and}$$
$$\widetilde{G}=(\widetilde{g}_1-\widetilde{g}_2\mathrm{u})-(g_3^{\dagger}+g_4^{\dagger}\mathrm{u})\mathbf{e}. \tag{17.31}$$

By the *spectral basis* of the geometric algebra $\mathbb{G}_{p+1,q+1}$ over the geometric algebra $\mathbb{G}_{p,q}$, we mean

$$\begin{pmatrix}1\\ \mathbf{e}\end{pmatrix}\mathrm{u}_+\begin{pmatrix}1 & \mathbf{e}\end{pmatrix}=\begin{pmatrix}\mathrm{u}_+ & \mathrm{u}_+\mathbf{e}\\ \mathbf{e}\mathrm{u}_+ & \mathrm{u}_-\end{pmatrix}. \tag{17.32}$$

In the spectral basis (17.32), the general element $G$ in (17.30) takes the form of a $2\times 2$ matrix over $\mathbb{G}_{p,q}$,

$$\begin{aligned}G&=\begin{pmatrix}1 & \mathbf{e}\end{pmatrix}\mathrm{u}_+\begin{pmatrix}1\\ \mathbf{e}\end{pmatrix}G\begin{pmatrix}1 & \mathbf{e}\end{pmatrix}\mathrm{u}_+\begin{pmatrix}1\\ \mathbf{e}\end{pmatrix}=\begin{pmatrix}1 & \mathbf{e}\end{pmatrix}\mathrm{u}_+\begin{pmatrix}G & G\mathbf{e}\\ \mathbf{e}G & \mathbf{e}G\mathbf{e}\end{pmatrix}\mathrm{u}_+\begin{pmatrix}1\\ \mathbf{e}\end{pmatrix}\\
&=\begin{pmatrix}1 & \mathbf{e}\end{pmatrix}\mathrm{u}_+\begin{pmatrix}g_1+g_2 & g_3+g_4\\ \mathbf{e}(g_3-g_4)\mathbf{e} & \mathbf{e}(g_1-g_2)\mathbf{e}\end{pmatrix}\begin{pmatrix}1\\ \mathbf{e}\end{pmatrix}\\
&=\begin{pmatrix}1 & \mathbf{e}\end{pmatrix}\mathrm{u}_+\begin{pmatrix}g_1+g_2 & g_3+g_4\\ g_3^{-}-g_4^{-} & g_1^{-}-g_2^{-}\end{pmatrix}\begin{pmatrix}1\\ \mathbf{e}\end{pmatrix},\end{aligned}$$

where $\mathbf{e}g\mathbf{e}=g^{-}$ for $g\in G_{p,q}$ is recognized as the operation of grade inversion defined in (17.31). We call

$$[G]=\begin{pmatrix}g_1+g_2 & g_3+g_4\\ g_3^{-}-g_4^{-} & g_1^{-}-g_2^{-}\end{pmatrix}=\begin{pmatrix}a & b\\ c & d\end{pmatrix} \tag{17.33}$$

the *matrix of* $G\in\mathbb{G}_{p+1,q+1}$ *over the geometric algebra* $\mathbb{G}_{p,q}$. Using (17.31), we also find that

$$[G^{\dagger}]=\begin{pmatrix}g_1^{\dagger}-g_2^{\dagger} & -g_3^{\dagger}-g_4^{\dagger}\\ -\widetilde{g}_3+\widetilde{g}_4 & \widetilde{g}_1+\widetilde{g}_2\end{pmatrix}=\begin{pmatrix}\widetilde{d} & -\widetilde{b}\\ -\widetilde{c} & \widetilde{a}\end{pmatrix}. \tag{17.34}$$

The $(p,q)$-*horosphere* is defined by

$$\mathscr{H}^{p,q}=\{\mathbf{x}_c|\ \mathbf{x}_c=\frac{1}{2}\mathbf{x}_a\mathbf{b}\mathbf{x}_a,\ \mathbf{x}_a=\mathbf{x}+\mathbf{a}\ \text{for}\ \mathbf{x}\in\mathbb{R}^{p,q}\}\subset\mathbb{R}^{p+1,q+1}. \tag{17.35}$$

By the *affine plane* $\mathscr{A}_a^{p,q}$, we mean

$$\mathscr{A}_a^{p,q}:=\{\mathbf{x}_a=\mathbf{x}+\mathbf{a}|\ \mathbf{x}\in\mathbb{R}^{p,q}\},$$

so the affine plane $\mathscr{A}_a^{p,q}\subset\mathbb{R}^{p+1,q+1}$ is just the pseudo-Euclidean space $\mathbb{R}^{p,q}$ displaced by the null vector $\mathbf{a}\in\mathbb{R}^{p+1,q+1}$. (In contrast, the affine plane $\mathscr{A}_h^n$ was obtained by displacing $\mathbb{R}^n$ by the unit vector $\mathbf{e}_{n+1}\in\mathbb{R}^{n+1}$, given in Definition 17.1.1.)

The reciprocal null vectors **a** and **b** have been previously specified in (17.29). Noting that $\mathbf{x}_a^2 = \mathbf{x}^2$ and expanding the point $\mathbf{x}_c \in \mathcal{H}^{p,q}$, we say that

$$\mathbf{x}_c = \frac{1}{2}\mathbf{x}_a\mathbf{b}\mathbf{x}_a = \frac{1}{2}(\mathbf{x}_a \cdot \mathbf{b} + \mathbf{x}_a \wedge \mathbf{b})\mathbf{x}_a = \mathbf{x}_a - \frac{1}{2}\mathbf{x}^2\mathbf{b}$$
$$= \exp\left(\frac{1}{2}\mathbf{x}\mathbf{b}\right)\mathbf{a}\exp(-\frac{1}{2}\mathbf{x}\mathbf{b}) \tag{17.36}$$

is the *conformal representant* of both $\mathbf{x} \in \mathbb{R}^{p,q}$ and $\mathbf{x}_a \in \mathcal{A}_a^{p,q}$. The conformal representant $\mathbf{x}_c$ is a null vector for all $\mathbf{x} \in \mathbb{R}^{p,q}$, since by (17.36)

$$\mathbf{x}_c^2 = \left[\exp\left(\frac{1}{2}\mathbf{x}\mathbf{b}\right)\mathbf{a}\exp\left(-\frac{1}{2}\mathbf{x}\mathbf{b}\right)\right]^2 = \mathbf{a}^2 = 0.$$

The points $\mathbf{x}_c$ are *homogeneous* in the sense that

$$\mathbf{x}_c = \frac{\alpha\mathbf{x}_c}{\alpha} = \frac{\alpha\mathbf{x}_c}{(\alpha\mathbf{x}_c)\cdot\mathbf{b}}$$

for all nonzero $\alpha \in \mathbb{R}$. Given $\mathbf{x}_c$, it is easy to get back $\mathbf{x}_a$ by the simple projection,

$$\mathbf{x}_a = (\mathbf{x}_c \wedge \mathbf{b})\cdot\mathbf{a}. \tag{17.37}$$

We can also retrieve the vector $\mathbf{x} \in \mathbb{R}^{p,q}$ from $\mathbf{x}_c$ by noting that

$$\mathbf{x} = (\mathbf{x}_c \wedge \mathbf{b} \wedge \mathbf{a})\cdot(\mathbf{b} \wedge \mathbf{a}). \tag{17.38}$$

The expression of the conformal representant $\mathbf{x}_c$ in (17.36) is interesting because it shows that all points on $\mathcal{H}^{p,q}$ can be obtained by a simple rotation of **a** in the plane of the bivector **xb**. The affine plane $\mathcal{A}_a^{p,q}$ and horosphere $\mathcal{H}^{p,q}$ are pictured in Fig. 17.9.

The horosphere has attracted a lot of attention in the computer science community because of its close relationship to the conformal group of $\mathbb{R}^{p,q}$. Any conformal transformation in $\mathbb{R}^{p,q}$ can be represented by an orthogonal transformation on the horosphere $\mathcal{H}^{p,q}$ in $\mathbb{R}^{p+1,q+1}$. We have already explored conformal transformations on $\mathbb{R}^{p,q}$ in Chap. 16, and we now show how we can connect up these ideas to orthogonal transformations in $\mathbb{R}^{p+1,q+1}$ restricted to the horophere $\mathcal{H}^{p,q}$.

For $\mathbf{x}_c, \mathbf{y}_c \in \mathcal{H}^{p,q}$,

$$\mathbf{x}_c \cdot \mathbf{y}_c = \left(\mathbf{x} + \mathbf{a} - \frac{\mathbf{x}^2}{2}\mathbf{b}\right)\cdot\left(\mathbf{y} + \mathbf{a} - \frac{\mathbf{y}^2}{2}\mathbf{b}\right)$$
$$= \mathbf{x}\cdot\mathbf{y} - \frac{1}{2}(\mathbf{x}^2 + \mathbf{y}^2) = -\frac{1}{2}(\mathbf{x} - \mathbf{y})^2, \tag{17.39}$$

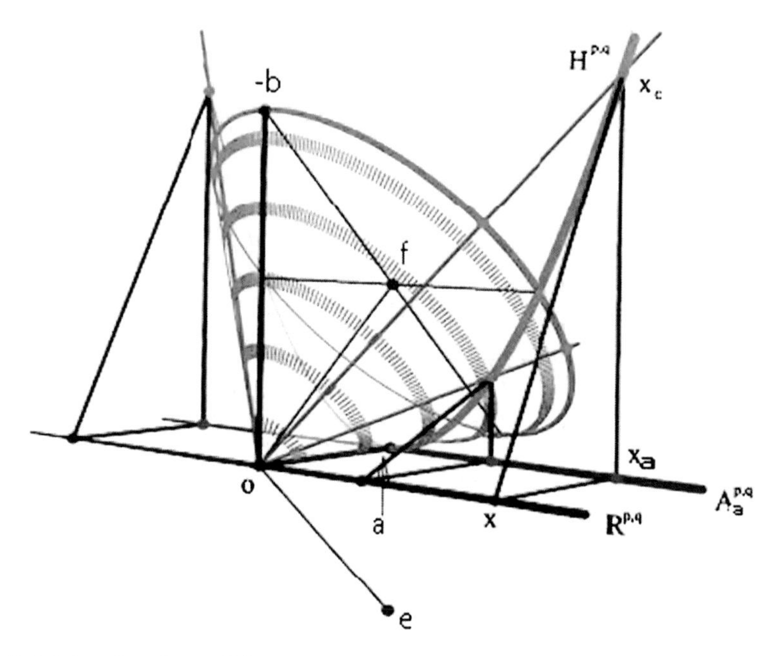


**Fig. 17.9** The affine plane $\mathscr{A}_a^{p,q}$ and horosphere $\mathscr{H}^{p,q}$

which is $-1/2$ the pseudo-Euclidean distance squared between the points $\mathbf{x}_c$ and $\mathbf{y}_c$. Thus, the pseudo-Euclidean structure of $\mathbb{R}^{p,q}$ is preserved in the form of the inner product $\mathbf{x}_c \cdot \mathbf{y}_c$ on the horosphere. The horosphere $\mathscr{H}^{p,q}$ is a nonlinear model of both the affine plane $\mathscr{A}_a^{p,q}$ and the pseudo-Euclidean space $\mathbb{R}^{p,q}$.

Let us now find the matrix $[\mathbf{x}_c]$, as given in (17.33), of the point $\mathbf{x}_c \in \mathscr{H}^{p,q}$. Noting that

$$\mathbf{x}_c = \mathbf{x} + \frac{1}{2}(1-\mathbf{x}^2)\mathbf{e} + \frac{1}{2}(1+\mathbf{x}^2)\mathbf{f}$$

when expanded in the standard basis of $\mathbb{R}^{p+1,q+1}$ and using (17.33) with

$$g_1 = \mathbf{x},\ g_2 = 0,\ g_3 = \frac{1}{2}(1-\mathbf{x}^2),\ g_4 = -\frac{1}{2}(1+\mathbf{x}^2),$$

we easily find that

$$[\mathbf{x}_c] = \begin{pmatrix} g_1 + g_2 & g_3 + g_4 \\ g_3^- - g_4^- & g_1^- - g_2^- \end{pmatrix} = \begin{pmatrix} \mathbf{x} & -\mathbf{x}^2 \\ 1 & -\mathbf{x} \end{pmatrix} = \begin{pmatrix} \mathbf{x} \\ 1 \end{pmatrix} \begin{pmatrix} 1 & -\mathbf{x} \end{pmatrix}. \tag{17.40}$$

Recalling (17.36), for any $\mathbf{y} \in \mathbb{R}^{p,q}$, define $S_{\mathbf{y}} = \exp\left(\frac{1}{2}\mathbf{y}\mathbf{b}\right)$. Calculating the matrix of $S_{\mathbf{y}}$, we find that

$$[S_{\mathbf{y}}] = \left[1 + \frac{1}{2}\mathbf{y}\mathbf{b}\right] = \begin{pmatrix} 1 & \mathbf{y} \\ 0 & 1 \end{pmatrix}.$$

Noting that the matrix of the null vector $[\mathbf{a}] = \begin{pmatrix} 0 & 0 \\ 1 & 0 \end{pmatrix}$, and using (17.34), we can reexpress the relationship (17.36) in the matrix form

$$[\mathbf{x}_c] = [S_{\mathbf{x}}\mathbf{a}S_{\mathbf{x}}^{\dagger}] = [S_{\mathbf{x}}][\mathbf{a}][S_{\mathbf{x}}^{\dagger}] = \begin{pmatrix} 1 & \mathbf{x} \\ 0 & 1 \end{pmatrix}\begin{pmatrix} 0 & 0 \\ 1 & 0 \end{pmatrix}\begin{pmatrix} 1 & -\mathbf{x} \\ 0 & 1 \end{pmatrix}.$$

More generally, with the help of (17.40), (17.33), and (17.34), any transformation of the form $L(\mathbf{x}_c) = G\mathbf{x}_c G^{\dagger}$ for $G \in \mathbb{G}_{p+1,q+1}$ can be expressed in the equivalent matrix form

$$[L(\mathbf{x}_c)] = [G\mathbf{x}_c G^{\dagger}] = \begin{pmatrix} a & b \\ c & d \end{pmatrix}\begin{pmatrix} \mathbf{x} \\ 1 \end{pmatrix}\begin{pmatrix} 1 & -\mathbf{x} \end{pmatrix}\begin{pmatrix} \widetilde{d} & -\widetilde{b} \\ -\widetilde{c} & \widetilde{a} \end{pmatrix}. \quad (17.41)$$

From the last equation, it is clear that for any transformation of the form $L(\mathbf{x}_c)$, we can work with the simpler *twistor form*

$$T(\mathbf{x}_c) = \begin{pmatrix} a & b \\ c & d \end{pmatrix}\begin{pmatrix} \mathbf{x} \\ 1 \end{pmatrix} = \begin{pmatrix} a\mathbf{x} + b \\ c\mathbf{x} + d \end{pmatrix} \quad (17.42)$$

of this equation without losing any information, since the full equation (17.41) can easily be recovered, [65, p.236].

The general transformation $L(\mathbf{x}_c)$, and its twistor equivalent $T(\mathbf{x}_c)$, is well defined for all values of $G \in \mathbb{G}_{p+1,q+1}$. However, for only certain values of $G$ will $L : \mathscr{H}^{p,q} \longrightarrow \mathscr{H}^{p,q}$. For these values of $G$, the twistor form $T(\mathbf{x}_c)$ becomes the conformal transformations of the Ahlfors–Vahlen matrices discussed in Chap. 16. The following table gives the equivalent expressions for the various kinds of conformal transformations.

| | $G$ | $[G]$ | $(a\mathbf{x}+b)(c\mathbf{x}+d)^{-1}$ |
|---|---|---|---|
| Translation | $e^{\frac{1}{2}\mathbf{y}\mathbf{b}}$ | $\begin{pmatrix} 1 & \mathbf{y} \\ 0 & 1 \end{pmatrix}$ | $\mathbf{x}+\mathbf{y}$ |
| Inversion | $\mathbf{e}$ | $\begin{pmatrix} 0 & 1 \\ 1 & 0 \end{pmatrix}$ | $\frac{1}{\mathbf{x}}$ |
| Dilation | $e^{\frac{1}{2}\phi\mathbf{u}}$ | $\begin{pmatrix} e^{\frac{1}{2}\phi} & 0 \\ 0 & e^{-\frac{1}{2}\phi} \end{pmatrix}$ | $e^{\phi}\mathbf{x}$ |
| Reflection | $\mathbf{y}$ | $\begin{pmatrix} \mathbf{y} & 0 \\ 0 & -\mathbf{y} \end{pmatrix}$ | $-\mathbf{y}\mathbf{x}\mathbf{y}^{-1}$ |
| Transversion | $e^{\mathbf{c}\mathbf{a}}$ | $\begin{pmatrix} 1 & 0 \\ -\mathbf{c} & 1 \end{pmatrix}$ | $\mathbf{x}(1-\mathbf{c}\mathbf{x})^{-1}$ |

From the table, it is clear that every orthogonal transformation $L(\mathbf{x}_c)$ in $\mathbb{R}^{p+1,q+1}$ acting on the horosphere $\mathscr{H}^{p,q}$ induces a corresponding conformal transformation on $\mathbb{R}^{p,q}$. Rewriting a transversion in the equivalent form

$$\mathbf{x}(1-\mathbf{cx})^{-1} = (\mathbf{x}^{-1}-\mathbf{c})^{-1},$$

it is clear that it is a composition of an inversion, a translation, and an inversion. In terms of the corresponding matrices, we see that

$$\begin{pmatrix} 0 & 1 \\ 1 & 0 \end{pmatrix} \begin{pmatrix} 1 & -\mathbf{c} \\ 0 & 1 \end{pmatrix} \begin{pmatrix} 0 & 1 \\ 1 & 0 \end{pmatrix} = \begin{pmatrix} 1 & 0 \\ -\mathbf{c} & 1 \end{pmatrix},$$

as expected.

## Exercises

1. Let $\mathbf{x}$ be a point in $\mathbb{R}^3$. This point can be represented by the vector $\mathbf{x}_a = \mathbf{x}+\mathbf{a}$ in the affine plane $\mathscr{A}_a^3 \subset \mathbb{R}^{4,1}$ and by the null vector $\mathbf{x}_c \in \mathscr{H}_c^3 \subset \mathbb{R}^{4,1}$. The point $\mathbf{x} \in \mathbb{R}^3$ can also be represented by the bivector $\mathbf{x}_c \wedge \mathbf{b} \in \mathbb{R}^{4,1}$, see Fig. 17.9. Show that $(\mathbf{x}_c \wedge \mathbf{b})^2 = 1$ and that the origin $0 \in \mathbb{R}^3$ is represented by the bivector $\mathbf{a} \wedge \mathbf{b} = -\mathrm{u}$.
2. (a) Let $\mathbf{x}_c \wedge \mathbf{y}_c \wedge \mathbf{b} = (\mathbf{x}_a \wedge \mathbf{y}_a) \wedge \mathbf{b}$ represent the line segment from the point $\mathbf{x}$ to the point $\mathbf{y}$ in $\mathbb{R}^2$. Show that

$$(\mathbf{x}_c \wedge \mathbf{y}_c \wedge \mathbf{b}) \cdot \mathbf{a} = \mathbf{x}_a \wedge \mathbf{y}_a$$

and

$$(\mathbf{x}_c \wedge \mathbf{y}_c \wedge \mathbf{b})^2 = (\mathbf{x}-\mathbf{y})^2.$$

(b) Show that the equation of a circle in $\mathbb{R}^2$ centered at the point $\mathbf{p} \in \mathbb{R}^2$ and with radius $r$ as represented in the horosphere $\mathscr{H}^2$ is

$$\{\mathbf{x}_c \in \mathscr{H}^2 | \quad (\mathbf{x}_c \wedge \mathbf{p}_c \wedge \mathbf{b})^2 = r^2\} \iff \left\{\mathbf{x}_c \in \mathscr{H}^2 | \quad \mathbf{x}_c \cdot \mathbf{p}_c = -\frac{1}{2}r^2\right\}.$$

Defining $\mathbf{s}_r = \mathbf{p}_c + \frac{1}{2}r^2\mathbf{b}$, the circle can also be represented can also be represented as all points $\mathbf{x}_c \in \mathscr{H}^2$ such that $\mathbf{x}_c \cdot \mathbf{s}_r = 0$.
3. Let $\mathbf{x}_c \wedge \mathbf{y}_c \wedge \mathbf{w}_c \wedge \mathbf{b}$ represent the triangular plane segment in $\mathbb{R}^2$ whose vertices are at the points $\mathbf{x}_c, \mathbf{y}_c, \mathbf{w}_c \in \mathscr{H}^2$. Show that

$$(\mathbf{x}_c \wedge \mathbf{y}_c \wedge \mathbf{w}_c \wedge \mathbf{b}) \cdot \mathbf{a} = \mathbf{x}_a \wedge \mathbf{y}_a \wedge \mathbf{w}_a$$

and that

$$(\mathbf{x}_c \wedge \mathbf{y}_c \wedge \mathbf{w}_c \wedge \mathbf{b})^2 = (\mathbf{x}\wedge\mathbf{y} - \mathbf{x}\wedge\mathbf{w} + \mathbf{y}\wedge\mathbf{w})^2 = -|(\mathbf{x}-\mathbf{w})\wedge(\mathbf{y}-\mathbf{w})|^2$$

where $|(\mathbf{x}-\mathbf{w})\wedge(\mathbf{y}-\mathbf{w})|$ is twice the area of the triangle in $\mathbb{R}^2$ with vertices at the points $\mathbf{x},\mathbf{y},\mathbf{w}\in\mathbb{R}^2$.

4. (a) Let $\mathbf{x}_c,\mathbf{y}_c,\mathbf{w}_c\in\mathscr{H}^2$. Show that $(\mathbf{x}_c\wedge\mathbf{y}_c\wedge\mathbf{w}_c)\cdot\mathbf{b}=(\mathbf{x}_c-\mathbf{y}_c)\wedge(\mathbf{y}_c-\mathbf{w}_c)$.
   (b) Let $\mathbf{x}_c$ and $\mathbf{y}_c$ be distinct points in $\mathscr{H}^1$, and let $\mathbf{w}_c\in\mathscr{H}^1$. Show that
   $$\mathbf{x}_c\wedge\mathbf{y}_c\wedge\mathbf{w}_c=0 \quad \text{iff} \quad \mathbf{w}_c=\mathbf{x}_c \text{ or } \mathbf{w}_c=\mathbf{y}_c.$$
   (c) More generally, show that $\mathbf{x}_c,\mathbf{y}_c,\mathbf{w}_c\in\mathscr{H}^3$ and $\mathbf{x}_c\wedge\mathbf{y}_c\neq 0$ then
   $$\mathbf{x}_c\wedge\mathbf{y}_c\wedge\mathbf{w}_c=0 \quad\Longleftrightarrow\quad \mathbf{w}_c=\mathbf{x}_c, \text{ or } \mathbf{w}_c=\mathbf{y}_c.$$
5. Given three distinct points, $\mathbf{x}_c,\mathbf{y}_c,\mathbf{w}_c\in\mathscr{H}^2$, show that there will always exist a fourth point $\mathbf{v}_c\in\mathscr{H}^2$ with the property that
   $$(\mathbf{x}_c\wedge\mathbf{y}_c)\cdot(\mathbf{w}_c\wedge\mathbf{v}_c)=0.$$
6. (a) Show that three distinct points, $\mathbf{x}_c,\mathbf{y}_c,\mathbf{w}_c\in\mathscr{H}^2$, determine the equation of a unique circle if $\mathbf{x}_a\wedge\mathbf{y}_a\wedge\mathbf{w}_a\neq 0$.
   (b) Show that fourth point, $\mathbf{d}_c\in\mathscr{H}^2$, will lie on this circle iff
   $$\mathbf{d}\wedge\mathbf{x}_c\wedge\mathbf{y}_c\wedge\mathbf{w}_c=0.$$
   (Without loss of generality, we may assume that the center of the circle $\mathbf{p}_c=\mathbf{a}$.)
   (c) Show that the radius $r$ of the circle determined by the $\mathbf{x}_c,\mathbf{y}_c,\mathbf{w}_c\in\mathscr{H}^2$ satisfies
   $$r^2=\frac{(\mathbf{x}_c\wedge\mathbf{y}_c\wedge\mathbf{w}_c)(\mathbf{w}_c\wedge\mathbf{y}_c\wedge\mathbf{x}_c)}{(\mathbf{b}\wedge\mathbf{x}_a\wedge\mathbf{y}_a\wedge\mathbf{w}_a)^2}.$$
7. (a) Prove Simpson's theorem 17.2.4 for points $\mathbf{x}_a,\mathbf{y}_a,\mathbf{w}_a$ in the affine plane $\mathscr{A}_a^2$. How does the proof differ from the proof given earlier in the affine plane $\mathscr{A}_h^2$?

# Chapter 18
# Lie Groups and Lie Algebras

*The universe is an enormous direct product of representations of symmetric groups.*

—Steven Weinberg

We have studied linear transformation on $\mathbb{R}^n$ using the traditional matrix formalism in Chap. 7 and more generally in Chaps. 8–10, using the machinery of geometric algebra. This chapter explains the *bivector* interpretation of a general linear operator and offers a new proof of the Cayley–Hamilton theorem based upon this interpretation. The bivector interpretation of a linear transformation leads naturally to the study of Lie algebras and their corresponding Lie groups, including an introductory discussion of the powerful Dynkin diagrams as a means of classifying all complex semisimple Lie algebras.

## 18.1 Bivector Representation

Let $\mathbb{G}_{n,n} = \mathbb{G}(\mathbb{R}^{n,n})$ be the $2^{2n}$ geometric algebra of neutral signature, which was basic to our study of the symmetric group in Chap. 12. The corresponding unitary geometric algebra $\mathbb{U}_{n,n} = \mathbb{G}_{n,n+1}$ was introduced in Chap.10. The bivector representation of a linear operator in $\mathbb{R}^n$ or $\mathbb{C}^n$ takes place in the $n$-dimensional *complimentary* or *dual null cones* $\mathscr{A}^n$ and $\mathscr{B}^n$. These null cones are defined by introducing the complimentary *Witt basis* of null vectors. Let $(\mathbf{e})_{(2n)}$ be the standard orthonormal basis of $\mathbb{G}_{n,n}$. Then the null cones $\mathscr{A}^n$ and $\mathscr{B}^n$ are specified by

$$\mathscr{A}^n = \text{span}(\mathbf{a})_{(n)} = \text{span}\left(\mathbf{a}_1 \ \dots \ \mathbf{a}_n\right) \quad \mathscr{B}^n = \text{span}(\mathbf{b})_{(n)}, \tag{18.1}$$

where $\mathbf{a}_i = \mathbf{e}_i + \mathbf{e}_{n+i}$ and $\mathbf{b}_i = \frac{1}{2}(\mathbf{e}_i + \mathbf{e}_{n+i})$ .

G. Sobczyk, *New Foundations in Mathematics: The Geometric Concept of Number*, DOI 10.1007/978-0-8176-8385-6_18,

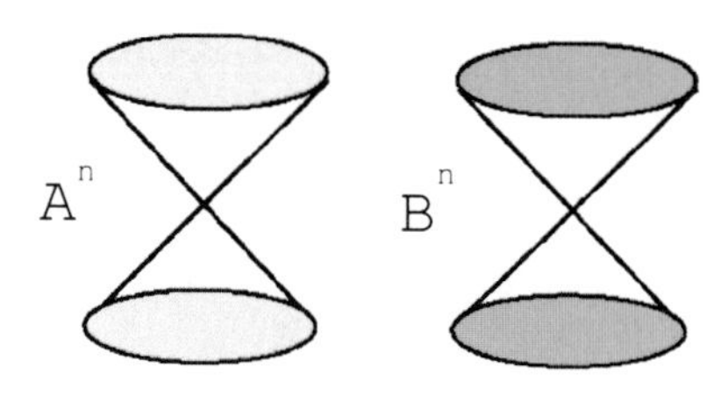


**Fig. 18.1** The null cones $\mathscr{A}^n$ and $\mathscr{B}^n$ make up a Witt basis of $\mathbb{R}^{n,n}$. It would take 6 dimensions to properly display dual 3-dimensional null cones. For clarity, the origin of $\mathscr{B}^n$ has been shifted to the right

The null vectors $(\mathbf{a})_{(n)}$ and $(\mathbf{b})_{(n)}$ satisfy the following basic properties, as is easily verified,

$$(\mathbf{a})^{\mathrm{T}}_{(n)} \cdot (\mathbf{a})_{(n)} = 0 = (\mathbf{b})^{\mathrm{T}}_{(n)} \cdot (\mathbf{b})_{(n)} \tag{18.2}$$

and

$$(\mathbf{a})^{\mathrm{T}}_{(n)} \cdot (\mathbf{b})_{(n)} = [1]_n, \tag{18.3}$$

where $[1]_n$ is the identity $n \times n$ matrix. The $n$-dimensional vector spaces defined by $\mathscr{A}^n$ and $\mathscr{B}^n$ are said to be *null cones* because they satisfy (18.2) and *reciprocal* to each other because they satisfy (18.3). As a consequence of these two relationships, we also have

$$(\mathbf{a})_{(n)} \cdot (\mathbf{b})^{\mathrm{T}}_{(n)} = \sum_{i=1}^{n} \mathbf{a}_i \cdot \mathbf{b}_i = n. \tag{18.4}$$

The null cones $\mathscr{A}^n$ and $\mathscr{B}^n$ have one point in common, the origin, and are pictured in Fig. 18.1.

By the *standard matrix basis* $\Omega^2_{n,n}$ of bivectors, we mean

$$\Omega^2_{n,n} = (\mathbf{a})^{\mathrm{T}}_{(n)} \wedge (\mathbf{b})_{(n)} = \begin{pmatrix} \mathbf{a}_1 \wedge \mathbf{b}_1 & \mathbf{a}_1 \wedge \mathbf{b}_2 & \dots & \mathbf{a}_1 \wedge \mathbf{b}_n \\ \mathbf{a}_2 \wedge \mathbf{b}_1 & \mathbf{a}_2 \wedge \mathbf{b}_2 & \dots & \mathbf{a}_2 \wedge \mathbf{b}_n \\ \dots & \dots & \dots & \dots \\ \dots & \dots & \dots & \dots \\ \mathbf{a}_n \wedge \mathbf{b}_1 & \mathbf{a}_n \wedge \mathbf{b}_2 & \dots & \mathbf{a}_n \wedge \mathbf{b}_n \end{pmatrix}. \tag{18.5}$$

The matrix $\Omega^2_{n,n}$ spans a $n^2$-dimensional linear space of bivectors, defined by the same symbol,

$$\Omega^2_{n,n} = \mathrm{span}\{\mathbf{a} \wedge \mathbf{b} |\ \mathbf{a} \in \mathscr{A}^n, \mathbf{b} \in \mathscr{B}^n\} = \{\mathbf{F} |\ \mathbf{F} = (\mathbf{a})_{(n)}[\mathbf{F}] \wedge (\mathbf{b})^{\mathrm{T}}_{(n)}\}$$

where $[\mathbf{F}]$ is a real (or complex) $n \times n$ matrix, called the *matrix* of the bivector $\mathbf{F}$. The *algebra* $\Omega_{n,n}$ is generated by taking all sums of geometric products of the elements of $\Omega^2_{n,n}$ and is an even subalgebra of the geometric algebra $\mathbb{G}_{n,n}$.

Each bivector $\mathbf{F} \in \Omega^2_{n,n}$ defines two linear operators, $\mathbf{F}: \mathscr{A}^n \to \mathscr{A}^n$ and $\mathbf{F}: \mathscr{B}^n \to \mathscr{B}^n$, specified by

$$\mathbf{F}(\mathbf{x}_a) := \mathbf{F} \cdot \mathbf{x}_a = \Big( (\mathbf{a})_{(n)}[\mathbf{F}] \wedge (\mathbf{b})^{\mathrm{T}}_{(n)} \Big) \cdot \mathbf{x}_a = (\mathbf{a})_{(n)}[\mathbf{F}](x_a)^{\mathrm{T}}_{(n)} \tag{18.6}$$

and

$$\mathbf{x}_b \cdot \mathbf{F} = \mathbf{x}_b \cdot \Big( (\mathbf{a})_{(n)}[\mathbf{F}] \wedge (\mathbf{b})^{\mathrm{T}}_{(n)} \Big) = (x_b)_{(n)}[\mathbf{F}](\mathbf{b})^{\mathrm{T}}_{(n)}.$$

The action of the linear transformation on the null cone $\mathscr{A}^n$ is completely determined by its bivector $\mathbf{F}$. The bivector representation of a linear transformation is the key idea to the study and classification of Lie algebras and Lie groups, as we discover below.

The space of bivectors $\Omega^2_{n,n}$, acting on $\mathscr{A}^n$, is isomorphic to the set of all endomorphisms of $\mathscr{A}^n$ into itself. Let $\mathbf{K} = (\mathbf{a})_{(n)} \wedge (\mathbf{b})^{\mathrm{T}}_{(n)}$, then we find that $[\mathbf{K}] = [1]_n$ is the identity matrix and $\mathbf{F} \cdot \mathbf{K} = \mathrm{tr}([\mathbf{F}])$ is the trace of the matrix $[\mathbf{F}]$ of $\mathbf{F}$. By the *rank* of a bivector $\mathbf{F} \in \Omega^2_{n,n}$, we mean the highest positive integer $k$ such that $\wedge^k \mathbf{F} \neq 0$. A bivector $\mathbf{F}$ is said to be *nonsingular* if $\wedge^n \mathbf{F} = \Big(\det[\mathbf{F}]\Big) \wedge^n \mathbf{K} \neq 0$.

We can provide a new proof of the Cayley–Hamilton theorem for a linear operator $f$ in terms of its bivector invariants $(\wedge^k \mathbf{K}) \cdot (\wedge^k \mathbf{F})$. The proof follows immediately by iterating the identity

$$\begin{aligned}
(\wedge^n \mathbf{F}) \cdot [(\wedge^n \mathbf{K})]\mathbf{x} &= (\wedge^n \mathbf{F}) \cdot [(\wedge^n \mathbf{K}) \cdot \mathbf{x}] = n[(\wedge^n \mathbf{F}) \cdot \mathbf{x}] \cdot (\wedge^{n-1} \mathbf{K}) \\
&= n^2 \big[(\wedge^{n-1} \mathbf{F}) \wedge (\mathbf{F} \cdot \mathbf{x})\big] \cdot (\wedge^{n-1} \mathbf{K}) \\
&= n^2 \Big[(\wedge^{n-1} \mathbf{F}) \cdot (\wedge^{n-1} \mathbf{K}) \mathbf{F} \cdot \mathbf{x} + (\wedge^{n-1} \mathbf{F}) \cdot [(\mathbf{F} \cdot \mathbf{x}) \cdot (\wedge^{n-1} \mathbf{K})]\Big] \\
&= n^2 \Big[(\wedge^{n-1} \mathbf{F}) \cdot (\wedge^{n-1} \mathbf{K}) \mathbf{F} \cdot \mathbf{x} - (n-1)[(\wedge^{n-1} \mathbf{F}) \cdot (\mathbf{F} \cdot \mathbf{x})] \cdot (\wedge^{n-2} \mathbf{K})\Big].
\end{aligned}$$

For example for $n = 3$, after iterating and simplifying, we get the relationship

$$\mathbf{F}^{(3)} : \mathbf{x} - (\mathbf{K} \cdot \mathbf{F})\mathbf{F}^{(2)} : \mathbf{x} + \frac{1}{4}[(\mathbf{K} \wedge \mathbf{K}) \cdot (\mathbf{F} \wedge \mathbf{F})](\mathbf{F} \cdot \mathbf{x}) - \frac{1}{36}[(\wedge^3 \mathbf{K}) \cdot (\wedge^3 \mathbf{F})]\mathbf{x} = 0,$$

where we have introduced the notation

$$f^k(\mathbf{x}) = f(f^{k-1}(\mathbf{x})) = \mathbf{F} \cdot (\mathbf{F}^{(k-1)} : \mathbf{x}) = \mathbf{F}^{(k)} : \mathbf{x}$$

for the repeated composition of the linear operator $f$ in terms of its bivector $\mathbf{F}$. This gives quite a different, but closely related, proof that a linear operator $f$ on $\mathbb{R}^3$ satisfies its characteristic polynomial and should be compared with the derivation (8.5) found in Chap. 8. Sometimes we use the alternative notation

$$\mathbf{F}^{(k)} : \mathbf{x} = \mathbf{F}^{(k)} \boxtimes \mathbf{x}, \tag{18.7}$$

since $\mathbf{F} \cdot \mathbf{x} = \mathbf{F} \boxtimes \mathbf{x}$.

## Exercises

1. (a) Find the bivector $\mathbf{K} \in \Omega^2_{2,2}$ of the identity matrix $[1] = \begin{pmatrix} 1 & 0 \\ 0 & 1 \end{pmatrix}$.
   (b) Let $f : \mathscr{A}^2 \to \mathscr{A}^2$ be defined by

$$f(\mathbf{a})_{(2)} = (\mathbf{a})_{(2)}[f]$$

   where $[f] = \begin{pmatrix} 1 & 3 \\ -1 & -2 \end{pmatrix}$. Find the bivector $\mathbf{F} \in \Omega^2_{2,2}$ of the matrix $[f]$.
   (c) Show that the trace of the matrix $[f]$ is given by $\mathrm{tr}[f] = \mathbf{F} \cdot \mathbf{K}$.
   (d) Prove the Cayley–Hamilton theorem for the matrix $[f]$ using its bivector representation.
2. (a) Find the bivector $\mathbf{K} \in \Omega^2_{3,3}$ of the identity matrix $[1] = \begin{pmatrix} 1 & 0 & 0 \\ 0 & 1 & 0 \\ 0 & 0 & 1 \end{pmatrix}$.
   (b) Let $f : \mathscr{A}^3 \to \mathscr{A}^3$ be defined by

$$f(\mathbf{a})_{(3)} = (\mathbf{a})_{(3)}[f]$$

   where $[f] = \begin{pmatrix} 1 & 2 & -1 \\ -1 & -2 & 2 \\ 1 & 0 & 1 \end{pmatrix}$. Find the bivector $\mathbf{F} \in \Omega^2_{3,3}$ of the matrix $[f]$.
   (c) Show that the trace of the matrix $[f]$ is given by $\mathrm{tr}[f] = \mathbf{F} \cdot \mathbf{K}$.
   (d) Prove the Cayley–Hamilton theorem for the matrix $[f]$ using its bivector representation.
3. Let $\mathbf{F} \in \Omega^2_{n,n}$.
   (a) Show that

$$\mathbf{F} \boxtimes \mathbf{K} = \mathbf{K} \boxtimes \mathbf{F} = 0.$$

   (b) Show that $\mathrm{e}^{\frac{t}{2}\mathbf{K}} \mathbf{F} \mathrm{e}^{-\frac{t}{2}\mathbf{K}} = \mathbf{F}$.
   (c) Show that $\mathrm{e}^{\frac{t}{2}\mathbf{F}} \mathbf{K} \mathrm{e}^{-\frac{t}{2}\mathbf{F}} = \mathbf{K}$.
4. Show that all of the bivectors of $\mathbb{G}^2_{n,n}$ are specified by

$$\mathbb{G}^2_{n,n} = \Omega^2_{n,n} \cup \{\mathbf{a}_i \wedge \mathbf{a}_j\}_{1 \le i < j \le n} \cup \{\mathbf{b}_i \wedge \mathbf{b}_j\}_{1 \le i < j \le n}.$$

5. Let $\mathbf{F} \in \Omega^2_{n,n}$, and let $f_t(\mathbf{x}_a) = \mathrm{e}^{\frac{t}{2}\mathbf{F}} \mathbf{x}_a \mathrm{e}^{-\frac{t}{2}\mathbf{F}}$ where $\mathbf{x}_a \in \mathscr{A}^n$ and $t \in \mathbb{R}$.
   (a) By expanding in a Taylor series around $t = 0$, show that

$$f_t(\mathbf{x}_a) = \mathbf{K} \cdot \mathbf{x}_a + t\mathbf{F} \cdot \mathbf{x}_a + \frac{t^2}{2!}\mathbf{F}^{(2)} : \mathbf{x}_a + \cdots + \frac{t^k}{k!}\mathbf{F}^{(k)} : \mathbf{x}_a + \cdots.$$

(b) Show that $f_s \circ f_t(\mathbf{x}_a) = f_{s+t}(\mathbf{x}_a)$. For this reason, we say that $f_t(\mathbf{x}_a)$ defines the *one parameter group*

$$\mathscr{G} = \{f_t| \quad t \in \mathbb{R}\}$$

with the group operation $f_s \circ f_t = f_{s+t}$.

(c) For $\mathbf{F} = \mathbf{a}_i \wedge \mathbf{b}_j$, $i \neq j$, show that

$$f_t(\mathbf{x}_a) = \mathbf{K} \cdot \mathbf{x}_a + t(\mathbf{a}_i \wedge \mathbf{b}_j) \cdot \mathbf{x}_a = \mathbf{x}_a + t(\mathbf{a}_i \wedge \mathbf{b}_j) \cdot \mathbf{x}_a.$$

(d) For $\mathbf{F} = \mathbf{a}_i \wedge \mathbf{b}_i$, show that

$$f_t(\mathbf{x}_a) = \mathbf{x}_a + (e^t - 1)(\mathbf{a}_i \wedge \mathbf{b}_i) \cdot \mathbf{x}_a.$$

## 18.2 The General Linear Group

The *general linear group* is the *Lie group* defined to be the set of bivectors

$$GL_n = \{\mathbf{F}| \ \mathbf{F} \in \Omega^2_{n,n} \text{ and } \wedge^n \mathbf{F} \neq 0\}, \tag{18.8}$$

together with the group operation $\mathbf{H} = \mathbf{F} \circ \mathbf{G}$, where

$$\mathbf{H} = \frac{1}{2}\mathbf{F} \cdot (\mathbf{G} \cdot \mathbf{x})\, \partial_{\mathbf{x}} = -\frac{1}{2}\partial_{\mathbf{x}}\, (\mathbf{x} \cdot \mathbf{G}) \cdot \mathbf{F}. \tag{18.9}$$

Recall that $\partial_{\mathbf{x}}$ is the vector derivative on $\mathbb{R}^{n,n}$.

http://en.wikipedia.org/wiki/General_linear_group

The vector derivative $\partial_{\mathbf{x}}$ on $\mathbb{G}_{n,n}$ can be broken down into two pieces with respect to the null cones $\mathscr{A}^n$ and $\mathscr{B}^n$. We write

$$\partial_{\mathbf{x}} = \partial_{\mathbf{x}_a} + \partial_{\mathbf{x}_b} \tag{18.10}$$

where

$$\partial_{\mathbf{x}_a} = (\mathbf{b})_{(n)}(\mathbf{a})^{\mathrm{T}}_{(n)} \cdot \partial_{\mathbf{x}}$$

and

$$\partial_{\mathbf{x}_b} = (\mathbf{a})_{(n)}(\mathbf{b})^{\mathrm{T}}_{(n)} \cdot \partial_{\mathbf{x}}.$$

Using the vector derivative on the null cone $\mathscr{A}^n$, the composition formula (18.9) takes the simpler form

$$\mathbf{H} = \Big(\mathbf{F} \cdot (\mathbf{G} \cdot \mathbf{x}_a)\Big) \wedge \partial_{\mathbf{x}_a} = -\partial_{\mathbf{x}_a} \wedge \Big((\mathbf{x}_a \cdot \mathbf{G}) \cdot \mathbf{F}\Big) \tag{18.11}$$

as follows from

$$\mathbf{F} = \frac{1}{2}(\mathbf{F}\cdot\mathbf{x})\wedge\partial_{\mathbf{x}} = \frac{1}{2}\Big((\mathbf{F}\cdot\mathbf{x}_a)\wedge\partial_{\mathbf{x}_a} + (\mathbf{F}\cdot\mathbf{x}_b)\wedge\partial_{\mathbf{x}_b}\Big) = (\mathbf{F}\cdot\mathbf{x}_a)\wedge\partial_{\mathbf{x}_a},$$

since by Problem 3(a) of the previous exercises

$$\mathbf{F}\boxtimes\mathbf{K} = [\mathbf{F}\cdot(\mathbf{a})_{(n)}]\wedge(\mathbf{b})^{\mathrm{T}}_{(n)} + (\mathbf{a})_{(n)}\wedge[\mathbf{F}\cdot(\mathbf{b})^{\mathrm{T}}_{(n)}] = 0,$$

which is equivalent to

$$\mathbf{F} = \partial_{\mathbf{x}_a}\wedge(\mathbf{x}_a\cdot\mathbf{F}) = (\mathbf{F}\cdot\mathbf{x}_a)\wedge\partial_{\mathbf{x}_a} = \partial_{\mathbf{x}_b}\wedge(\mathbf{x}_b\cdot\mathbf{F}).$$

Let $\mathbf{F}\in\Omega^2_{n,n}$ and

$$f_t(\mathbf{x}_a) = \mathrm{e}^{\frac{t}{2}\mathbf{F}}\mathbf{x}_a\mathrm{e}^{-\frac{t}{2}\mathbf{F}} = \mathbf{K}\cdot\mathbf{x}_a + t\mathbf{F}\cdot\mathbf{x}_a + \frac{t^2}{2!}\mathbf{F}^{(2)}:\mathbf{x}_a + \cdots + \frac{t^k}{k!}\mathbf{F}^{(k)}:\mathbf{x}_a + \cdots. \quad (18.12)$$

Clearly, for each $t\in\mathbb{R}$,

$$f_s\circ f_t(\mathbf{x}_a) = f_s(f_t(\mathbf{x}_a)) = f_{s+t}(\mathbf{x}_a),$$

so $f_t$ defines a one parameter group. Now define

$$\mathbf{G}_t = -\partial_{\mathbf{x}_a}\wedge\Big(\exp\Big(\frac{t}{2}\mathbf{F}\Big)\mathbf{x}_a\exp\Big(-\frac{t}{2}\mathbf{F}\Big)\Big) = -\partial_{\mathbf{x}_a}\wedge f_t(\mathbf{x}_a). \quad (18.13)$$

Using (18.12) and expanding $\mathbf{G}_t$ in a Taylor series around $t=0$, we find that

$$\mathbf{G}_t = \mathbf{K} + t\mathbf{F} + \frac{t^2}{2!}\Big(\mathbf{F}\boxtimes(\mathbf{F}\boxtimes\mathbf{x}_a)\Big)\wedge\partial_{\mathbf{x}_a} + \cdots + \frac{t^k}{k!}\Big(\mathbf{F}^{(k)}\boxtimes\mathbf{x}_a\Big)\wedge\partial_{\mathbf{x}_a} + \cdots. \quad (18.14)$$

It follows that

$$\mathbf{G}_t\in GL_n \quad\text{and}\quad \mathbf{G}_0 = \mathbf{K},$$

where $\mathbf{K}\in GL_n$ is the identity element.

From (18.14), we see that the set $\Omega^2_{n,n}$ of bivectors makes up the *tangent space of bivectors* of the Lie group $GL_n$ at the identity element $\mathbf{K}\in GL_n$. We make the following

**Definition 18.2.1.** The general Lie algebra $gl_n = \Omega^2_{n,n}$ is the tangent space of bivectors to $GL_n$ at the identity element $\mathbf{K}\in GL_n$, together with the Lie algebra product. Given $\mathbf{A},\mathbf{B}\in gl_n$,

$$\mathbf{A}\boxtimes\mathbf{B} = \frac{1}{2}(\mathbf{AB}-\mathbf{BA}).$$

The Lie algebra product satisfies the famous *Jacobi identity*

$$\mathbf{A}\boxtimes(\mathbf{B}\boxtimes\mathbf{C}) = (\mathbf{A}\boxtimes\mathbf{B})\boxtimes\mathbf{C} + \mathbf{B}\boxtimes(\mathbf{A}\boxtimes\mathbf{C}), \quad (18.15)$$

which for bivectors is always valid.

We have the following important

**Theorem 18.2.2.** *For each* $\mathbf{F} \in gl_n$,

$$\mathrm{e}^{\frac{t}{2}\mathbf{F}} \mathbf{x}_a \mathrm{e}^{-\frac{t}{2}\mathbf{F}} = (\mathbf{a})_{(n)} \mathrm{e}^{t[\mathbf{F}]} (x_a)^{\mathrm{T}}_{(n)}, \tag{18.16}$$

*where* $\mathbf{x}_a = (\mathbf{a})_{(n)} (x_a)^{\mathrm{T}}_{(n)}$ *and* $[\mathbf{F}]$ *is the matrix of* $\mathbf{F}$.

*Proof.* It is sufficient to show that

$$\mathrm{e}^{\frac{t}{2}\mathbf{F}} (\mathbf{a})_{(n)} \mathrm{e}^{-\frac{t}{2}\mathbf{F}} = (\mathbf{a})_{(n)} \mathrm{e}^{t[\mathbf{F}]} \quad \text{or} \quad \mathrm{e}^{\frac{t}{2}\mathbf{F}} (\mathbf{a})_{(n)} \mathrm{e}^{-\frac{t}{2}\mathbf{F}} \wedge (\mathbf{b})^{\mathrm{T}}_{(n)} = (\mathbf{a})_{(n)} \mathrm{e}^{t[\mathbf{F}]} \wedge (\mathbf{b})^{\mathrm{T}}_{(n)}.$$

The proof follows immediately by noting that both sides are identical when expanded in a Taylor series around $t = 0$. □

The expression

$$\mathbf{G}_t = \mathrm{e}^{\frac{t}{2}\mathbf{F}} (\mathbf{a})_{(n)} \mathrm{e}^{-\frac{t}{2}\mathbf{F}} \wedge (\mathbf{b})^{\mathrm{T}}_{(n)} = (\mathbf{a})_{(n)} \mathrm{e}^{t[\mathbf{F}]} \wedge (\mathbf{b})^{\mathrm{T}}_{(n)} \in GL_n \tag{18.17}$$

shows that the group element $\mathbf{G}_t \in GL_n$ can either be defined by the left or right sides of (18.16). We say that the left side of the equation is a *double covering* of the group element $\mathbf{G}$, since

$$\mathbf{G}_t = \mathrm{e}^{\frac{t}{2}\mathbf{F}} (\mathbf{a})_{(n)} \mathrm{e}^{-\frac{t}{2}\mathbf{F}} \wedge (\mathbf{b})^{\mathrm{T}}_{(n)} = \left( -\mathrm{e}^{\frac{t}{2}\mathbf{F}} \right) (\mathbf{a})_{(n)} \left( -\mathrm{e}^{-\frac{t}{2}\mathbf{F}} \right) \wedge (\mathbf{b})^{\mathrm{T}}_{(n)}.$$

The right side of (18.16) relates the group element $\mathbf{G}_t \in GL_n$, defined by the bivector $\mathbf{F} \in gl_n$, to exponential of the matrix $[\mathbf{F}]$ of $\mathbf{F}$.

The general linear group $GL_n$ has many important subgroups. The *special linear group* $SL_n$ is the subgroup of $GL_n$ defined by

$$SL_n := \{\mathbf{G} \in GL_n | \quad \wedge^n \mathbf{G} = \wedge^n \mathbf{K}\} \quad \iff \quad \{\mathbf{G} \in GL_n | \quad \det \mathbf{G} = 1\}. \tag{18.18}$$

Just as for the general linear group $GL_n$, the special linear group $SL_n$ has a corresponding *special linear Lie algebra* defined to be the tangent space of bivectors to $SL_n$ at the identity element $\mathbf{K}$. For $\mathbf{F} \in gl_n$, consider the one parameter subgroup in $SL_n$ defined by (18.13). Using (18.14) and imposing the condition (18.18), we find that

$$0 = \frac{\mathrm{d}}{\mathrm{d}t} (\wedge^n \mathbf{G}_t)_{t=0} = n\mathbf{F} \wedge^{n-1} \mathbf{K} = \mathrm{tr}(\mathbf{F}) \wedge^n \mathbf{K}, \tag{18.19}$$

so that $\mathrm{tr}(\mathbf{F}) = \mathbf{K} \cdot \mathbf{F} = 0$. Thus, we have found that

$$sl_n = \{\mathbf{F} \in gl_n | \ \mathbf{F} \cdot \mathbf{K} = 0\} \subset gl_n.$$

## Exercises

1. Let $\mathbf{A},\mathbf{B} \in GL_2$ be given by

$$\mathbf{A} = \mathbf{a}_1 \wedge \mathbf{b}_1 + 2\mathbf{a}_1 \wedge \mathbf{b}_2 + 3\mathbf{a}_2 \wedge \mathbf{b}_2, \quad \mathbf{B} = -\mathbf{a}_1 \wedge \mathbf{b}_1 - \mathbf{a}_2 \wedge \mathbf{b}_1 + \mathbf{a}_2 \wedge \mathbf{b}_2.$$

   (a) Find the matrices $[\mathbf{A}]$ and $[\mathbf{B}]$ of $\mathbf{A}$ and $\mathbf{B}$, respectively.
   (b) Find the bivector $\mathbf{C} \in GL_2$ such that $\mathbf{C} = \mathbf{B} \circ \mathbf{A}$, and show that $[\mathbf{C}] = [\mathbf{B}][\mathbf{A}]$.

2. More generally, for $\mathbf{A} = (\mathbf{a})_{(n)} \wedge [\mathbf{A}](\mathbf{b})^{\mathrm{T}}_{(n)}$ and $\mathbf{B} = (\mathbf{a})_{(n)} \wedge [\mathbf{B}](\mathbf{b})^{\mathrm{T}}_{(n)}$ in $GL_n$, show that $[\mathbf{C}] = [\mathbf{B} \circ \mathbf{A}] = [\mathbf{B}][\mathbf{A}]$.
3. (a) Show that

$$\left(\mathrm{e}^{\frac{t}{2}\mathbf{a}_1 \wedge \mathbf{b}_2}\, \mathbf{x}_a\, \mathrm{e}^{-\frac{t}{2}\mathbf{a}_1 \wedge \mathbf{b}_2}\right) \wedge \partial_{\mathbf{x}_a} = \mathbf{K} + t\mathbf{a}_1 \wedge \mathbf{a}_2.$$

   (b) Calculate

$$\left(\mathrm{e}^{\frac{t}{2}\mathbf{a}_1 \wedge \mathbf{b}_1}\, \mathbf{x}_a\, \mathrm{e}^{-\frac{t}{2}\mathbf{a}_1 \wedge \mathbf{b}_1}\right) \wedge \partial_{\mathbf{x}_a}.$$

4. Prove the Jacobi identity (18.15) for bivectors.
5. Find the one parameter group in $SL_2$ specified by

$$\left(\mathrm{e}^{\frac{t}{2}(\mathbf{a}_1 \wedge \mathbf{b}_1 - \mathbf{a}_2 \wedge \mathbf{b}_2)}\, \mathbf{x}_a\, \mathrm{e}^{-\frac{t}{2}(\mathbf{a}_1 \wedge \mathbf{b}_1 - \mathbf{a}_2 \wedge \mathbf{b}_2)}\right) \wedge \partial_{\mathbf{x}_a}.$$

   *Suggestion:* Use Problem 5(d) of Sect. 18.1.
6. (a) Show that

$$\mathrm{e}^{\frac{1}{2}\mathbf{a}_1 \wedge \mathbf{b}_2}\mathbf{a}_1 \wedge \mathbf{b}_1 \mathrm{e}^{-\frac{1}{2}\mathbf{a}_1 \wedge \mathbf{b}_2} = \mathbf{a}_1 \wedge \mathbf{b}_1 - t\mathbf{a}_1 \wedge \mathbf{b}_2.$$

   (b) Show that

$$\mathrm{e}^{\frac{1}{2}\mathbf{a}_1 \wedge \mathbf{b}_2}\mathbf{a}_1 \wedge \mathbf{b}_2 \mathrm{e}^{-\frac{1}{2}\mathbf{a}_1 \wedge \mathbf{b}_2} = \mathbf{a}_1 \wedge \mathbf{b}_2.$$

   (c) Show thats

$$\mathrm{e}^{\frac{1}{2}\mathbf{a}_1 \wedge \mathbf{b}_2}\mathbf{a}_2 \wedge \mathbf{b}_1 \mathrm{e}^{-\frac{1}{2}\mathbf{a}_1 \wedge \mathbf{b}_2} = \mathbf{a}_2 \wedge \mathbf{b}_1 + t(\mathbf{a}_1 \wedge \mathbf{b}_1 - \mathbf{a}_2 \wedge \mathbf{b}_2) - t^2\mathbf{a}_1 \wedge \mathbf{b}_2.$$

7. Let $\mathbf{A},\mathbf{B} \in gl_n$. Define the exponential mapping $\mathrm{e}^{t\mathbf{A}\boxtimes} : gl_n \to gl_n$ by

$$\mathrm{e}^{t\mathbf{A}\boxtimes}(\mathbf{B}) = \mathbf{B} + t\mathbf{A} \boxtimes \mathbf{B} + \cdots + \frac{t}{k!}\mathbf{A}^{(k)} \boxtimes \mathbf{B} + \cdots.$$

   Show that $\mathrm{e}^{t\mathbf{A}^{(k)}\boxtimes}(\mathbf{B})$ is just the Taylor series expansion of $f_t(\mathbf{B}) = \mathrm{e}^{\frac{t}{2}\mathbf{A}}\mathbf{B}\mathrm{e}^{\frac{-t}{2}\mathbf{A}}$ around $t = 0$.

## 18.3 The Algebra $\Omega_{n,n}$

The space of bivectors $\Omega^2_{n,n}$ and the corresponding algebra $\Omega_{n,n}$ have many interesting properties. Consider $\mathbf{W} \in \Omega^2_{n,n}$ to be a bivector variable. Noting that $(\mathbf{a}_i \wedge \mathbf{b}_j) \cdot (\mathbf{a}_j \wedge \mathbf{b}_i) = 1$ for each $\mathbf{a}_i \wedge \mathbf{b}_j \in \Omega^2_{n,n}$, it is natural to define the *bivector derivative* $\partial_{\mathbf{W}}$ by requiring that for all $\mathbf{A} \in \Omega^2_{n,n}$,

$$\mathbf{A} \cdot \partial_{\mathbf{W}}\, \mathbf{W} = \mathbf{A} = \partial_{\mathbf{W}} \mathbf{W} \cdot \mathbf{A}, \tag{18.20}$$

which is simply saying that $\mathbf{A} \cdot \partial_{\mathbf{W}}$ is the $\mathbf{A}$-directional derivative in the direction of the bivector $\mathbf{A}$. We can express (18.20) in terms of its component partial $(\mathbf{a}_i \wedge \mathbf{b}_j)$-derivatives $\partial_{ij} = (\mathbf{a}_i \wedge \mathbf{b}_j) \cdot \partial_{\mathbf{W}}$. We find that

$$\partial_{\mathbf{W}} = (\mathbf{a})_{(n)} [\partial_{\mathbf{W}}]^{\mathrm{T}} \wedge (\mathbf{b})^{\mathrm{T}}_{(n)}, \tag{18.21}$$

where $\mathbf{W} = (\mathbf{a})_{(n)} [\mathbf{W}] \wedge (\mathbf{b})^{\mathrm{T}}_{(n)}$ for $[\mathbf{W}] = [w_{ij}]$ and $[\partial_{\mathbf{W}}] = [\partial_{ij}]$.

Let $\mathbf{F} = (\mathbf{a})_{(n)} [\mathbf{F}] \wedge (\mathbf{b})^{\mathrm{T}}_{(n)}$ and $\mathbf{G} = (\mathbf{a})_{(n)} [\mathbf{G}] \wedge (\mathbf{b})^{\mathrm{T}}_{(n)}$. Since $\mathbf{F}$ and $\mathbf{G}$ are bivectors in $\Omega^2_{n,n}$, we have $\mathbf{F}\mathbf{G} = \mathbf{F} \odot \mathbf{G} + \mathbf{F} \boxtimes \mathbf{G}$ where the symmetric and skew-symmetric parts are given by

$$\mathbf{F} \odot \mathbf{G} = \frac{1}{2}(\mathbf{F}\mathbf{G} + \mathbf{G}\mathbf{F}) = \mathbf{F} \cdot \mathbf{G} + \mathbf{F} \wedge \mathbf{G}, \text{ and } \mathbf{F} \boxtimes \mathbf{G} = \frac{1}{2}(\mathbf{F}\mathbf{G} - \mathbf{G}\mathbf{F}) = <\mathbf{F}\mathbf{G}>_2 .$$

In terms of the matrices $[\mathbf{F}]$, $[\mathbf{G}]$ of the respective components of $\mathbf{F}$ and $\mathbf{G}$, we find that

$$\begin{aligned} \mathbf{F} \cdot \mathbf{G} &= \left[(\mathbf{a})_{(n)} [\mathbf{F}] \wedge (\mathbf{b})^{\mathrm{T}}_{(n)}\right] \cdot \left[(\mathbf{a})_{(n)} [G] \wedge (\mathbf{b})^{\mathrm{T}}_{(n)}\right] \\ &= \left((\mathbf{a})_{(n)} [F][G]\right) \cdot (\mathbf{b})^{\mathrm{T}}_{(n)} = trace([\mathbf{F}][\mathbf{G}]) = \mathbf{G} \cdot \mathbf{F}, \end{aligned} \tag{18.22}$$

$$\mathbf{F} \wedge \mathbf{G} = \sum_{i<j} \left[ \det \begin{pmatrix} f_{ii} & f_{ij} \\ g_{ji} & g_{jj} \end{pmatrix} + \det \begin{pmatrix} g_{ii} & g_{ij} \\ f_{ji} & f_{jj} \end{pmatrix} \right] \mathbf{a}_i \wedge \mathbf{b}_i \wedge \mathbf{a}_j \wedge \mathbf{b}_j, \tag{18.23}$$

and

$$\begin{aligned} \mathbf{F} \boxtimes \mathbf{G} &= \left[(\mathbf{a})_{(n)} [\mathbf{F}] \wedge (\mathbf{b})^{\mathrm{T}}_{(n)}\right] \boxtimes \left[(\mathbf{a})_{(n)} [\mathbf{G}] \wedge (\mathbf{b})^{\mathrm{T}}_{(n)}\right] \\ &= (\mathbf{a})_{(n)} \left([\mathbf{F}][\mathbf{G}] - [\mathbf{G}][\mathbf{F}]\right) \wedge (\mathbf{b})^{\mathrm{T}}_{(n)}. \end{aligned} \tag{18.24}$$

The last relationship (18.24) relates the Lie algebra $gl_n(\mathbb{R})$ to the usual *Lie bracket* of the matrices $[\mathbf{F}]$ and $[\mathbf{G}]$ of the bivectors $\mathbf{F}, \mathbf{G} \in \Omega^2_{n,n}$. If we consider $\mathbf{F}$ and $\mathbf{G}$ to be complex bivectors in $\mathbb{G}_{n,n+1}$, we have the corresponding complex Lie algebra $gl_n(\mathbb{C})$.

As a simple application of the identities (18.22)–(18.24), we use them to calculate the bivector derivative (18.21) of the bivector variable $\mathbf{W}$. We find that $\partial_{\mathbf{W}}\mathbf{W} = \partial_{\mathbf{W}} \odot \mathbf{W} + \partial_{\mathbf{W}} \boxtimes \mathbf{W}$, where

$$\partial_{\mathbf{W}} \bigodot \mathbf{W} = \partial_{\mathbf{W}} \cdot \mathbf{W} + \partial_{\mathbf{W}} \wedge \mathbf{W} = n^2 - 2\sum_{i<j} \mathbf{a}_i \wedge \mathbf{b}_i \wedge \mathbf{a}_j \wedge \mathbf{b}_j = n^2 - \mathbf{K} \wedge \mathbf{K} \quad (18.25)$$

and

$$\partial_{\mathbf{W}} \boxtimes \mathbf{W} = 0. \quad (18.26)$$

We give here a number of other basic bivector derivative identities which will be needed:

$$\partial_{\mathbf{W}}\ \mathbf{W} \cdot \mathbf{F} = \mathbf{F}, \ \text{ and } \ \partial_{\mathbf{W}} \cdot (\mathbf{W} \wedge \mathbf{F}) = (n-1)^2 \mathbf{F} \quad (18.27)$$

and

$$\partial_{\mathbf{W}} \wedge \mathbf{W} \wedge \mathbf{F} = -\mathbf{K} \wedge \mathbf{K} \wedge \mathbf{F}. \quad (18.28)$$

For the skew-symmetric part $\partial_{\mathbf{W}}\ \mathbf{W} \boxtimes \mathbf{F}$, we find that

$$\partial_{\mathbf{W}}\ \mathbf{W} \boxtimes \mathbf{F} = 2(n-1)\mathbf{F} - (\mathbf{K} \wedge \mathbf{K}) \cdot \mathbf{F} = 2n\mathbf{F} - 2(\mathbf{F} \cdot \mathbf{K})\mathbf{K}. \quad (18.29)$$

The last equality in (18.29) implies the interesting identity

$$(\mathbf{K} \wedge \mathbf{K}) \cdot \mathbf{F} = 2(\mathbf{F} \cdot \mathbf{K})\mathbf{K} - 2\mathbf{F}. \quad (18.30)$$

Another relationship which we will occasionally use is

$$\mathbf{K} \wedge \mathbf{K} \wedge (\mathbf{X} \boxtimes \mathbf{Y}) = 0. \quad (18.31)$$

## Exercises

1. Prove the identities (18.25)–(18.31).
2. Show that for a bivector $\mathbf{A} \in \Omega^2_{n,n}$,

$$(\mathbf{A} \boxtimes \partial_{\mathbf{W}}) \boxtimes \mathbf{W} = \partial_{\mathbf{W}} \boxtimes (\mathbf{W} \boxtimes \mathbf{A}).$$

3. For $\mathbf{A} \in \Omega^2_{n,n}$, show that

$$\partial_{\mathbf{W}} \mathbf{A} \mathbf{W} = (n^2 - 4n + 2)\mathbf{A} + 2(\mathbf{A} \cdot \mathbf{K})\mathbf{K} - \mathbf{K} \wedge \mathbf{K} \wedge \mathbf{A}.$$

4. For $\mathbf{A} \in \Omega^2_{n,n}$, calculate

$$\partial_{\mathbf{W}} \boxtimes \left( e^{\frac{t}{2}\mathbf{A}}\ \mathbf{W}\ e^{-\frac{t}{2}\mathbf{A}} \right).$$

5. Let $\mathbf{A} = \mathbf{a}_1 \wedge \mathbf{b}_1 + 2\mathbf{a}_2 \wedge \mathbf{b}_1 - \mathbf{a}_2 \wedge \mathbf{b}_2$ and $\mathbf{B} = 2\mathbf{a}_1 \wedge \mathbf{b}_1 - \mathbf{a}_1 \wedge \mathbf{b}_2 + 3\mathbf{a}_2 \wedge \mathbf{b}_2$.
   (a) Find the matrices $[\mathbf{A}]$ and $[\mathbf{B}]$ of $\mathbf{A}$ and $\mathbf{B}$.
   (b) Verify that $\mathbf{A} \cdot \mathbf{B} = \mathrm{tr}([\mathbf{A}][\mathbf{B}])$.
   (c) Verify that $\mathbf{A} \boxtimes \mathbf{B} = \mathbf{a}_{(2)}([\mathbf{A}][\mathbf{B}] - [\mathbf{B}][\mathbf{A}]) \wedge (\mathbf{b})^{\mathrm{T}}_{(2)}$.
   (d) Calculate $\mathbf{A} \wedge \mathbf{B}$ using (18.23).
6. Let $\mathbf{A} \in \Omega^2_{n,n}$ and $[\mathbf{A}]$ be the matrix of $\mathbf{A}$. Show that

$$(\mathbf{a})_{(n)}[\mathbf{A}]^k = \mathbf{A}^{(k)} \boxtimes (\mathbf{a})_{(n)}.$$

7. Let $\mathbf{A}, \mathbf{B} \in gl_n = \Omega^2_{n,n}$. By the *Killing form* $\kappa$ on $gl_n$, we mean

$$\kappa(\mathbf{A}, \mathbf{B}) := \partial_{\mathbf{W}} \cdot \Big( (\mathbf{W} \boxtimes \mathbf{A}) \boxtimes \mathbf{B} \Big).$$

   (a) Show that $\kappa(\mathbf{A}, \mathbf{B}) = 2n\mathbf{A} \cdot \mathbf{B} - 2(\mathbf{K} \cdot \mathbf{A})(\mathbf{K} \cdot \mathbf{B}) = \kappa(\mathbf{B}, \mathbf{A})$.
   (b) Show that the Killing form reduces to $\kappa(\mathbf{A}, \mathbf{B}) = 2n\mathbf{A} \cdot \mathbf{B}$ on the Lie algebra $sl_n$.

## 18.4 Orthogonal Lie Groups and Their Lie Algebras

Let $n = p + q$ and $(\mathbf{e})_{(p,q)} = (\mathbf{e}_1, \ldots, \mathbf{e}_p, \mathbf{e}_{n+1}, \ldots, \mathbf{e}_{n+q})$ be the standard orthonormal basis of $\mathbb{R}^{p,q} \subset \mathbb{R}^{n,n}$. Define

$$I_{p,q} = \wedge(\mathbf{e})_{(p,q)}, \qquad I^{-1}_{p,q} = (-1)^q I^{\dagger}_{p,q}, \tag{18.32}$$

and the corresponding projections $P_{p,q} : \mathbb{G}_{n,n} \to \mathbb{G}_{p,q}$ by

$$P_{p,q}(\mathbf{A}_r) = \mathbf{A}_r \cdot I_{p,q} I^{-1}_{p,q} \tag{18.33}$$

for each $\mathbf{A}_r \in \mathbb{G}^r_{p,q}$. Acting on the null cone $\mathscr{A}^n$, the projection $P_{p,q} : \mathscr{A}^n \to \mathbb{R}^{p,q}$. In addition, $P_{p,q} : \Omega^2_{n,n} \to \mathbb{G}^2_{p,q}$. More precisely,

$$P_{p,q}(\Omega^2_{n,n}) = \frac{1}{2}(\mathbf{e})^{\mathrm{T}}_{(p,q)} \wedge (\mathbf{e})_{(p,q)}[g] = \begin{pmatrix} 0 & \mathbf{e}_{12} & \ldots & \mathbf{e}_{1\,n+q} \\ -\mathbf{e}_{12} & 0 & \ldots & \mathbf{e}_{2\,n+q} \\ \ldots & \ldots & \ldots & \ldots \\ \ldots & \ldots & \ldots & \ldots \\ \ldots & \ldots & \ldots & \ldots \\ -\mathbf{e}_{1\,n+q} & -\mathbf{e}_{2\,n+q} & \ldots & 0 \end{pmatrix} [g],$$

where $[g] = (\mathbf{e})^{\mathrm{T}}_{(p,q)} \cdot (\mathbf{e})_{(p,q)}$.

Let $\mathbf{F} = (\mathbf{a})_{(n)} \wedge [\mathbf{F}](\mathbf{b})^{\mathrm{T}}_{(n)}$, where $[\mathbf{F}] = \mathbf{F} \cdot \left((\mathbf{a})^{\mathrm{T}}_{(n)} \wedge (\mathbf{b})_{(n)}\right)$. Projecting $\mathbf{F}$ onto $\mathbb{G}^2_{p,q}$, we find that

$$\mathbf{F}_{p,q} := P_{p,q}(\mathbf{F}) = \frac{1}{2}(\mathbf{e})_{(p,q)} \wedge [\mathbf{F}][g](\mathbf{e})^{\mathrm{T}}_{(p,q)}. \tag{18.34}$$

The matrix $[\mathbf{F}_{p,q}]$ with respect to the basis $(\mathbf{e})_{(p,q)}$ is defined by

$$\mathbf{F}_{p,q} \cdot (\mathbf{e})_{(p,q)} = (\mathbf{e})_{(p,q)}[\mathbf{F}_{p,q}] \iff [\mathbf{F}_{p,q}] = -[g]\left((\mathbf{e})^{\mathrm{T}}_{(p,q)} \wedge (\mathbf{e})_{(p,q)}\right) \cdot \mathbf{F}_{p,q}. \tag{18.35}$$

Dotting both sides of (18.34) on the right by $\mathbf{x} = (\mathbf{e})_{(p,q)}(x)^{(p,q)}$ gives

$$\mathbf{F}_{p,q} \cdot \mathbf{x} = \frac{1}{2}(\mathbf{e})_{(p,q)}\left([\mathbf{F}] - [g][\mathbf{F}]^{\mathrm{T}}[g]\right)(x)^{(p,q)}.$$

If the matrix $[\mathbf{F}]$ satisfies the property $[\mathbf{F}]^{\mathrm{T}} = -[g][\mathbf{F}][g]$, then this last equation simplifies to

$$\mathbf{F}_{p,q} \cdot \mathbf{x} = (\mathbf{e})_{(p,q)}[\mathbf{F}](x)^{(p,q)} \iff \mathbf{F}_{p,q} \cdot (\mathbf{e})_{(p,q)} = (\mathbf{e})_{(p,q)}[\mathbf{F}]. \tag{18.36}$$

Comparing (18.35) with (18.36), we see that

$$(\mathbf{e})_{(p,q)}[\mathbf{F}] = (\mathbf{e})_{(p,q)}[\mathbf{F}_{p,q}] \iff [\mathbf{F}] = [\mathbf{F}_{p,q}]$$

in the case when $[\mathbf{F}]^{\mathrm{T}} = -[g][\mathbf{F}][g]$.

We are now ready to make the following

**Definition 18.4.1.** The Lie algebra $so_{p,q}$ is the Lie subalgebra of bivectors of $gl_n$, specified by

$$so_{p,q} := \{\mathbf{A} \in gl_n | \ [\mathbf{A}] = -[g][\mathbf{A}]^{\mathrm{T}}[g]\} \tag{18.37}$$

where $\mathbf{A} = (\mathbf{e})_{(p,q)} \wedge [\mathbf{A}](\mathbf{e})^{\mathrm{T}}_{(p,q)}$.

The Lie algebra $so_{p,q}$ can be projected onto the corresponding *spin Lie algebra*,

$$\mathrm{spin}_{p,q} := \{\mathbf{A}_{p,q} = P_{p,q}(\mathbf{A}) | \ \mathbf{A} \in gl_n\} = \{\mathbf{A}_{p,q} \in \mathbb{G}^2_{p,q}\}. \tag{18.38}$$

Comparing (18.37) and (18.38), we see that

$$\mathrm{spin}_{p,q} = P_{p,q}\left(so_{p,q}\right) = P_{p,q}\left(gl_n\right) = P_{p,q}\left(sl_n\right). \tag{18.39}$$

We can also view the corresponding Lie group $SO_{p,q}$ as a subgroup of $GL_n$. We first define the *orthogonal Lie group* $O_{p,q}$.

**Definition 18.4.2.** The Lie group $O_{p,q}$ is the subgroup of $GL_n$ specified by

$$O_{p,q} := \{\mathbf{F} \in GL_n |\ P_{p,q}\Big(\mathbf{F}\cdot\mathbf{x}_a\Big)\cdot P_{p,q}\Big(\mathbf{F}\cdot\mathbf{y}_a\Big) = P_{p,q}(\mathbf{x}_a)\cdot P_{p,q}(\mathbf{y}_a)\}$$

for all $\mathbf{x}_a, \mathbf{y}_a \in \mathscr{A}^n$.

Using (18.6), this is equivalent to saying that

$$\Big((\mathbf{e})_{(p,q)}[\mathbf{F}](x_a)^{\mathrm{T}}_{(n)}\Big)\cdot\Big((\mathbf{e})_{(p,q)}[\mathbf{F}](y_a)^{\mathrm{T}}_{(n)}\Big) = \Big((\mathbf{e})_{(p,q)}(x_a)^{\mathrm{T}}_{(n)}\Big)\cdot\Big((\mathbf{e})_{(p,q)}(y_a)^{\mathrm{T}}_{(n)}\Big)$$

for all $\mathbf{x}_a, \mathbf{y}_a \in \mathscr{A}^n$ or, even more simply,

$$[\mathbf{F}]^{\mathrm{T}}[g][\mathbf{F}] = [g] \quad \Longleftrightarrow \quad [g][\mathbf{F}]^{\mathrm{T}}[g] = [\mathbf{F}]^{-1} \tag{18.40}$$

where the metric tensor $[g] = (\mathbf{e})^{\mathrm{T}}_{(p,q)}\cdot(\mathbf{e})_{(p,q)}$. In establishing (18.40), we are using the fact that (18.6) is equivalent to the statement that

$$\mathbf{F}\cdot(\mathbf{a})_{(n)} = (\mathbf{a})_{(n)}[\mathbf{F}].$$

The Lie group $SO_{p,q}$ of the Lie algebra $so_{p,q}$ is the subgroup of $O_{p,q}$ defined by

$$SO_{p,q} := \{\mathbf{G} \in O_{p,q}| \quad \det[\mathbf{G}] = 1. \tag{18.41}$$

There is one other important class of classical Lie groups that needs to be mentioned, and that is the *sympletic Lie groups* $\mathrm{Sp}_{2n} \subset \mathrm{GL}_{2n}$. First, we define

$$\mathbf{J} = (\mathbf{a})_{(n)} \wedge (\mathbf{b})^{\mathrm{T}}_{(n')} - (\mathbf{a})_{(n')} \wedge (\mathbf{b})^{\mathrm{T}}_{(n)} \in \mathbb{G}^2_{2n,2n}$$

where $(\mathbf{a})_{(n)} = (\mathbf{a}_1,\ldots,\mathbf{a}_n)$ and $(\mathbf{b})_{(n')} = (\mathbf{b}_{n+1},\ldots,\mathbf{b}_{2n})$. Next, we note that $P_{2n}(\mathscr{A}^{2n})$ $= \mathbb{R}^{2n}$ and define

$$\mathbf{J}_{2n} := P_{2n}(\mathbf{J}) = (\mathbf{e})_{(n)} \wedge (\mathbf{e})^{\mathrm{T}}_{(n')},$$

where

$$(\mathbf{e})_{(n)} = (\mathbf{e}_1,\ldots,\mathbf{e}_n), \text{ and } (\mathbf{e})_{(n')} = (\mathbf{e}_{n+1},\ldots,\mathbf{e}_{2n}).$$

We can now give

**Definition 18.4.3.** The sympletic group, $\mathrm{Sp}_{2n} \subset \mathrm{GL}_{2n}$, is the subgroup of $\mathrm{GL}_{2n}$ specified by

$$\mathrm{Sp}_{2n} := \{\mathbf{F} \in \mathrm{GL}_{2n}|\ \mathbf{J}_{2n}\cdot(\mathbf{x}_a \wedge \mathbf{y}_a) = \mathbf{J}_{2n}\cdot\Big((\mathbf{F}\cdot\mathbf{x}_a)\wedge(\mathbf{F}\cdot\mathbf{y}_a)\Big) \text{ for all } \mathbf{x}_a,\mathbf{y}_a \in \mathscr{A}^{2n}\}. \tag{18.42}$$

The condition that

$$\mathbf{J}_{2n} \cdot (\mathbf{x}_a \wedge \mathbf{y}_a) = \mathbf{J}_{2n} \cdot \Big( (\mathbf{F} \cdot \mathbf{x}_a) \wedge (\mathbf{F} \cdot \mathbf{y}_a) \Big)$$

for all $\mathbf{x}_a, \mathbf{y}_a \in \mathscr{A}^{2n}$ means that the *skew-symmetric bilinear form* $\mathbf{J}_{2n} \cdot (\mathbf{x}_a \wedge \mathbf{y}_b)$ is preserved under the action of any bivector $\mathbf{F} \in \mathrm{Sp}_{2n}$. Recalling that the matrix $[\mathbf{F}]$ of the bivector $\mathbf{F} \in \mathrm{GL}_{2n}$ is defined by

$$\mathbf{F} \cdot (\mathbf{a})_{(2n)} = (\mathbf{a})_{(2n)} [\mathbf{F}],$$

the condition (18.42) takes the alternative form

$$\mathbf{J}_{2n} \cdot \Big( (\mathbf{a})^{\mathrm{T}}_{(2n)} \wedge (\mathbf{a})_{(2n)} \Big) = \mathbf{J}_{2n} \cdot \Big( [\mathbf{F}]^{\mathrm{T}} (\mathbf{a})^{\mathrm{T}}_{(2n)} \wedge (\mathbf{a})_{(2n)} [\mathbf{F}] \Big). \tag{18.43}$$

But the matrix

$$[\mathbf{J}_{2n}] = \Big( (\mathbf{e})_{(2n)} \wedge (\mathbf{e})^{\mathrm{T}}_{(2n)} \Big) \cdot \Big( (\mathbf{a})^{\mathrm{T}}_{(2n)} \wedge (\mathbf{a})_{(2n)} \Big) = \begin{pmatrix} 0 & I_n \\ -I_n & 0 \end{pmatrix}$$

where $I_n = [1]_n$ is the identity $n \times n$ matrix. Using this relationship on the right side of (18.43) gives the classical result that $\mathbf{F} \in \mathrm{Sp}_{2n}$ iff

$$[\mathbf{J}_{2n}] = [\mathbf{F}]^{\mathrm{T}} [\mathbf{J}_{2n}] [\mathbf{F}]. \tag{18.44}$$

We now find the corresponding Lie algebra $sp_{2n}$ of dimension $n(2n+1)$ by using the property (18.17). Suppose that $\mathbf{A} \in sp_{2n}$ and that $\mathbf{A} \cdot (\mathbf{a})_{(2n)} = (\mathbf{a})_{(2n)} [\mathbf{A}]$ and that $\mathbf{F}_t \in \mathrm{Sp}_{2n}$ is a one parameter group at the identity element $\mathbf{K}_{2n} \in \mathrm{GL}_{2n}$. Then using (18.17),

$$\mathbf{F}_t = (\mathbf{a})_{(2n)} \mathrm{e}^{t[\mathbf{A}]} \wedge (\mathbf{b})^{\mathrm{T}}_{(2n)} \in \mathrm{GL}_{2n} \quad \Longleftrightarrow \quad [\mathbf{F}_t]^{\mathrm{T}} [\mathbf{J}_{2n}] [\mathbf{F}_t] = [\mathbf{J}_{2n}].$$

Differentiating this last relationship, and evaluating at $t = 0$, gives the result

$$\mathbf{A} \in sp_{2n} \quad \Longleftrightarrow \quad [\mathbf{A}]^{\mathrm{T}} [\mathbf{J}_{2n}] + [\mathbf{J}_{2n}] [\mathbf{A}]^{\mathrm{T}} = 0.$$

A different treatment of the classical groups in geometric algebra can be found in [19]. See also [55, p.220] for a discussion of the Lipschitz, Pin, and Spin groups. Also, it should be pointed out, there are Lie groups which are isomorphic with no subgroups of the general linear group $\mathrm{GL}_n$, [8].

## Exercises

The *generators* of the Lie algebra $gl_n$ are the bivectors of $\Omega^2_{n,n}$.

1. (a) Use (18.39) to find the bivector generator of the Lie algebra

$$\text{spin}_2 = P_2\Big(\Omega_{2,2}\Big).$$

   (b) What is the matrix of the generator of $\text{spin}_2$?
   (c) Use the matrix found in (b) to write down the bivector generator of $so_2$.
2. (a) Use (18.39) to find the generators of the Lie algebra

$$\text{spin}_{1,1} = P_{1,1}\Big(\Omega_{2,2}\Big).$$

   (b) What is the matrix of the generator of $\text{spin}_{1,1}$?
   (c) Use the matrix found in b) to write down the bivector generator of $so_{1,1}$.
3. (a) Use (18.39) to find the bivector generators of the Lie algebra

$$\text{spin}_3 = P_3\Big(\Omega_{3,3}\Big).$$

   (b) Find the matrices of the three generators of $\text{spin}_3$.
   (c) Use the matrices found in b) to write down the three bivector generators of $so_3$.
   (d) Show that $\text{spin}_3 \equiv so_3$ by showing that their Lie algebras are isomorphic. (Their generators have the same Lie algebra multiplication table.)
4. (a) Use (18.39) to find the generators of the Lie algebra

$$\text{spin}_{2,1} = P_{2,1}\Big(\Omega_{3,3}\Big).$$

   (b) Find the matrices of the three generators of $\text{spin}_{2,1}$.
   (c) Use the matrices found in (b) to write down the three generators of $so_{2,1}$.
   (d) Show that $\text{spin}_{2,1} \equiv so_{2,1}$ by showing that their Lie algebras are isomorphic. (Their generators have the same Lie algebra multiplication table.)
5. (a) Recalling the definition of the unitary space $\mathscr{H}^{p,q}$ and its geometric algebra $\mathscr{U}_{p,q}$, given in Sect. 10.3 of Chap. 10, we can define the projection operator (18.33)

$$P_{p,q+1} : \mathscr{A}^n(\mathbb{C}) \to \mathscr{H}^{p,q},$$

   where $\mathscr{A}^n(\mathbb{C})$ is the complexified null cone and $i = e_{n+q+1}$. The *unitary Lie algebra* $u_{p,q}$ is defined by

$$u_{p,q} := \{A \in gl_n(\mathbb{C}) |\ \ [\mathbf{A}] = -[g][A]^*[g]\}$$

where $[\mathbf{A}]^* = [\overline{\mathbf{A}}]^{\mathrm{T}}$. Alternatively, $u_{p,q}$ can be defined by

$$u'_{p,q} := \{\mathbf{A}_{p,q} = P_{p,q+1}(\mathbf{A}) | \ \mathbf{A} \in gl_n(\mathbb{C})\} = \{\mathbf{A}_{p,q} \in \mathscr{U}^2_{p,q}\}.$$

By a similar analysis given to establish (18.39), show that

$$su'_{p,q} = P_{p,q+1}\Big(su_{p,q}\Big) = P_{p,q+1}\Big(gl_n(\mathbb{C})\Big) = P_{p,q+1}\Big(sl_n(\mathbb{C})\Big).$$

(b) The unitary Lie group $U_{p,q}$, analogous to Definition 18.4.2, can be defined by

$$U_{p,q} := \{\mathbf{F} \in GL_n(\mathbb{C})^2 | \ P_{p,q+1}\Big(\mathbf{F}\cdot\mathbf{x}_a\Big)\cdot P_{p,q+1}\Big(\mathbf{F}\cdot\mathbf{y}_a\Big) = P_{p,q+1}(\mathbf{x}_a)\cdot P_{p,q+1}(\mathbf{y}_a) \tag{18.45}$$

for all $\mathbf{x}_a, \mathbf{y}_a \in \mathscr{A}^n(\mathbf{C})$. Show that $\mathbf{F} \in GL_n(\mathbb{C})^2$ iff

$$[\mathbf{F}]^{-1} = [g][\mathbf{F}]^*[g] = [g][\overline{\mathbf{F}}]^{\mathrm{T}}[g].$$

The *special unitary group* is the subgroup of $U_{p,q}$ defined by

$$SU_{p,q} := \{\mathbf{G} \in U_{p,q} | \quad \det[\mathbf{G}] = 1\}. \tag{18.46}$$

6. (a) Define the *Pauli bivectors* by

$$\sigma_1 = (\mathbf{a})_{(2)} \wedge \begin{pmatrix} 0 & 1 \\ 1 & 0 \end{pmatrix} (\mathbf{b})^{\mathrm{T}}_{(2)}, \sigma_2 = (\mathbf{a})_{(2)} \wedge \begin{pmatrix} 0 & -\mathrm{i} \\ \mathrm{i} & 0 \end{pmatrix} (\mathbf{b})^{\mathrm{T}}_{(2)}$$

and

$$\sigma_3 = (\mathbf{a})_{(2)} \wedge \begin{pmatrix} 1 & 0 \\ 0 & -1 \end{pmatrix} (\mathbf{b})^{\mathrm{T}}_{(2)},$$

where $\mathrm{i} = \mathbf{e}_5$. Show that the group $SU_2$ consists of all complex bivectors

$$\mathbf{F} = (\mathbf{a})_{(2)} \wedge [\mathbf{F}](\mathbf{b})^{\mathrm{T}}_{(2)}$$

where the matrix $[\mathbf{F}]$ of $\mathbf{F}$ has determinant one and is of the form

$$[\mathbf{F}] = a[1]_3 + \mathrm{i}b[\sigma_1] + \mathrm{i}c[\sigma_2] + \mathrm{i}d[\sigma_3],$$

$a, b, c, d \in \mathbb{R}$ and $[1]_2$ is the identity matrix.

(b) Define the rotation bivectors $\mathbf{E}_i := (\mathbf{a})_{(3)} \wedge [\mathbf{E}_i](\mathbf{b})_{(3)}$ of $SO_3$ by

$$[\mathbf{E}_1] = \begin{pmatrix} 1 & 0 & 0 \\ 0 & 0 & 1 \\ 0 & -1 & 0 \end{pmatrix}, [\mathbf{E}_2] = \begin{pmatrix} 0 & 0 & -1 \\ 0 & 1 & 0 \\ 1 & 0 & 0 \end{pmatrix}, [\mathbf{E}_3] = \begin{pmatrix} 0 & -1 & 0 \\ 1 & 0 & 0 \\ 0 & 0 & 1 \end{pmatrix}.$$

Show that there is a group homomorphism between the group bivectors of $SO_3$ and the Pauli bivectors of $SU_2$.

(c) Show that the Lie algebras $so_3$ and $su_2$ are algebraically isomorphic.

7. Calculate the Killing form $\kappa(\mathbf{A},\mathbf{B})$ on the Lie algebras $so_n$, $su_n$, and $sp_n$.

## 18.5 Semisimple Lie Algebras

Whereas not all Lie groups are isomorphic to a matrix group [8], it is well known that all Lie algebras can be faithfully represented by matrix Lie algebras on $\mathbb{R}^n$. Since we have seen that each linear transformation on $\mathbb{R}^n$ can be represented by a corresponding bivector, it follows that every Lie algebra can be represented by a corresponding Lie algebra of bivectors. Lie algebras arise in all kinds of different settings, so the problem of their classification is an important one. While there is no general way to classify all Lie algebras, the so-called classical Lie algebras of orthogonal transformations, or simple Lie algebras, lend themselves to a neat classification by the Dynkin diagrams of their *root systems*. Whereas it is impossible to deal with this beautiful subject in any detail, it is the purpose of the remainder of this chapter to give the reader a general idea about how the structure of $gl_n(\mathbb{C})$ and its *semisimple Lie subalgebras* can be elegantly and concisely formulated in the language of geometric algebra.

Let $h$ be a Lie subalgebra of a Lie algebra $g$. We say that $h$ is an *ideal* of $g$ if

$$\mathbf{B}\boxtimes\mathbf{A}\in h \quad \text{for all} \quad \mathbf{B}\in g, \mathbf{A}\in h. \tag{18.47}$$

An ideal $h$ is said to be *nontrivial* if $h\neq\{0\}$. An ideal $h$ of a Lie algebra $g$ is said to be an *Abelian ideal* if $\mathbf{A}\boxtimes\mathbf{B}=0$ for all $\mathbf{A},\mathbf{B}\in h$. A Lie algebra $g\neq 0$ is said to be *simple* if it has no nontrivial ideals, and $g$ is said to be *semisimple* if it has no nontrivial Abelian ideals. It follows that a simple Lie algebra is also semisimple.

There are several more definitions that must be given if we are going to be able gain entry into the vast edifice of Lie algebras and Lie groups that has been constructed over the last century. The *commutator subalgebra* $\mathscr{D}g$ of a Lie algebra $g$ is defined by

$$\mathscr{D}g=\{\mathbf{A}\boxtimes\mathbf{B}| \quad \text{for all} \quad \mathbf{A},\mathbf{B}\in g.\} \tag{18.48}$$

The commutator subalgebra $\mathscr{D}g\subset g$ is an ideal Lie subalgebra in $g$. A Lie algebra $g$ is said to be *nilpotent* if there is an integer $k$ such that

$$\mathbf{A}_1\boxtimes(\mathbf{A}_2\boxtimes(\ldots\boxtimes(\mathbf{A}_k\boxtimes\mathbf{B})\ldots))=0. \tag{18.49}$$

Furthermore, the Lie algebra $g$ is said to be *solvable* if its commutatior subalgebra $\mathscr{D}g$ is a nilpotent Lie algebra. It follows that a solvable Lie algebra is only slightly more general than a nilpotent Lie algebra.

http://en.wikipedia.org/wiki/Solvable_Lie_algebra

We give some standard examples. Let $v_n$ be the Lie subalgebra of $gl_n$ defined by

$$v_n = \{\mathbf{A} |\ \mathbf{A} \in gl_n \ \text{and} \ [\mathbf{A}] \text{ is strictly upper triangular}\},$$

then $v_n$ is a nilpotent Lie algebra. Now define the Lie subalgebra $\tau_n$ of $gl_n$,

$$\tau_n = \{\mathbf{A} |\ \mathbf{A} \in gl_n \ \text{and} \ [\mathbf{A}] \text{ is upper triangular}\},$$

then $\tau_n$ is a solvable Lie algebra with the nilpotent commutator subalgebra $\mathscr{D}\tau_n = v_n$. It is also interesting to note that a Lie algebra $g$ is semisimple if and only if its Killing form is nonsingular, $g$ is nilpotent if and only if its Killing form is zero, and $g$ is solvable if and only if its Killing form vanishes on $\mathscr{D}g$.

The key to the classification of a semisimple complex Lie algebra $g$ is that each such Lie algebra contains a *Cartan subalgebra.* A Cartan subalgebra $h \subset g$ of a semisimple Lie algebra $g$ is a maximal Abelian subalgebra with the property that if $\mathbf{A} \in g$ and $\mathbf{H} \boxtimes \mathbf{A} = 0$ for all $\mathbf{H} \in h$, then $\mathbf{A} \in h$.

http://en.wikipedia.org/wiki/Cartan_subalgebra

To understand how the classification scheme works, we now define what is meant by a *root system* of a semisimple complex Lie algebra $g$ with a Cartan subalgebra $h$. In this case, the Killing form $\kappa$ can be chosen so that $\kappa(\mathbf{A},\mathbf{B}) = \mathbf{A} \cdot \mathbf{B}$ for all $\mathbf{A},\mathbf{B} \in g$. A bivector $h_\alpha \in h$ determines a *root* if there exists a nonzero element $\mathbf{A} \in g$ such that

$$h \boxtimes \mathbf{A} = (h_\alpha \cdot h)\mathbf{A} \quad \text{for all} \quad h \in h.$$

Sometimes, we will refer to the bivectors $h_\alpha$ as *cobivectors* because they are being used to determine the *dual form* $(h_\alpha \cdot h)$.

**Definition 18.5.1.** Let $g$ be a semisimple complex Lie algebra with the with Cartan subalgebra $h$. A finite set $\mathscr{S} \subset h$ of nonzero cobivectors in $h$, called *roots*, form a root system in $g$ if:

1. The set $\mathscr{S}$ spans an $n$-dimensional space of bivectors $\mathscr{S}^n = \text{span}(\mathscr{S})$ over the real numbers.
2. If $h_\alpha \in \mathscr{S}$ is a root cobivector, then $-h_\alpha \in \mathscr{S}$ is also a root cobivector.
3. If $h_{\alpha_1}, h_{\alpha_2} \in \mathscr{S}$ are roots, then

$$h_{\alpha_2} - 2(h_{\alpha_2} \cdot \hat{h}_{\alpha_1})\hat{h}_{\alpha_1} \in \mathscr{S}$$

   is also a root in $\mathscr{S}$, where $\hat{h}_\alpha = \frac{h_\alpha}{\sqrt{h_\alpha \cdot h_\alpha}}$.
4. If $h_{\alpha_1}, h_{\alpha_2} \in \mathscr{S}$, then

$$2\frac{h_{\alpha_1} \cdot h_{\alpha_2}}{h_{\alpha_1} \cdot h_{\alpha_1}} = 2\hat{h}_{\alpha_1} \cdot \hat{h}_{\alpha_2} \frac{|h_{\alpha_2}|}{|h_{\alpha_1}|} \in Z$$

   where $Z$ is the set of integers. It follows that $4\cos^2\theta \in Z$ where $\hat{h}_{\alpha_1} \cdot \hat{h}_{\alpha_2} = \cos\theta$.

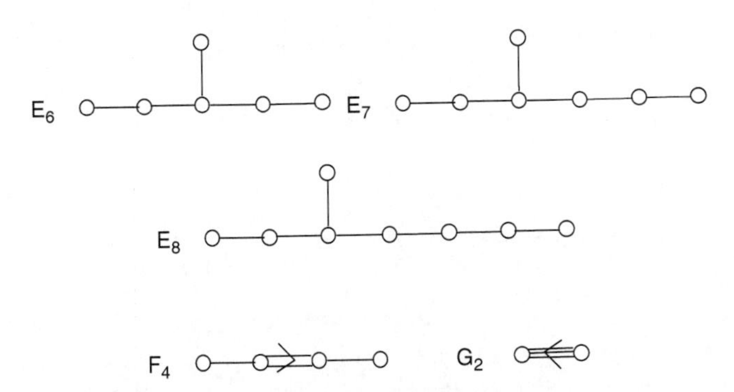


**Fig. 18.2** The Dynkin digrams represent the only possible simple Lie algebras of finite dimension over the complex numbers. Each diagram contains all the information necessary to construct the given Lie algebra

Given a root system $\mathscr{S} \subset h$ of a semisimple complex Lie algebra $g$, it is always possible to find a subset $\mathscr{S}^+ \subset \mathscr{S}$ called the *positive roots* of $\mathscr{S}$. A positive root subset $\mathscr{S}^+$ is characterized by the following two properties:

1. For each root $h_\alpha \in \mathscr{S}$, either $h_\alpha$ or $-h_\alpha$ is in $\mathscr{S}^+$ but not both.
2. If $h_\alpha, h_\beta \in \mathscr{S}^+$, $h_\alpha \neq h_\beta$, and $h_\alpha + h_\beta \in \mathscr{S}$, then $h_\alpha + h_\beta \in \mathscr{S}^+$.

An element $h_\alpha \in \mathscr{S}^+$ is *indecomposable* or *simple* if $h_\alpha \neq h_\beta + h_\gamma$ for some $h_\beta, h_\gamma \in \mathscr{S}^+$. The set $\mathscr{R} \subset \mathscr{S}^+$ of simple roots makes up a basis of $\mathscr{S}^n$, and any positive root $h_\alpha \in \mathscr{S}^+$ is then a linear combination of the simple roots with positive coefficients. The *rank* of a root system $\mathscr{S}$ is $n$, the dimension of $\mathscr{S}^n$. A more complete discussion of root systems, and the proof of the last statement, can be found online at

http://en.wikipedia.org/wiki/Root_system

See also [7, 26] and many other references.

We give here a brief description of the irreducible *Dynkin diagrams* for all possible root systems for simple complex Lie algebras, Fig. 18.2. Some explanation of the meaning of the above Dynkin diagrams is required for even a rudimentary understanding of the structure that they reveal. The vertices of a diagram represent the positive roots $h_\alpha$ of the particular root system in $\mathscr{S}^n$ that they characterize. A line segment is drawn between each non-orthogonal pairs of positive roots; it is undirected if the cobivectors make an angle of $120^\circ = \frac{2\pi}{3}r$, a directed double line segment connects them if they make an angle of $135^\circ = \frac{3\pi}{4}$, and a directed trible line segment means they make an angle of $150^\circ = \frac{5\pi}{6}r$. A "directed arrow" in the diagram points to the shorter cobivector.

It can be shown from the basic properties of a root system that if $h_\alpha$ and $h_\beta$ are positive simple roots, then

$$h_\alpha, h_\alpha + h_\beta, \ldots, h_\alpha + s h_\beta \tag{18.50}$$

are all positive roots, where $s = -2\frac{h_\alpha \cdot h_\beta}{h_\beta \cdot h_\beta}$ and $h_\alpha + (s+1)h_\beta$ is not a root. For a detailed description of the significance of these Dynkin diagrams, see [7,26]. For a detailed account of the automorphism groups in Clifford algebras, see [55,64].

## 18.6 The Lie Algebras $A_n$

We have seen that the bivectors in (18.5), under the Lie algebra product (18.24), make up the Lie algebra $gl_n$. All finite dimensional Lie algebras can be represented as Lie subalgebras of $gl_n$. In particular, we are interested in all finite dimensional Lie algebras that are characterized by their root systems as represented by their Dynkin diagrams. We begin by considering those *simple Lie algebras* whose root systems are characterized by the Dynkin diagram $A_n$ where $n = 1, 2, \ldots$. The matrices of the bivectors which characterize the Lie algebra with the root system $A_n$ are all those matrices which are trace-free, called the *special linear Lie algebra* $sl_n$. We can thus write

$$sl_n(\mathbb{C}) = \{B \in gl_n | K \cdot B = 0\} \subset gl_n. \tag{18.51}$$

The Lie algebra $sl_{n+1}$ has the root structure of the Dynkin diagram of $A_n$. This means that $sl_{n+1}$ has *n positive diagonal root elements* $h_{\alpha_i} \in sl_{n+1}$. These elements, when *normalized*, give the elements

$$h_i = 2\frac{h_{\alpha_i}}{h_{\alpha_i} \cdot h_{\alpha_i}}, \tag{18.52}$$

of the *Cartan subalgebra* $\mathscr{H}_n = (h)_{(n)} = \big(h_1, \ldots, h_n\big)$ of $sl_{n+1}$. The *Cartan matrix* $[\mathscr{H}_n]$ of $sl_n$ is then defined by

$$[\mathscr{H}_n] = (h)^{\mathrm{T}}_{(n)} \cdot (h_\alpha)_{(n)}. \tag{18.53}$$

To see how all this works in practice, let us consider the cases for $n = 1, 2, 3$. The Lie algebra $sl_2$ has one simple positive root $h_{\alpha_1}$, which we may choose to be the element $h_{\alpha_1} = \mathbf{a}_1 \wedge \mathbf{b}_1 - \mathbf{a}_2 \wedge \mathbf{b}_2 \in sl_2$. Using (18.52), we see that the corresponding normalized diagonal element $h_1 = h_{\alpha_1}$, which then gives the Cartan matrix

$$[\mathscr{H}_1] = [h_1 \cdot h_{\alpha_1}] = [2].$$

The Lie algebra $sl_2 = \text{span}\{h_1, X_{12}, Y_{21}\}$, where $X_{12} = \mathbf{a}_1 \wedge \mathbf{b}_2$ and $Y_{21} = \mathbf{a}_2 \wedge \mathbf{b}_1$. In general, the symbol $X_{ij} = \mathbf{a}_i \wedge \mathbf{b}_j$, where $i < j$, is used to represent the *positive root bivectors* that lie *above* the diagonal elements in (18.5). Similarly, the symbol $Y_{ji} = \mathbf{a}_j \wedge \mathbf{b}_i$, where $i < j$ is used to represent the *negative root bivectors* that lie *below* the diagonal elements in (18.5). The Lie algebra $sl_2$ is then completely determined by the Lie algebra products

$$h_1 \boxtimes X_{12} = (h)^{\mathrm{T}}_{(1)} \cdot (h_\alpha)_{(1)} X_{12} = 2X_{12},\ h_1 \boxtimes Y_{12} = (h)^{\mathrm{T}}_{(1)} \cdot (-h_\alpha)_{(1)} Y_{21} = -2Y_{21},$$

$$\text{and } X_{12} \boxtimes Y_{21} = h_1. \tag{18.54}$$

The cases for $n = 2$ and $n = 3$ are more instructive. For $n = 2$, the case of $sl_3$, we have two simple positive roots, which may be chosen as $h_{\alpha_1} = \mathbf{a}_1 \wedge \mathbf{b}_1 - \mathbf{a}_2 \wedge \mathbf{b}_2$ and $h_{\alpha_2} = \mathbf{a}_2 \wedge \mathbf{b}_2 - \mathbf{a}_3 \wedge \mathbf{b}_3$. Using (18.52) to normalize these roots, gives the *Abelian* Cartan subalgebra

$$\mathscr{H}_2 = (h)_{(2)} = span\begin{pmatrix} h_1 & h_2 \end{pmatrix}$$

and the Cartan matrix

$$[\mathscr{H}_2] = (h)^{\mathrm{T}}_{(2)} \cdot (h_\alpha)_{(2)} = \begin{pmatrix} h_1 \cdot h_{\alpha_1} & h_1 \cdot h_{\alpha_2} \\ h_2 \cdot h_{\alpha_1} & h_2 \cdot h_{\alpha_2} \end{pmatrix} = \begin{pmatrix} 2 & -1 \\ -1 & 2 \end{pmatrix},$$

where in this case once again $h_i = h_{\alpha_i}$ for $i = 1, 2$.

We now select the positive root bivectors $X_{12}, X_{23}$, and $X_{13} = X_{12} \boxtimes X_{23}$, which satisfy the general *positive Lie eigenvalue equation*

$$(h)^{\mathrm{T}}_{(2)} \boxtimes \begin{pmatrix} X_{12} & X_{23} & X_{13} \end{pmatrix}$$

$$= \begin{pmatrix} h_1 \\ h_2 \end{pmatrix} \cdot \begin{pmatrix} h_{\alpha_1} & h_{\alpha_2} & h_{\alpha_1} + h_{\alpha_2} \end{pmatrix} \begin{pmatrix} X_{12} & 0 & 0 \\ 0 & X_{23} & 0 \\ 0 & 0 & X_{13} \end{pmatrix}, \tag{18.55}$$

where

$$\begin{pmatrix} h_1 \\ h_2 \end{pmatrix} \cdot \begin{pmatrix} h_{\alpha_1} & h_{\alpha_2} & h_{\alpha_1} + h_{\alpha_2} \end{pmatrix} = \begin{pmatrix} 2 & -1 & 1 \\ -1 & 2 & 1 \end{pmatrix} \tag{18.56}$$

is the general *Lie algebra eigenmatrix* of the Lie eigenvalue equation (18.55). Note that the last column of this matrix is just the sum of the first two columns. This property is a direct consequence of the root structure of $A_2$.

Similarly, it can be directly verified that negative root bivectors $Y_{21}, Y_{32}$, and $Y_{31} = -Y_{21} \boxtimes Y_{32}$ satisfy the corresponding *negative Lie eigenvalue equation*

$$(h)^{\mathrm{T}}_{(2)} \boxtimes \begin{pmatrix} Y_{21} & Y_{32} & Y_{31} \end{pmatrix} = -\begin{pmatrix} 2 & -1 & 1 \\ -1 & 2 & 1 \end{pmatrix} \begin{pmatrix} Y_{21} & 0 & 0 \\ 0 & Y_{32} & 0 \\ 0 & 0 & Y_{31} \end{pmatrix}. \tag{18.57}$$

To completely determine the Lie algebra $sl_3$, calculate the additional Lie algebra products

$$\begin{pmatrix} X_{12} \\ X_{23} \\ X_{13} \end{pmatrix} \boxtimes \begin{pmatrix} Y_{21} & Y_{32} & Y_{31} \end{pmatrix} = \begin{pmatrix} h_1 & 0 & -Y_{32} \\ 0 & h_2 & Y_{21} \\ -X_{23} & X_{12} & h_1 + h_2 \end{pmatrix}.$$

We have now completely determined the structure of $sl_3$ from the structure of its root system $A_2$.

For the case when $n = 3$, $sl_4$ has the root structure of $A_3$. In this case, we take as the three simple positive roots $h_{\alpha_i} = \mathbf{a}_i \wedge \mathbf{b}_i - \mathbf{a}_{i+1} \wedge \mathbf{b}_{i+1}$, for $i = 1,2,3$. By applying (18.52) to normalize $h_{\alpha_i}$, we again find that $h_i = h_{\alpha_i}$, giving the Cartan subalgebra $\mathscr{H}_3 = (h)_{(3)}$ and the Cartan matrix

$$[\mathscr{H}_3] = (h)^{\mathrm{T}}_{(3)} \cdot (h_\alpha)_{(3)} = \begin{pmatrix} 2 & -1 & 0 \\ -1 & 2 & -1 \\ 0 & -1 & 2 \end{pmatrix}$$

which completely determines the root structure of $sl_3$. We now choose the positive root bivectors $X_{12}, X_{23}, X_{34}$, which further determine the positive root bivectors $X_{13} = X_{12} \boxtimes X_{23}$, $X_{24} = X_{23} \boxtimes X_{34}$, and $X_{14} = X_{13} \boxtimes X_{34}$.

The general *positive Lie eigenvalue equation* for $sl_4$ is then given by

$$(h)^{\mathrm{T}}_{(3)} \boxtimes \begin{pmatrix} X_{12} & X_{23} & X_{34} & X_{13} & X_{24} & X_{14} \end{pmatrix}$$

$$= \begin{pmatrix} h_1 \\ h_2 \\ h_3 \end{pmatrix} \cdot \begin{pmatrix} h_{\alpha_1} & h_{\alpha_2} & h_{\alpha_3} & h_{\alpha_1} + h_{\alpha_2} & h_{\alpha_2} + h_{\alpha_3} & h_{\alpha_1} + h_{\alpha_2} + h_{\alpha_3} \end{pmatrix} [X], \tag{18.58}$$

where

$$[X] = \begin{pmatrix} X_{12} & 0 & 0 & 0 & 0 & 0 \\ 0 & X_{23} & 0 & 0 & 0 & 0 \\ 0 & 0 & X_{34} & 0 & 0 & 0 \\ 0 & 0 & 0 & X_{13} & 0 & 0 \\ 0 & 0 & 0 & 0 & X_{24} & 0 \\ 0 & 0 & 0 & 0 & 0 & X_{14} \end{pmatrix},$$

and

$$\left[\mathscr{H}_3 \middle| \begin{pmatrix} 1 & -1 & 1 \\ 1 & 1 & 0 \\ -1 & 1 & 1 \end{pmatrix}\right]$$

$$= \begin{pmatrix} h_1 \\ h_2 \\ h_3 \end{pmatrix} \cdot \left(h_{\alpha_1}\ h_{\alpha_2}\ h_{\alpha_3}\ h_{\alpha_1}+h_{\alpha_2}\ h_{\alpha_2}+h_{\alpha_3}\ h_{\alpha_1}+h_{\alpha_2}+h_{\alpha_3}\right)$$

$$= \begin{pmatrix} 2 & -1 & 0 & 1 & -1 & 1 \\ -1 & 2 & -1 & 1 & 1 & 0 \\ 0 & -1 & 2 & -1 & 1 & 1 \end{pmatrix} \tag{18.59}$$

is the general *Lie algebra eigenmatrix* of the Lie eigenvalue equation (18.55). Note how the last three columns of this matrix are formed augmenting the Cartan matrix with the appropriate sums of the three columns of the Cartan matrix. This property is a direct consequence of Cartan matrix and the root structure of $A_3$.

Similarly, it can be directly verified that negative root bivectors

$$Y_{21}, Y_{32}, Y_{43}, Y_{31}, Y_{42}, Y_{41},$$

where $Y_{31} = -Y_{21} \boxtimes Y_{32}$, $Y_{42} = -Y_{32} \boxtimes Y_{43}$, and $Y_{41} = -Y_{31} \boxtimes Y_{43}$ satisfy the corresponding *negative Lie eigenvalue equation*

$$(h)^{\mathrm{T}}_{(3)} \boxtimes (Y_{j>i})_{(6)} = -\left[\mathscr{H}_3 \middle| \begin{pmatrix} 1 & -1 & 1 \\ 1 & 1 & 0 \\ -1 & 1 & 1 \end{pmatrix}\right][Y]. \tag{18.60}$$

To completely determine the Lie algebra $sl_3$, we calculate the additional Lie algebra products

$$(X_{i<j})^{\mathrm{T}}_{(6)} \boxtimes (Y_{j>i})_{(6)}$$

$$= \begin{pmatrix} h_1 & 0 & 0 & -Y_{32} & 0 & -Y_{42} \\ 0 & h_2 & 0 & Y_{21} & -Y_{43} & 0 \\ 0 & 0 & h_3 & 0 & Y_{32} & Y_{31} \\ -X_{23} & X_{12} & 0 & h_1+h_2 & 0 & -Y_{43} \\ 0 & -X_{34} & X_{23} & 0 & h_2+h_3 & Y_{21} \\ -X_{24} & 0 & X_{13} & -X_{34} & X_{12} & h_1+h_2+h_3 \end{pmatrix}.$$

We have now completely determined the structure of $sl_3$ from the structure of its root system $A_2$.

# References

1. Ablamowicz, R., Sobczyk, G.: Lectures on Clifford (Geometric) Algebras and Applications. Birkhäuser, Boston (2004)
2. Ahlfors, L.V.: Complex Analysis, 3rd edn. McGraw-Hill, New York (1979)
3. Baylis, W.E., Huschilt, J., Jiansu, W.: Why i? Am. J. Phys. **60**(9), 788 (1992)
4. Baylis, W.E.: Electrodynamics: A Modern Geometric Approach (Progress in Mathematical Physics). Birkhäuser, Boston (1998)
5. Baylis, W.E., Sobczyk, G.: Relativity in clifford's geometric algebras of space and spacetime. Int. J. Theor. Phys. **43**(10), 1386–1399 (2004)
6. Bayro Corrochano, E., Sobczyk, G. (eds.): Geometric Algebra with Applications in Science and Engineering. Birkhäuser, Boston (2001)
7. Belinfante, J.G.F., Kolman, B.: A Survey of Lie Groups and Lie Algebras with Applications and Computational Methods. Society for Industrial and Applied Mathematics, Pennsylvania (1972)
8. Birkhoff, G.: Lie groups isomorphic with no linear group. Bull. Am. Math. Soc., **42**, 882–888 (1936)
9. Born, M.: Einstein's Theory of Relativity, rev. edn. Dover, New York (1962)
10. Brackx, F., Delanghe R., Sommen, F.: Clifford Analysis. Research Notes in Mathematics, vol. 76. Pitman Advanced Publishing Program, Boston (1982)
11. Brackx, F., De Schepper, H., Sommen, F.: The hermitian clifford analysis toolbox. Adv. Appl. Clifford Algebras **18**, 451–487 (2008)
12. Clifford, W.K.: Applications of grassmann's extensive algebra. Am. J. Math. **1**, 350–358 (1878)
13. Clifford, W.K.: On the classification of geometric algebras, In: R. Tucker (ed.) Mathematical Papers by William Kingdon Clifford, pp. 397–401. Macmillan, London (1882) (Reprinted by Chelsea, New York, 1968)
14. Crowe, M.J.: A History of Vector Analysis. Chapter 6. Dover, New York (1985)
15. Cullen, C.G.: Matrices and Linear Transformations, 2nd edn. Dover, New York (1972)
16. Curtis, C.W.: Pioneers of Representation Theory: Frobenius, Burnside, Schur, and Brauer, AMS and the London Mathematical Society (1999). http://www.ams.org/bookstore-getitem/item=HMATH-15-S
17. Dantzig, T.: NUMBER: The Language of Science, 4th edn. Free Press, New York (1967)
18. Davis, P.J.: Interpolation and Approximation. Dover, New York (1975)
19. Doran, C., Hestenes, D., Sommen, F., Van Acker, N.: Lie groups as spin groups. J. Math. Phys., **34**(8), 3642–3669 (1993)
20. Dorst, L., Doran, C., Lasenby, J. (eds.): Applications of Geometric Algebra in Computer Science and Engineering. Birkhäuser, Boston (2002)

G. Sobczyk, *New Foundations in Mathematics: The Geometric Concept of Number*, DOI 10.1007/978-0-8176-8385-6,

21. Einstein, A., Lorentz, H.A., Minkowski, H., Weyl, H.: On the Electrodynamics of Moving Bodies. In: The Principle of Relativity, pp. 37–65. Dover, New York (1923). Translated from "Zur Elektrodynamik bewegter Körper", Annalen der Physik, 17 (1905)
22. Fishback, W.T.: Projective & Euclidean Geometry, 2nd edn. Wiley, New York (1969)
23. Fjelstad, P.: Extending relativity via the perplex numbers. Am. J. Phys. **54**(5), 416 (1986)
24. Flanders, H.: Differential Forms with Applications to the Physical Sciences. Dover, New York (1989)
25. French, A.P.: Special Relativity. Norton, New York (1968)
26. Fulton, W., Harris, J.: Representation Theory: A First Course. Springer, New York (1991)
27. Friedberg, S.H., Insel, A.J., Spence, L.E.: Linear Algebra. Prentice-Hall, Englewood Cliffs (1979)
28. Gallian, J.A.: Contemporary Abstract Algebra, 6th edn. Houghton Mifflin Company, Boston (2006)
29. Gantmacher, F.R.: Theory of Matrices, translated by Hirsch, K.A. Chelsea Publishing, New York (1959)
30. Gel'fand, I.M., Shilov, G.E.: Generalized Functions. Properties and Operations, vol. 1. Academic, New York (1964)
31. Havel, T.F.: Geometric Algebra: Parallel Processing for the Mind (Nuclear Engineering) (2002). http://www.garretstar.com/secciones/clases/MT318/lect1.pdf, http://web.mit.edu/tfhavel/www/
32. Heath, T.L: Euclid's Elements, vol. 2, p. 298, 2nd edn. Dover, New York (1956)
33. Hestenes, D.: Space Time Algebra. Gordon and Breach, New York (1966)
34. Hestenes, D.: Proper particle mechanics. J. Math. Phys. **15**, 1768–1777 (1974)
35. Hestenes, D.: The design of linear algebra and geometry. Acta Appl. Math. vol. 23, pp. 65–93. Kluwer Academic, Dordrecht (1991)
36. Hestenes, D.: New Foundations for Classical Mechanics, 2nd edn. Kluwer, Dordrecht (1999)
37. Hestenes, D.: Point groups and space groups in geometric algebra, In: Doerst, L., Doran, C., Lasen, J. (eds.) Applications of Geometric Algebra with Applications in Computer Science and Engineering, pp. 3–34. Birkhauser, Boston (2002)
38. Hestenes, D.: Spacetime physics with geometric algebra. Am. J. Phys. **71**(6), pp. 691–714 (2003)
39. Hestenes, D.: Gauge Theory Gravity with Geometric Calculus, Foundations of Physics, 35(6):903–970 (2005)
40. Hestenes, D., Holt, J.: The crystallographic space groups in geometric algebra. J. Math. Phys. **48**, 023514 (2007)
41. Hestenes, D.: Grassmann's Legacy. In: Grassmann Bicentennial Conference (1809-1877) September 16–19, (2009) Potsdam Szczecin (DE PL). http://geocalc.clas.asu.edu/pdf/GrassmannLegacy2.pdf
42. Hestenes, D., Reany, P., Sobczyk, G.: Unipodal algebra and roots of polynomials. Adv. Appl. Clifford Algebras **1**(1), 31–51 (1991)
43. Hestenes D., Sobczyk. G.: Clifford Algebra to Geometric Calculus: A Unified Language for Mathematics and Physics, 2nd edn. Kluwer, Dordrecht (1992)
44. Hestenes, D., Ziegler, R.: Projective geometry with Clifford algebra, Acta Applicandae Mathematicae, vol. 23, p. 25–63, Kluwer Academic, Dordrecht (1991)
45. Hicks, N.J.: Notes on Differential Geometry. Van Nostrand Company, Princeton (1965)
46. Horn, R., Johnson, C.R.: Matrix Analysis. Cambridge University Press, New York (1990)
47. Jackson, J.D.: Classical Electrodynamics. Wiley, New York (1962)
48. James, G., Liebeck, M.: Representations and Characters of Groups, 2nd edn. Cambridge University Press, Cambridge (2001)
49. Klein, F.: Elementary Mathematics From an Advanced Standpoint, vol. 1, 3rd edn. Dover, New York (1924)
50. Lam, T.Y.: Representations of finite groups: a hundred years. Part I Notices of the AMS **45**(3), 361–372 (1998)

51. Lasenby, A., Doran, C., & Gull, S.: Gravity, gauge theories and geometric algebra, Phil. Trans. R. Lond. A 356: 487–582 (1998)
52. Lee, J.M.: Manifolds and Differential Geometry, Graduate Studies in Mathematics, vol. 107. American Mathematical Society, Providence, Rhode Island (2009)
53. Linz, P.: Theoretical Numerical Analysis. Wiley, New York (1979)
54. Lounesto, P.: Clical Algebra Calculator and user manual, Helsinki University of Technology of Mathematics, Research Report **248**, (1994) http://users.tkk.fi/ppuska/mirror/Lounesto/CLICAL.htm
55. Lounesto, P.: Clifford Algebras and Spinors, 2nd edn. Cambridge University Press, Cambridge (2001)
56. Millman, R.S., Parker, G.D.: Elements of Differential Geometry. Prentice-Hall, Englewood Cliffs (1977)
57. Nash, J.: C1 isometric imbeddings. Ann. Math. **60**(3), 383–396 (1954)
58. Nash, J.: The imbedding problem for riemannian manifolds. Ann. Math. **63**(1), 20–63 (1956)
59. Nahin, P.: An Imaginary Tale: The story of the Square Root of Minus One. Princeton University Press, Princeton (1998)
60. Nering, E.: Linear Algebra and Matrix Theory (Paperback). Wiley, New York (1976)
61. Niven, I.N., Zuckerman, H.S., Montgomery, H.L.: An Introduction to the Theory of Numbers, 5th edn. Wiley, New York (1991)
62. Oziewicz, Z.: How do you add relative velocities? In: Pogosyan, G.S., Vicent, L.E., Wolf, K.B. (eds.) Group Theoretical Methods in Physics. Institute of Physics, Bristol (2005)
63. Pontryagin, L.S.: Hermitian operators in a space with indefinite metric. Izv. Akad. Nauk SSSR Ser. Mat. **8**, 243–280 (1944)
64. Porteous, I.R.: Clifford Algebras and the Classical Groups. Cambridge University Press, Cambridge (1995)
65. Pozo, J., Sobczyk, G.: Geometric algebra in linear algebra and geometry. Acta Appl. Math. **71**, 207–244 (2002)
66. Shilov, G.E.: Linear Algebra. Dover, New York (1977)
67. Sobczyk, G.: Mappings of Surfaces in Euclidean Space using Geomtric Algebra. Ph.D dissertation, Arizona State University (1971). http://www.garretstar.com/secciones/publications/publications.html
68. Sobczyk, G.: Spacetime vector analysis. Phys. Lett. **84A**, 45–49 (1981)
69. Sobczyk, G.: Conjugations and hermitian operators in spacetime. Acta Phys. Pol. **B12**(6), 509–521 (1981)
70. Sobczyk, G.: A complex gibbs-heaviside vector algebra for space-time. Acta Phys. Pol. **B12**(5), 407–418 (1981)
71. Sobczyk, G.: Unipotents, idempotents, and a spinor basis for matrices. Adv. Appl. Clifford Algebras **2**(1), 51–64 (1992)
72. Sobczyk, G.: Noncommutative extensions of number: an introduction to clifford's geometric algebra. Aportaciones Mat. Comun. **11**, 207–218 (1992)
73. Sobczyk, G.: Simplicial calculus with geometric algebra. In: Micali, A., et al. (eds.) Clifford Algebras and their Applications in Mathematical Physics, p. 279–292. Kluwer, the Netherlands (1992)
74. Sobczyk, G.: Linear transformations in unitary geometric algebra. Found. Phys. **23**(10), 1375–1385 (1993)
75. Sobczyk, G.: Jordan form in associative algebras. In: Oziewicz, Z., et al. (eds.) Clifford Algebras and Quantum Deformations, pp. 357–364. Kluwer, the Nethelands (2003)
76. Sobczyk, G.: Jordan form in clifford algebra. In: Bracks, F., et al. (eds.) Clifford Algebras and their Applications in Mathematical Physics, pp. 33–41. Kluwer, the Netherlands (2003)
77. Sobczyk, G.: Hyperbolic number plane. College Math. J. **26**(4), 268–280 (1995)
78. Sobczyk, G.: The generalized spectral decomposition of a linear operator. College Math. J. **28**(1), 27–38 (1997)
79. Sobczyk, G.: Spectral Integral Domains in the Classroom. Aportaciones Matematicas. Serie Comunicaciones, vol. 20, pp. 169–188. Sociedad Matemática Mexicana, Mexico (1997)

80. Sobczyk, G.: The missing spectral basis in algebra and number theory. The American Mathematical Monthly, vol. 108, pp. 336–346 (2001)
81. Sobczyk, G.: Generalized Vandermonde determinants and applications. Aportaciones Matematicas, Serie Comunicaciones, vol. 30, pp. 203–213. Sociedad Matemática Mexicana, Mexico (2002)
82. Sobczyk, G.: Clifford geometric algebras in multilinear algebra and non-euclidean geometries. Byrnes, J., (ed.) Computational Noncommutative Algebra and Applications: NATO Science Series, pp. 1–28. Kluwer, Dordrecht (2004)
83. Sobczyk, G.: Quantum Hermite Interpolation Polynomials. Aportaciones Matematicas, Parametric Optimization and Related Topics VII 18, Sociedad Matemática Mexicana, Mexico, pp. 105–112 (2004)
84. Sobczyk, G.: Structure of Factor Algebras and Clifford Algebra. Linear Algebra and Its Applications, vol. 241–243, pp. 803–810, Elsevier Science, New York (1996)
85. Sobczyk, G.: The spectral basis and rational interpolation. Proceedings of "Curves and Surfaces." Avignon, France, arXiv:math/0602405v1 (2006)
86. Sobczyk, G.: Geometric matrix algebra. Lin. Algebra Appl. **429**, 1163–1173 (2008)
87. Sobczyk, G., Yarman, T.: Unification of Space-Time-Matter-Energy, Appl. Comput. Math. **7**, No. 2, pp. 255–268 (2008)
88. Sobczyk, G., León Sanchez, O.: The fundamental theorem of calculus. Adv. Appl. Clifford Algebras **21**, 221–231 (2011)
89. Sobczyk, G.: Conformal mappings in geometric algebra. Not. AMS. **59**(2), 264–273 (2012)
90. Sobczyk, G.: Unitary geometric algebra. In: Ablamowicz, R., Vaz, J. (eds.) Special Volume of Advances in Applied Clifford Algebras in Memory of Prof. Jaime Keller, pp. 283–292. Springer Basel AG (2012). http://link.springer.com/article/10.1007/s00006-011-0277-5
91. Spiegel, M.R.: Vector Analysis and an introduction to Tensor Analysis. Schaum's Outline Series. Schaum Publishing, New York (1959)
92. Spivak, M.S.: Calculus on Manifolds. W.A. Benjamin, New York (1965)
93. Stoer, J., Bulirsch, R.: Introduction to Numerical Analysis, 2nd edn. Translated by Bartels, R., Gautschi, W., Witzgall, C. Springer, New York (1993)
94. Struik. D.J.: A Concise History of Mathematics. Dover, New York (1967)
95. Thomas, G.B., Finney, R.L.: Calculus and Analytic Geometry, 8th edn. Addison-Wesley, Reading, MA (1996)
96. Verma, N.: Towards an Algorithmic Realization of Nash's Embedding Theorem. CSE, UC San Diego. http://cseweb.ucsd.edu/~naverma/manifold/nash.pdf
97. Whitney, H.: Differentiable manifolds. Ann. Math. **37**, 645–680 (1936)
98. Yarman, T.: The End Results of General Relativity Theory via just Energy Conservation and Quantum Mechanics, Foundations of Physics Letters, **19**(7), pp. 675–694 (2006)
99. Young, J.W.: Projective Geometry. The Open Court Publishing Company, Chicago (1930)

# Symbols

$gcd$ Greatest common denominator: (1.1), (1.2)

$\mathbb{Z}_h$ Modular numbers modulo($h$): (1.3)

$(a_{m-1}\ a_{m-2}\ \ldots a_1\ a_0)_p$ $p$-adic number basis: (1.7)

$\mathbb{K}[x]_h$ Modular polynomials modulo($h(x)$): (1.11)

$(a_{m-1}\ a_{m-2}\ \ldots a_1\ a_0)_{x-x_0}$ $(x-x_0)$-adic number basis: (1.15)

$z = x + \mathrm{i}y$, $w = x + \mathrm{u}y$ Complex number $z$ and hyperbolic number $w$:

$|w|_h \equiv \sqrt{|ww^-|}$ Hyperbolic modulus of $w$: (2.1)

$z = r(\cos\theta + \mathrm{i}\sin\theta) \equiv r\exp\mathrm{i}\theta$ Polar form of $z$. (2.5)

$w = \pm\rho(\cosh\phi + \mathrm{u}\sinh\phi) \equiv \pm\rho\exp\mathrm{u}\phi$ Hyperbolic polar form of $w$: (2.6)

$\langle\overline{z_1}z_2\rangle_0$ Inner product of $z_1$ and $z_2$: (2.8)

$\langle\overline{z_1}z_2\rangle_i$ Outer product of $z_1$ and $z_2$: (2.8)

$\langle w_1^- w_2\rangle_0$ Inner product of $w_1$ and $w_2$: (2.9)

$\langle w_1^- w_2\rangle_u$ Outer product of $w_1$ and $w_2$: (2.9)

$\{u_+, u_-\}$ Idempotent basis: (2.11)

$|X_1 - X_2|_h$ Spacetime distance between two events: (2.22)

$V = \mathrm{d}X/\mathrm{d}t = c + \mathrm{u}v$ Spacetime velocity: (2.25)

$P = \gamma mV$ Spacetime momentum: (2.25)

$E = \gamma mc^2$ Total energy: (2.27)

$\mathbf{ab} = \mathbf{a}\cdot\mathbf{b} + \mathbf{a}\wedge\mathbf{b}$ Geometric product: (3.6), (3.13), (3.29)

$\mathbf{ab} = |\mathbf{a}||\mathbf{b}|\mathrm{e}^{\mathrm{i}\theta}$ Euler form: (3.8)

G. Sobczyk, *New Foundations in Mathematics: The Geometric Concept of Number*, DOI 10.1007/978-0-8176-8385-6,

$\mathbb{G}_2 = \mathbb{G}(\mathbb{R}^2)$ Geometric algebra of $\mathbb{R}^2$: (3.9)

$\mathbb{G}_3 = \mathbb{G}(\mathbb{R}^3)$ Geometric algebra of $\mathbb{R}^3$: (3.11)

$I = \mathbf{e}_{123}$ Pseudoscalr of $\mathbb{G}_3$: (3.11)

$Z(\mathbb{G}_3)$ Center of $\mathbb{G}_3$: (3.12)

$\mathbf{a} \times \mathbf{b}$ Vector cross product: (3.15)

$\mathbf{a} \cdot (\mathbf{b} \wedge \mathbf{c})$, $\mathbf{a} \wedge (\mathbf{b} \wedge \mathbf{c})$ Inner and outer products: (3.16), (3.17)

$\mathbf{A} \boxtimes \mathbf{B} = \frac{1}{2}(\mathbf{AB} - \mathbf{BA})$ Anti-symmetric product:

$\sqrt{\hat{\mathbf{a}}\hat{\mathbf{b}}}$ Square root: (3.23)

$(\mathbf{e})_{(n)}$, $(\mathbf{e})^{\mathrm{T}}_{(n)}$ Row and column of vectors: (3.26), (4.1), (4.2)

$\mathbf{a}^2 = |\mathbf{a}|^2$ Magnitude of vector: (3.28)

$\mathbb{G}_n = \mathrm{span}\{\mathbf{e}_{\lambda_{(k)}}\}_{k=0}^{n}$ Standard basis of $\mathbb{G}_n$: (3.30)

$A^\dagger$, $A^-$, $\widetilde{A}$ 3 conjugations: (3.31), (3.32), (3.33)

$\mathbf{a} \cdot \mathbf{B}_k$, $\mathbf{a} \wedge \mathbf{B}_k$ Inner and outer products: (3.40), (3.39)

$|\mathbf{A}_k| = \sqrt{|\mathbf{A}_k \cdot \mathbf{A}_k^\dagger|}$ Magnitude of $k$-vector: (3.44)

$F_{\mathbf{a}}(\mathbf{x})$ Directional derivative: (3.45)

$\partial_{\mathbf{x}} = \sum_{i=1}^{n} \mathbf{e}_i \frac{\partial}{\partial x^i}$ Vector derivative: (3.46)

$[\mathbf{b} \cdot \partial_{\mathbf{x}}, \mathbf{a} \cdot \partial_{\mathbf{x}}] = \mathbf{b} \cdot \partial_{\mathbf{x}}\, \mathbf{a} \cdot \partial_{\mathbf{x}} - \mathbf{a} \cdot \partial_{\mathbf{x}}\, \mathbf{b} \cdot \partial_{\mathbf{x}}$ Bracket operation: (3.55)

$[\mathbf{b}, \mathbf{a}] = \mathbf{b} \cdot \partial_{\mathbf{x}} \mathbf{a} - \mathbf{a} \cdot \partial_{\mathbf{x}} \mathbf{b}$ Bracket operation: (3.56)

$g^{\mathbf{e}_1} := \mathbf{e}_1 g \mathbf{e}_1$ e-conjugate: (4.10)

$[g]$ Matrix of the geometric number $g$: (4.12), (4.13)

$g^-$, $g^\dagger$, $\widetilde{g}$ 3 conjugations: (4.14), (4.15), (4.16)

$\mathbf{v}_{(k)} = \wedge(\mathbf{v})_{(k)}$ $k$-vector notation: (5.1), (5.4)

$\mathbb{G}_n^k$ Space of $k$-vectors: (5.5)

$\langle M \rangle_k$ $k$-vector part: (5.7)

$\underline{f}(\mathbf{v}_1 \wedge \cdots \wedge \mathbf{v}_k) = f(\mathbf{v}_1) \wedge \cdots \wedge f(\mathbf{v}_k)$ Outermorphism rule: Definition 7.1.1

$\partial_{(k)}$ Simplicial $k$-derivative: (7.3)

$\det f = \partial_{(n)} \cdot f_{(n)}$ Determinant of $f$: (7.5)

$f(\mathbf{e})_{(n)} = (f\mathbf{e})_{(n)} = (\mathbf{e})_{(n)}[f]$ Matrix of $f$: (7.7)

$(\mathbf{e})^{(n)}$ Column of vectors: (7.9)

$[1]_n$ Identity matrix: (7.10)

$\overline{f}$ Adjoint of $f$: (7.18)

$End(\mathbb{R}^n)$ Endomorphisms: defined before (8.1)

$\varphi_f(\lambda)$ Characteristic polynomial of $f$: (8.1)

$\psi_{\mathbf{v}}(\lambda)$ Minimal polynomial of $f$: (8.7)

$q^{(m-1)}(\mathbf{a})$ Jordan chain: (8.11)

$\mathbf{e}^{\mathrm{i}}(\mathbf{e}_j) = \delta^i_j$ Dual basis: (9.1)

$\mathbb{R}^{p,q}$ Pseudo-Euclidean space: (10.2)

$\mathbf{e}^{\mathrm{i}}$ Reciprocal basis vector: (10.4)

$(\mathbf{e})^{(n)}$ Reciprocal basis: (10.5)

$\mathbb{G}_{p,q}$ Geometric algebra of signature $(p,q)$: (10.10)

$\partial_{\mathbf{x}} = \sum_{i=1}^n \mathbf{e}^{\mathrm{i}} \frac{\partial}{\partial x^i}$ Vector derivative of signature $(p,q)$: (10.11)

$(\mathbf{a})^{(n)}$ Reciprocal basis: (10.18)

$\det A = \mathbf{a}^{\dagger}_{(n)} \cdot \mathbf{a}_{(n)}$ Gram determinant: (10.17)

$\mathbb{H}^{p,q}$, $\mathbb{U}^{p,q}$ Unitary geometric algebra: (10.19)

$\mathbf{x} \cdot \mathbf{y}$, $\mathbf{x} \wedge \mathbf{y}$ Hermitian inner and outer products: (10.20)

$(\mathbf{a})_{(\overline{k})}$ Hermitian row of vectors: (10.28), (10.29)

$\mathbf{a}_{(\overline{k})}$ Complex $k$-vectors: (10.30)

$f^*$ Hermitian adjoint of $f$: (10.35)

$\mathscr{B}_j = \{1, \mathbf{e}'_1, \mathbf{e}'_2, \mathbf{e}'_1\mathbf{e}'_2\}$ Relative basis: (11.15)

$\mathbb{R}^2_j$ Relative plane: (11.16)

$\mathscr{H}^+$ Relative positive oriented bivectors: (11.17)

$g_1 \odot g_2$, $g_1 \boxtimes g_2$ Relative inner and outer products: (11.19)

$\mathbb{G}_{1,2}$ Standard basis: (11.31)

$a \circ b$ Twisted product: (12.1)

$a^+$, $a^-$ Special symbols: (12.2)

$\mathbb{G}_{n,n}$ Geometric algebra: (12.18)

$u^{\pm\pm\ldots\pm}_{1\,2\ldots n}$ Primitive idempotent: (12.19)

$tangle(a_1,a_2;b_1,b_2)$ Entanglement table: Definition 12.3.1

$\mathscr{R}$ Group algebra: (12.43), (12.46), (12.58)

$\beta(R)$, $\beta(M)$ Boundary of chain: (13.1), (13.4)

$\mathbf{x}_i = \frac{\partial \mathbf{x}}{\partial s^i}$ Tangent vectors: (13.2)

$\mathbf{x}^i$ Reciprocal vectors: (13.5)

$\dot{g}(\mathbf{x})\dot{\partial}_{\mathbf{x}}\dot{f}(\mathbf{x})$ Two sided vector derivative: (13.7)

$\int_M g(\mathbf{x})\mathrm{d}\mathbf{x}_{(m)}f(\mathbf{x})$ Directed integral: (13.11)

$\mathscr{D}(M) = \int_M \mathrm{d}\mathbf{x}_{(m)}$ Directed content: (13.12)

$\int_{\beta(M)} g(\mathbf{x})\mathrm{d}\mathbf{x}_{(m-1)}f(\mathbf{x})$ Directed integral over boundary: (13.13)

$s(t) = \int_{\mathbf{x}} |d\mathbf{x}|$ Arc length of curve: (14.1)

$\mathscr{T}_{\mathbf{x}}^1$, $\mathscr{T}_{\mathbf{x}}$ Tangent space and algebra: (15.2)

$I_{\mathbf{x}} = \wedge(\mathbf{x})_{(k)}$ Pseudoscalar on $k$-surface: (15.3)

$\mathbf{n}_{\mathbf{x}} = \widehat{I}_{\mathbf{x}}i^{-1}$ Unit normal to $k$-surface: (15.5)

$[g] = [g_{ij}]$ Metric tensor of $k$-surface: (15.6)

$\partial_{\mathbf{x}} = \sum_{i=1}^{k} \mathbf{x}^i \frac{\partial}{\partial s^i}$ Vector derivative on $k$-surface: (15.7)

$F_{\mathbf{a}}(\mathbf{x}) = \mathbf{a} \cdot \partial_{\mathbf{x}} F(\mathbf{x})$ Directional derivative on $k$-surface: (15.8)

$F_{\mathbf{a},\mathbf{b}}(\mathbf{x})$ Second derivative on $k$-surface: (15.12)

$H_{\mathbf{a}}(g)$ Directional derivatives of $g$-tensor: (15.13), (15.15)

$P_{\mathbf{x}}$ Projection operator: Definition 15.1.4

$(\mathbf{x}_{ij}) = (\mathbf{K}_{ij}) + (\mathbf{L}_{ij})$ Matrix of second derivatives: (15.21), (15.22)

$\mathbf{L}_{\mathbf{a},\mathbf{b}} = P_{\mathbf{a}}(P_{\mathbf{x}}(\mathbf{b}))$ Derivatives of normal vectors: (15.24)

$L(g) = \dot{\partial}_{\mathbf{x}}\dot{P}_{\mathbf{x}}(g)$ Shape operator: (15.26)

$\underline{\mathbf{n}}(\mathbf{a}) = -P_{\mathbf{a}}(\mathbf{n})$ Weingarten map: (15.30)

$\frac{\mathrm{d}^2\mathbf{x}}{\mathrm{d}s^2} = \kappa_g\mathbf{S} + \kappa_n\mathbf{K}$ Geodesic and normal curvature: (15.31), (15.32)

$\nabla_{\mathbf{x}}F(\mathbf{x})$, $F_{/\mathbf{a}}(\mathbf{x})$ Coderivatives: (15.39), (15.40)

$K_G$, $\kappa_m$ Gausian and mean curvature: (15.42), (15.43)

$L_{\mathbf{a},\mathbf{b}} = \mathbf{L}_{\mathbf{a},\mathbf{b}} \cdot \mathbf{n}$ Second fundamental form: (15.44)

$R(\mathbf{a} \wedge \mathbf{b}) = \partial_{\mathbf{v}} \wedge P_{\mathbf{a}}P_{\mathbf{b}}(\mathbf{v})$ Riemann curvature bivector: Definition 15.5.1

$R_{ijkl} = R(\mathbf{x}_i \wedge \mathbf{y}_j) \cdot (\mathbf{x}_k \wedge \mathbf{x}_l)$ Classical Riemann curvature tensor: (15.51)

$(\mathbf{x}')_{(k)} = \underline{f}(\mathbf{x})_{(k)}$ Push forward: (16.2)

$\overline{f}(\partial'_{\mathbf{x}'}) = \partial_{\mathbf{x}}$ Pull back: (16.4)

$\underline{f}_{\mathbf{a}}(\mathbf{b}) = \underline{f}_{\mathbf{b}}(\mathbf{a})$ Integrability: (16.7)

$\underline{f}_{\not{b}}(\mathbf{a}),\ \underline{f}_{\not{a},\not{b}}$ Intrinsic coderivatives: Definition 16.1.1

$L'(A) = \dot{\partial}_{\mathbf{x}'}\dot{\underline{f}}(A)$ Generalized shape operator: (16.16)

$\Phi'(A_r) = \underline{f}(\Phi(A_r))$ Shape divergence: (16.17)

$\mathscr{W}_4(\mathbf{a}_{(2)}) = R(\mathbf{a}_{(2)}) \wedge \mathbf{a}_{(2)}$ Conformal Weyl 4-vector: (16.28)

$\mathscr{W}_C(\mathbf{a}_{(2)})$ Classical conformal Weyl tensor: (16.29)

$\mathbf{a} \cdot \delta_{\mathbf{x}} = \mathbf{a} \cdot \nabla_{\mathbf{x}} + \underline{h}_{\mathbf{x}}^{-1} \underline{h}_{\mathbf{a}}$ Affine directional $h$-coderivative: (16.37)

$H_{\not{a}}(\mathbf{b}) = \mathbf{K}_{\not{a}}(\mathbf{b}) + \mathbf{T}_{\not{a}}(\mathbf{b})$ Affine tensor: (16.44)

$\mathscr{A}_h^n$ Affine plane: Definition 17.1.1

$S^*_{\mathbf{x}_h}$ $S_{\mathbf{x}_h}$ Rays and representants: (17.1), (17.2)

$S_{A_r} \cap S_{B_s}$ $S_{A_r} \cup S_{B_s}$, Meet and join: (17.3), (17.4)

$\Pi^n$ Projective plane: (17.11)

$r(\mathbf{a}, \mathbf{b}, \mathbf{c}, \mathbf{d})$ Cross ratio: (17.15)

$\mathscr{H}(\mathbf{a}, \mathbf{b}, \mathbf{c}, \mathbf{d})$ Harmonically related: (17.16)

$\mathscr{H}^{p,q}$ $(p,q)$-horosphere: (17.35)

$\mathbf{x}_a,\ \mathbf{x}_c,\ [\mathbf{x}_c]$ Homogeneous and conformal representants: (17.36), (17.40)

$\mathscr{A}^n,\ \mathscr{B}^n$ Null cones: (18.1)

$\Omega^2_{n,n}$ Standard basis of bivectors: (18.5)

$\mathbf{F}^{(k)} : \mathbf{x} = \mathbf{F}^{(k)} \boxtimes \mathbf{x}$ Standard basis of bivectors: (18.7)

$GL_n$ General linear group: (18.8)

$\partial_{\mathbf{x}} = \partial_{\mathbf{x}_a} + \partial_{\mathbf{x}_b}$ Decomposition of vector derivative: (18.10)

$gl_n = \Omega^2_{n,n}$ General Lie algebra: Definition 18.2.1

$\partial_{\mathbf{W}}$ Bivector derivative: (18.20), (18.21)

$I_{p,q} = \wedge(\mathbf{e})_{(p,q)}$ Pseudoscalar: (18.32)

$P_{p,q}$ Projection operator: (18.33)

$\mathbf{F}_{p,q} = P_{p,q}(\mathbf{F})$ Projection of bivector: (18.34)

$so_{p,q}$ Lie algebra $so_{p,q}$: Definition 18.4.1

$spin_{p,q}$ Lie algebra $spin_{p,q}$: Definition 18.39

$O_{p,q}$ Orthogonal Lie group $O_{p,q}$: Definition 18.4.2

$SO_{p,q}$ Lie algebra $SO_{p,q}$: (18.41)

$Sp_{2n}$ Lie group $Sp_{2n}$: Definition 18.42, (18.44)

$SO_{p,q}$ Lie algebra $SO_{p,q}$: (18.41)

$U_{p,q}$ Unitary Lie group $U_{p,q}$: (18.45)

$SU_{p,q}$ Special unitary Lie group $SU_{p,q}$: (18.46)

$\mathscr{D}g$ Commutator subalgebra: (18.48)

$\mathscr{S} \subset h$ Root system: Definition 18.5.1

$sl_n(\mathbb{C})$ Special linear Lie algebra: (18.51)

$[\mathscr{H}_n]$ Cartan matrix: (18.53)

# Index



G. Sobczyk, *New Foundations in Mathematics: The Geometric Concept of Number*, DOI 10.1007/978-0-8176-8385-6,

**C**



**D**

## Q



## R



## S



## T

Printed by Publishers' Graphics LLC
SO20121029.19.19.2